Methods in Molecular Biology

Series Editor
John M. Walker
School of Life and Medical Sciences
University of Hertfordshire
Hatfield, Hertfordshire, AL10 9AB, UK

For further volumes:
http://www.springer.com/series/7651

Extracellular Vesicles

Methods and Protocols

Edited by

Winston Patrick Kuo

CloudHealth Genomics, Ltd., Shanghai, China
Westchester Biotech Project, Asbury Park, NJ, USA

Shidong Jia

Predicine, Inc., Hayward, CA, USA

Humana Press

Editors
Winston Patrick Kuo
CloudHealth Genomics, Ltd.
Shanghai, China

Westchester Biotech Project
Asbury Park, NJ, USA

Shidong Jia
Predicine, Inc.
Hayward, CA, USA

ISSN 1064-3745 ISSN 1940-6029 (electronic)
Methods in Molecular Biology
ISBN 978-1-4939-7251-7 ISBN 978-1-4939-7253-1 (eBook)
DOI 10.1007/978-1-4939-7253-1

Library of Congress Control Number: 2017947900

Printed on acid-free paper

This Humana Press imprint is published by Springer Nature
The registered company is Springer Science+Business Media LLC
The registered company address is: 233 Spring Street, New York, NY 10013, U.S.A.

Preface

The study of extracellular vesicles (EVs), including exosomes and microvesicles, is an emerging field across many disciplines that involve many research and development efforts on their biogenesis, role in intercellular communication by transporting biomolecules between close and distant cells, and their role in normal and disease physiology.

In both healthy and pathological states, the functions of EVs are for removal of protein waste and communication across cells by way of immune response activation and/or deactivation, cellular proliferation, and, in cancer, tumor metastasis. More recently, in the era of Precision Medicine, the clinical utility of EVs is its minimally to noninvasive approach (liquid biopsy) to enable screening, assessing tumor heterogeneity, monitoring therapeutic responses, and minimal residual disease detection to EV-based therapeutics. About 25–50% of tissue biopsies have sufficient material for any downstream analysis (genomic or proteomic), which has made the EV field very attractive, hence the growing research interest in this space.

What makes EVs very interesting is that they are present in many biological fluids (blood (serum and plasma), urine, saliva, breast milk, CSF, follicular fluid, semen, lung lavage, and tears) and contain unique characteristics as nucleic acids (DNA, mRNA, miRNA molecules, noncoding RNA), proteins, and lipids identical to the original cell. Many researchers are also studying EVs derived from stem cells, cell cultures, and parasites, among many other areas. As this field grows, there is a demand for established methods and protocols to isolate and characterize EVs among other techniques including purification, imaging, biofluid-specific and cell-specific isolation, and downstream genomic, metabolomic and proteomic profiling approaches.

In this book, we have brought together an international group of leading scientists with domain knowledge/expertise in the area of EVs across many basic and clinical disciplines. We have attempted to include a variety of different techniques related to the growing EV applications, as at times; using only one technique or two is insufficient to address the question at hand.

We would like to express our greatest appreciation and gratitude to all the contributing authors, as without their time and effort in putting their chapters together along with their notes, this book would not be possible. In addition, we would like to give special thanks to Professor John M. Walker, Professor Emeritus, School of Life and Medical Sciences, University of Hertfordshire, for the invitation, as we wouldn't have thought to organize a book on methods and protocols related to extracellular vesicles.

Shanghai, China *Winston Patrick Kuo*

Contents

The original version of this chapter was revised. An erratum to this chapter can be found at DOI 10.1007/978-1-4939-7253-1_36

Contributors

TAKANORI AKAGI • *Department of Materials Engineering, School of Engineering, The University of Tokyo, Tokyo, Japan*

ÁNGEL AYUSO-SACIDO • *Instituto Madrileño de Estudios Avanzados, IMDEA Nanociencia, Madrid, Spain; Fundación de Investigación HM Hospitales, Hospital de Madrid Group, Madrid, Spain; Instituto de Medicina Molecular Aplicada (IMMA), School of Medicine, San Pablo-CEU University, Madrid, Spain*

CRISTOBAL BELDA-INIESTA • *Fundación de Investigación HM Hospitales, Hospital de Madrid Group, Madrid, Spain*

PETER N. BROWN • *Drug Discovery Programme, Beatson Institute for Cancer Research, Glasgow, UK*

VIRGINIA CAMACHO • *Beth Israel Deaconess Medical Center, Harvard Medical School, Boston, MA, USA*

PAULINE CARNELL-MORRIS • *Malvern Instruments Ltd, Amesbury, UK*

CESAR M. CASTRO • *Cancer Program, MGH Center for Systems Biology, Massachusetts General Hospital, Harvard Medical School, Boston, MA, USA; Division of Hematology-Oncology, Massachusetts General Hospital, Harvard Medical School, Boston, MA, USA*

CHIHCHEN CHEN • *Institute of Nanoengineering and Microsystems, National Tsing Hua University, Hsinchu, Taiwan; Department of Power Mechanical Engineering, National Tsing Hua University, Hsinchu, Taiwan*

ZIXIN CHEN • *Vascular Biology Center, Department of Medicine, Medical College of Georgia, Augusta University, Augusta, GA, USA; The First Clinical Medical College, Guangzhou University of Chinese Medicine, Guangzhou Shi, Guangdong Sheng, China*

GEORGE M. CHURCH • *Department of Genetics, Harvard Medical School, Boston, MA, USA; Wyss Institute for Biologically Inspired Engineering, Harvard University, Boston, MA, USA*

PETR CIZMAR • *Imaging Division, Image Sciences Institute, University Medical Centre Utrecht, Utrecht, The Netherlands*

EMANUELE COCUCCI • *Division of Pharmaceutics and Pharmaceutical Chemistry, College of Pharmacy, The Ohio State University, Columbus, OH, USA*

UTKAN DEMIRCI • *Department of Radiology, Bio-Acoustic MEMS in Medicine (BAMM) Laboratory, Canary Center at Stanford for Cancer Early Detection, Stanford School of Medicine, Palo Alto, CA, USA; Department of Electrical Engineering, Stanford University, Stanford, CA, USA*

REBECCA DRAGOVIC • *Nuffield Department of Obstetrics & Gynaecology, University of Oxford, Oxford, UK*

LUIS FILGUEIRA • *Department of Medicine, Unit of Anatomy, University of Fribourg, Fribourg, Switzerland*

BAO JUN GAO • *CloudHealth Genomics, Ltd, Shanghai, China*

NOEMÍ GARCÍA-ROMERO • *Instituto Madrileño de Estudios Avanzados, IMDEA Nanociencia, Madrid, Spain; Fundación de Investigación HM Hospitales, Hospital de Madrid Group, Madrid, Spain*

IONITA GHIRAN • *Beth Israel Deaconess Medical Center, Harvard Medical School, Boston, MA, USA*

JIANG HE • *Department of Radiology and Medical Imaging, University of Virginia, Charlottesville, VA, USA*
MITJA L. HEINEMANN • *Institute of Laboratory Medicine, Clinical Chemistry and Molecular Diagnostics, University Hospital Leipzig, Leipzig, Germany*
MICHELLE M. HILL • *The University of Queensland Diamantina Institute, The University of Queensland, Brisbane, QLD, Australia*
PATRICK HOLE • *Malvern Instruments Ltd, Amesbury, UK*
YI-HSING HSIAO • *Institute of Nanoengineering and Microsystems, National Tsing Hua University, Hsinchu, Taiwan; Institute of Biomedical Engineering and Nanomedicine, National Health Research Institutes, Miaoli, Taiwan*
TAO HUANG • *Department of Radiology and Medical Imaging, University of Virginia, Charlottesville, VA, USA*
BO HUANG • *Biochemistry and Molecular Biology, Tongji Medical College, Huazhong University of Science and Technology, Wuhan, China; State Key Laboratory of Medical Molecular Biology and Department of Immunology, Institute of Basic Medical Sciences, Chinese Academy of Medical Sciences and Peking Union Medical College, Beijing, China*
STEPHANIE N. HURWITZ • *Department of Biomedical Sciences, Florida State University College of Medicine, Tallahassee, FL, USA*
TAKANORI ICHIKI • *Department of Materials Engineering, School of Engineering, The University of Tokyo, Tokyo, Japan*
HYUNGSOON IM • *Cancer Program, MGH Center for Systems Biology, Massachusetts General Hospital, Harvard Medical School, Boston, MA, USA*
FATIH INCI • *Department of Radiology, Bio-Acoustic MEMS in Medicine (BAMM) Laboratory, Canary Center at Stanford for Cancer Early Detection, Stanford School of Medicine, Palo Alto, CA, USA*
ALEXANDER R. IVANOV • *Department of Chemistry and Chemical Biology, Barnett Institute of Chemical and Biological Analysis, Northeastern University, Boston, MA, USA*
KAZUYA IWAI • *Division of Protein Engineering, Cancer Institute, Japanese Foundation for Cancer Research, Tokyo, Japan; Department of Oral & Maxillofacial Implantology, Tokyo Dental College, Tokyo, Japan*
SEUNGA JEONG • *Division of Bioengineering, Incheon National University, Incheon, South Korea*
JOHN S. JI • *Environmental Health Science, Duke Kunshan University, Shanghai, China*
JASON GANG JIN • *CloudHealth Genomics, Ltd, Shanghai, China*
CHENGWEI JU • *Vascular Biology Center, Department of Medicine, Medical College of Georgia, Augusta University, Augusta, GA, USA; Department of Cardiology, Zhongda Hospital, Medical School of Southeast University, Nanjing, People's Republic of China*
HENRIKE JULICH-HAERTEL • *Department of Medicine II, Saarland University Medical Center, Saarland University, Homburg, Germany*
HONGJUN KANG • *Department of Critical Care Medicine, Chinese PLA General Hospital, Beijing, China*
SHLOMIT KENIGSBERG • *CreATe Fertility Centre, Toronto, ON, Canada*
BENTE KIERULF • *Thermo Fisher Scientific, Oslo, Norway*
IL-MAN KIM • *Vascular Biology Center, Department of Medicine, Medical College of Georgia, Augusta University, Augusta, GA, USA*

DARREN KORBIE • *Centre for Personalised Nanomedicine, Australian Institute for Bioengineering and Nanotechnology, The University of Queensland, Brisbane, QLD, Australia*
MIROSLAW KORNEK • *Department of Medicine II, Saarland University Medical Center, Saarland University, Homburg, Germany*
STELLA KOUREMBANAS • *Division of Newborn Medicine & Department of Medicine, Boston Children's Hospital, Boston, MA, USA; Department of Pediatrics, Harvard Medical School, Boston, MA, USA*
EMMA J.K. KOWAL • *Department of Genetics, Harvard Medical School, Boston, MA, USA; Wyss Institute for Biologically Inspired Engineering, Harvard University, Boston, MA, USA*
SIMION KREIMER • *Department of Chemistry and Chemical Biology, Barnett Institute of Chemical and Biological Analysis, Northeastern University, Boston, MA, USA*
WINSTON PATRICK KUO • *CloudHealth Genomics, Ltd, Shanghai, China; Westchester Biotech Project, Asbury Park, NJ, USA*
CHARLES PIN-KUANG LAI • *Institute of Biomedical Engineering, National Tsing Hua University, Hsinchu, Taiwan; Institute of Atomic and Molecular Sciences, Academia Sinica, Taipei, Taiwan*
REBECCA E. LANE • *Centre for Personalised Nanomedicine, Australian Institute for Bioengineering and Nanotechnology, The University of Queensland, Brisbane, QLD, Australia*
HAKHO LEE • *Cancer Program, MGH Center for Systems Biology, Massachusetts General Hospital, Harvard Medical School, Boston, MA, USA*
LANJUAN LI • *State Key Laboratory for Diagnosis and Treatment of Infectious Diseases, First Affiliated Hospital, College of Medicine, Zhejiang University, Hangzhou, Zhejiang Province, China; Collaborative Innovation Center for Diagnosis and Treatment of Infectious Diseases, Hangzhou, Zhejiang Province, China*
YONGJUN LI • *Vascular Biology Center, Department of Medicine, Medical College of Georgia, Augusta University, Augusta, GA, USA; Department of Cardiology, Zhongda Hospital, Medical School of Southeast University, Nanjing, People's Republic of China*
LI-GUO LIANG • *State Key Laboratory for Diagnosis and Treatment of Infectious Diseases, First Affiliated Hospital, College of Medicine, Zhejiang University, Hangzhou, Zhejiang Province, China; Collaborative Innovation Center for Diagnosis and Treatment of Infectious Diseases, Hangzhou, Zhejiang Province, China; Institute for Translational Medicine, Zhejiang University, Hangzhou, Zhejiang Province, China*
CLIFFORD L. LIBRACH • *CreATe Fertility Centre, Toronto, ON, Canada; Department of Obstetrics and Gynecology, University of Toronto, Toronto, ON, Canada; Department of Gynecology, Women's College Hospital, Toronto, ON, Canada*
YUTAO LIU • *Vascular Biology Center, Department of Medicine, Medical College of Georgia, Augusta University, Augusta, GA, USA*
JING LIU • *Department of Gastroenterology, Zhongnan Hospital of Wuhan University, Wuhan, China; The Hubei Clinical Center & Key Laboratory of Bowel Diseases, Wuhan, China; Biochemistry and Molecular Biology, Tongji Medical College, Huazhong University of Science and Technology, Wuhan, China*
RICHARD LOBB • *Tumour Microenvironment Laboratory, QIMR Berghofer Medical Research Institute, Herston, QLD, Australia*
VERONIKA LUKACS-KORNEK • *Department of Medicine II, Saarland University Medical Center, Saarland University, Homburg, Germany*

ANDREAS MÖLLER • *Tumour Microenvironment Laboratory, QIMR Berghofer Medical Research Institute, Herston, QLD, Australia; School of Medicine, University of Queensland, Brisbane, QLD, Australia*
GENSHAN MA • *Department of Cardiology, Zhongda Hospital, Medical School of Southeast University, Nanjing, People's Republic of China*
JINGWEI MA • *Department of Immunology, Tongji Medical College, Huazhong University of Science and Technology, Wuhan, China*
PIERRE-YVES MANTEL • *Department of Medicine, Unit of Anatomy, University of Fribourg, Fribourg, Switzerland*
SMART MBAGWU • *Department of Medicine, Unit of Anatomy, University of Fribourg, Fribourg, Switzerland*
DAVID G. MECKES JR • *Department of Biomedical Sciences, Florida State University College of Medicine, Tallahassee, FL, USA*
S. ALEX MITSIALIS • *Division of Newborn Medicine & Department of Medicine, Boston Children's Hospital, Boston, MA, USA; Department of Pediatrics, Harvard Medical School, Boston, MA, USA*
FATEMEH MOMEN-HERAVI • *Columbia College of Dental Medicine, New York, NY, USA*
TAKU MURAKAMI • *Hitachi Chemical Co. America, Ltd. R & D Center, Irvine, CA, USA*
AXL NEURAUTER • *Thermo Fisher Scientific, Oslo, Norway*
SHU-WING NG • *Gynecologic Oncology Division, Department of Obstetrics, Gynecology and Reproductive Biology, Brigham and Women's Hospital, Harvard Medical School, Boston, MA, USA*
MAKIYA NISHIKAWA • *Department of Biopharmaceutics and Drug Metabolism, Graduate School of Pharmaceutical Sciences, Kyoto University, Kyoto, Japan*
KETIL W. PEDERSEN • *Thermo Fisher Scientific, Oslo, Norway*
GORJANA RACKOV • *Instituto Madrileño de Estudios Avanzados, IMDEA Nanociencia, Madrid, Spain; Fundación de Investigación HM Hospitales, Hospital de Madrid Group, Madrid, Spain*
AVIV REGEV • *Broad Institute of MIT and Harvard, Cambridge, MA, USA; Department of Biology, MIT, Cambridge, MA, USA*
WON JONG RHEE • *Division of Bioengineering, Incheon National University, Incheon, South Korea*
MINGYI SHANG • *Department of Radiology, Shanghai Tongren Hospital, Shanghai, China*
YAN SHEN • *Vascular Biology Center, Department of Medicine, Medical College of Georgia, Augusta University, Augusta, GA, USA*
YE-FENG SHENG • *State Key Laboratory for Diagnosis and Treatment of Infectious Diseases, First Affiliated Hospital, College of Medicine, Zhejiang University, Hangzhou, Zhejiang Province, China; Collaborative Innovation Center for Diagnosis and Treatment of Infectious Diseases, Hangzhou, Zhejiang Province, China; Institute for Translational Medicine, Zhejiang University, Hangzhou, Zhejiang Province, China*
KIYOTAKA SHIBA • *Division of Protein Engineering, Cancer Institute, Japanese Foundation for Cancer Research, Tokyo, Japan*
JULIANO C. DA SILVEIRA • *Department of Veterinary Medicine, Faculty of Animal Sciences and Food Engineering, University of São Paulo, São Paulo, Brazil*
AGNIESZKA SIUPA • *Malvern Instruments Ltd, Amesbury, UK*
CHAO SONG • *CloudHealth Genomics, Ltd, Shanghai, China*
YUKI TAKAHASHI • *Department of Biopharmaceutics and Drug Metabolism, Graduate School of Pharmaceutical Sciences, Kyoto University, Kyoto, Japan*

YOSHINOBU TAKAKURA • *Department of Biopharmaceutics and Drug Metabolism, Graduate School of Pharmaceutical Sciences, Kyoto University, Kyoto, Japan*
YAOLIANG TANG • *Vascular Biology Center, Department of Medicine, Medical College of Georgia, Augusta University, Augusta, GA, USA*
KE TANG • *Biochemistry and Molecular Biology, Tongji Medical College, Huazhong University of Science and Technology, Wuhan, China*
DIONNE TANNETTA • *Department of Food and Nutritional Sciences, University of Reading, Reading, UK*
DMITRY TER-OVANESYAN • *Department of Genetics, Harvard Medical School, Boston, MA, USA; Wyss Institute for Biologically Inspired Engineering, Harvard University, Boston, MA, USA; Department of Molecular and Cellular Biology, Harvard University, Cambridge, MA, USA; Broad Institute of MIT and Harvard, Cambridge, MA, USA*
JOHN C. TIGGES • *Beth Israel Deaconess Medical Center, Harvard Medical School, Boston, MA, USA; Flow Cytometry Core, Beth Israel Deaconess Medical Center, Boston, MA, USA*
VASILIS TOXAVIDIS • *Beth Israel Deaconess Medical Center, Harvard Medical School, Boston, MA, USA*
MATT TRAU • *Centre for Personalised Nanomedicine, Australian Institute for Bioengineering and Nanotechnology, The University of Queensland, Brisbane, QLD, Australia; School of Chemistry and Molecular Biosciences, The University of Queensland, Brisbane, QLD, Australia*
MICHAEL TU • *School of Dentistry, University of California, Los Angeles, CA, USA*
SABINE KATHARINA URBAN • *Department of Medicine II, Saarland University Medical Center, Saarland University, Homburg, Germany*
JODY VYKOUKAL • *McCombs Institute for the Early Detection and Treatment of Cancer, The University of Texas MD Anderson Cancer Center, Houston, TX, USA*
MICHAEL WALCH • *Department of Medicine, Unit of Anatomy, University of Fribourg, Fribourg, Switzerland*
XIN WANG • *Center for Biomimetic Medicine, Houston Methodist Research Institute, Houston, TX, USA*
SHUQI WANG • *State Key Laboratory for Diagnosis and Treatment of Infectious Diseases, First Affiliated Hospital, College of Medicine, Zhejiang University, Hangzhou, Zhejiang Province, China; Collaborative Innovation Center for Diagnosis and Treatment of Infectious Diseases, Hangzhou, Zhejiang Province, China; Institute for Translational Medicine, Zhejiang University, Hangzhou, Zhejiang Province, China; Department of Radiology, Bio-Acoustic MEMS in Medicine (BAMM) Laboratory, Canary Center at Stanford for Cancer Early Detection, Stanford School of Medicine, Palo Alto, CA, USA*
CHRIS WANG • *School of Dentistry, University of California, Los Angeles, CA, USA*
AUSTIN WANG • *School of Dentistry, University of California, Los Angeles, CA, USA*
FANG WEI • *School of Dentistry, University of California, Los Angeles, CA, USA*
NEAL L. WEINTRAUB • *Vascular Biology Center, Department of Medicine, Medical College of Georgia, Augusta University, Augusta, GA, USA*
GARETH R. WILLIS • *Division of Newborn Medicine & Department of Medicine, Boston Children's Hospital, Boston, MA, USA; Department of Pediatrics, Harvard Medical School, Boston, MA, USA*
DAVID T.W. WONG • *School of Dentistry, University of California, Los Angeles, CA, USA*
ANTHONY YAN-TANG WU • *Institute of Biomedical Engineering, National Tsing Hua University, Hsinchu, Taiwan; Institute of Atomic and Molecular Sciences, Academia Sinica, Taipei, Taiwan*

Brandon A. Wyse • *CreATe Fertility Centre, Toronto, ON, Canada*
Cindy M. Yamamoto • *Hitachi Chemical Co. America, Ltd. R & D Center, Irvine, CA, USA*
Satoshi Yamamoto • *Division of Protein Engineering, Cancer Institute, Japanese Foundation for Cancer Research, Tokyo, Japan; Department of Oral & Maxillofacial Implantology, Tokyo Dental College, Tokyo, Japan*
Katherine Yang • *Cancer Program, MGH Center for Systems Biology, Massachusetts General Hospital, Harvard Medical School, Boston, MA, USA*
Hang Yin • *Department of Chemistry & Biochemistry, University of Colorado Boulder, Boulder, CO, USA; BioFrontiers Institute, University of Colorado Boulder, Boulder, CO, USA*
Mitsutaka Yoshida • *Division of Protein Engineering, Cancer Institute, Japanese Foundation for Cancer Research, Tokyo, Japan; Department of Oral & Maxillofacial Implantology, Tokyo Dental College, Tokyo, Japan*
Hong Yu • *Department of Cardiology, Second Affiliated Hospital, College of Medicine, Zhejiang University, Hangzhou, People's Republic of China*
Yuana Yuana • *Imaging Division, Image Sciences Institute, University Medical Centre Utrecht, Utrecht, The Netherlands*
Natasa Zarovni • *Exosomics Siena SpA, Siena, Italy*
Sherry Zhou • *Department of Radiology, Bio-Acoustic MEMS in Medicine (BAMM) Laboratory, Canary Center at Stanford for Cancer Early Detection, Stanford School of Medicine, Palo Alto, CA, USA*
Davide Zocco • *Exosomics Siena SpA, Siena, Italy*

Chapter 1

Extracellular Vesicles: A Brief Overview and Its Role in Precision Medicine

Mingyi Shang, John S. Ji, Chao Song, Bao Jun Gao, Jason Gang Jin, Winston Patrick Kuo, and Hongjun Kang

Abstract

Precision medicine has emerged as an approach to tailor therapies for an individual at the time of diagnosis and/or treatment. This emergence has been fueled by the ability to profile nucleic acids, along with proteins and lipids isolated from biofluids, a method called "liquid biopsy," either by or in combination of one of the following components: circulating tumor cells (CTCs), cell-free DNA (cfDNA), and/or extracellular vesicles (EVs). EVs are membrane-surrounded structures released by cells in an evolutionarily conserved manner. EVs have gained much attention from both the basic and clinical research areas, as EVs appear to play a role in many diseases; however, the well-known case is cancer. The hallmark of EVs in cancer is their role as mediators of communication between cells both at physiological and pathophysiological levels; hence, EVs are thought to contribute to the creation of a microenvironmental niche that promotes cancer cell survival, as well as reprogramming distant tissue for invasion. It is important to understand the mechanistic and functional aspects at the basic science level of EVs to get a better grasp on their role in healthy and disease states. EVs range from 30–1000 nm membrane-enclosed vesicles that are released by many mammalian cell types and present in a variety of biofluids. EVs have emerged as an area of clinical interest in the era of Precision Medicine, from their role in liquid biopsy (tissue biopsy free) approach for screening, assessing tumor heterogeneity, monitoring therapeutic responses, and minimal residual disease detection to EV-based therapeutics. EVs' diagnostic and therapeutic exploitation is under intense investigation in various indications. This chapter highlights EV biogenesis, composition of EVs, and their potential role in liquid biopsy diagnostics and therapeutics in the area of cancer.

Key words Exosomes, Biogenesis, Extracellular vesicles, Precision medicine, Liquid biopsy, Diagnostics, Therapeutics

1 Tissue Biopsy versus Liquid Biopsy

Many initial clinical and translational applications of EVs have been in cancer; however, EVs do play a role in many other pathophysiological processes. The current management of cancer relies on a combination of imaging, tissue biopsy, and appropriate blood work for the diagnosis, monitoring and molecular classification-based patient stratification for appropriate treatment (Fig. 1), where the

Winston Patrick Kuo and Shidong Jia (eds.), *Extracellular Vesicles: Methods and Protocols*, Methods in Molecular Biology, vol. 1660, DOI 10.1007/978-1-4939-7253-1_1, © Springer Science+Business Media LLC 2017

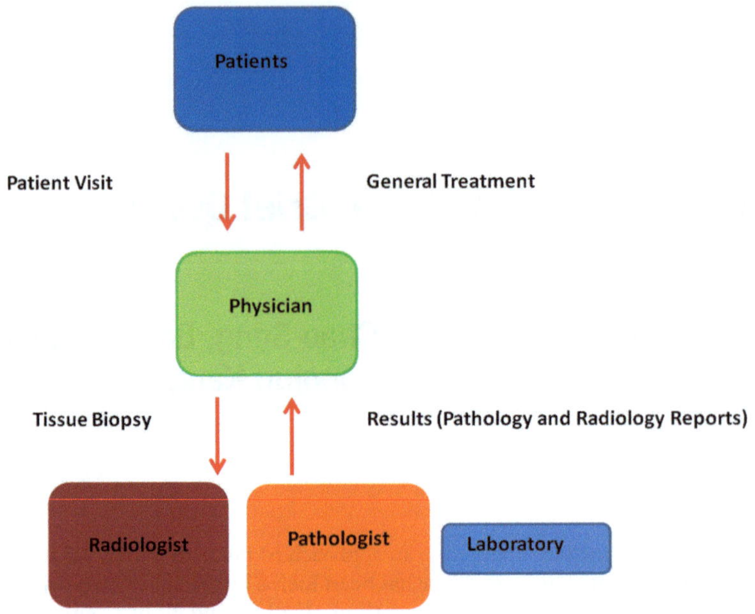

Fig. 1 Current standard-of-care process: the patient visits physician, the physician requests a radiograph, a tissue biopsy, and a blood workup. The pathologist and radiologist play an important role in diagnosis and treatment, and the laboratory plays a small role

emphasis is on the reports provided by the pathologist and radiologist. Conventional standard of care when a patient is diagnosed with cancer, be it Stage II or later stages, a tissue biopsy is generally performed, an invasive procedure, harvesting the tumor sample, which is often challenging in patients with advanced disease. Given the well-recognized intratumor genetic heterogeneity [1, 2], biopsy of small tumor fragments does not necessarily represent all genetic aberrations in the tumor, but sampling the entire tumor in each patient is not realistic either in most cases (Fig. 2). In a tissue biopsy, one might detect a mutation, and mistakenly, miss other mutations, thus delivering the inappropriate therapy, resulting in a relapse of the tumor. Moreover, tumors evolve all the time from local to advanced stages and by adapting to selective pressure from treatment. For diagnosing and better monitoring of molecular tumor changes, liquid biopsy is emerging as a more convenient alternative to tissue biopsy, with its less risk to the patient and potentially lower cost. Cancer is currently being genetically characterized from a blood sample by next-generation sequencing (NGS), where the detection of genetic aberrations can come from any of these components, CTCs, cfDNA, and/or EVs. The future paradigm in accessing a patient will have more emphasis on the molecular profiling of the circulating biomarkers and less on the pathologist and radiologist; however, their roles will complement

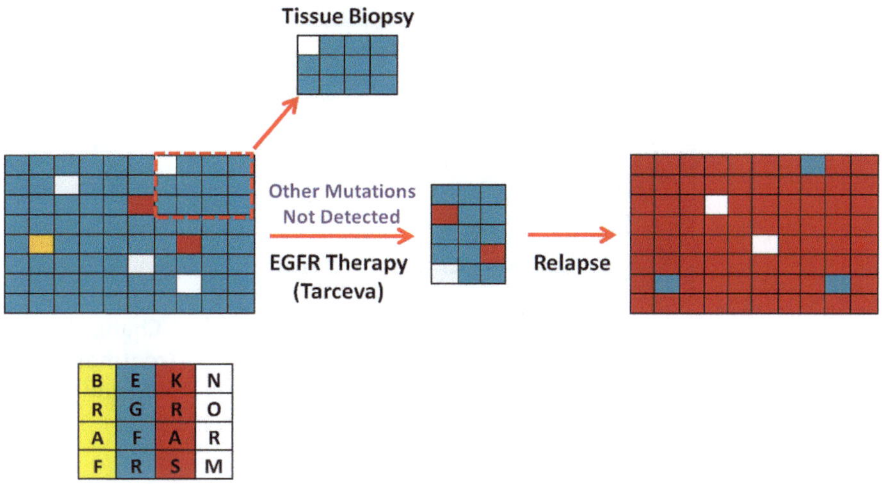

Fig. 2 Tissue biopsy issue: A tumor is present with several mutations (EGFR, KRAS and BRAF), tissue biopsy might not pick up all the mutations or even a resistance to a drug in this example. Based on the tissue biopsy, a mutation is detected (EGFR), missing the BRAF and KRAS mutations, but based on the mutation detected, the appropriate drug is given, however not accounting for the other mutations, and a relapse occurs

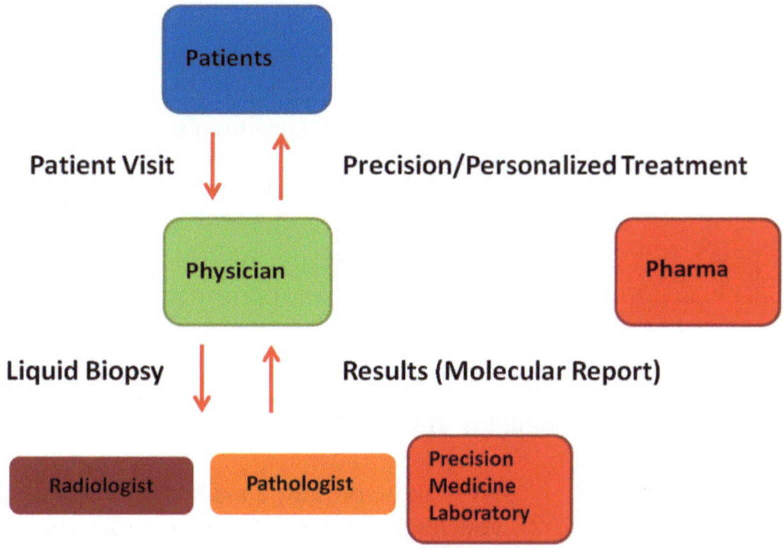

Fig. 3 Future standard-of-care: the radiologist and pathologist play a smaller role, whereas the laboratory and pharma are the major players. Instead of a tissue biopsy, a liquid biopsy is collected and processed in the lab for genetic testing and the results (mutations) are correlated with actionable drug therapies that are available, thus providing precision medicine

each other and have more value after a diagnosis is detected by way of liquid biopsy, minimizing surgical risk (Fig. 3). Blood samples are easily accessible and therefore save the cost for time in operating room and the ability to potentially detect mutations as compared to

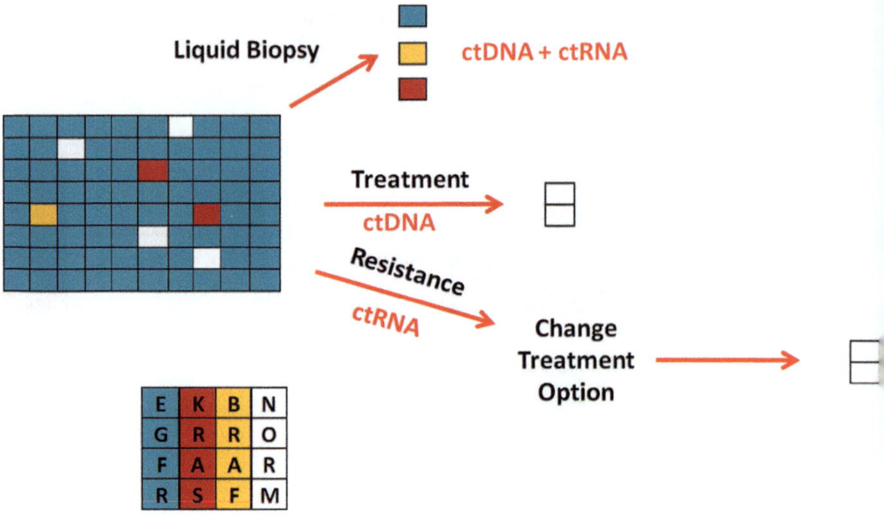

Fig. 4 Liquid biopsy concept: A blood sample is collected versus a tissue biopsy, where you can detect all the mutations (EGFR, BRAF, and KRAS), in addition, you can detect a drug resistance/sensitivity by way of ctRNA and give the appropriate treatment

a tissue biopsy to provide the appropriate treatment, even if a drug resistance or sensitivity is detected (Fig. 4). Another advantage is the serial blood specimen's collection at different stages of cancer management, thus enabling real-time tumor monitoring. The concept of Precision Medicine is of special interest for the global health-care ecosystem in selecting the appropriate personalized treatment/therapy, of which the goal is to provide "the right patient with the appropriate drug at the right dose at the right time" [3].

2 What are Extracellular Vesicles?

EVs can be broadly classified into three main subgroups: (a) microvesicles/microparticles/ectosomes that are produced by outward budding and fission of the plasma membrane; (b) exosomes that are formed within the endosomal network and released upon fusion of multivesicular bodies with the plasma membrane; and (c) apoptotic bodies are released as blebs of cells undergoing apoptosis. EVs contain lipids, membrane and cytosolic proteins, along with nucleic acids (miRNA, RNA, and DNA), which make them a key player in intercellular communication. Released EVs provide unconventional methods of cellular communication and exchange of bioactive molecules between cells. EVs allow long-distance communication between cells without the need for direct cell–cell contact, permitting cells having new surface molecules that may help alter the cellular signaling in response to particular stimuli; such biological properties of EVs make them excellent candidates for therapeutics.

The first report of nanometer-sized vesicles came from the analysis of normal and neoplastic cell lines that demonstrated populations of EVs 40 nm in diameter released constitutively by cells [4]. Around the same time, others demonstrated the presence of EVs using gel chromatography and ultracentrifugation [5, 6]. Electron microscopy of vesicles released during erythrocyte transferrin receptor exocytosis showed the potential secreted contents of late endosomes released into the extracellular space through EV-like vesicular structures [7]. Thus, the function of EVs was initially thought to be solely the secretion of transferrin receptors and other unwanted membrane proteins during the maturation of erythrocytes; however, in the early 1980s, intracellular late endosomes, spherical cellular structures containing many lumenal vesicles were considered part of the predegenerative complex process. The results were therefore criticized as being membranes shed by dying cells and less attention on EVs at that time, until the recent emergence of several publications demonstrating the extracellular release of these vesicles through fusion [8] between multivesicular bodies (MVB) containing late endosomes and plasma membranes of cytotoxic T lymphocytes [9], dendritic cells (DCs) [10], B lymphocytes [11], mast cells [12], neuronal cells [13], sperm [14], melanoma cells [15], ovarian carcinoma cells [16], and salivary gland epithelial cells [17]. These studies support EVs as entities released by living cells by secretory mechanisms that are distinct from degenerative processes; therefore, most cells in the human body release EVs by their involvement of the endocytic pathway followed by MVB with the plasma membrane.

3 Biogenesis of EVs

The formation of EVs occurs via the classic endocytic pathway [18–21]. Reverse budding occurs constitutively as the plasma membrane begins to stretch inward, whereby the vesicle pinches off into the intracellular space and continues down the endocytic pathway, encountering early endosomes, intraluminal vesicles (ILVs), and multivesicular bodies (MVBs) (late endosomes) [22, 23]. The path diverges, either releasing a transformed version of the initial vesicle back into the extracellular space as an EV or being digested and degraded in a lysosome (Fig. 5).

While this is the accepted pathway by which exosomes form, it is important to note that other pathways exist for the formation of other microvesicles. Ectosomes, for example, form in response to various signals on the surface of the cell membrane such as complement activation. The complement system is composed of over twenty soluble and membrane bound proteins with critical roles

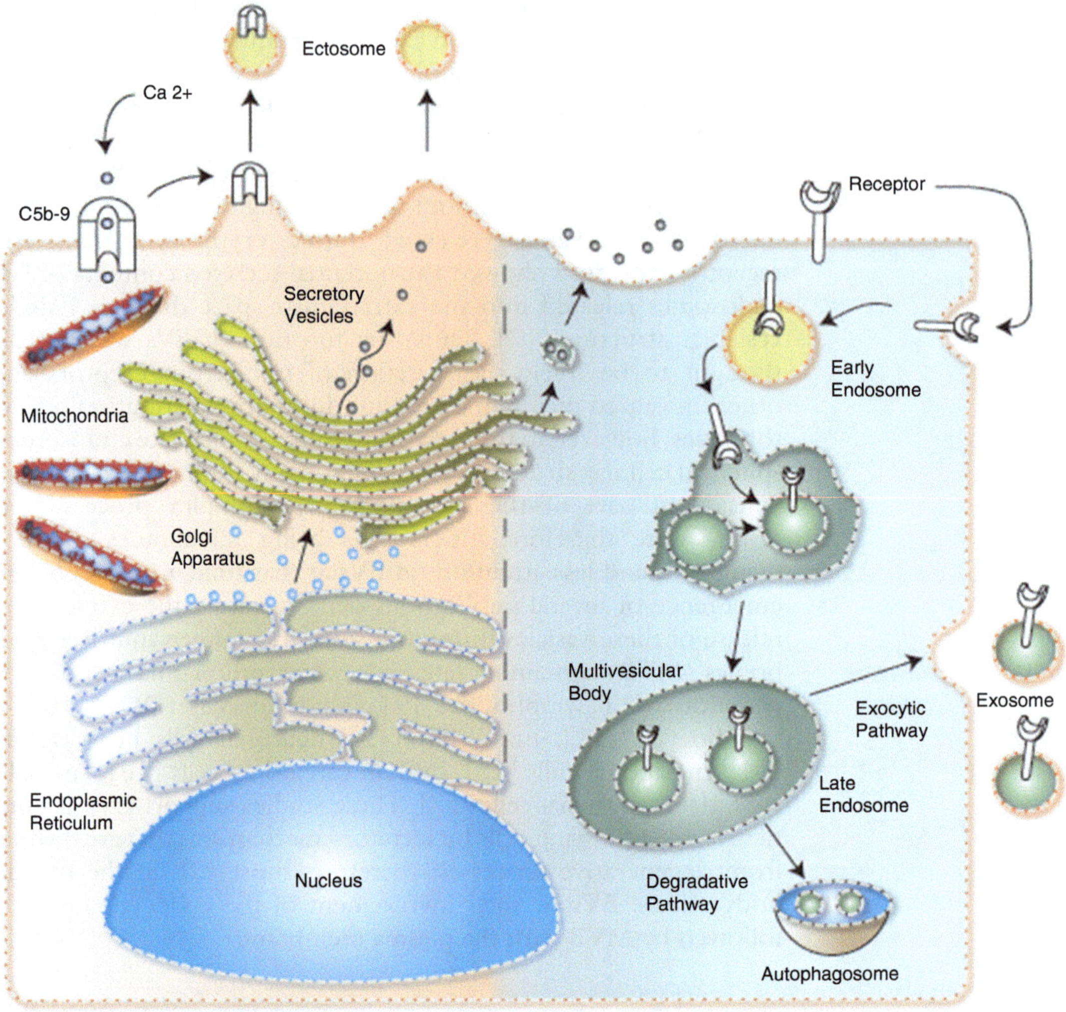

Fig. 5 Biogenesis of exosomes. Exosomes are formed by the endocytic/exocytic pathway from intraluminal vesicles and when MVBs interact with the plasma membrane of the cell. Shedding vesicles, ectosomes, and other particles form by either intracellular means or are released during complement-activated attack

in recognizing, binding, and removal of foreign particles as well as initiating and regulating innate and acquired immune responses. Activation of the complement system can occur through any of the classical, alternative, or lectin pathways, and culminates in homo-oligomerization and insertion of the membrane attack complex (MAC) pore components in the plasma membrane.

Many proteins are important for the formation of EVs. Vesicles initially form from specialized coated regions on the plasma membrane by pinching of budding vesicles, or transport vesicles, mediated by clathrin-lined pits [24]. Clathrin, a protein responsible for the formation of coated vesicles, is known to mediate shuttling of intracellular vesicles from the plasma membrane, Golgi apparatus, and early and late endosomes. Clathrin is a coated protein that begins to cover the outside membrane of inward budding vesicles and controls their size and shape [25]. Adaptor protein (AP) complexes linked with clathrin create selective attachment of membranes and molecules in the creation of the transport vesicle [26]. Different adaptor proteins, such as AP-2, AP-1A, AP-3A, and AP-4, are associated with various stages of endocytosis [26]. Dynamin, a GTPase protein, surrounds the neck of the forming vesicle and helps the newly formed membranes to fuse [27, 28].

Rab proteins, a subfamily of monomeric GTPases, direct transport vesicles to specific spots on the target membrane [23, 29]. As part of the endocytic pathway, the transport vesicle will use the Rab proteins to bind to an early endosome [21, 22].

Early endosomes are located near the plasma membrane and travel into the cell till the membrane pinches inward forming an ILV. Ubiquitin, a very important protein complex that operates at a number of transport steps, is required to form ILVs and help guide extracellular proteins to form clathrin-coated vesicles (Fig. 6). Ubiquitin molecules are added to proteins in order to sort them in the endocytic clathrin-dependent pathway [29]. The role of oligo-ubiquitination machinery is to provide the sorting signals and increase MVB sorting efficiency [30].

The ubiquitin tags are also recognized by the endosomal sorting complex required for transport (ESCRT), a cytosolic protein complex required to mediate the sorting process of ILVs [19] (Fig. 6). The ESCRT protein complex comprises four protein complexes; ESCRT-0, ESCRT-I, ESCRT-II, and ESCRT-III; their role is to pass protein cargo from the early endosome to the site of transmembrane protein sorting process and ILV formation [20, 21, 25, 30, 31]. In order for the ESCRT complex to bind to the membrane, these various complexes require ubiquitinated proteins as well as phosphatidylinositol-3 phosphate (PI(3)P) created when a lipid kinase phosphorylates phosphatidylinositol [29]. Once this process is accomplished, the ESCRT complex breaks off forming the late endosome, or MVBs, via adenosine triphosphatase vacuolar protein sorting 4 (Vps4), an AAA-type ATPase, and is recycled back into the cytoplasm [29, 30, 32].

As ILVs begin to accumulate in the endosome, the vesicle begins to mature from an early endosome to a late endosome. The maturation of an early endosome into a late endosome is generally characterized by the presence of MVBs [33]. MVBs are endosomes that have accumulated ILVs [21, 25]. The presence of the various Rab proteins has also been used to identify the maturation of endosomes [25, 34]. Rab 4, 5, and 11 proteins have been found in early endosomes in contrast to Rab 7 and 9, which have been identified on late endosomes [20, 25, 34]. As mentioned, Rab5 can be found in early endosomes and mediates endocytosis as well as the fusion of transport vesicles with the membrane of the

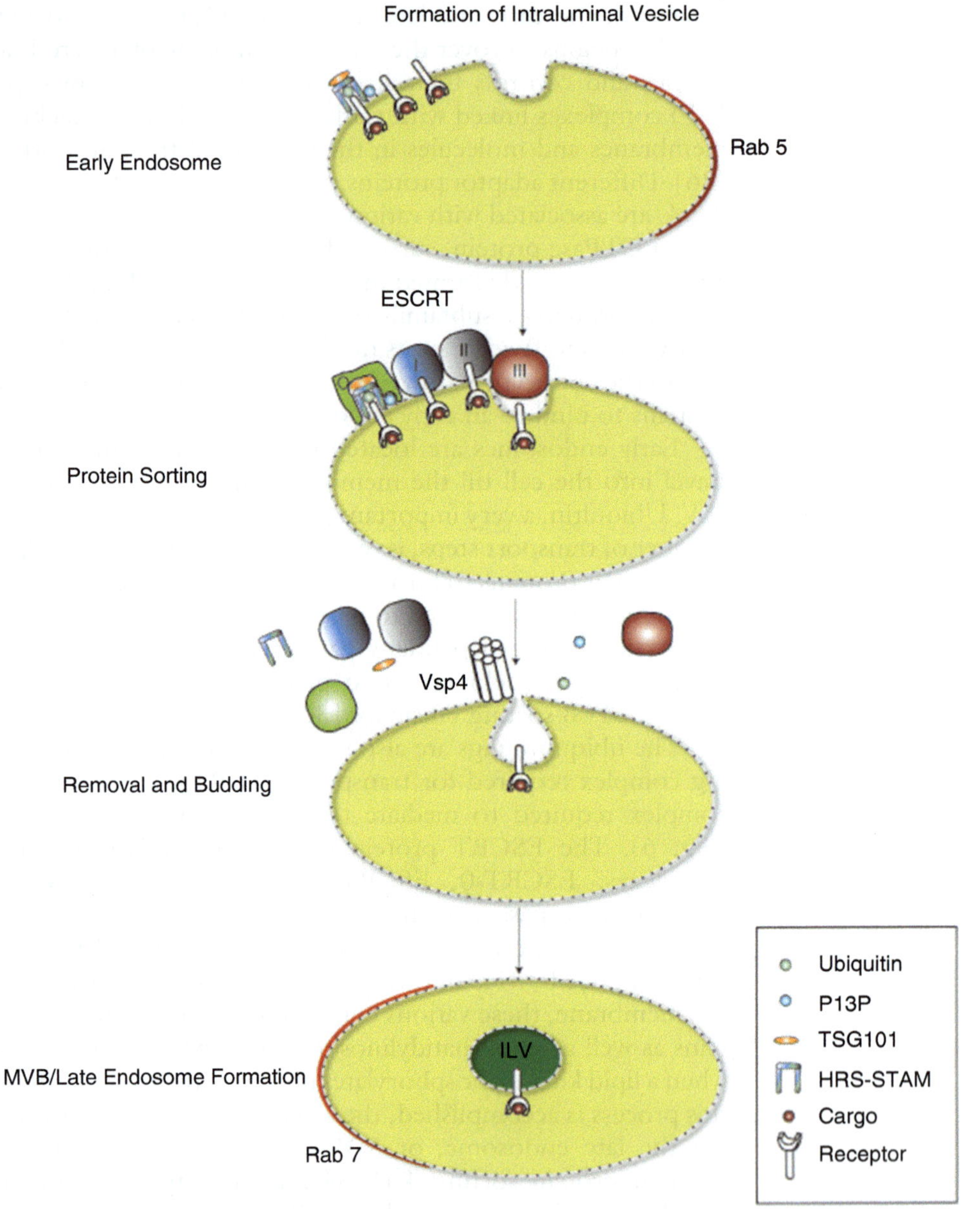

Fig. 6 Formation of intraluminal vesicles. Ubiquitin-tagged proteins are guided into clathrin-coated pits via the four parts of the endosomal sorting complex required for transport (ESCRT). Adenosine triphosphate vacuolar protein sorting (Vsp-4) mediates the dissociation of the ESCRT as the intraluminal vesicle (ILV) is formed in the MVB. Rab proteins indicate the maturation of the endosome

early endosomes [20, 25, 34]. An increase in EV secretion in Rab11 overexpressing cells with high intracellular calcium can be observed, indicating an involvement of Rab11 and calcium metabolism in the process of EV biogenesis [35]. Other evidence indicates that the accumulation of cholesterol in endosome membranes regulates Rab7 expression, which in turn controls the release of EVs [36].

MVBs (late endosomes) can eventually continue down one of two pathways, degradative or exocytotic pathways [37]. In the degradative pathway, the MVB fuses with a lysosome, becoming an autophagosome, and the MVB and its cargo are degraded [20] and serves as a means of downregulating cell receptors such as activated growth factor receptors [38] (Fig. 5). In the exocytic pathway, MVBs fuse with the plasma membrane of the host cell, leading to the release of ILVs into the extracellular space as EVs [19, 20, 30].

4 Diagnostic Potential of EVs

The presence of biomarker-containing EVs in serum [39], plasma [40], urine [41], cerebrospinal fluid [42], saliva [43], and bronchoalveolar lavage fluid [44] along with a variety of molecules that are found within them, such as nucleic acids, proteins, and lipids, can yield molecular signature patterns that are informative of the state of the human physiology or disease condition which can be exploited as detection tools. Fragile molecules such as RNA are rendered resistant to degradation by RNases in body fluids when packaged within EVs.

An area of intense investigation is the identification of EV biomarkers released by tumors in order to facilitate cancer diagnosis. Initially, it was difficult to distinguish EVs per se from other EVs for them to have clinical utility in "cancer diagnostics." Recent studies clearly demonstrate the feasibility of this specific approach [20, 45]. The ability to detect cancer early in its progression is necessary in order to dramatically improve the patient's odds for successful treatment, survival, and quality of life. It is generally thought that tumor-derived EVs express a component of the protein antigens and nucleic acids (DNA, mRNAs, and miRNAs) that are expressed in the original tumor. While EV release is a common component across all proliferating cell types, it is enhanced in tumor cells and generally in response to cellular stress [46]. The mRNA extracted from the tumor EVs appears to contain a snapshot of the tumor transcriptome, with serum EVs from glioblastoma patients contain the EGFRvIII mutant transcript as well as upregulated mRNA and miRNA associated with the tumor [47]. In order to advance the application of EV-based diagnostics to prostate cancer, researchers have investigated the detection of two well-described prostate-specific biomarkers, PCA-3 and TMPRSS2-ERG fusion transcripts [48, 49]. Both PCA-3 and TMPRSS2-ERG chimeric mRNA transcripts were found in EVs isolated from urine [50]. In addition, the identification of key transcripts within EVs in urine has the advantages of being both noninvasive and potentially informative as to the overall tumor genotype and malignancy status of the tumor. Generally, miRNAs are considered to be more stable than mRNAs in serum or whole blood. miRNAs such as miR-21, miR-106a, miR-146, miR-155, miR-191, miR-192, miR-203, miR-205, miR-210, miR-212, and miR-214 are present in plasma EVs from lung cancer patients [46]. A comparison between peripheral circulation miRNA-derived from EVs and miRNA-derived from the lung tumors indicated that the miRNA signatures were not significantly different. However, significant differences were observed between the mean EV-derived miRNA concentrations of the lung adenocarcinoma group and the control group.

miR-21, miR-141, miR-200a, miR-200c, miR-200b, miR-203, miR-205, and miR-214 are useful markers to detect various stages of ovarian cancer in tumor-derived EVs using anti-EpCAM [51]. According to the authors, these ovarian cancer-specific miRNAs profiles could not be isolated in healthy individuals. The increase of EV production during oncogenesis and the elevated expression of miRNA in EVs of cancer patients are excellent features for the development of noninvasive diagnostic assays. Additional studies and further validation are necessary to identify the appropriate mRNA and miRNA signatures and confirm their utility of EV isolation in the prospective screening of patients.

5 Therapeutic Potential of EVs

EVs are an alternative to liposomes in the delivery of therapeutic agents as they are composed of natural nonsynthetic components, and their small size and flexibility enable them to cross major biological membranes. Their lipid bilayer structure protects the cargo from degradation, facilitating delivery to its target [52–54]. EVs are naturally occurring secreted membrane vesicles, are less toxic, and are better tolerated in the body as evidenced by their ubiquitous presence in biological fluids [53]. EV delivery systems are recognized by the immune system as "self," thus avoiding rejection. Investigations using EVs have been used to deliver anti-inflammatory agents, such as curcumin, to activated myeloid cells in vivo, demonstrating both the stability and bioavailability of the agents in the blood [55]. Curcumin has been reported to inhibit the suppressive activity of T-cells by downregulation of TGF-β and IL-10 and, in addition, by inducing effector T-cells to destroy cancer cells [55].

EVs can be used as therapeutic vehicles to deliver RNA interference (siRNA) and miRNA regulatory molecules to suppress the growth of cancer cells [56]. For example, EVs containing let-7a miRNA were targeted against EGFR-expressing breast cancerous cells by intravenous injection in RAG2(−/−) mice where the donor cells were instructed to express the transmembrane domain of platelet-derived growth factor receptor fused to the GE11 peptide

[57]. In addition, EVs have been considered as a potential therapeutic tool in modulating neovascularization. Endothelial-derived EVs containing proteins such as Delta-like 4 (a transmembrane ligand for Notch receptors—expressed in arterial blood vessels and sprouting endothelial cells) and matrix metalloproteinases promote angiogenesis [58–60].

DC-derived EVs has been proposed, and promising results have resulted in two Phase I clinical trials as cell-free vaccines [61]. DC-derived EVs combined therapy has been successfully utilized as a maintenance immunotherapy in cancer patients with nonoperable non-small cell lung cancer in Phase II clinical trials [62]. All this data essentially indicates the potential of successful delivery of cytokines, nucleic acids, vaccine adjuvants, fluorescent labels, costimulatory signals, and gene therapy vectors using EVs to combat different pathological entities.

6 Conclusion

In summary, EVs are multifunctional biological entities that are secreted from many mammalian cells and play an active role in major biological processes such as regulation of immune response, antigen presentation, transfer of bioactive molecules between cells, and transfer of viruses and prions. The dramatic increase in understanding of EV biogenesis and biology strongly suggests the huge potential of EVs for the development of new diagnostic assays and therapeutic approaches. EVs provide a practical liquid biopsy approach to develop assays for screening, assessing tumor heterogeneity, monitoring therapeutic responses, and minimal residual disease detection. As EVs have gained attention for their role in intercellular cell communications and insights into disease processes, data has suggested that EVs may provide an efficient tool for therapeutic delivery; however, much work remains to be done to ensure their safety and effective use. EVs are attractive over existing drug delivery systems because of their small size, lack of toxicity, and target specificity. In addition, the range of therapeutic applications is quite promising from the use in immunotherapeutics using checkpoint inhibitors, to vaccines. The interaction of EVs with their biological targets still remains largely unknown, and hence, more research is required to achieve successful EV delivery to target cells. In conclusion, the potential impact to Precision Medicine initiatives warrants a systematic undertaking to develop new methods to interrogate EV populations across biological fluids and understanding their therapeutic potential.

References

1. Marusyk A, Almendro V, Polyak K (2012) Intra-tumour heterogeneity: a looking glass for cancer? Nat Rev Cancer 12(5):323–334. doi:10.1038/nrc3261
2. Janku F (2014) Tumor heterogeneity in the clinic: is it a real problem? Ther Adv Med Oncol 6(2):43–51. doi:10.1177/1758834013517414
3. Sadee W, Dai Z (2005) Pharmacogenetics/geno-mics and personalized medicine. Hum Mol Genet 14(2):R207–R214. doi:10.1093/hmg/ddi261
4. Trams EG, Lauter CJ, Salem N Jr, Heine U (1981) Exfoliation of membrane ecto-enzymes in the form of micro-vesicles. Biochim Biophys Acta 645(1):63–70
5. Taylor DD, Chou IN, Black PH (1983) Isolation of plasma membrane fragments from cultured murine melanoma cells. Biochem Biophys Res Commun 113(2):470–476
6. Taylor DD, Doellgast GJ (1979) Quantitation of peroxidase-antibody binding to membrane fragments using column chromatography. Anal Biochem 98(1):53–59
7. Harding C, Heuser J, Stahl P (1983) Receptor-mediated endocytosis of transferrin and recycling of the transferrin receptor in rat reticulocytes. J Cell Biol 97(2):329–339
8. Meckes DG Jr, Shair KH, Marquitz AR, Kung CP, Edwards RH, Raab-Traub N (2010) Human tumor virus utilizes exosomes for intercellular communication. Proc Natl Acad Sci U S A 107(47):20370–20375. doi:10.1073/pnas.1014194107
9. Peters PJ, Geuze HJ, Van der Donk HA, Slot JW, Griffith JM, Stam NJ, Clevers HC, Borst J (1989) Molecules relevant for T cell-target cell interaction are present in cytolytic granules of human T lymphocytes. Eur J Immunol 19 (8):1469–1475. doi:10.1002/eji.1830190819
10. Zitvogel L, Regnault A, Lozier A, Wolfers J, Flament C, Tenza D, Ricciardi-Castagnoli P, Raposo G, Amigorena S (1998) Eradication of established murine tumors using a novel cell-free vaccine: dendritic cell-derived exosomes. Nat Med 4(5):594–600
11. Raposo G, Nijman HW, Stoorvogel W, Liejendekker R, Harding CV, Melief CJ, Geuze HJ (1996) B lymphocytes secrete antigen-presenting vesicles. J Exp Med 183 (3):1161–1172
12. Raposo G, Tenza D, Mecheri S, Peronet R, Bonnerot C, Desaymard C (1997) Accumulation of major histocompatibility complex class II molecules in mast cell secretory granules and their release upon degranulation. Mol Biol Cell 8(12):2631–2645
13. Faure J, Lachenal G, Court M, Hirrlinger J, Chatellard-Causse C, Blot B, Grange J, Schoehn G, Goldberg Y, Boyer V, Kirchhoff F, Raposo G, Garin J, Sadoul R (2006) Exosomes are released by cultured cortical neurones. Mol Cell Neurosci 31(4):642–648. doi:10.1016/j.mcn.2005.12.003
14. Sullivan R, Saez F, Girouard J, Frenette G (2005) Role of exosomes in sperm maturation during the transit along the male reproductive tract. Blood Cells Mol Dis 35(1):1–10. doi:10.1016/j.bcmd.2005.03.005
15. Mears R, Craven RA, Hanrahan S, Totty N, Upton C, Young SL, Patel P, Selby PJ, Banks RE (2004) Proteomic analysis of melanoma-derived exosomes by two-dimensional polyacrylamide gel electrophoresis and mass spectrometry. Proteomics 4(12):4019–4031. doi:10.1002/pmic.200400876
16. Safaei R, Larson BJ, Cheng TC, Gibson MA, Otani S, Naerdemann W, Howell SB (2005) Abnormal lysosomal trafficking and enhanced exosomal export of cisplatin in drug-resistant human ovarian carcinoma cells. Mol Cancer Ther 4(10):1595–1604. doi:10.1158/1535-7163.MCT-05-0102
17. Kapsogeorgou EK, Abu-Helu RF, Moutsopoulos HM, Manoussakis MN (2005) Salivary gland epithelial cell exosomes: a source of autoantigenic ribonucleoproteins. Arthritis Rheum 52(5):1517–1521. doi:10.1002/art.21005
18. Castellana D, Toti F, Freyssinet JM (2010) Membrane microvesicles: macromessengers in cancer disease and progression. Thromb Res 125(Suppl 2):S84–S88. doi:10.1016/S0049-3848(10)70021-9
19. Cocucci E, Racchetti G, Meldolesi J (2009) Shedding microvesicles: artefacts no more. Trends Cell Biol 19(2):43–51. doi:10.1016/j.tcb.2008.11.003
20. Mathivanan S, Lim JW, Tauro BJ, Ji H, Moritz RL, Simpson RJ (2010) Proteomics analysis of A33 immunoaffinity-purified exosomes released from the human colon tumor cell line LIM1215 reveals a tissue-specific protein signature. Mol Cell Proteomics 9(2):197–208. doi:10.1074/mcp.M900152-MCP200
21. Sorkin A, Goh LK (2009) Endocytosis and intracellular trafficking of ErbBs. Exp Cell Res 315(4):683–696
22. Keller S, Sanderson MP, Stoeck A, Altevogt P (2006) Exosomes: from biogenesis and

secretion to biological function. Immunol Lett 107(2):102–108. doi:10.1016/j.imlet.2006.09.005
23. Thery C, Zitvogel L, Amigorena S (2002) Exosomes: composition, biogenesis and function. Nat Rev Immunol 2(8):569–579. doi:10.1038/nri855
24. Nazarenko I, Rana S, Baumann A, McAlear J, Hellwig A, Trendelenburg M, Lochnit G, Preissner KT, Zoller M (2010) Cell surface tetraspanin Tspan8 contributes to molecular pathways of exosome-induced endothelial cell activation. Cancer Res 70(4):1668–1678. doi:10.1158/0008-5472.CAN-09-2470
25. Sorkin A, von Zastrow M (2009) Endocytosis and signalling: intertwining molecular networks. Nat Rev Mol Cell Biol 10(9):609–622. doi:10.1038/nrm2748
26. Ohno H (2006) Clathrin-associated adaptor protein complexes. J Cell Sci 119(Pt 18):3719–3721. doi:10.1242/jcs.03085
27. Praefcke GJ, Ford MG, Schmid EM, Olesen LE, Gallop JL, Peak-Chew SY, Vallis Y, Babu MM, Mills IG, McMahon HT (2004) Evolving nature of the AP2 alpha-appendage hub during clathrin-coated vesicle endocytosis. EMBO J 23(22):4371–4383. doi:10.1038/sj.emboj.7600445
28. van Dam EM, Stoorvogel W (2002) Dynamin-dependent transferrin receptor recycling by endosome-derived clathrin-coated vesicles. Mol Biol Cell 13(1):169–182. doi:10.1091/mbc.01-07-0380
29. Lakkaraju A, Rodriguez-Boulan E (2008) Itinerant exosomes: emerging roles in cell and tissue polarity. Trends Cell Biol 18(5):199–209. doi:10.1016/j.tcb.2008.03.002
30. van Niel G, Wubbolts R, Ten Broeke T, Buschow SI, Ossendorp FA, Melief CJ, Raposo G, van Balkom BW, Stoorvogel W (2006) Dendritic cells regulate exposure of MHC class II at their plasma membrane by oligoubiquitination. Immunity 25(6):885–894. doi:10.1016/j.immuni.2006.11.001
31. Piper RC, Luzio JP (2001) Late endosomes: sorting and partitioning in multivesicular bodies. Traffic 2(9):612–621
32. Yeo SC, Xu L, Ren J, Boulton VJ, Wagle MD, Liu C, Ren G, Wong P, Zahn R, Sasajala P, Yang H, Piper RC, Munn AL (2003) Vps20p and Vta1p interact with Vps4p and function in multivesicular body sorting and endosomal transport in Saccharomyces cerevisiae. J Cell Sci 116(Pt 19):3957–3970. doi:10.1242/jcs.00751
33. Thery C, Ostrowski M, Segura E (2009) Membrane vesicles as conveyors of immune responses. Nat Rev Immunol 9(8):581–593. doi:10.1038/nri2567
34. Rink J, Ghigo E, Kalaidzidis Y, Zerial M (2005) Rab conversion as a mechanism of progression from early to late endosomes. Cell 122(5):735–749. doi:10.1016/j.cell.2005.06.043
35. Savina A, Vidal M, Colombo MI (2002) The exosome pathway in K562 cells is regulated by Rab11. J Cell Sci 115(Pt 12):2505–2515
36. Feng D, Zhao WL, Ye YY, Bai XC, Liu RQ, Chang LF, Zhou Q, Sui SF (2010) Cellular internalization of exosomes occurs through phagocytosis. Traffic 11(5):675–687. doi:10.1111/j.1600-0854.2010.01041.x
37. Trajkovic K, Hsu C, Chiantia S, Rajendran L, Wenzel D, Wieland F, Schwille P, Brugger B, Simons M (2008) Ceramide triggers budding of exosome vesicles into multivesicular endosomes. Science 319(5867):1244–1247. doi:10.1126/science.1153124
38. Futter CE, Pearse A, Hewlett LJ, Hopkins CR (1996) Multivesicular endosomes containing internalized EGF-EGF receptor complexes mature and then fuse directly with lysosomes. J Cell Biol 132(6):1011–1023
39. Logozzi M, De Milito A, Lugini L, Borghi M, Calabro L, Spada M, Perdicchio M, Marino ML, Federici C, Iessi E, Brambilla D, Venturi G, Lozupone F, Santinami M, Huber V, Maio M, Rivoltini L, Fais S (2009) High levels of exosomes expressing CD63 and caveolin-1 in plasma of melanoma patients. PLoS One 4(4):e5219. doi:10.1371/journal.pone.0005219
40. Caby MP, Lankar D, Vincendeau-Scherrer C, Raposo G, Bonnerot C (2005) Exosomal-like vesicles are present in human blood plasma. Int Immunol 17(7):879–887. doi:10.1093/intimm/dxh267
41. Pisitkun T, Shen RF, Knepper MA (2004) Identification and proteomic profiling of exosomes in human urine. Proc Natl Acad Sci U S A 101(36):13368–13373. doi:10.1073/pnas.0403453101
42. Vella LJ, Greenwood DL, Cappai R, Scheerlinck JP, Hill AF (2008) Enrichment of prion protein in exosomes derived from ovine cerebral spinal fluid. Vet Immunol Immunopathol 124(3–4):385–393. doi:10.1016/j.vetimm.2008.04.002
43. Sharma S, Rasool HI, Palanisamy V, Mathisen C, Schmidt M, Wong DT, Gimzewski JK (2010) Structural-mechanical characterization of nanoparticle exosomes in human saliva, using correlative AFM, FESEM, and force spectroscopy. ACS Nano 4(4):1921–1926. doi:10.1021/nn901824n

44. Qazi KR, Torregrosa Paredes P, Dahlberg B, Grunewald J, Eklund A, Gabrielsson S (2010) Proinflammatory exosomes in bronchoalveolar lavage fluid of patients with sarcoidosis. Thorax 65(11):1016–1024. doi:10.1136/thx.2009.132027
45. Welton JL, Khanna S, Giles PJ, Brennan P, Brewis IA, Staffurth J, Mason MD, Clayton A (2010) Proteomics analysis of bladder cancer exosomes. Mol Cell Proteomics 9 (6):1324–1338. doi:10.1074/mcp.M000063-MCP201
46. Rabinowits G, Gercel-Taylor C, Day JM, Taylor DD, Kloecker GH (2009) Exosomal microRNA: a diagnostic marker for lung cancer. Clin Lung Cancer 10(1):42–46. doi:10.3816/CLC.2009.n.006
47. Skog J, Wurdinger T, van Rijn S, Meijer DH, Gainche L, Sena-Esteves M, Curry WT Jr, Carter BS, Krichevsky AM, Breakefield XO (2008) Glioblastoma microvesicles transport RNA and proteins that promote tumour growth and provide diagnostic biomarkers. Nat Cell Biol 10(12):1470–1476. doi:10.1038/ncb1800
48. de Kok JB, Verhaegh GW, Roelofs RW, Hessels D, Kiemeney LA, Aalders TW, Swinkels DW, Schalken JA (2002) DD3(PCA3), a very sensitive and specific marker to detect prostate tumors. Cancer Res 62(9):2695–2698
49. Tomlins SA, Rhodes DR, Perner S, Dhanasekaran SM, Mehra R, Sun XW, Varambally S, Cao X, Tchinda J, Kuefer R, Lee C, Montie JE, Shah RB, Pienta KJ, Rubin MA, Chinnaiyan AM (2005) Recurrent fusion of TMPRSS2 and ETS transcription factor genes in prostate cancer. Science 310(5748):644–648. doi:10.1126/science.1117679
50. Nilsson J, Skog J, Nordstrand A, Baranov V, Mincheva-Nilsson L, Breakefield XO, Widmark A (2009) Prostate cancer-derived urine exosomes: a novel approach to biomarkers for prostate cancer. Br J Cancer 100(10):1603–1607. doi:10.1038/sj.bjc.6605058
51. Taylor DD, Gercel-Taylor C (2008) MicroRNA signatures of tumor-derived exosomes as diagnostic biomarkers of ovarian cancer. Gynecol Oncol 110(1):13–21. doi:10.1016/j.ygyno.2008.04.033
52. El Andaloussi S, Lakhal S, Mager I, Wood MJ (2013) Exosomes for targeted siRNA delivery across biological barriers. Adv Drug Deliv Rev 65(3):391–397. doi:10.1016/j.addr.2012.08.008
53. Lai RC, Yeo RW, Tan KH, Lim SK (2013) Exosomes for drug delivery–a novel application for the mesenchymal stem cell. Biotechnol Adv 31(5):543–551. doi:10.1016/j.biotechadv.2012.08.008
54. Vickers KC, Remaley AT (2012) Lipid-based carriers of microRNAs and intercellular communication. Curr Opin Lipidol 23(2):91–97. doi:10.1097/MOL.0b013e328350a425
55. Sun D, Zhuang X, Xiang X, Liu Y, Zhang S, Liu C, Barnes S, Grizzle W, Miller D, Zhang HG (2010) A novel nanoparticle drug delivery system: the anti-inflammatory activity of curcumin is enhanced when encapsulated in exosomes. Mol Ther 18(9):1606–1614. doi:10.1038/mt.2010.105
56. O'Loughlin AJ, Woffindale CA, Wood MJ (2012) Exosomes and the emerging field of exosome-based gene therapy. Curr Gene Ther 12(4):262–274
57. Ohno S, Takanashi M, Sudo K, Ueda S, Ishikawa A, Matsuyama N, Fujita K, Mizutani T, Ohgi T, Ochiya T, Gotoh N, Kuroda M (2013) Systemically injected exosomes targeted to EGFR deliver antitumor microRNA to breast cancer cells. Mol Ther 21(1):185–191. doi:10.1038/mt.2012.180
58. Karlsson MR, Kahu H, Hanson LA, Telemo E, Dahlgren UI (2000) Tolerance and bystander suppression, with involvement of CD25-positive cells, is induced in rats receiving serum from ovalbumin-fed donors. Immunology 100(3):326–333
59. Karlsson MR, Kahu H, Hanson LA, Telemo E, Dahlgren UI (2002) An established immune response against ovalbumin is suppressed by a transferable serum factor produced after ovalbumin feeding: a role of CD25+ regulatory cells. Scand J Immunol 55(5):470–477
60. Martinez MC, Andriantsitohaina R (2011) Microparticles in angiogenesis: therapeutic potential. Circ Res 109(1):110–119. doi:10.1161/CIRCRESAHA.110.233049
61. Chaput N, Flament C, Viaud S, Taieb J, Roux S, Spatz A, Andre F, LePecq JB, Boussac M, Garin J, Amigorena S, Thery C, Zitvogel L (2006) Dendritic cell derived-exosomes: biology and clinical implementations. J Leukoc Biol 80(3):471–478. doi:10.1189/jlb.0206094
62. Viaud S, Ploix S, Lapierre V, Thery C, Commere PH, Tramalloni D, Gorrichon K, Virault-Rocroy P, Tursz T, Lantz O, Zitvogel L, Chaput N (2011) Updated technology to produce highly immunogenic dendritic cell-derived exosomes of clinical grade: a critical role of interferon-gamma. J Immunother 34 (1):65–75. doi:10.1097/CJI.0b013e3181fe535b

Chapter 2

Red Blood Cells: A Source of Extracellular Vesicles

Winston Patrick Kuo, John C. Tigges, Vasilis Toxavidis, and Ionita Ghiran

Abstract

During their lifetime, like all other cell types, red blood cells (RBCs) release both exosomes and plasma membrane derived EVs (ectosomes). RBC exosomes are formed only during the development of RBCs in bone marrow, and are released following the fusion of microvesicular bodies (MVB) with the plasma membrane. On the other hand, RBC EVs are generated during normal aging of RBCs in circulation by budding of the plasma membrane due to complement-mediated calcium influx, followed by vesicle shedding. This makes red blood cells and stored red cells a reliable source of EVs for basic and clinical research.

Key words Red blood cells, Extracellular vesicles, Diagnostics, Stored blood

1 Introduction

The most studied RBC-derived EVs are formed during RBC storage [1–3]. During blood storage, RBCs undergo metabolic, biochemical, and structural changes that significantly affect RBC membrane integrity, deformability and oxygen-carrying capacity, which can lead to poor tissue oxygenation and decreased posttransfusion recipient survival due to maladaptive inflammatory response to various transfused blood components [4–8]. The degree of loss of RBC function depends on several factors such as prestorage blood preparation techniques, additives to the solutions used for RBC storage, and length of storage [9]. In addition, the increase in generation of oxygen radicals during blood storage accelerates RBC storage lesions, significantly decreasing RBC storage time, and exacerbating post-transfusion responses to RBC transfusion [5, 10, 11]. One of the characteristics of RBC storage lesions is the formation of RBC-derived EVs, leading to a gradual decrease in RBC size and progressive decrease in membrane deformability. Identification of EVs in stored blood supernatant can be done by any of the contrast enhancement techniques such as dark-field microscopy, phase contrast or DIC Nomarski.

Winston Patrick Kuo and Shidong Jia (eds.), *Extracellular Vesicles: Methods and Protocols*, Methods in Molecular Biology, vol. 1660, DOI 10.1007/978-1-4939-7253-1_2, © Springer Science+Business Media LLC 2017

2 The Mechanisms of EV Generation During RBC Storage

Ongoing depletion of ATP resources during blood storage leads to a steady decrease in activity of plasma membrane Ca^{++} pumps resulting in a continual increase in intra-RBC [Ca^{++}] concentration. Prolonged increase in intra RBC [Ca^{++}] and depletion of RBC ATP reserves alter the activity of key plasma membrane enzymes involved in maintaining membrane phospholipid asymmetry such as flippase, floppase, and especially scramblase. Ca^{++}-dependent activation of RBC scramblase leads to increased exposure of anionic phospholipids on the external leaflet of plasma membrane, particularly phosphatidylserine (PS), followed by vesiculation and release of EVs. The precise molecular mechanisms and signaling pathways responsible for formation and release of RBC-derived EVs are hitherto not known. As a direct result of the continuous formation of EVs, aside from the gradual decrease in RBC size, both stored RBCs as well as RBC-derived EVs, display increasing levels of PS on their surface. Importantly, recent studies have started to shed light on the biological activities of EVs, implicating them in modulating immune response by down regulating the inflammatory response of monocytes and macrophages post phagocytosis, as well as in promoting thrombosis.

3 Functional Consequences of EV Formation During Storage

RBC-derived EVs are enriched in DAG, cholesterol, CR1(Complement receptor 1, CD35) and GPI-anchored proteins such as CD55, CD59, and acetylcholinesterase compared to their relative distribution in RBC plasma membrane [12]. Notably, abundant RBC plasma membrane proteins such as glycophorin A (GPA) were found to be present in EVs at a significantly lower concentration compared to RBC plasma membrane, strongly suggesting that vesiculation of the plasma membrane is not a random process. EV formation results in a gradual but significant loss of complement regulatory particles such as CR1, CD55 and CD59 from stored RBCs. In addition to the gradual loss of complement regulatory function of stored RBCs, loss of CR1 leads to a steady decrease in the ability of stored RBCs to bind and clear complement-opsonized particles.

Immune clearance evolved as an efficient means to bind and remove inflammatory particles from the intravascular space, a process critical in preventing activation of circulating leukocytes and vascular endothelium. Notably, most mammals mark pathogens and immune complexes for clearance with complement opsonins and immobilize them with the C3b binding protein, factor H, to adherent platelets, which signal for their removal by macrophages in the liver and spleen. Primates, however, have a different clearance system that is based on RBC CR1 and complement fragments C3b, C4b MBL and C1q [13–21]. Nonhuman primate RBCs have GPI-anchored CR1 [22] whereas humans have a transmembrane form of CR1, which makes the human clearance system unique. Although resting neutrophils express at least two to three hundred times more CR1 at the cell surface than RBC, opsonized particles are preferentially bound to CR1 RBC, which has been ascribed to increase RBC CR1 avidity for immune complexes compared to neutrophils (PMN) CR1, likely due to ability of CR1 to cluster [23]. Recently, complement-opsonized PMN and platelet-derived microparticles were found to interact and bind RBCs, a finding that underscores the important role of EVs as conveyers of cell-cell communication.

4 EVs Generated by RBC in Circulation by Complement

The complement system is comprised of over 20 soluble and membrane bound proteins with critical roles in recognizing, binding, and removal of foreign particles as well as initiating and regulating innate and acquired immune responses. Activation of the complement system can occur through any of the classical, alternative, or lectin pathways, and culminates homo-oligomerization and insertion of the membrane attack complex (MAC) pore components in the plasma membrane. This process leads to an unregulated water and Ca^{++} influx followed by cell lysis and death. Interestingly, this mechanism of nonspecific cell lysis is shared across distant species from bacterial exotoxin streptolysin O, to peptide melittin from bee venom, and cytolytic granule component perforin from cytotoxic lymphocytes [24–27]. Nucleated cells utilize various strategies against MAC-dependent death, from constitutive and regulated expression of complement regulatory proteins at the plasma membrane such as CD46, CR1, CD55 and CD59 that prevent MAC formation, to exocytosis and endocytosis of the MAC-containing areas of the plasma membrane [28, 29]. The result of MAC-induced exocytosis is the formation and release in circulation of EVs, which contains, in addition to the proteins and lipids mentioned above, components of the MAC complex.

RBCs have severely limited abilities to defend against MAC lysis due to the lack of endocytosis, mitochondria (as an efficient source of ATP) or a full complement of cytoskeletal proteins. Therefore, unlike nucleated cells, few copies of MAC seem to be effective in inducing RBC lysis. Circulating RBCs can remove MACs from the plasma membrane by vesiculation resulting in the formation of EVs. This mechanism seems to require Ca^{++}, calpain activation and local disruption of the spectrin skeleton. As a result of EV formation during the 120-day life cycle of circulating RBC, the size and protein composition of RBC varies among circulating

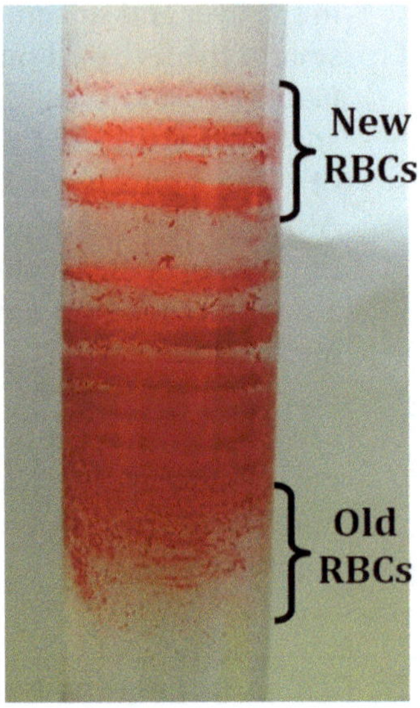

Fig. 1 Layers of various age RBCs obtained by centrifugation of whole blood in Percoll. Fresh RBCs were washed twice in HBSS and layered over Percoll and centrifuged for 20 min at 15,000 × *g*. Top layers contain less dense, new RBCs, whereas older cells form the bottom layers. Layers that are in between the top and bottom layers represent RBCs that are in between the two extreme populations

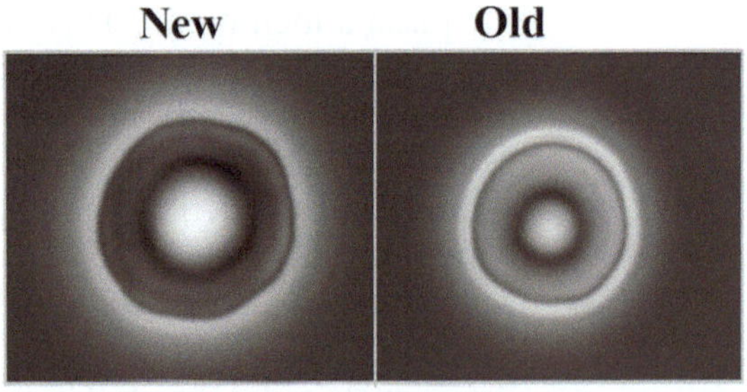

Fig. 2 New RBCs are morphologically distinct from old RBCs. Fresh RBCs were washed twice in HBSS and fractionated on Percoll gradient at 4 °C. Less dense (new RBCs) and more dense (old RBCs) fractions were collected, washed twice in PBS and analyzed by phase contrast microscopy

RBCs (Fig. 1), with new RBCs being larger with a full complement of membrane proteins and old RBCs smaller and denser with significantly fewer membrane proteins (Fig. 2). To illustrate the role of complement in red blood cell-derived EV formation, red blood cells were incubated with all the required complement components for MAC formation, but C7: C5b6, C8 and C9 (Fig. 3a), or with all complement components, specifically C5b6, C7, C8 and C9 (Fig. 3b). The supernatant analyzed by nano-flow cytometry showed that complement activation promoted a robust formation

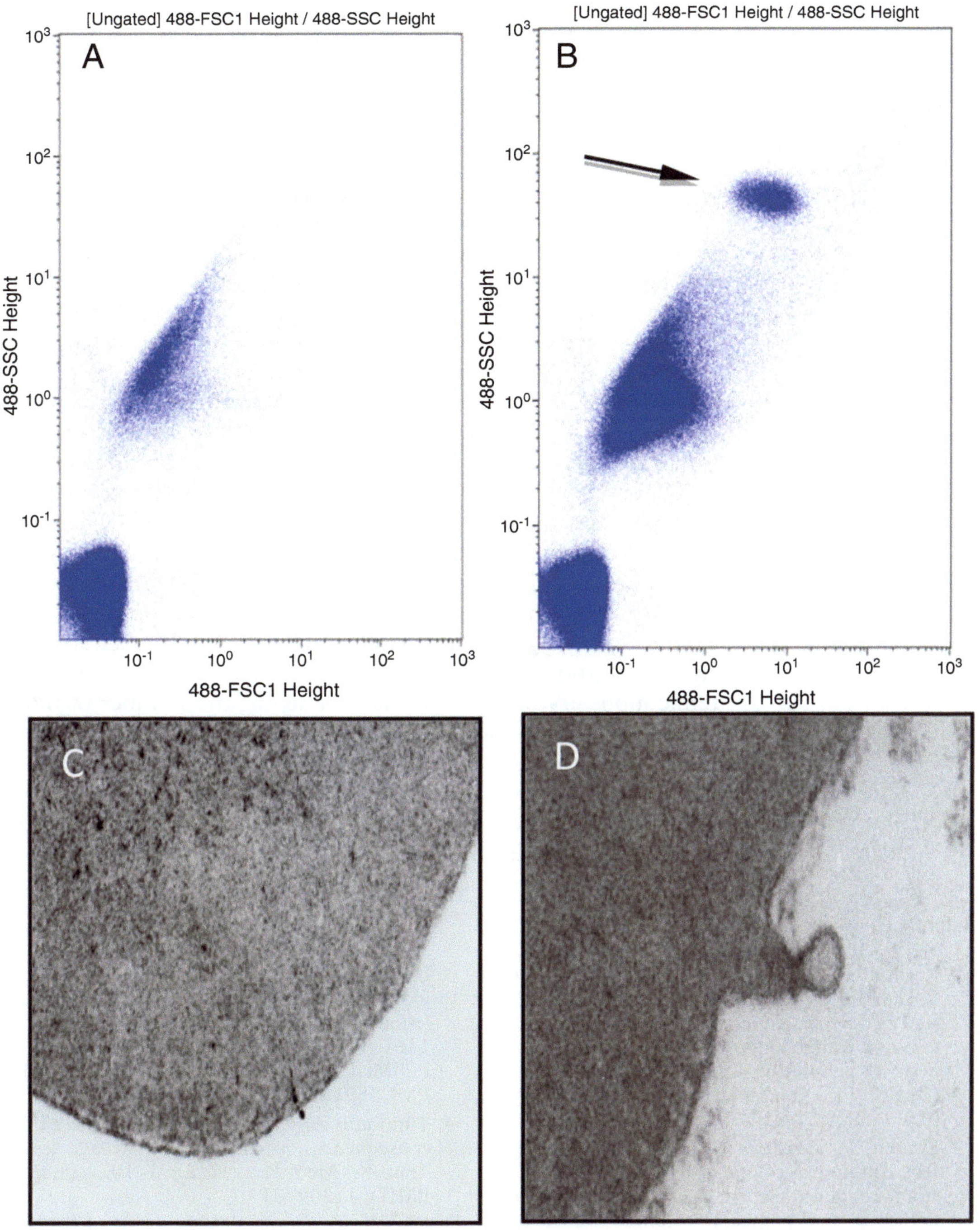

Fig. 3 Complement activation generates RBC EVs. Fresh RBCs were incubated with MAC-required complement components in the absence of C7 (**a**) or presence (**b**). EVs were analyzed by nano-flow cytometry. RBCs in serum in the presence of buffer (*left*) or 5 μM cobra venom serum (*right*), for 1 h at 37 °C. Cells were then prepared for standard TEM analysis. *Arrow* indicates a late stage on RBC EV formation. (**c**) TEM of RBCs in the absence of C7 and (**d**) TEM of the RBCs in the presence of C7

of small (250 nm) EVs in the absence of cell lysis. In a separate experiment, the alternative pathway of complement activation was triggered by addition of cobra venom factor in the absence (Fig. 3a) or presence (Fig. 3b) of calcium chelator EGTA. The cells were then examined by electron microscopy (Fig. 3c, d). The results show complement induced the formation of small (160–180 nm) vesicle budding from plasma membrane, without cell lysis. EVs of various cellular origins with vastly different compositions are found circulating in blood at any given time [30]. Circulating EVs have a half-life of 5–10 min and are removed by tissue resident macrophages from liver spleen lungs [31]. If EVs are proven to be relevant prognostic, diagnostic and treatment efficacy markers, this could be an advantage as information conveyed by cell- or tissue-specific EV either as protein, lipid or nucleic acid composition would be up to date. Although in the case of cancer, a marker for tumor genesis is a significant up-regulation of complement-regulatory proteins at the plasma membrane, and a consequent increased resistance to complement attack [32–35]. Therefore, MAC-mediated formation of tumor-specific EVs could be substantially decreased in cancer suggesting that intracellularly formed EVs, would be the more relevant for diagnostic purposes.

As expected, during pathological situations characterized by significant and dysregulated complement activation, such as systemic lupus erythematosus (SLE) or sepsis, the number of blood EVs is significantly increased (not shown). Recently several reports have shown that RBCs contain microRNA (miRNA) species that have specific biological activities [36, 37]. Whether RBC-derived EVs are also enriched in miRNA and have signaling properties remains to be determined.

References

1. Burnouf T, Chou ML, Goubran H, Cognasse F, Garraud O, Seghatchian J (2015) An overview of the role of microparticles/microvesicles in blood components: are they clinically beneficial or harmful? Transfus Apher Sci 53 (2):137–145. doi:10.1016/j.transci.2015.10.010. PubMed PMID: 26596959
2. Keuren JF, Magdeleyns EJ, Govers-Riemslag JW, Lindhout T, Curvers J (2006) Effects of storage-induced platelet microparticles on the initiation and propagation phase of blood coagulation. Br J Haematol 134(3):307–313. doi:10.1111/j.1365-2141.2006.06167.x. PubMed PMID: 16848773
3. Donadee C, Raat NJ, Kanias T, Tejero J, Lee JS, Kelley EE, Zhao X, Liu C, Reynolds H, Azarov I, Frizzell S, Meyer EM, Donnenberg AD, Qu L, Triulzi D, Kim-Shapiro DB, Gladwin MT (2011) Nitric oxide scavenging by red blood cell microparticles and cell-free hemoglobin as a mechanism for the red cell storage lesion. Circulation 124(4):465–476. doi:10.1161/CIRCULATIONAHA.110.008698. PubMed PMID: 21747051; PMCID: PMC3891836
4. Tinmouth A, Chin-Yee I (2001) The clinical consequences of the red cell storage lesion. Transfus Med Rev 15(2):91–107. PubMed PMID: 11309731
5. Chin-Yee I, Arya N, d'Almeida MS (1997) The red cell storage lesion and its implication for transfusion. Transfus Sci 18(3):447–458. doi: S095538869700043X [pii]. Epub 1997/08/05. PubMed PMID: 10175158
6. Spiess BD (2007) Red cell transfusions and guidelines: a work in progress. Hematol Oncol Clin North Am 21(1):185–200. doi:10.1016/j.hoc.2006.11.006. Epub

2007/01/30. S0889-8588(06)00190-0 [pii]. PubMed PMID: 17258127

7. Solheim BG, Hess JR (2009) In: Simon TL, Snyder EI, Solheim BG, Strauss RG, Peride M (eds) Rossi's principles of transfusion medicine, 4th edn. Blackwell Publishing, New Jersey
8. Hess JR (2006) An update on solutions for red cell storage. Vox Sang 91(1):13–19. doi:10.1111/j.1423-0410.2006.00778.x. Epub 2006/06/08. VOX778 [pii]. PubMed PMID: 16756596
9. Gyongyossy-Issa MI, Weiss SL, Sowemimo-Coker SO, Garcez RB, Devine DV (2005) Pre-storage leukoreduction and low-temperature filtration reduce hemolysis of stored red cell concentrates. Transfusion 45(1):90–96. doi:10.1111/j.1537-2995.2005.04061.x. Epub 2005/01/14. TRF04061 [pii]. PubMed PMID: 15647023
10. Yoshida T, AuBuchon JP, Tryzelaar L, Foster KY, Bitensky MW (2007) Extended storage of red blood cells under anaerobic conditions. Vox Sang 92(1):22–31. doi:10.1111/j.1423-0410.2006.00860.x. Epub 2006/12/22. VOX860 [pii]. PubMed PMID: 17181587
11. Yoshida T, AuBuchon JP, Dumont LJ, Gorham JD, Gifford SC, Foster KY, Bitensky MW (2008) The effects of additive solution pH and metabolic rejuvenation on anaerobic storage of red cells. Transfusion 48 (10):2096–2105. doi:10.1111/j.1537-2995.2008.01812.x. Epub 2008/07/18. TRF01812 [pii]. PubMed PMID: 18631166
12. Pascual M, Lutz HU, Steiger G, Stammler P, Schifferli JA (1993) Release of vesicles enriched in complement receptor 1 from human erythrocytes. J Immunol 151(1):397–404. PubMed PMID: 8326133
13. Nelson RAJ (1953) The immune adherence phenomenon: an immunologically specfic reaction between microorganisms and erythrocytes leading to enhanced phagocytosis. Science 118:733–737
14. Pilsczek FH, Nicholson-Weller A, Ghiran I (2005) Phagocytosis of Salmonella montevideo by human neutrophils: immune adherence increases phagocytosis, whereas the bacterial surface determines the route of intracellular processing. J Infect Dis 192(2):200–209. doi:10.1086/430947. Epub 2005/06/18. JID31399 [pii]. PubMed PMID: 15962214
15. Klickstein LB, Bartow TJ, Miletic B, Rabson LD, Smith JA, Fearon DT (1988) Identification of distinct C3b and C4b recognition sites in the human C3b/C4b receptor (CR1, CD35) by deletion mutagenesis. J Exp Med 168:1699–1717
16. Krych M, Hourcade D, Atkinson JP (1991) Sites within the complement C3b/C4b receptor important for the specificity of ligand binding. Proc Natl Acad Sci U S A 88 (10):4353–4357
17. Makrides SC, Scesney SM, Ford PJ, Evans KS, Carson GR, Marsh HCJ (1992) Cell surface expression of the C3b/C4b receptor (CR1) protects Chinese hamster ovary cells from lysis by human complement. J Biol Chem 267:24754–24761
18. Klickstein LB, Barbashov S, Liu T, Jack RM, Nicholson-Weller A (1997) Complement receptor type 1 (CR1, CD35) is a receptor for C1q. Immunity 7:345–355
19. Tas SW, Klickstein LB, Nicholson-Weller A (1999) C1q and C4b are additive ligands for CR1 and mediate erythrocyte adhesion. J Immunol 163:5056–5063
20. Ghiran I, Barbashov SF, Klickstein LB, Tas SW, Jensenius JC, Nicholson-Weller A (2000) Complement receptor 1/CD35 is a receptor for mannan-binding lectin. J Exp Med 192:1797–1808
21. Fearon DT (1980) Identification of the membrane glycoprotein that is the C3b receptor of the human erythrocyte, polymorphonuclear leukocyte, B lymphocyte and monocyte. J Exp Med 152:20–30
22. Cornacoff JB, Hebert LA, Smead WL, Van Aman ME, Birmingham DJ, Waxman FJ (1983) Primate erythrocyte-immune complex-clearing mechanism. J Clin Invest 71:236–247
23. Paccaud JP, Carpentier J-L, Schifferli JA (1990) Difference in the clustering of complement receptor 1 (CR1) on polymorphonuclear leukocytes and erythrocytes: effect on immune adherence. Eur J Immunol 20:283–289
24. Laine RO, Morgan BP, Esser AF (1988) Comparison between complement and melittin hemolysis: anti-melittin antibodies inhibit complement lysis. Biochemistry 27 (14):5308–5314. PubMed PMID: 2458761
25. Laine RO, Esser AF (1989) Identification of the discontinuous epitope in human complement protein C9 recognized by anti-melittin antibodies. J Immunol 143(2):553–557. PubMed PMID: 2472443
26. Black SM, Schott ME, Batdorf BH, Benson BA, Rutherford MS, Levay-Young BK, Dalmasso AP (2010) IL-4 induces protection of vascular endothelial cells against killing by complement and melittin through lipid biosynthesis. Eur J Immunol 40(3):803–812. doi:10.1002/eji.200939488. PubMed PMID: 20017192

27. Benson BA, Vercellotti GM, Dalmasso AP (2015) IL-4 and IL-13 induce protection from complement and melittin in endothelial cells despite initial loss of cytoplasmic proteins: membrane resealing impairs quantifying cytotoxicity with the lactate dehydrogenase permeability assay. Xenotransplantation 22 (4):295–301. doi:10.1111/xen.12172. PubMed PMID: 26031609; PMCID: PMC4519407

28. Hetland G, Johnson E, Eskeland T (1987) Formation of the membrane attack complex of complement (MAC) on erythrocytes from monocyte-produced terminal complement components. Scand J Immunol 25 (6):571–577. PubMed PMID: 3602934

29. Esser AF (1991) Big MAC attack: complement proteins cause leaky patches. Immunol Today 12(9):316–318. doi:10.1016/0167-5699(91)90006-F. discussion 21. PubMed PMID: 1721818

30. Kreimer S, Belov AM, Ghiran I, Murthy SK, Frank DA, Ivanov AR (2015) Mass-spectrometry-based molecular characterization of extracellular vesicles: lipidomics and proteomics. J Proteome Res 14(6):2367–2384. doi:10.1021/pr501279t. PubMed PMID: 25927954

31. Choi H, Lee DS (2016) Illuminating the physiology of extracellular vesicles. Stem Cell Res Ther 7(1):55. doi:10.1186/s13287-016-0316-1. PubMed PMID: 27084088; PMCID: PMC4833943

32. Ziller F, Macor P, Bulla R, Sblattero D, Marzari R, Tedesco F (2005) Controlling complement resistance in cancer by using human monoclonal antibodies that neutralize complement-regulatory proteins CD55 and CD59. Eur J Immunol 35(7):2175–2183. doi:10.1002/eji.200425920. PubMed PMID: 15971270

33. Pilzer D, Gasser O, Moskovich O, Schifferli JA, Fishelson Z (2005) Emission of membrane vesicles: roles in complement resistance, immunity and cancer. Springer Semin Immunopathol 27(3):375–387. doi:10.1007/s00281-005-0004-1. PubMed PMID: 16189651

34. Lopatina T, Gai C, Deregibus MC, Kholia S, Camussi G (2016) Cross talk between cancer and mesenchymal stem cells through extracellular vesicles carrying nucleic acids. Front Oncol 6:125. doi:10.3389/fonc.2016.00125. PubMed PMID: 27242964; PMCID: PMC4876347

35. Desrochers LM, Antonyak MA, Cerione RA (2016) Extracellular vesicles: satellites of information transfer in cancer and stem cell biology. Dev Cell 37(4):301–309. doi:10.1016/j.devcel.2016.04.019. PubMed PMID: 27219060; PMCID: PMC4995598

36. Azzouzi I, Moest H, Wollscheid B, Schmugge M, Eekels JJ, Speer O (2015) Deep sequencing and proteomic analysis of the microRNA-induced silencing complex in human red blood cells. Exp Hematol 43(5):382–392. doi:10.1016/j.exphem.2015.01.007. PubMed PMID: 25681748

37. Wang ZY, Yang FM, Liu J, Li R, Li XP, Jing ZH (2015) Correlation between the expression of microRNA 451 in red blood cells and chronic mountain sickness. Zhongguo Shi Yan Xue Ye Xue Za Zhi 23(2):481–484. doi:10.7534/j.issn.1009-2137.2015.02.036. PubMed PMID: 25948209

Part I

Isolation of Extracellular Vesicles

Chapter 3

Isolation of Extracellular Vesicles by Ultracentrifugation

Fatemeh Momen-Heravi

Abstract

Extracellular vesicles (EVs) represent a group of heterogeneous vesicles that can be obtained from almost all biofluids. EVs, including microvesicles, exosomes, and apoptotic bodies, can deliver bioactive cargos and signaling molecules. Various physiological roles and pathophysiological roles for EVs in diseases such as cancer, infectious diseases, endocrine diseases, and neurodegenerative disorders have been recognized. These observations highlight EVs as potential novel biomarkers and targets for therapeutic intervention. One of the major limitations in the use of EVs for diagnosis and therapeutic purposes is the lack of standardization of isolation techniques. Here, we describe protocols for ultracentrifugation and sucrose gradient isolation methods, which are the current gold standard, and are the most studied methods for EV isolation.

Key words Exosomes, Extracellular vesicles, Ultracentrifugation, Sucrose gradient, Isolation

1 Introduction

Extracellular vesicles are heterogeneous vesicles that play pivotal roles in cellular communications [1, 2]. Emerging evidence indicates that EVs are present in biofluids and carry different biomacromolecules such as lipids, proteins, and nucleic acids that provide a snapshot of the parental cells at the time of secretion [2, 3]. EVs and their molecular cargoes gained attention as disease biomarkers as well as drug delivery vehicles. One of the major limitations in use of EVs for diagnosis and therapeutic purposes is the lack of standardization of isolation techniques [4]. Here, we describe ultracentrifugation and sucrose gradient methods, which currently are the most commonly used and most studied methods for EV isolation. Differential centrifugation separation is primarily based on the size of different particles. This type of isolation has been used extensively for the isolation of EVs and is considered the gold standard for isolation, specifically for the isolation of smaller subpopulations of EVs or exosomes [5]. Although centrifugation works solely based on the size of different EVs, after resuspension of the pallet

Winston Patrick Kuo and Shidong Jia (eds.), *Extracellular Vesicles: Methods and Protocols*, Methods in Molecular Biology, vol. 1660, DOI 10.1007/978-1-4939-7253-1_3,

it can be coupled with immune magnetic beads isolation which can lead to the isolation of subpopulations of EVs based on surface markers [6]. In this way the lipoprotein particle and viruses that share the same size range with EVs can be efficiently separated from the EVs. In this chapter, we focus primarily on ultracentrifugation and sucrose gradient centrifugation. However, the readers should consider merging these basic methods with size exclusion, immune magnetic bead isolation and microfluidics as needed.

2 Materials

Ethylenediaminetetraacetic acid (EDTA), Conditioned medium based on cells, Phosphate-buffered saline (PBS), Tris-buffered saline (TBS), refrigerated centrifuge, centrifuges, tubes, ultracentrifuge, fixed-angle or swinging-bucket rotor, nanoparticle tracking analysis, TEM microscope, Western blot equipment and reagents/antibodies, exosome-depleted FBS, −80 °C freezer, OptiPrep™ (60% (w/v) solution of iodixanol), refractometer. Prepare stock of Tris/sucrose/D2O solution by using 30 g protease-free sucrose, 2.4 g Tris base, 50 ml D2O, increase volume to 100 ml with D2O. The solution can be stored for up to 2 months at 4 °C. Iodixanol gradient medium can be prepared in a 0.02 M HEPES [4-(2-hydroxyethyl)-1-piperazine ethanesulfonic acid]/NaOH buffer.

3 Methods

3.1 Differential Centrifugation/Ultracentrifugation Protocols

For consistency and reproducibility, all samples to be compared should be spun at the same speed and using the same rotor type. All centrifugations should be performed at 4 °C.

1. Dilute the biofluid (serum, plasma, saliva or milk) in at least 50% in the phosphate buffered saline in an appropriate RNase/DNase-free tube. This step reduces the viscosity and increases the efficiency of vesicle isolation [7] (*see* **Note 1**). In the case of culture media skip this step. For the isolation of EVs from the cultured cells, it is crucial that the cells were cultured in a exosome-depleted fetal bovine serum (FBS) (*see* **Note 2**).
2. Add 0.3 μl of EDTA/ml to the whole volume and vortex briefly. This will prevent the aggregation of EVs (*see* **Note 3**).
3. Centrifuge the mixture for 10 min at 1500 × *g* to remove dead cells. Discard the pellet and transfer the supernatant to a new collection tube.
4. If isolation of exosomes, large apoptotic bodies, and microparticles are intended, a single centrifugation step with the forces range of 10,000–20,000 × *g* for 20–30 min should be done.

The pallet contains the large apoptotic bodies and large microvesicles. Collect the supernatant and continue with the next step for isolation of smaller subpopulation of EVs (*see* **Note 4**).

5. Select the appropriate rotor (*see* **Note 5**). Determine the rotor and k-factor and the required time for ultracentrifugation. The k-factor can be defined as a scale of the time taken for a particle to sediment through a particular medium via ultracentrifugation [4, 8]. The k-factor indicates the relative pelleting efficiency of a given rotor at a maximum rotation speed and can be used to determine the time t (in hours) which is required for the sedimentation of EVs using different rotors [4]. The correlation between k-factor, time, and s (the sedimentation coefficient, in Svedbergs) can be represented by the following formula:

$$t = \frac{\mathrm{k}}{s}$$

The maximum angular velocity (ω) of a centrifuge (in rad/s) and the minimum and maximum radius r of the rotor determine the value of k-factor [9]. The k-factor can be calculated from the following formula where $\mathrm{r_{max}}$ is the maximum radius from the axis of rotation in centimeters and $\mathrm{r_{min}}$ is minimum radius from axis of rotation in centimeters:

$$k = \frac{\ln\left(r_{\max}/r_{\min}\right)}{\omega^2} \times \frac{10^{13}}{3600}$$

The following equation can be used to calculate k factors in the cases where the centrifuge speed is determined as revolutions per minute (RPM):

$$k = \frac{2.53.10^5 \times \ln\left(r_{\max}/r_{\min}\right)}{\left(\mathrm{RPM}/1000\right)^2}$$

Most efficient rotors for the isolation of EVs have the lowest *k*-factor value and operate at a relatively high centrifugal force (RCF) or *g*, and have a low sedimentation path length [4].

Performance of different rotors can be compared by the following formula, which can be used to calculate the time required for the sedimentation of EVs in one specific rotor compared to another rotor. The k-factor and centrifugation times (t) for two specific rotors (A and B) can be concerted by the following formula:

$$\frac{T\mathrm{a}}{K\mathrm{a}} = \frac{T\mathrm{b}}{K\mathrm{b}}$$

Where Ta, is the time for sedimentation in rotor A and Ka is the k factor of that rotor. Kb is the k factor of the other rotor and Tb is the time required for sedimentation of rotor B. These calculations can be performed by an online tool available at www.beckmancoulter.com.

6. Select the appropriate centrifuge tube based on the rotor. Ideally, clear tubes and chemical resistant tubes are preferred over opaque tubes and tubes with poor resistance. Consider the maximum allowed speed and volume which is indicated on the tube packaging.
7. Transfer biofluid to the ultracentrifugation tube and ensure closure of tube after loading the volume of biofluid.
8. For the isolation of smaller subpopulations of EVs (exosomes, small microvesicles and apoptotic bodies), ultracentrifugation should be performed at high speed (100,000–120,000 × *g*) for a set amount of time based on rotor type (at least 70 min) (*see* **Note 6**).
9. The pellet of EVs is suspended in phosphate buffered saline (PBS) based on the size of pellet (20–100 μl) of fresh PBS. The suspension should be aliquoted for the characterization step via transmission electron microscope (TEM), atomic-force microscopy (AFM), western blot for exosomal markers, and Nanoparticle Tracking Analysis (NTA) [10]. The remaining suspended pellet should be stored at −80 °C for further characterization and analysis. For western blotting, common EVs markers such as CD63, CD9, CD81, and phosphatidylserin can be used to confirm the successful isolation. TEM can be used for assessing the size and morphology of the EVs as well as the purity of the samples. The relative amount of protein aggregates can be seen in the TEM images (*see* **Note 7**). If a special subpopulation of EVs is of interest, the surface markers can be used for selective isolation (*see* **Note 8**).

3.2 Sucrose Gradient Separation Protocol

As discussed previously, one of the problems associated with ultracentrifugation is contamination with other nonvesicles. Density gradient enables separation of particles by density and can be considered continuation of ultracentrifugation protocol when extra purity is needed. Protein and/or protein–RNA aggregates can be efficiently separated from the exosomes which they aggregate at the bottom of the tube. The density of EVs is reported to be between 1.13 and 1.19 g/ml [11].

1. Load Tris/sucrose/D2O solution at the bottom of a 4–15 ml centrifuge tube to make a 30% cushion. Alternatively, OptiPrep™ which is a 60% (w/v) solution of iodixanol in water can be used to make the gradient medium (*see* **Note 9**).
2. Load pellet on top of the sucrose cushion or iodixanol gradient without disturbing the interface before ultracentrifugation for

2–12 h. The sucrose cushion should be kept in the cold room or refrigerator (4 °C). This step will facilitate the separation.

3. Centrifuge for 75 min up to 24 h at 75,000–160,000 × *g* at 4 °C (*see* **Note 10**).
4. With a syringe or pipette collect the top phase of the Tris/sucrose/D2O cushion, which contains exosomes. The density of this phase can be assessed by a refractometer.
5. Transfer the exosomes to a new ultracentrifuge tube. Add 40–60 ml PBS based on the tube capacity. Centrifuge for 70 min–24 h at 100,000 × *g*, 4 °C.
6. Resuspend the pellet in 35–100 μl of PBS (*see* **Note 11**).

4 Notes

1. Viscosity is an important parameter to consider when working with a biofluid. In the case of equal amounts of EVs, a lower viscosity biofluid yields more EVs in the isolated pellet [7]. Viscosities of different biofluids should be standardized and specimens have to be diluted to reach similar viscosity values to use similar protocols. If not, longer ultracentrifugation time and higher speed are needed for more viscous biofluids [7].
2. Alternative is to culture cells without serums. Although, specific cells can survive without serum for a few days, this change will inevitably induce a stress response which may lead to the different EVs signature. No studies compared the effect of exosome-depleted serum or serum free culture system [12].
3. EDTA has been shown to prevent EV aggregation, as well as the formation of insoluble siRNAs in the process of electroporation of EVs with siRNA [13, 14].
4. It is possible to replace **steps 3** and **4** of differential centrifugation/ultracentrifugation protocols with a serial filtration step; 0.8 μm, 0.4 μm, and 0.22 μm filters can be used for ultimate goal of smaller subpopulation/exosome isolation; this will eliminate dead cells and larger vesicles while keeping small vesicles for downstream isolation by ultracentrifugation.
5. Two types of rotors have been used for the isolation of EVs: swinging bucket and fixed-angle. In the swinging bucket rotors, at the time of centrifugation the buckets swing out a horizontal position. This type of rotor is less efficient in pelleting compared to fixed angle rotors. However, it has a larger path length and is suitable for separating EVs and does not physically damage EVs. Fixed angle rotors have a high efficiency for pelleting. Fixed-angle rotor sediments EVs against the wall of the centrifuge tube and gradually forces this material

down the wall to the bottom of the tube which can lead to aggregation.

6. Increased ultracentrifugation times can lead to a greater yield of exosomal RNA and protein. In contrast, very long ultracentrifugation times may result in coaggregation of proteins in the EV pellet [8].
7. The coaggregation of proteins and lipids with EVs isolated by ultracentrifugation is a common finding, which can interfere with separation by individual vesicle size. Another high-speed centrifugation should be considered to reduce the contamination of proteins and lipids [4, 15]. However, it has the drawback of decreasing the efficiency and its cost and benefits based on the down-stream aims should be assessed.
8. For example, for the isolation of the exosome subpopulation, immune magnetic beads can be used to isolate the exosome subpopulation based on exosomal surface markers [6].
9. Iodixanol ($C_{35}H_{44}I_6N_6O_{15}$) is an iodinated compound which has a lower viscosity than sucrose. Iodixanol gradient has been used in both viruses and EVs studies [16, 17]. In the virus isolation studies iodixanol showed better resolution of viruses compared to the sucrose gradient method [18, 19]. In a report on isolation of EVs from saliva using iodixanol, EV-related signals were concentrated in second fraction. However, the saliva was pretreated with sonication and filtration [19]. Iodixanol showed superior ability to isolate HIV virus from EVs compared to sucrose gradient [18].
10. The time can be varied based on the type of EVs, rotor type, and tube length. For example, in long tubes contaminating aggregates may not reach the lower fraction of the tube and coaggregate together with EVs in the upper fractions [12]. Fixed angle rotor has more efficacy in the isolation based on sucrose gradient separation and can be used to reduce the required ultracentrifugation time [19]. The time can vary for different type of EVs, as some studies showed that it can take over 60 h for vesicles to reach the equilibrium [20].
11. Although in the classic literature this method was described for isolation of exosome subtraction, there is an overlapping density between different type of EVs [11]. Isolated EVs should be checked for biological activity, specific EVs markers with western blot or flow cytometry, morphology and characteristics (TEM and AFM), and number (NTA). Since the density of different subpopulation of EVs based on type of biofluid can be different, it is recommended that every fraction be evaluated for specific EV markers.

References

1. Yanez-Mo M, Siljander PR, Andreu Z, Zavec AB, Borras FE, Buzas EI, Buzas K, Casal E, Cappello F, Carvalho J, Colas E, Cordeiro-da Silva A, Fais S, Falcon-Perez JM, Ghobrial IM, Giebel B, Gimona M, Graner M, Gursel I, Gursel M, Heegaard NH, Hendrix A, Kierulf P, Kokubun K, Kosanovic M, Kralj-Iglic V, Kramer-Albers EM, Laitinen S, Lasser C, Lener T, Ligeti E, Line A, Lipps G, Llorente A, Lotvall J, Mancek-Keber M, Marcilla A, Mittelbrunn M, Nazarenko I, Nolte-'t Hoen EN, Nyman TA, O'Driscoll L, Olivan M, Oliveira C, Pallinger E, Del Portillo HA, Reventos J, Rigau M, Rohde E, Sammar M, Sanchez-Madrid F, Santarem N, Schallmoser K, Ostenfeld MS, Stoorvogel W, Stukelj R, Van der Grein SG, Vasconcelos MH, Wauben MH, De Wever O (2015) Biological properties of extracellular vesicles and their physiological functions. J Extracell Vesicles 4:27066. doi:10.3402/jev.v4.27066
2. Raposo G, Stoorvogel W (2013) Extracellular vesicles: exosomes, microvesicles, and friends. J Cell Biol 200(4):373–383. doi:10.1083/jcb.201211138
3. Jia S, Zocco D, Samuels ML, Chou MF, Chammas R, Skog J, Zarovni N, Momen-Heravi F, Kuo WP (2014) Emerging technologies in extracellular vesicle-based molecular diagnostics. Expert Rev Mol Diagn 14(3):307–321. doi:10.1586/14737159.2014.893828
4. Momen-Heravi F, Balaj L, Alian S, Mantel PY, Halleck AE, Trachtenberg AJ, Soria CE, Oquin S, Bonebreak CM, Saracoglu E, Skog J, Kuo WP (2013) Current methods for the isolation of extracellular vesicles. Biol Chem 394(10):1253–1262. doi:10.1515/hsz-2013-0141
5. Mathivanan S, Fahner CJ, Reid GE, Simpson RJ (2012) ExoCarta 2012: database of exosomal proteins, RNA and lipids. Nucleic Acids Res 40(Database issue):D1241–D1244. doi:10.1093/nar/gkr828
6. Bukong TN, Momen-Heravi F, Kodys K, Bala S, Szabo G (2014) Exosomes from hepatitis C infected patients transmit HCV infection and contain replication competent viral RNA in complex with Ago2-miR122-HSP90. PLoS Pathog 10(10):e1004424. doi:10.1371/journal.ppat.1004424
7. Momen-Heravi F, Balaj L, Alian S, Trachtenberg AJ, Hochberg FH, Skog J, Kuo WP (2012) Impact of biofluid viscosity on size and sedimentation efficiency of the isolated microvesicles. Front Physiol 3:162. doi:10.3389/fphys.2012.00162
8. Cvjetkovic A, Lotvall J, Lasser C (2014) The influence of rotor type and centrifugation time on the yield and purity of extracellular vesicles. J Extracell Vesicles 3. doi:10.3402/jev.v3.23111
9. Langer K, Balthasar S, Vogel V, Dinauer N, von Briesen H, Schubert D (2003) Optimization of the preparation process for human serum albumin (HSA) nanoparticles. Int J Pharm 257(1–2):169–180
10. Momen-Heravi F, Balaj L, Alian S, Tigges J, Toxavidis V, Ericsson M, Distel RJ, Ivanov AR, Skog J, Kuo WP (2012) Alternative methods for characterization of extracellular vesicles. Front Physiol 3:354. doi:10.3389/fphys.2012.00354
11. van der Pol E, Boing AN, Harrison P, Sturk A, Nieuwland R (2012) Classification, functions, and clinical relevance of extracellular vesicles. Pharmacol Rev 64(3):676–705. doi:10.1124/pr.112.005983
12. Witwer KW, Buzas EI, Bemis LT, Bora A, Lasser C, Lotvall J, Nolte-'t Hoen EN, Piper MG, Sivaraman S, Skog J, Thery C, Wauben MH, Hochberg F (2013) Standardization of sample collection, isolation and analysis methods in extracellular vesicle research. J Extracell Vesicles 2. doi:10.3402/jev.v2i0.20360
13. Momen-Heravi F, Bala S, Bukong T, Szabo G (2014) Exosome-mediated delivery of functionally active miRNA-155 inhibitor to macrophages. Nanomedicine 10(7):1517–1527. doi:10.1016/j.nano.2014.03.014
14. Kooijmans SA, Stremersch S, Braeckmans K, de Smedt SC, Hendrix A, Wood MJ, Schiffelers RM, Raemdonck K, Vader P (2013) Electroporation-induced siRNA precipitation obscures the efficiency of siRNA loading into extracellular vesicles. J Control Release 172(1):229–238. doi:10.1016/j.jconrel.2013.08.014
15. Thery C, Amigorena S, Raposo G, Clayton A (2006) Isolation and characterization of exosomes from cell culture supernatants and biological fluids. Curr Protoc Cell Biol Chapter 3:Unit 3 22. doi:10.1002/0471143030.cb0322s30
16. Baietti MF, Zhang Z, Mortier E, Melchior A, Degeest G, Geeraerts A, Ivarsson Y, Depoortere F, Coomans C, Vermeiren E, Zimmermann P, David G (2012) Syndecan-syntenin-ALIX regulates the biogenesis of exosomes. Nat Cell Biol 14(7):677–685. doi:10.1038/ncb2502

17. Tauro BJ, Mathias RA, Greening DW, Gopal SK, Ji H, Kapp EA, Coleman BM, Hill AF, Kusebauch U, Hallows JL, Shteynberg D, Moritz RL, Zhu HJ, Simpson RJ (2013) Oncogenic H-ras reprograms Madin-Darby canine kidney (MDCK) cell-derived exosomal proteins following epithelial-mesenchymal transition. Mol Cell Proteomics 12 (8):2148–2159. doi:10.1074/mcp.M112.027086
18. Cantin R, Diou J, Belanger D, Tremblay AM, Gilbert C (2008) Discrimination between exosomes and HIV-1: purification of both vesicles from cell-free supernatants. J Immunol Methods 338(1–2):21–30. doi:10.1016/j.jim.2008.07.007
19. Iwai K, Minamisawa T, Suga K, Yajima Y, Shiba K (2016) Isolation of human salivary extracellular vesicles by iodixanol density gradient ultracentrifugation and their characterizations. J Extracell Vesicles 5:30829. doi:10.3402/jev.v5.30829
20. Palma J, Yaddanapudi SC, Pigati L, Havens MA, Jeong S, Weiner GA, Weimer KM, Stern B, Hastings ML, Duelli DM (2012) MicroRNAs are exported from malignant cells in customized particles. Nucleic Acids Res 40 (18):9125–9138. doi:10.1093/nar/gks656

Chapter 4

Sequential Filtration: A Gentle Method for the Isolation of Functional Extracellular Vesicles

Mitja L. Heinemann and Jody Vykoukal

Abstract

A prevalent challenge in isolating extracellular vesicles (EVs) from biological fluids is the reliable depletion of abundant contaminants—including free proteins and biomolecules, as well as nontarget vesicle subpopulations and other nanoparticulates—from the sample matrix while maximizing recovery. Sequential Filtration is a recently published approach for the size-based isolation of exosomes that is ideally suited for large-volume biofluid samples such as ascites, urine, lavage fluid, or cell-conditioned media. We describe a straightforward, three-step protocol comprising back-to-back steps of dead-end (normal) filtration, tangential-flow filtration, and track-etched membrane filtration that can be applied to yield a homogeneous population of exosome-sized extracellular vesicles. The approach is scalable and employs relatively gentle manipulation forces to fractionate and concentrate extracellular vesicles with good purity and functional integrity.

Key words Extracellular vesicles, Exosomes, Exosome isolation, Sequential filtration, Early detection, Functional exosomes, Cancer biology

1 Introduction

Exosomes and microvesicles are of increasing interest in current medical and biomolecular research. To fully understand extracellular vesicle composition, functions, and medical relevance, adequate isolation methods are crucial.

Exosomes and microvesicles can be purified by a variety of different methods. Currently employed methods rely on ultracentrifugation, chemical precipitation, antibody-coated beads, size-exclusion chromatography, or membrane filtration [1–3].

Each of these principles has its own set of merits and limitations, and thus, many alterations and combinations of these methods are utilized in order to optimize the isolation process to achieve specific desired results [4].

We have developed a novel, scalable isolation method that most notably gently isolates functional exosomes of uniform size and morphology from a wide range of input volumes [5]. The method

Winston Patrick Kuo and Shidong Jia (eds.), *Extracellular Vesicles: Methods and Protocols*, Methods in Molecular Biology, vol. 1660, DOI 10.1007/978-1-4939-7253-1_4, © Springer Science+Business Media LLC 2017

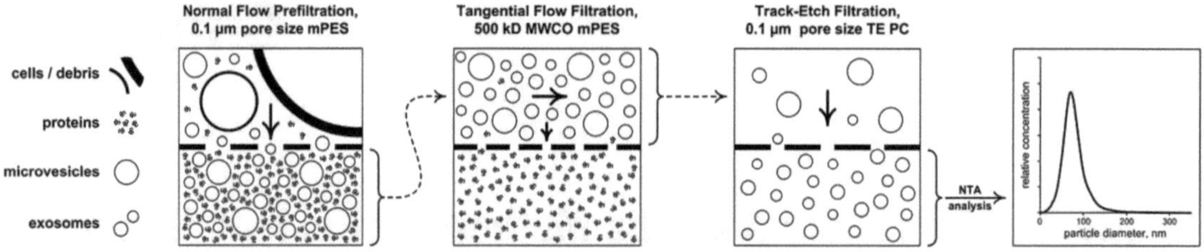

Fig. 1 Schematic of the filtration protocol for the isolation of exosomes from cell culture supernatants and patient-derived biofluids. In step 1, cells and cell debris are filtered out. Step 2 depletes free protein and concentrates the sample. In step 3, microvesicles are removed by low-pressure filtration through a filter with clearly defined pore sizes. Size distribution is confirmed by Nanotracking Analysis. (Reprinted from *Journal of Chromatography A*, 1371, M.L. Heinemann et al., Benchtop isolation and characterization of functional exosomes by sequential filtration, 125–135, 2014, with permission from Elsevier)

relies on three sequential filtration steps (Fig. 1): First, dead-end (normal) filtration is used to remove cells, cell debris, and large (*greater than nominal 100 nm diameter*) extracellular vesicles. Subsequently, tangential flow filtration (TFF) is employed to concentrate exosomes and remove non-exosome-associated proteins and biomolecules and small (*less than approximately 10 nm diameter*) nonexosomal particles. Finally, filtration through a track-etched membrane having a uniform, specific pore size (e.g., Whatman/GE *Nucleopore* 50, 80, 100 or 200 nm diameter) enables size-defined fractionation of exosomes and nonexosomal particles such as microvesicles from the concentrated sample. In order to minimize physical stress on exosomes and microvesicles [6], pressure is monitored and controlled by modification of filtration speed and by use of adjustable tubing clamps.

Due to the minimal physical and chemical alterations involved in this isolation workflow, full functional integrity of the extracellular vesicle yield is be tter ensured. Suggested applications of this isolation method are in vitro and in vivo research investigations of exosome functions; large-scale production of cell culture-derived exosomes (e.g., for drug delivery applications [3]); or exosome purification from patient-derived biofluids such as ascites, urine or lavage fluids [7].

2 Materials

Prepare all solutions with deionized water and analytical grade reagents. Store all reagents following the manufacturer's instructions. Store the cleaning solution at 4 °C.

2.1 Prefiltration

1. Vacuum filtration unit with receiver flask, nominal 0.1 µm pore size, polyethersulfone (PES) membrane, 150–1000 mL capacity (e.g., Millipore *Stericup* or Corning *Vacuum Filter System*).
2. Phosphate Buffered Saline (PBS).

2.2 Tangential Flow Filtration (TFF) System

1. Peristaltic pump (e.g., Spectrum Labs *KrosFlo Research II* or Cole-Parmer *Masterflex*).
2. Pressure Monitor (e.g., Spectrum Labs *KrosFlo Digital Pressure Monitor* or Sartorius *Vivaflow* mechanical inline *Pressure Indicator*).
3. Holders for filter module and reservoir bottle (lab provided or as suggested by TFF system manufacturer).
4. 500 kDa MWCO (theoretical 10 nm pore dimension) hollow fiber filter module (Spectrum Labs *MicroKros* or *MidiKros*) or 100 kDa MWCO (theoretical 6 nm pore dimension) crossflow filtration cassette (Sartorius *Vivaflow 50*, *50R* or *200*).
5. Pressure Transducers (3×) (for use with Digital Pressure Monitor).
6. 0.8 mm silicone tubing with Luer fittings (or as provided by TFF system manufacturer).
7. Cable binders.
8. 3-port conical reservoir bottle (250 mL) with dip tubes.
9. Adjustable clamps.
10. Three-way Luer valves.
11. 220 nm syringe filters.
12. Waste container (e.g., 1000 mL Erlenmeyer flask).
13. Phosphate Buffered Saline (PBS).
14. Deionized water.
15. Cleaning solution: 1 N NaOH (add 4 g NaOH to 1 liter deionized water). You may optionally add 0.05% sodium azide. This cleaning solution enables reuse of the tangential flow filtration system (as specified by manufacturer) by removing protein residues from within the system. Not all hollow fiber filter modules are sold for multiple use applications; deviation from the manufacturer's recommendations could impair filtration performance.

2.3 Track Etch Filtration

1. 25 mm diameter track-etched membrane filters; 0.03, 0.05, 0.08, or 0.10 μm pore size (e.g., Whatman/GE *Nucleopore*). We recommend using filters with 0.10 μm pore sizes (*see* **Note 1**).
2. Autoclavable filter holders, 25 mm (Whatman/GE *Swin-Lok Filter Holder*): Place track-etched membrane filter in filter holder (and autoclave, if desired).
3. Syringe, 20 mL.
4. Syringe pump.
5. Pressure transducer (if using digital pressure monitor).

6. Pressure Monitor (e.g., Spectrum Labs *KrosFlo Digital Pressure Monitor*).
7. 0.8 mm silicon tube with Luerlock fitting.

3 Methods

3.1 Prefiltration of Biofluid

1. Add up to 200 mL cell culture supernatant or other large-volume biofluid (*see* **Note 2**) to a sterile bottle-top filter and filter at low speed with gentle vacuum.
2. After complete draining of the sample, add an additional 50 mL of buffer on top of the filter in order to increase the yield by washing residual EVs out of the filter into the filtrate.
3. Transfer the filtrate into a 250 mL 3-port conical-bottom reservoir bottle.

3.2 Tangential Flow Filtration

1. Set up the Tangential Flow Filtration System as indicated in (Fig. 2). Suggested equipment is listed in the Materials section. Tubing lengths should be kept as short as possible to minimize holdup volume.
2. Connect all parts of the system with Luer fittings (*see* **Note 3**). Close the lower filtrate outlet of the filter module.
3. Place pressure transducers at the suggested locations and connect to the pressure monitor. Equip all air inlets with 0.22 μm syringe filters to avoid contamination with air pollutants during the filtration.
4. Attach the adjustable clamps at the suggested locations, but keep the clamps loose until calibration of the pressure monitor is performed.
5. Do not connect the conical-bottom reservoir bottle yet, as the system needs to be primed first. Instead, connect the two hoses leading to the reservoir bottle to one other using Luer fittings.
6. Wet the filter module membranes by adjusting the 3-way valves such that the deionized water supply is open and connections to the air inlet, PBS supply, and conical reservoir bottle are closed (Fig. 3).
7. Start the peristaltic pump and tighten the clamps in order to fill the filter module with water. Make sure that the pressure never exceeds 20 PSI. Once all the membranes are wetted (you will see the water level rise in the filter), stop the pump, close the water supply and open the PBS supply (*see* **Note 4**).
8. Start the pump again and thus replace the water in the system with PBS (*see* **Note 5**). Stop the pump again.
9. Attach the conical reservoir bottle with the filtrate from **step 1** to the previously prepared TFF system.

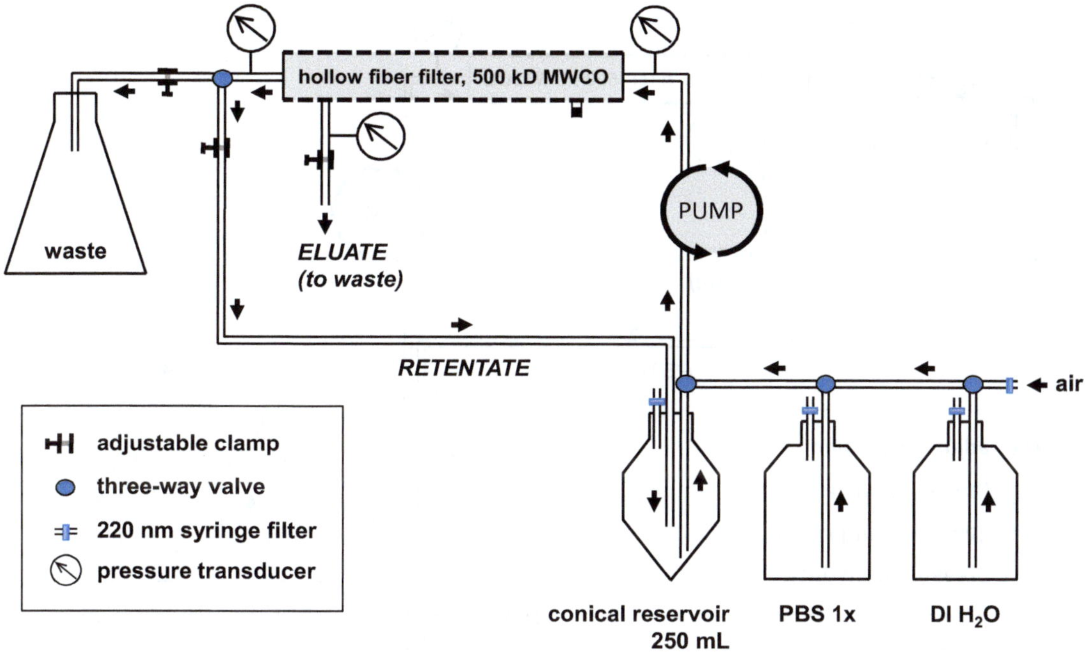

Fig. 2 Suggested setup for tangential flow filtration. *Arrows* indicate direction of flow. Filtrate from step 1 of the protocol is loaded into conical reservoir bottle. A peristaltic pump drives sample through a hollow fiber filter module (pore size: 500 kDa) and back into the reservoir bottle. Small molecules, such as free proteins, can pass the filter pores, elute from the system, and are discarded. Larger particles, such as exosomes and microvesicles, stay in circulation. Pressure is constantly displayed on a pressure monitor and maintained on a low level by adjustable clamps. Retained exosome-enriched sample is transferred to step 3 of the protocol. (Reprinted from *Journal of Chromatography A*, 1371, M.L. Heinemann et al., Benchtop isolation and characterization of functional exosomes by sequential filtration, 125–135, 2014, with permission from Elsevier)

10. Close the valve leading to the PBS and DI water supplies so as to only filter the exosome sample.
11. Before starting the pump, make sure all valves are adjusted to direct a closed circulation-loop from the reservoir bottle through the filter and back into the bottle.
12. Start the filtration slowly. Increase the pump speed while adjusting the clamps to provide a stable trans-membrane pressure of approx. 1.5–2.0 PSI (*see* **Notes 6** and 7).
13. When the volume of the reservoir bottle is decreased to approx. 50 mL, add 200 mL of PBS to the system. This can be done in two different ways:
    (a) Loosen the clamps, open the PBS supply valve and start the peristaltic pump until volume in reservoir bottle reaches 250 mL again. Be careful to stop the pump before the bottle overflows (*see* **Note 8**).

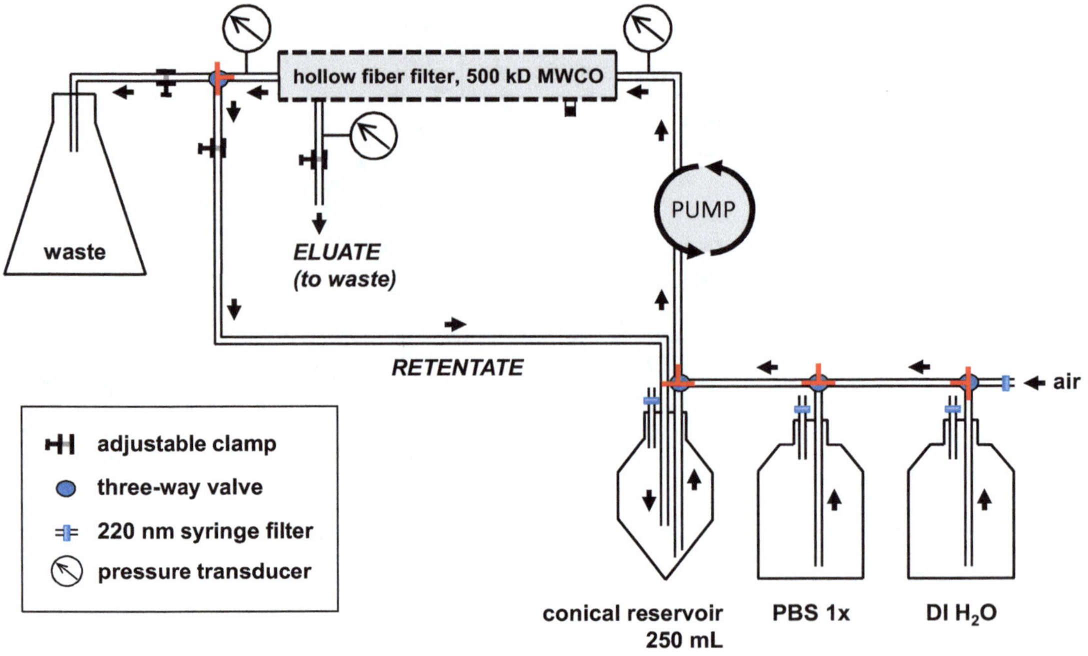

Fig. 3 Suggested setup for preparation of tangential flow filtration system. Filter membranes need to be flushed with deionized water and PBS before performing exosome enrichment. For initial flushing with deionized water, adjust three way valves as indicated. After flushing with deionized water, pause the peristaltic pump, close the deionized water supply, and open the PBS supply for subsequent flushing with PBS. The system is then ready for the exosome-containing sample

(b) Disconnect the air filter from the reservoir bottle and manually inject 200 mL PBS in to the bottle with a syringe.

14. Repeat the previous two steps four more times for a total of five diafiltrations (*see* **Note 9**) During the last filtration, reduce the volume until the reservoir bottle is empty and only the holdup volume of the system remains (depending on the specific equipment and setup, this volume should be between 1 and 5 mL).
15. Open the air supply by adjusting the valves as indicated in (Fig. 4) and drain the system into the reservoir bottle (*see* **Note 10**).
16. Detach the reservoir bottle and connect the two hoses that were attached to the bottle to each other again (as in **step 1**).
17. Transfer the contents of the reservoir bottle into a 20 mL syringe. TFF is now complete. If reusing system and filter, complete **steps 18** and **19**, below.
18. Loosen the clamps and flush the system with at least 200 mL deionized water by adjusting valves to direct flow through all fluid paths.

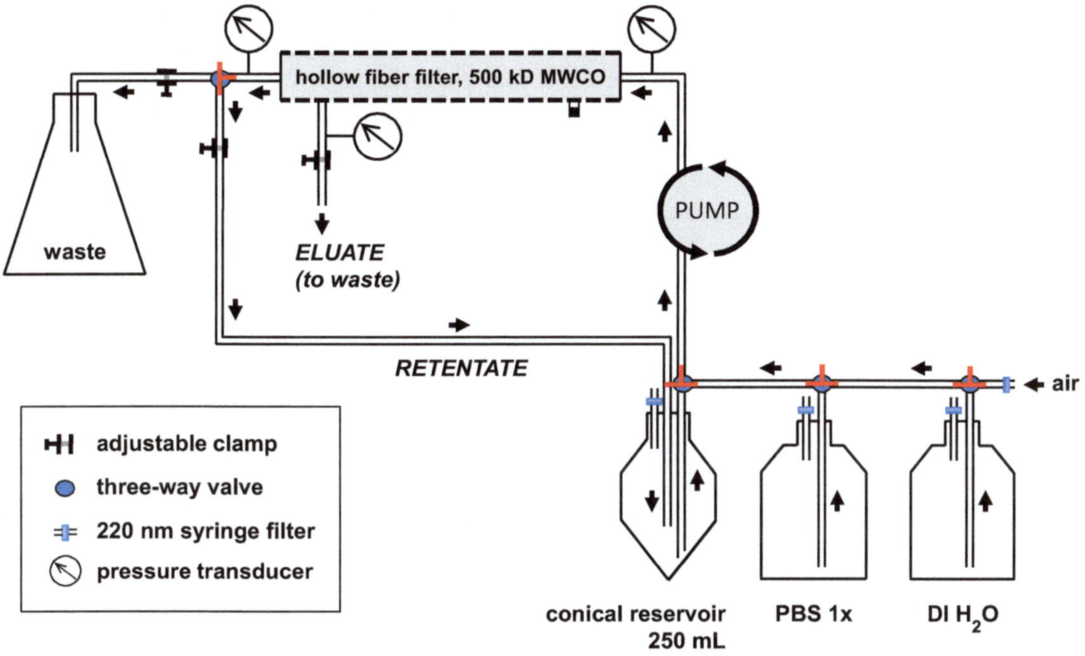

Fig. 4 Suggested adjustments to the tangential flow filtration system for system drainage. When sample is reduced down to the holdup volume of the system, stop the pump and adjust three way valves as indicated. Flushing the system with a few ml of PBS is suggested in order to increase exosome recovery

19. Fill the reservoir bottle with cleaning solution and reattach it to the system. Start the pump and tighten the clamps until a stable trans-membrane pressure of 1.5–2.5 PSI is reached. Stop the pump before the system is drained completely. Store the filter filled with cleaning solution until reusing it. Before the next filtration, flush the filter with deionized water at low pressure (as described **step 2** of this protocol).

3.3 Track-Etch Filtration

1. After completion of TFF, load the remaining contents of the reservoir bottle into a 20 mL syringe, as indicated in **step** 7 of the TFF protocol.
2. Connect the parts of the system in the following order: syringe, pressure transducer, syringe filter, hose. Place the syringe in a syringe pump and make sure the hose rests in a conical-bottom tube to catch the filtrate. Attach the transducer to a pressure monitor (*see* **Note 11**).
3. Start the filtration while monitoring trans-membrane pressure (*see* **Note 11**). Try to stay below 3.0 PSI. Do not allow the trans-membrane pressure to exceed 3.5 PSI (0.25 bar). The filtration may take from 30 min to up to a few hours, depending on both the concentration of particles in the sample and on the selected filter membrane (*see* **Note 12**).

4. At the end of the filtration, rinsing the filter is an optional step. Since the filter holder has a small holdup volume, rinsing will increase the exosomal yield: Load 1–2 mL PBS in the syringe and reconnect it to syringe pump and pressure transducer. Filter as described above.

4 Notes

1. We recommend performing prefiltration, track-etch filtration and all transferring steps within a biological safety hood to avoid contamination. This is especially crucial if downstream functional assays involving cell cultures are planned.
2. To process more than 200 mL of cell culture supernatant, or other large-volume biofluid, simply use a filtration unit with a larger receiver flask. In this case, the Tangential Flow setup will need to be modified (*see* Subheading 3.2).
3. If the connections between the Luer fittings and hoses are not tight enough, cable binders or tubing clamps can be used to create a more secure connection. If using a filter module without Luerlock connections (such as Spectrum Labs *MidiKros*), use an appropriate-sized tube and tie it to the filter module directly, using cable binders.

If processing copious amounts of media, a larger conical reservoir bottle than suggested in this protocol will be required. Alternatively, additional media can periodically be injected into the standard volume reservoir during filtration via a 3-way valve.

4. Always stop the pump before adjusting the three-way valves.
5. Use of deionized water or other nonphysiologic buffer during the filtration process could compromise vesicle integrity or functional properties. PBS or other buffer of approximately 290 mOsmol/L is recommended.
6. If the transmembrane pressure increases rapidly, immediately stop filtration. The most common causes of rapid pressure increase are saturated air filters or incorrectly adjusted valves.
7. To achieve consistent results, it is important to ensure that the tangential flow filtration system does not completely drain or overflow.
8. If the air filters in the TFF system become wet, they will impede airflow and will need to be replaced.
9. One can save time by performing fewer than five diafiltrations. However, doing so will limit the depletion of free protein from the exosome sample.

10. The system may be flushed with 1–3 mL of PBS to increase the total exosome yield. However, performing this additional step will decrease the final net exosome concentration as the harvest volume is increased.
11. Here, we define "transmembrane pressure" as simply 50% of the premembrane pressure, as we do not attach a second transducer behind the filter for the calculation of an absolute transmembrane pressure. However, this approximation is acceptable for monitoring pressure and identifying pressure peaks.
12. The rate of pressure gain during rack etch filtration depends on the particle concentration and the filter. The density of holes in the membrane varies, resulting in varying prefiltration pressure.

References

1. Théry C, Amigorena S, Raposo G (2006) Isolation and characterization of exosomes from cell culture supernatants and biological fluids. Curr Protoc Cell Biol. Chapter 3, Unit 3.22
2. Rekker K, Saare M, Roost AM et al (2014) Comparison of serum exosome isolation methods for microRNA profiling. Clin Biochem 47:135–138
3. Vader P, Mol EA, Pasterkamp G et al (2016) Extracellular vesicles for drug delivery. Adv Drug Deliv Rev 106:1–9
4. Witwer KW, Buzás EI, Bemis LT et al (2013) Standardization of sample collection, isolation and analysis methods in extracellular vesicle research. J Extracell Vesicles 2:18389
5. Heinemann ML, Ilmer M, Silva LP et al (2014) Benchtop isolation and characterization of functional exosomes by sequential filtration. J Chromatogr A 1371:125–135
6. Yellon DM, Davidson SM (2014) Exosomes: nanoparticles involved in cardioprotection? Circ Res 114:325–332
7. Pocsfalvi G, Stanly C, Fiume I et al (2016) Chromatography and its hyphenation to mass spectrometry for extracellular vesicle analysis. J Chromatogr A 1439:26–41

Chapter 5

Paper-Based for Isolation of Extracellular Vesicles

Yi-Hsing Hsiao and Chihchen Chen

Abstract

Paper-based devices chemically functionalized with capturing molecules enable the isolation and characterization of extracellular vesicles (EVs) from samples of limited amount. Here, we describe the isolation of EV subpopulations from human serum samples. The morphology, content, and amount of captured EVs can be assessed using scanning electron microscopy (SEM), transcriptome analysis, and paper-based enzyme-linked immunosorbent assays (pELISA), respectively. A colorimetric readout can be detected from 10 μL serum within 10 min.

Key words Extracellular vesicles, Exosomes, Cellulose paper, Microfluidics, Paper ELISA, Chemical conjugation

1 Introduction

Most of cell types release extracellular vesicles (EVs) through different biogenesis routes into the extracellular space including plasma [1, 2], urine [3, 4], milk [2, 5], saliva [6], amniotic fluid [7], malignant effusions [8] and aqueous humor [9]. EVs are heterogeneous membranous particles that range in size from 30 to 5000 nm and contain exclusive and particular subsets of nucleic acids and proteins inherited from parental cells. EVs are increasingly recognized to involve in physiological as well as disease processes, such as cell communication [10], angiogenesis [11], metastasis [12], and eye diseases [9]. A huge interest has been sparked in assessing their efficiency and applications on medical and biological studies. The common procedure for purifying EVs involves a series of centrifugations [13] and/or filtration [14] to remove large debris and cellular contaminants, followed by a final high-speed sucrose density gradient ultracentrifugation [15, 16]. Other methods, such as utilizing magnetic beads [17], polymeric precipitation [18–20], two-aqueous phase separation [21] and microfluidic techniques [22–24] have also been developed. We have developed a

Winston Patrick Kuo and Shidong Jia (eds.), *Extracellular Vesicles: Methods and Protocols*, Methods in Molecular Biology, vol. 1660, DOI 10.1007/978-1-4939-7253-1_5, © Springer Science+Business Media LLC 2017

paper-based immunoaffinity device that provides a simple, time- and cost-effective, and efficient way to isolate and characterize subgroups of EVs [24, 25]. Cellulose paper cut into a defined shape can be arranged and laminated using two plastic sheets with the registered through-holes. In contrast to the general strategy to define the fluid boundary in paper-based devices by printing hydrophobic wax or polymers, these patterns of laminated paper are resistant to many organic solvents [26–28]. Paper test zones are chemically modified [29–31] to provide stable and dense coverage of capturing molecules which have a high affinity for specific surface markers on EV subgroups [32, 33]. Biological samples can be pipetted directly onto the paper test zones, and EVs are isolated after the rinse step; scanning electron microscope (SEM), enzyme-linked immunosorbent assay (ELISA), or transcriptomic analysis can be performed to characterize isolated EVs.

2 Materials

2.1 Collection of Serum Samples

1. Serum separation tubes and conical tubes.
2. 0.8 μm filters (*see* **Note 1**).

2.2 Fabrication of Paper Devices

1. Whatman cellulose filter paper No. 1 (*see* **Note 2**).
2. Polystyrene sheets.
3. A thermal laminator.
4. A manual hole punch or a CAD-automated laser cutter.
5. A plasma chamber.
6. Silane solution: 4% (v/v) 3-mercaptopropyl trimethoxysilane in ethanol (200 proof) (*see* **Note 3**).
7. *N*-γ-maleimidobutyryloxy succinimide ester (GMBS) solution: 0.01 μmol/mL GMBS in ethanol (200 proof) (*see* **Note 3**).
8. Blocking solution: 1% (w/v) bovine serum albumin (BSA) in phosphate buffered saline (PBS). Store at 4 °C.
9. NeutrAvidin solution: 10 μg/mL NeutrAvidin solution in PBS (*see* **Note 4**). Store at 4 °C and use within 4 weeks.

2.3 Immunopurification

1. Biotinylated capturing molecule solution: the capturing molecule of choice can be anti-CD63 antibodies, annexin V, or other molecules that would bind to EVs (*see* **Note 5**).
2. Rinse solution: use blocking buffer to rinse off unbound antibodies while use annexin V binding buffer to rinse off unbound annexin V molecules.

2.4 Scanning Electromicrographs (SEM)

1. Cacodylate buffer: 0.2 M sodium cacodylate trihydrate solution, pH 7.4 (*see* **Note 6**).
2. Karnovsky's fixative containing 5% glutaraldehyde and 4% paraformaldehyde in 0.08 M cacodylate buffer, pH 7.4 (*see* **Note 7**).

2.5 RNA Isolation

1. MirVana RNA isolation kit.
2. Ambion plant RNA isolation aid.
3. MinElute RNA cleanup kit.
4. RNase-free microcentrifuge tubes and pipette tips.

2.6 Paper-Based ELISA

1. Primary antibody solution: 1:1000 primary antibody (e.g., rabbit anti-human CD9) in PBS.
2. Secondary antibody solution: 1:1000 secondary antibody (e.g., horseradish peroxidase (HRP)-conjugated goat anti-rabbit antibody) in PBS.
3. Substrate solution: Mix 1:1 (v/v) of hydrogen peroxide and 3,3′,5,5′-tetramethylbenzidine (TMB).
4. A desktop scanner.

3 Methods

Carry out all procedures at room temperature unless otherwise specified.

3.1 Collection of Serum Samples

1. Collect ~8.5 mL of peripheral blood by venipuncture into each serum separation tube and gently invert the tube five times. Keep the tube in a vertical position for 30 min.
2. Centrifuge the tube at 1200 × *g* for 15 min. Transfer the top, clear, serum layer into a clean conical tube.
3. Centrifuge the conical tube at 3000 × *g* for 30 min. Pass the supernatant through a 0.8 μm filter. The resulting filtered serum can be used immediately or be store at −80 °C.

3.2 Fabrication of Paper Devices

1. Cut cellulose filter paper into circles of 5 mm in diameter.
2. Create an array of registered through-holes in two polystyrene sheets. Each through-hole exhibits a diameter of <5 mm (*see* **Note 8**).
3. Place one polystyrene sheet on the bench, and overlay a single paper disc on top of each corresponding through-hole. Sandwich the paper discs by registering and overlaying the second sheet of polystyrene. Check the overall alignment before bonding the sandwich in a thermal laminator (Fig. 1).
4. Insert the laminated device into a plasma chamber and activate the paper surface with oxygen plasma (100 mW, 1% oxygen, 30 s).

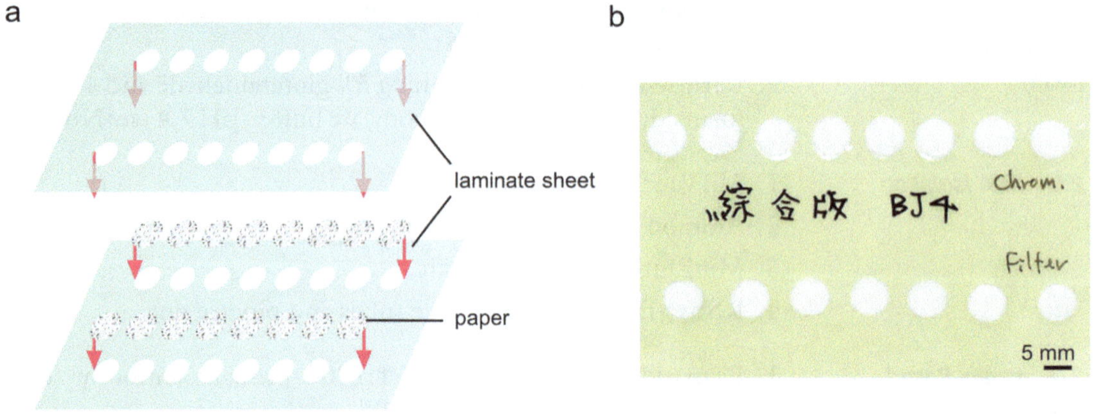

Fig. 1 The fabrication of the paper-based device. (**a**) Filter paper is cut into the desired shape and laminated. (**b**) Image of the paper-based device

5. Upon release of plasma chamber vacuum, immediately transfer the paper device into the silane solution and incubate for 30 min. Rinse with 100% ethanol.
6. Incubate the device with a GMBS solution for 15 min. Rinse with 100% ethanol.
7. Incubate the device with NeutrAvidin solution for 1 h at 4 °C. The device can then proceed to the following steps, or be stored at 4 °C for 4 weeks if needed (Fig. 2).

3.3 Immunopurification

1. Pipette 10 μL blocking solution onto each paper disc and incubate for 10 min. Repeat two more times.
2. Pipette 10 μL biotinylated capturing molecule solution onto each paper disc and incubate for 10 min. Repeat two more times.
3. Pipette each paper disc with 10 μL rinse solution and incubate for 1 min. Repeat two more times.
4. Pipette serum samples onto each paper disc at a rate of 5 μL/min.
5. Pipette each paper disc with 10 μL rinse solution and incubate for 1 min. Repeat two more times.

3.4 SEM Analysis

1. Pipette 10 μL 0.5×Karnovsky's fixative onto each paper disc and incubate for 10 min to fix EVs isolated.
2. Rinse the paper device twice with ample PBS for 5 min.
3. Dehydrate the paper device with increasing concentrations of ethanol—one 35% ethanol incubation, two 50% ethanol incubations, two 70% ethanol incubations, two 95% ethanol incubations, and four 100% ethanol incubations. Each incubation step lasts 10 min.

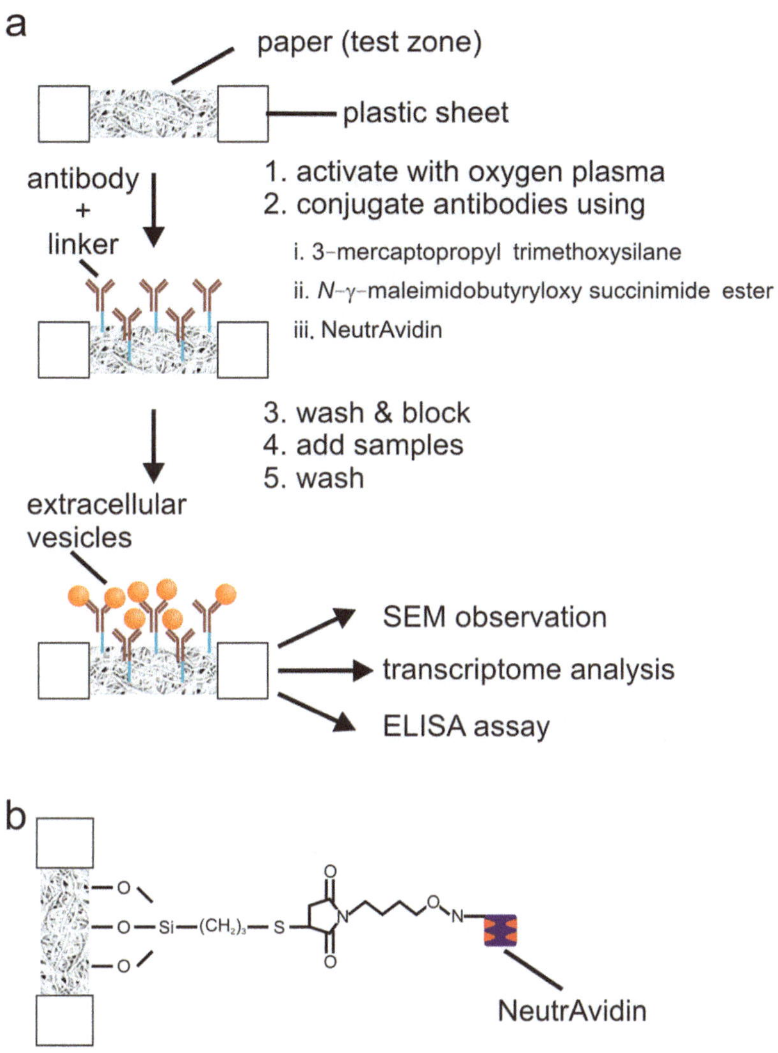

Fig. 2 The operation procedure of paper-based assays for isolation and characterization of extracellular vesicles. (**a**) The surface of the paper was first activated with a brief treatment of oxygen plasma and reacted with the following molecules, 3-mercaptopropyl trimethoxysilane, *N*-γ-maleimidobutyryloxy succinimide ester, and NeutrAvidin, in that order. The schematic structure of the linker formed on the surface was shown in (**b**). Biotinylated capturing molecules, e.g., antibodies, were immobilized via the strong affinity between biotin and NeutrAvidin molecules. Extracellular vesicles isolated on the functionalized paper surface could be characterized by using SEM, transcriptome and ELISA assays (reproduced from ref. 24 with permission of Springer)

4. Critical dry the paper device and sputter-coat it with palladium/gold.
5. Perform SEM at low electron acceleration voltage (~5 kV) to prevent excess charging of the paper device (Fig. 3a–e).

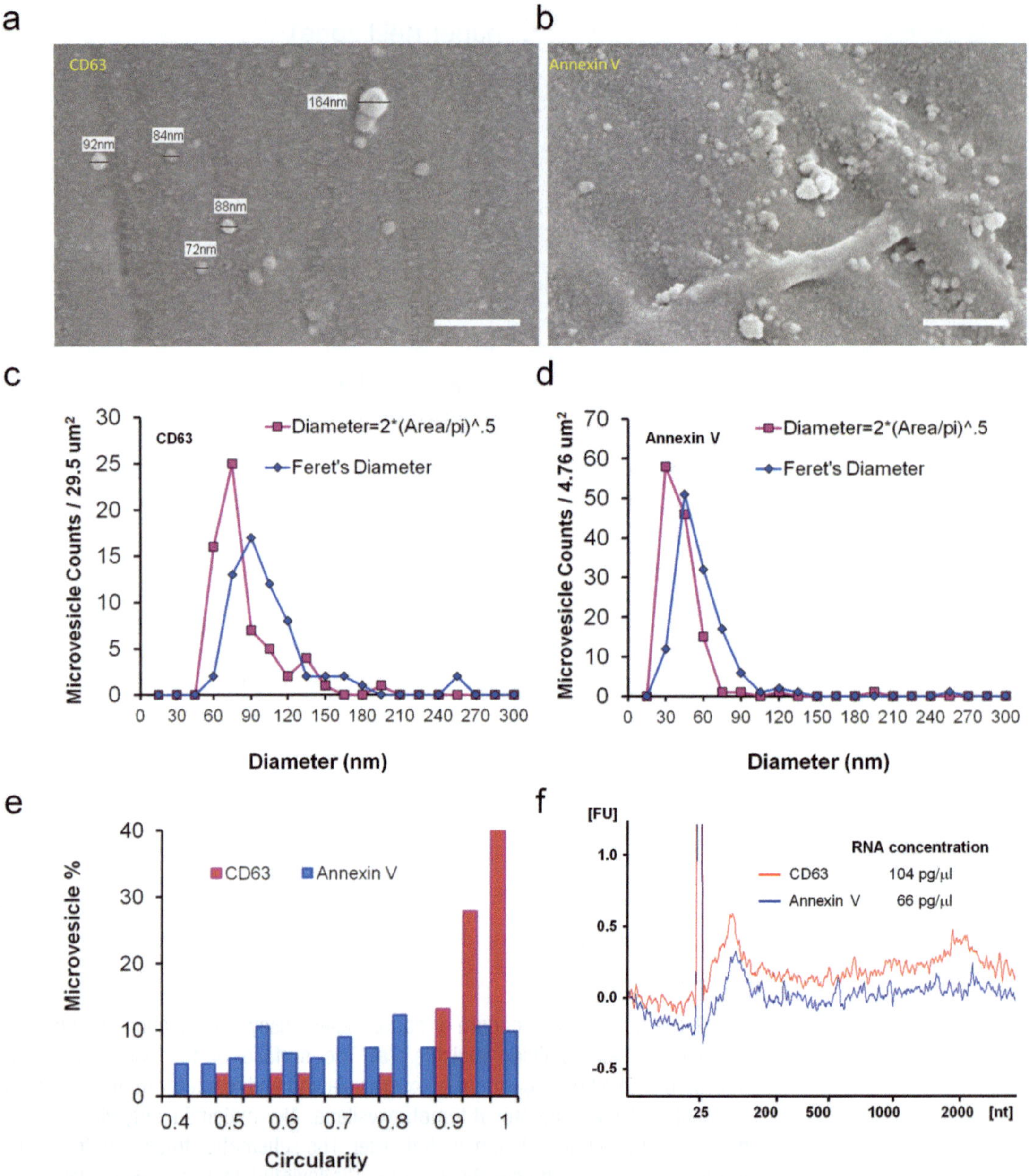

Fig. 3 EVs were captured on functionalized filter paper. SEM images showing the EVs isolated on the filter paper coated with anti-CD63 antibodies (**a**) and annexin V molecules (**b**), respectively (scale bars: 400 nm). The distributions of both the projected area diameters and Feret's diameters of EVs in image (**a**) and (**b**) are shown in (**c**) and (**d**), respectively. Projected area diameters were calculated from the projected area assuming circular geometry, while Feret's diameter were determined as the greatest distance possible between any two points along the boundary of a region of interest. (**e**) Profiles of the circularity of EVs captured on the anti-CD63 antibody- and annexin V-coated papers. (**f**) Bioanalyzer data showing the intensities (*y*-axis, in arbitrary units) and size (*x*-axis, in nucleotides (nt)) distributions of fluorescently labeled total RNA extracted from EVs captured on anti-CD63 antibody- or annexin V-coated filter papers. The migrating peak at 25 nt represents an internal standard (reproduced from ref. 24 with permission of Springer)

3.5 RNA Isolation

1. Place each paper disc into an RNase-free microcentrifuge tube (*see* **Note 9**).
2. Lyse EVs by adding 35 μL polyvinylpyrrolidone-based RNA isolation aid and 265 μL lysis/binding buffer to each tube. Mix vigorously by using a vortex mixer.
3. Add 30 μL homogenate provided with the isolation kit to the lysate. Vortex vigorously and incubate the tube on ice for 10 min.
4. Add 330 μL acid phenol/chloroform to the lysate.
5. Vortex vigorously for at least 15 s per tube.
6. Centrifuge at 10,000 × *g* for 5 min.
7. Transfer the RNA-containing, top, aqueous phase into a fresh tube and record the volume removed (V).
8. Precipitate the total RNA with 1.25 × V of 100% ethanol and mix thoroughly.
9. Add 700 μL sample into the filter cartridge provided with the isolation kit. Centrifuge at 10,000 × *g* for 15 s. Discard the flow-through. Repeat this step using the same filter cartridge if the sample volume is more than 700 μL.
10. Add 700 μL washing solution #1 provided with the isolation kit into the filter cartridge. Centrifuge at 10,000 × *g* for 15 s. Discard the flow-through.
11. Add 500 μL washing solution #2/3 provided with the isolation kit into the filter cartridge. Centrifuge at 10,000 × *g* for 15 s. Discard the flow-through. Repeat one more time.
12. Centrifuge at 10,000 × *g* for 1 min to remove residual ethanol. Discard the flow-through.
13. Place the filter column in a fresh tube.
14. Add 33 μL elution solution preheated to 95 °C into the filter cartridge. Close the cap and wait for 20 s. Repeat two more times (*see* **Note 10**).
15. Centrifuge at maximum speed for 30 s. Save the flow-through RNA sample. Store at −80 °C if needed, or proceed to the following steps if further concentrating or purifying RNAs is deemed necessary.
16. Add 350 μL RTL buffer provided with the cleanup kit into the RNA sample. Mix well.
17. Add 250 μL ethanol into RNA sample. Mix well by pipetting.
18. Transfer the RNA sample into the spin column. Close the cap gently. Centrifuge at 8000 × *g* for 15 s. Discard the flow-through.

Table 1
Extraction of total RNA from sera from healthy volunteers and analyzing with a Bioanalyzer ($n = 4$ for all conditions, reproduced from ref. 24 with permission of Springer)

Platform	Capturing molecule	Serum vol. (μL)	Average RNA extracted (ng)	Average RNA extracted per serum (pg/μL)	Reference
Microfluidic device	anti-human CD63	400	4.14	10.4	[22]
Paper device	anti-human CD63	72	1.49 ± 0.08	20.7	
Paper device	annexin V	72	0.99 ± 0.26	13.8	
Serum	–	72	1.23 ± 0.58	17.0	

19. Add 500 μL RPE buffer provided with the cleanup kit into the spin column. Close the cap gently. Centrifuge at 8000 × *g* for 15 s. Discard the flow-through.
20. Add 500 μL 80% ethanol into the spin column. Close the cap gently. Centrifuge at 8000 × *g* for 2 min. Discard the flow-through.
21. Place the spin column in a fresh tube.
22. Keep the lid open. Centrifuge at 8000 × *g* for 2 min. Discard the flow-through.
23. Place the spin column in a fresh tube.
24. Add 14 μL RNase-free water into the spin column. Close the cap gently.
25. Centrifuge at maximum speed for 1 min. Save the flow-through RNA sample. Store at −80 °C if needed (Fig. 3f and Table 1).

3.6 Paper-Based ELISA

1. Add 5 μL primary antibody solution onto each paper disc and incubate for 1 min.
2. Rinse paper devices with 30 mL PBS shaken at 100 r.p.m. for 30 s.
3. Add 5 μL secondary antibody solution onto each paper disc and incubate for 1 min.
4. Rinse paper devices with 30 mL PBS shaken at 100 r.p.m. for 30 s.
5. Add 5 μL substrate solution onto each paper disc and scan the resulting change in spot color intensity using a desktop scanner (Fig. 4).

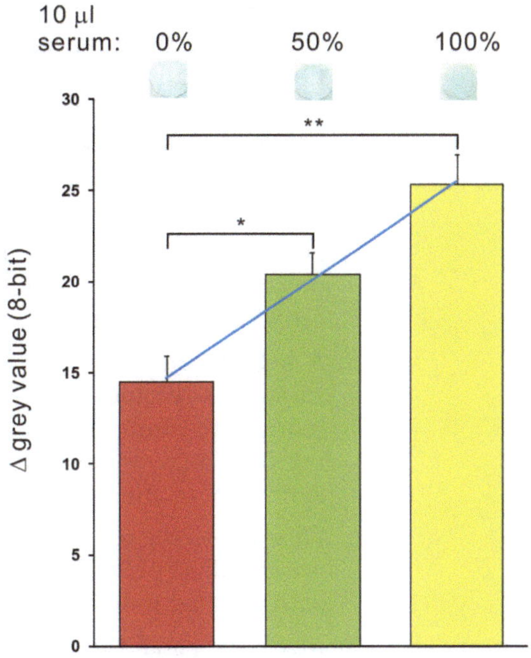

Fig. 4 P-ELISA for the detection of EVs. *Grey* value changes for paper devices coated with 10 μL samples of PBS, 50% serum (diluted with PBS) and 100% serum samples were 14.5 ± 1.4, 20.4 ± 1.2, and 25.3 ± 1.6, respectively ($n = 2$ for each group, $^{**}p = 0.02$ and $^{*}p = 0.05$, Student's *t*-test). The overlaid linear regression line, $Y = 10.8X + 14.7$ with a coefficient of determination $R2 = 0.95$, shows the linear change in *grey* values produced by the enzymatic reaction of horseradish peroxidase (HRP) in the P-ELISA assay within the range of serum samples tested (reproduced from ref. 24 with permission of Springer)

4 Notes

1. Other sized filters may be used based on the research interest. EVs can be commonly classified into exosomes (30–100 nm in diameter), microvesicles (50–1000 nm in diameter), and apoptotic bodies (800–5000 nm in diameter) [34]. However, classification of EVs is still under discussion and investigation.
2. Cellulose paper grade 1 was chosen for its low protein adsorption and relatively large average pore size (11 μm). Other porous materials, such as nitrocellulose may be used based on the research interest [35].
3. Prepare 3-mercaptopropyl trimethoxysilane and GMBS solutions immediately before use. Special caution must be taken, since they are both moisture sensitive and toxic. Wear protective gloves and avoid inhalation or contact with skin and eyes.

Keep the stock bottle tightly closed and allow it to equilibrate to room temperature before opening to avoid water condensation. Alternatively, open the stock bottle inside a nitrogen-filled glove bag or box. Aliquot into small vials to avoid frequent opening of the stock bottle.

4. NeutrAvidin and avidin both have a very strong affinity for biotin. We choose NeutrAvidin for its near-neutral isoelectric point of pH 6.3 to reduce nonspecific interactions with charged objects. In contrast, avidin with an isoelectric point of 10.5 carries positive charges at neutral pH.
5. A higher concentration of antibody is used due to the large surface-to-volume ratio of the paper device. We use 20 μg/mL biotinylated anti-CD63 antibody in blocking solution containing 0.09% (w/v) sodium azide, and 1:20 (v/v) biotinylated annexin V in annexin V binding buffer.
6. Dissolve 4.28 g sodium cacodylate trihydrate in 80 mL distilled water. Adjust pH to 7.4 with HCl and then make up to 100 mL with distilled water. Store at −20 °C.
7. Mix 2.0 g paraformaldehyde with 25 mL of distilled water in a 125 mL Erlenmeyer flask and heat to 60 °C on a hot plate. Add 2–4 drops of 1.0 M sodium hydroxide and stir until the solution clears. Cool down to room temperature. Add 5 mL 50% EM grade glutaraldehyde and 20 mL 0.2 M cacodylate buffer, pH 7.4. Store at −20 °C.
8. Make paper discs and through-holes on polystyrene sheets using either a manual hole punch or a CAD-automated laser cutter. The diameter of the though-hole is slightly smaller than that of the paper disc.
9. Take individual paper discs from the bulk device by cutting away regions of excess polystyrene laminate using scissors or a razor blade.
10. Elution solution may be nuclease-free water or nuclease-free 0.1 mM EDTA.

Acknowledgment

This work was supported in part by the Taiwan National Science Council grants—NSC 99-2320-B-007-005-MY2 (CC) and NSC 101-2628-E-007-011-MY3 (CMC), and the Veterans General Hospitals and University System of Taiwan Joint Research Program (CC).

References

1. Caby MP, Lankar D, Vincendeau-Scherrer C, Raposo G, Bonnerot C (2005) Exosomal-like vesicles are present in human blood plasma. Int Immunol 17(7):879–887
2. Lässer C, Alikhani VS, Ekstrom K, Eldh M, Paredes PT, Bossios A, Sjostrand M, Gabrielsson S, Lötvall J, Valadi H (2011) Human saliva, plasma and breast milk exosomes contain RNA: uptake by macrophages. J Transl Med 9:9. doi:10.1186/1479-5876-9-9
3. Wiggins RC, Glatfelter A, Kshirsagar B, Brukman J (1986) Procoagulant Activity in Nornal human-urine associated with subcellular particles. Kidney Int 29(2):591–597
4. Raj DAA, Fiume I, Capasso G, Pocsfalvi G (2012) A multiplex quantitative proteomics strategy for protein biomarker studies in urinary exosomes. Kidney Int 81(12):1263–1272
5. Admyre C, Johansson SM, Qazi KR, Filen JJ, Lahesmaa R, Norman M, Neve EPA, Scheynius A, Gabrielsson S (2007) Exosomes with immune modulatory features are present in human breast milk. J Immunol 179 (3):1969–1978
6. Keller S, Ridinger J, Rupp AK, Janssen JWG, Altevogt P (2011) Body fluid derived exosomes as a novel template for clinical diagnostics. J Transl Med 9:86. doi:10.1186/1479-5876-9-86
7. Asea A, Jean-Pierre C, Kaur P, Rao P, Linhares IM, Skupski D, Witkin SS (2008) Heat shock protein-containing exosomes in mid-trimester amniotic fluids. J Reprod Immunol 79 (1):12–17
8. Bard MP, Hegmans JP, Hemmes A, Luider TM, Willemsen R, Severijnen LAA, van Meerbeeck JP, Burgers SA, Hoogsteden HC, Lambrecht BN (2004) Proteomic analysis of exosomes isolated from human malignant pleural effusions. Am J Respir Cell Mol Biol 31(1):114–121
9. Perkumas KM, Hoffman EA, McKay BS, Allingham RR, Stamer WD (2007) Myocilin-associated exosomes in human ocular samples. Exp Eye Res 84(1):209–212
10. Anderson HC, Mulhall D, Garimella R (2010) Role of extracellular membrane vesicles in the pathogenesis of various diseases, including cancer, renal diseases, atherosclerosis, and arthritis. Lab Invest 90(11):1549–1557
11. Montecalvo A, Larregina AT, Shufesky WJ, Stolz DB, Sullivan MLG, Karlsson JM, Baty CJ, Gibson GA, Erdos G, Wang ZL, Milosevic J, Tkacheva OA, Divito SJ, Jordan R, Lyons-Weiler J, Watkins SC, Morelli AE (2012) Mechanism of transfer of functional microRNAs between mouse dendritic cells via exosomes. Blood 119(3):756–766
12. Grange C, Tapparo M, Collino F, Vitillo L, Damasco C, Deregibus MC, Tetta C, Bussolati B, Camussi G (2011) Microvesicles released from human renal cancer stem cells stimulate angiogenesis and formation of lung premetastatic niche. Cancer Res 71(15):5346–5356
13. Théry C, Clayton A, Amigorena S, Raposo G (2006) Isolation and characterization of exosomes from cell culture supernatants and biological fluids. In: Morgan K (ed) Current protocols in cell biology. John Wiley, New York, NY, p 3.22
14. Rood IM, Deegens JKJ, Merchant ML, Tamboer WPM, Wilkey DW, Wetzels JFM, Klein JB (2010) Comparison of three methods for isolation of urinary microvesicles to identify biomarkers of nephrotic syndrome. Kidney Int 78(8):810–816
15. Lamparski HG, Metha-Damani A, Yao JY, Patel S, Hsu DH, Ruegg C, Le Pecq JB (2002) Production and characterization of clinical grade exosomes derived from dendritic cells. J Immunol Methods 270(2):211–226
16. Cantin R, Diou J, Belanger D, Tremblay AM, Gilbert C (2008) Discrimination between exosomes and HIV-1: purification of both vesicles from cell-free supernatants. J Immunol Methods 338(1-2):21–30
17. Taylor DD, Gercel-Taylor C (2008) MicroRNA signatures of tumor-derived exosomes as diagnostic biomarkers of ovarian cancer. Gynecol Oncol 110(1):13–21
18. Fernandez-Llama P, Khositseth S, Gonzales PA, Star RA, Pisitkun T, Knepper MA (2010) Tamm-Horsfall protein and urinary exosome isolation. Kidney Int 77(8):736–742
19. Alvarez ML, Khosroheidari M, Ravi RK, DiStefano JK (2012) Comparison of protein, microRNA, and mRNA yields using different methods of urinary exosome isolation for the discovery of kidney disease biomarkers. Kidney Int 82(9):1024–1032
20. Witwer KW, Buzás EI, Bemis LT, Bora A, Lässer C, Lötvall J, Nolte-'t Hoen EN, Piper MG, Sivaraman S, Skog J, Théry C, Wauben MH, Hochberg F (2013) Standardization of sample collection, isolation and analysis methods in extracellular vesicle research. J Extracel Vesicles 2:20360
21. Shin H, Han C, Labuz JM, Kim J, Kim J, Cho S, Gho YS, Takayama S, Park J (2015) High-yield isolation of extracellular vesicles using aqueous two-phase system. Sci Rep 5:13103

22. Chen C, Skog J, Hsu CH, Lessard RT, Balaj L, Wurdinger T, Carter BS, Breakefield XO, Toner M, Irimia D (2010) Microfluidic isolation and transcriptome analysis of serum microvesicles. Lab Chip 10(4):505–511
23. Davies RT, Kim J, Jang SC, Choi EJ, Gho YS, Park J (2012) Microfluidic filtration system to isolate extracellular vesicles from blood. Lab Chip 12(24):5202–5210
24. Chen C, Lin BR, Wang HK, Fan ST, Hsu MY, Cheng CM (2014) Paper-based immunoaffinity devices for accessible isolation and characterization of extracellular vesicles. Microfluid Nanofluid 16(5):849–856
25. Chen C, Lin BR, Hsu MY, Cheng CM (2015) Paper-based devices for isolation and characterization of extracellular vesicles. J Vis Exp 98: e52722
26. Carrilho E, Martinez AW, Whitesides GM (2009) Understanding wax printing: a simple micropatterning process for paper-based microfluidics. Anal Chem 81(16):7091–7095
27. Dungchai W, Chailapakul O, Henry CS (2009) Electrochemical detection for paper-based microfluidics. Anal Chem 81(14):5821–5826
28. Glavan AC, Martinez RV, Subramaniam AB, Yoon HJ, Nunes RMD, Lange H, Thuo MM, Whitesides GM (2014) Omniphobic "R-F Paper" produced by silanization of paper with fluoroalkyltrichlorosilanes. Adv Funct Mater 24(1):60–70
29. Usami S, Chen HH, Zhao YH, Chien S, Skalak R (1993) Design and construction of a linear shear-stress flow chamber. Ann Biomed Eng 21 (1):77–83
30. Cras JJ, Rowe-Taitt CA, Nivens DA, Ligler FS (1999) Comparison of chemical cleaning methods of glass in preparation for silanization. Biosens Bioelectron 14(8-9):683–688
31. Murthy SK, Sin A, Tompkins RG, Toner M (2004) Effect of flow and surface conditions on human lymphocyte isolation using microfluidic chambers. Langmuir 20 (26):11649–11655
32. Scanu A, Molnarfi N, Brandt KJ, Gruaz L, Dayer JM, Burger D (2008) Stimulated T cells generate microparticles, which mimic cellular contact activation of human monocytes: differential regulation of pro- and anti-inflammatory cytokine production by high-density lipoproteins. J Leukoc Biol 83 (4):921–927
33. Inal JM, Ansa-Addo EA, Stratton D, Kholia S, Antwi-Baffour SS, Jorfi S, Lange S (2012) Microvesicles in health and disease. Arch Immunol Ther Exp (Warsz) 60(2):107–121
34. Crescitelli R, Lässer C, Szabo TG, Kittel A, Eldh M, Dianzani I, Buzás EI, Lötvall J (2013) Distinct RNA profiles in subpopulations of extracellular vesicles: apoptotic bodies, microvesicles and exosomes. J Extracel Vesicles 2:20677
35. Fridley GE, Holstein CA, Oza SB, Yager P (2013) The evolution of nitrocellulose as a material for bioassays. MRS Bull 38 (4):326–330

Chapter 6

Filter-Based Extracellular Vesicle mRNA Isolation and High-Throughput Gene Expression Analysis

Cindy M. Yamamoto, Taku Murakami, and Shu-Wing Ng

Abstract

Extracellular vesicles (EVs) are a heterogeneous group of membrane-encapsulated particles with different ranges of size, density, and cargo. Various types of RNA including mRNA are enclosed within EVs and can serve as novel biomarkers for disease detection and patient management. Ultracentrifugation, precipitation, antibody-based capture and filter-based methods are available as in-house laboratory procedures or commercially available kits to isolate EVs. Here, we describe a filter-based method for EV mRNA isolation that is designed for parallel processing of large sample numbers.

Key words Exosomes, Microvesicles, Urine, Ascites, Extracellular vesicles, Filter, mRNA, Isolation, High-throughput, Gene expression

1 Introduction

EVs are cell-derived membrane vesicles released into the extracellular space. They encapsulate components such as proteins, lipids, DNAs, mRNAs, and noncoding RNAs and can reflect the molecular profiles of parental cells [1]. Major EV constituents include exosomes (~30–100 nm), which are derived from multivesicular bodies (MVB) and microvesicles (~100–1000 nm), which originate from direct budding of the plasma membrane. Size, biogenesis and cell surface markers for each EV type, however, have not been clearly defined and these are the challenges in EV isolation and separation. Most currently available EV isolation methods, including widely used differential ultracentrifugation, polymer-based precipitation and filtration are unable to completely separate each type of EV due to the overlap in these characteristics [2–4]. The method described here uses a depth filter which has been shown to bind exosomes and microvesicles [5, 6].

Biofluids such as urine or plasma have been reported to contain EVs in the range of 10^{8-11} particles/mL with RNA quantities in the range of pg–ng/mL, dependent on the method of isolation [7–9].

Winston Patrick Kuo and Shidong Jia (eds.), *Extracellular Vesicles: Methods and Protocols*, Methods in Molecular Biology, vol. 1660, DOI 10.1007/978-1-4939-7253-1_6, © Springer Science+Business Media LLC 2017

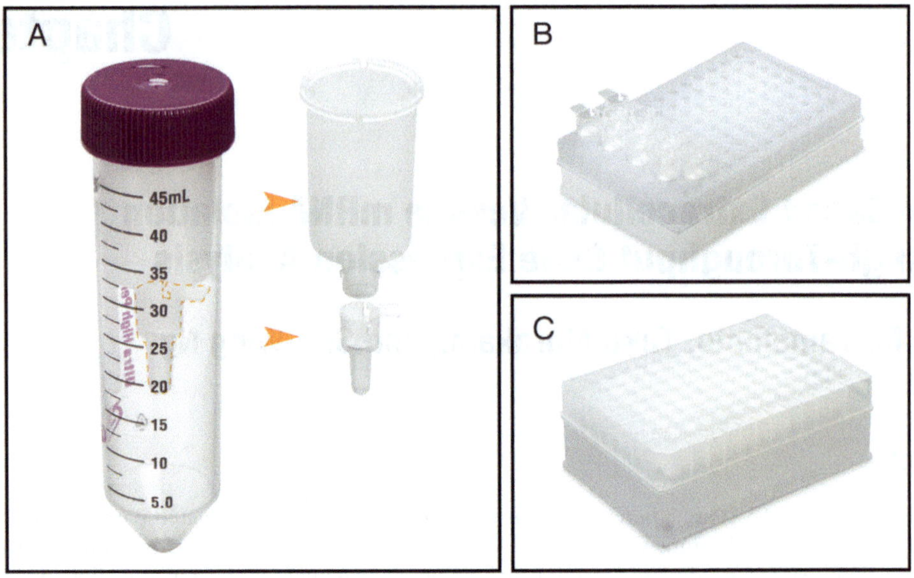

Fig. 1 Filter-based platform technologies. (**a**) The Exosome Collection Tube accommodates volumes up to 12.5 mL. After sample is applied and centrifuged, EVs are collected on the filter tip which can be detached from the collection tube. (**b**) The filter tips are then placed on a filter tip holder over the mRNA Capture Plate for lysis and mRNA hybridization. (**c**) The 96-well Exosome Filterplate handles volumes from 100 to 400 μL

Enrichment methods are frequently necessary because although the number of EV particles is large, RNA quantities are limiting. Sample volume required for a study will depend on downstream applications. In the method described here, high porosity of the filter allows large volumes to be applied without clogging, while EVs are concentrated onto the filter during low-speed centrifugation ($2000 \times g$). Two platforms based on the same filter technology are available. The 96-well filterplate is intended for applications when smaller volumes (100–400 μL) are sufficient, while the single filter tubes accommodate larger volumes (≤12.5 mL) (Fig. 1). Both of these platforms can be aligned to a 96-well oligo(dT)-coated microplate for mRNA hybridization. This configuration enables cDNA to be prepared for quantitative PCR analysis directly on this plate [6]. Alternatively, the mRNA can be isolated for downstream applications such as transcriptome (mRNA) next generation sequencing (NGS).

Increasing numbers of studies demonstrate that EVs are not only involved in cell-to-cell communication and cellular homeostasis in healthy individuals, but are also associated with pathological conditions such as diabetes and cancer [10–12]. The use of EVs and their components as potential biomarkers are being intensively investigated. EV isolation methods designed for simultaneous large scale sample processing would facilitate these studies. The methods here describe urine, plasma and ascites sample preparation with a focus on EV mRNA isolation and gene expression analysis from a 96-well platform.

2 Materials

Materials supplied by Hitachi Chemical Diagnostics, Inc. (HCD) include the ExoComplete™ Exosome Isolation Tubes, filter tip holder, mRNA Capture Plate, Lysis Buffer, Wash Buffer A, Wash Buffer B, RT dilution buffer and 10× RT buffer.

2.1 Urine Collection and Processing

1. 4 oz. specimen containers and 50 mL sterile tubes.
2. Refrigerated swinging bucket centrifuge that can accommodate 50 mL tubes.
3. 25× PBS (ThermoFisher).
4. Vortex.

2.2 Ascites Collection

1. Yankauer suction tip and drainage bag.
2. Refrigerated swinging bucket centrifuge that can accommodate 50 mL tubes.

2.3 Plasma Collection

1. Blood collection tube with EDTA or sodium citrate anticoagulant.
2. Refrigerated swinging bucket centrifuge that can accommodate 96-well plates.

2.4 Extracellular Vesicle Capture

1. ExoComplete™ Exosome Isolation Tubes.
2. ExoComplete™ Exosome Filterplate.

2.5 mRNA Hybridization and Isolation

1. Filter tip holder.
2. mRNA Capture Plate.
3. Working Lysis Buffer: Lysis Buffer (HCD) with 1% 2-mercaptoethanol and 500 μg/mL proteinase K.
4. Wash Buffers A and B (HCD).
5. DNase/RNase-free water or DEPC water.
6. Single or 8-well channel pipettor.
7. Single or 8-well channel aspirator.
8. Adhesive aluminum foil seal.
9. Crushed ice and ice pan.
10. Reservoirs.
11. Vacuum desiccator (optional).

2.6 cDNA Synthesis

1. cDNA Master Mix: 1× reverse transcriptase (RT) buffer, 10 μM random hexamers, 1.25 mM dNTP, 0.1 U/μL RNase Inhibitor and 2.5 U/μL MMLV reverse transcriptase. Bring up to desired volume with RT dilution buffer (HCD).
2. Incubator.

2.7 Real-Time PCR Analysis

1. PCR Master Mix: 1× SsoAdvanced Universal SYBR Green Supermix (Bio-Rad), 1 μM sense and antisense primer pair.
2. 384-well plates.
3. Optical sealing tape.
4. ViiA™ 7 real-time PCR system (ThermoFisher Scientific).

3 Methods

3.1 Urine Collection and Processing

1. Upon informed consent, collect urine in specimen cup.
2. Transfer collected urine into 50 mL tubes.
3. Urine can either be processed immediately to next step or stored frozen at −80 °C until further use.
   (a) If frozen, place in 37 °C water bath until thawed and proceed to next step.
4. Centrifuge at 800 × *g* for 15 min at 4 °C to pellet cells and large debris.
5. Transfer 10 mL supernatant to a 50 mL tube.
6. Add 2.5 mL 25× PBS to the 10 mL urine supernatant and vortex briefly (*see* **Note 1**).
7. Proceed to Subheading 3.4, **step 1**.

3.2 Ascites Collection and Processing

1. Upon informed consent, ascites are collected from peritoneal cavity using Yankauer suction connected to a drainage bag. The volume of the ascites ranges from 500 mL to 2 L. Ascites from smaller volumes can be collected using bulb suction.
2. Transfer ascites into 50 mL tubes.
3. Centrifuge at 2000 × *g* for 15 min at 4 °C to get rid of cell debris.
4. Ascites can either be processed immediately (proceed to Subheading 3.4, **step 1**) or stored frozen at −80 °C until further use.
5. If frozen, place at 37 °C until thawed.
   (a) Recentrifuge at 2000 × *g* for 15 min at 4 °C.
   (b) Transfer supernatant to a sterile tube and proceed to Subheading 3.4, **step 1**.

3.3 Plasma Collection and Processing

1. Upon informed consent, collect blood in 8 mL blood collection tube.
2. Centrifuge blood at 2000 × *g* for 15 min at 4 °C.
3. Transfer plasma into microfuge tubes.

4. Plasma can either be processed immediately (proceed to Subheading 3.4, **step 1** or Subheading 3.5, **step 1**) or stored frozen at −80 °C until further use.
5. If frozen, place at 37 °C until thawed.
   (a) Recentrifuge at 2000 × *g* for 15 min at 4 °C.
   (b) Transfer supernatant to a sterile tube and proceed to Subheading 3.4, **step 1** or Subheading 3.5, **step 1**.

3.4 Collection Tube Extracellular Vesicle Capture and mRNA Isolation

1. Apply sample to the Exosome Collection Tube (maximum volume is 12.5 mL, *see* **Notes 2** and **3**).
2. Centrifuge at 2500 × *g* for 10 min at 4 °C.
3. Place mRNA Capture Plate under the filter tip holder.
4. Remove filter tip from the Exosome Collection Tube and place into the filter tip holder (*see* **Note 4**).
   (a) Maximum of 48 filter tips in a checkerboard pattern can be placed in the filter tip holder at one time. If there are more than 48 samples, use a second tip holder and place over a 96-well microplate.
5. Apply 80 μL Working Lysis Buffer to each well and cover with aluminum seal and lid.
6. Incubate at 37 °C for 10 min.
7. Centrifuge at 2500 × *g* for 5 min.
8. Repeat **step 7** if there is a second filter tip holder (*see* **Note 5**).
9. Incubate the mRNA Capture Plate with lid at 2–8 °C overnight (>16 h, *see* **Note 6**).
10. Following overnight incubation, place mRNA Capture Plate, Wash Buffers A and B on ice in preparation for wash step (*see* **Note 7**). All wash steps should be performed on ice.
11. Aspirate EV lysates from the mRNA Capture Plate.
12. Apply 100 μL of cold Wash Buffer A to each well and aspirate.
13. Repeat two more times.
14. Apply 150 μL of cold Wash Buffer B to each well and incubate 3 min.
15. Aspirate and repeat two more times.
16. Remove remaining residual liquid (*see* **Note 8**).
17. If cDNA synthesis is the downstream application desired, skip to Subheading 3.6. (For mRNA isolation, continue to **step 18**).
18. Apply 80 μL RNase/DNase-free water to each well and cover tightly with aluminum seal.
19. Heat to 65 °C for 5 min. and then place on ice.
20. Centrifuge at 1500 × *g* for 1 min at 4 °C.
21. Transfer mRNA to sterile 8-well strips and store at −80 °C (*see* **Note 9**).

3.5 Filterplate Extracellular Vesicle Capture and mRNA Isolation

1. Place Exosome Filterplate over the 96-well deep well plate.
2. Apply sample to the Exosome Filterplate (maximum volume is 400 μL, *see* **Note 3**).
3. Centrifuge at 2500 × *g* for 5 min at 4 °C.
4. Place mRNA Capture Plate under the Exosome Filterplate.
5. Apply 60 μL Working Lysis Buffer to each well and cover with aluminum seal and lid.
6. Incubate at 37 °C for 10 min.
7. Centrifuge at 2500 × *g* for 5 min.
8. Incubate the mRNA Capture Plate with lid at 2–8 °C overnight (>16 h, *see* **Note 6**).
9. Following overnight incubation, place mRNA Capture Plate, Wash Buffers A and B on ice in preparation for wash step (*see* **Note 7**). All wash steps should be performed on ice.
10. Aspirate EV lysates from the mRNA Capture Plate.
11. Apply 100 μL of cold Wash Buffer A to each well and aspirate.
12. Repeat two more times.
13. Apply 150 μL of cold Wash Buffer B to each well and incubate 3 min.
14. Aspirate and repeat two more times.
15. Remove remaining residual liquid (*see* **Note 8**).
16. If cDNA synthesis is the downstream application desired, skip to Subheading 3.6. (For mRNA isolation, continue to **step 17**).
17. Apply 60 μL RNase/DNase-free water to each well and cover tightly with aluminum seal.
18. Heat to 65 °C for 5 min. and then place on ice.
19. Centrifuge at 1500 × *g* for 1 min at 4 °C.
20. Transfer mRNA to sterile 8-well strips and store at −80 °C (*see* **Note 9**).

3.6 cDNA Synthesis

1. Apply desired volume of cDNA Master Mix to each well and cover with aluminum seal.
2. Centrifuge at 1500 × *g* for 1 min at 4 °C.
3. Incubate at 37 °C for 2 h.
4. Briefly mix solution by plate vortex.
5. Centrifuge at 1500 × *g* for 1 min at 4 °C.
6. Store at 4 °C.

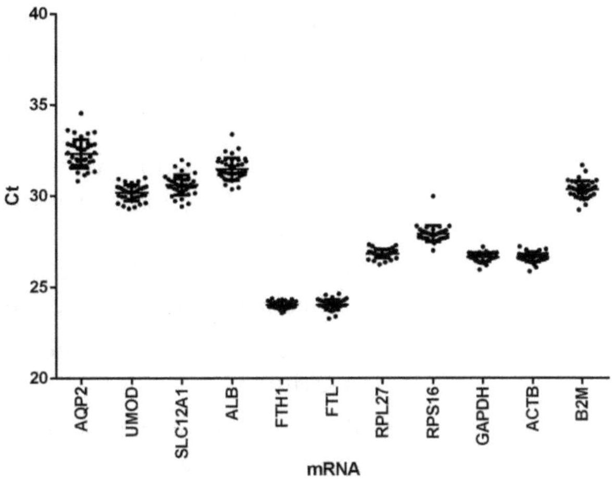

Fig. 2 Urinary EV gene expression analysis. First morning, urine from three donors were pooled, aliquoted, and kept at −80 °C until processing. Using standard procedure, cDNA was prepared in the mRNA Capture Plate following the application of 10 mL pooled urine to the Exosome Collection Tubes. Real-time PCR analysis of kidney-specific genes (AQP2, UMOD, SLC12A1 and ALB) and mRNA previously identified to be highly expressed in urinary EVs (FTH1, FTL, RPL27, RPS16, GAPDH, ACTB and B2M) are shown. Data represent mean ± standard deviation of $n = 40$

3.7 Real-Time PCR Quantitation and Analysis

1. Prepare PCR Master Mix for each target gene for final 5 μL PCR reaction volume.
2. Add 3 μL PCR master mix into designated position of 384-well plate.
3. Add 2 μL cDNA into appropriate wells.
4. Seal plate with optical tape.
5. Centrifuge at 1500 × *g* for 1 min at 4 °C.
6. Insert plate into ViiA7 real-time PCR instrument and initiate program.
   (a) Step 1: 95 °C 10 min 1×; Step 2: 95 °C 0.5 min, 65 °C 1 min, 40×; Step 3 (melting curve program): 95 °C 0.25 min, 60 °C 1 min, 60 °C to 95 °C at 0.05 °C/s, 95 °C 0.25 min
7. Export data to Excel format for processing (Fig. 2).

4 Notes

1. Apply 1/4 volume of 25× PBS to total sample volume for urine samples. Samples such as ascites, plasma and cell culture supernatants do not require addition of 25× PBS before applying to filter.

2. Sample volumes >12.5 mL can be applied to the Exosome Collection Tube by repeated application and centrifugation of the sample. Urine volumes up to 30 mL have been applied to the Exosome Collection Tube without clogging.
3. For smaller sample volumes between 100 and 400 μL, the Exosome Filterplate (96-well format) can be used for high-throughput sample processing (Fig. 1b, c). For plasma, volumes of 800 μL have been applied by repeated application and centrifugation of the sample.
4. If nozzle of filter tip is not dry, wipe with Kimwipes or briefly centrifuge.
5. Remove first filter tips/filter tip holder from mRNA Capture Plate. Then place the second filter tips/filter tip holder over the mRNA Capture Plate.
6. Overnight incubation of EV lysate in mRNA capture plate at 4 °C is recommended to maximize hybridization efficiency.
7. Wash Buffers A and B can be stored at 2–8 °C if desired.
8. To completely remove remaining residual liquid from mRNA Capture Plate after final wash step, a brief centrifugation and repeat aspiration may be performed. A 10-min vacuum drying step can also be applied if desired.
9. If sequencing is the downstream application, a DNase treatment of the isolated mRNA can be performed (Fig. 3).

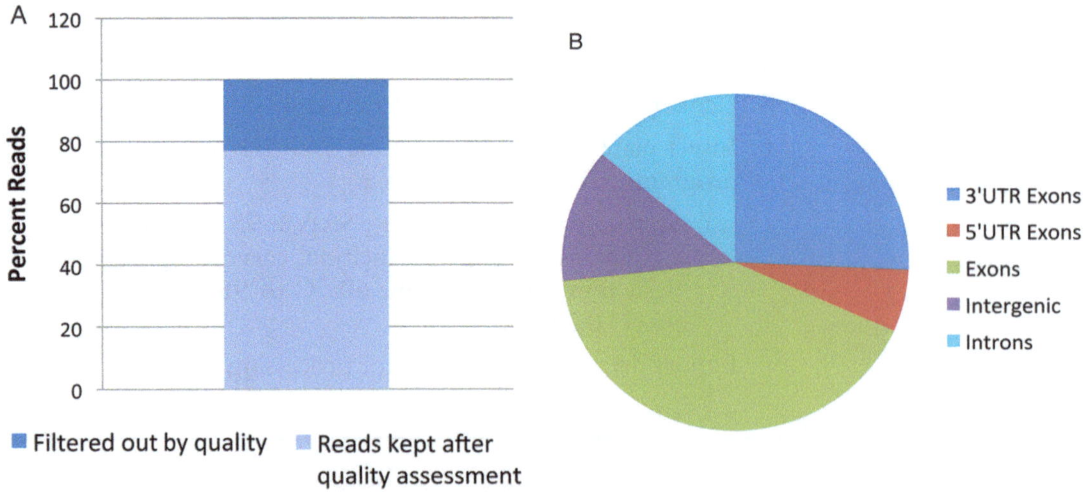

Fig. 3 Next generation mRNA sequencing. EV mRNA from a 10 mL urine sample was isolated for next generation mRNA sequencing using Clontech Low Input library prep kit. The total number of reads was 25,141,528 and percentage of reads aligned was 79.44% with 6500 genes detected. The trimmed and filtered reads were aligned to the UCSC reference genome hg19 assembly by TopHat. (**a**) The sequence read quality assessment shows >75% of reads kept after quality assessment. (**b**) The distribution of reads mapping to genome annotations indicates >80% exon and intergenic sequences

References

1. Miranda KC, Bond DT, McKee M, Skog J, Paunescu TG, Da Silva N et al (2010) Nucleic acids within urinary exosomes/microvesicles are potential biomarkers for renal disease. Kidney Int 78(2):191–199. doi:10.1038/ki.2010.106
2. Thery C, Amigorena S, Raposo G, Clayton A (2006) Isolation and characterization of exosomes from cell culture supernatants and biological fluids. Curr Protoc Cell Biol Chapter 3, Unit 3.22
3. Rider MA, Hurwitz SN, Meckes Jr DG (2016) ExtraPEG: a polyethylene glycol-based method for enrichment of extracellular vesicles. Sci Rep 6:23978. doi:10.1038/srep23978
4. Grant R, Ansa-Addo E, Stratton D, Antwi-Baffour S, Jorfi S et al (2011) A filtration-based protocol to isolate human plasma membrane-derived vesicles and exosomes from blood plasma. J Immunol Methods 371 (1-2):143–151
5. Aoki J, Ohashi K, Mitsuhashi M, Murakami T, Oakes M, Kobayashi T, Doki N, Kakihana K, Sakamaki H (2014) Posttransplantation bone marrow assessment by quantifying hematopoietic cell-derived mRNAs in plasma exosomes/microvesicles. Clin Chem 60 (4):675–682
6. Murakami T, Oakes M, Ogura M, Tovar V, Yamamoto C, Mitsuhashi M (2014) Development of glomerulus-, tubule-, and collecting duct-specific mRNA assay in human urinary exosomes and microvesicles. PLoS One 9(10): e109074. doi:10.1371/journal.pone.0109074
7. Miranda KC, Bond DT, Levin JZ, Adiconis X, Sivachenko A et al (2014) Massively parallel sequencing of human urinary exosome/microvesicle RNA reveals a predominance of non-coding RNA. PLoS One 9(5):e96094. doi:10.1371/journal.pone.0096094
8. Enderle D, Spiel A, Coticchia CM, Berghoff E, Mueller R, Schumberger M et al (2015) Characterization of RNA from exosomes and other extracellular vesicles isolated by a novel spin column-based method. PLoS One 10(8): e0136133. doi:10.1371/journal.pone.0136133
9. Oosthuyzen W, Sime NEL, Ivy JR, Turtle EJ, Street JM, Pound J et al (2013) Quantification of human urinary exosomes by nanoparticle tracking analysis. J Physiol 591 (23):5833–5842
10. Zubiri I, Posada-Ayala M, Sanz-Maroto A, Calvo E, Martin-Lorenzo M, Gonzalez-Calero L, de la Cuesta F et al (2014) Diabetic nephropathy induces changes in the proteome of human urinary exosomes as revealed by label-free comparative analysis. J Proteomics 96:92–102
11. Tang MK, Wong AS (2015) Exosomes: emerging biomarkers and targets for ovarian cancer. Cancer Lett 367(1):26–33
12. Tkach M, Clotilde T (2016) Communication by extracellular vesicles: where we are and where we need to go. Cell 164:1226–1232

Chapter 7

Specific and Generic Isolation of Extracellular Vesicles with Magnetic Beads

Ketil W. Pedersen, Bente Kierulf, and Axl Neurauter

Abstract

This chapter covers magnetic bead-based isolation and analysis of the smallest members of extracellular vesicles (EVs), the exosomes (30–150 nm), generally regarded to originate from the multivesicular bodies (MVBs). Also included, are descriptions of how to prepare samples prior to isolations. The magnetic bead-based isolation workflow is dramatically shortened both by omitting the pre-enrichment step and providing an option for a very short capture time. Three direct exosome isolation strategies are described: (1) "Specific and Direct," (2) "Semi Generic and Direct" and (3) "Generic and Direct" as well as exosome release from the magnetic beads. Detailed description of downstream exosome analysis is included covering flow cytometry, Western blot and electron microscopy. Finally, a description of exosome isolation from more complex starting material including urine and serum/plasma is discussed.

Key words Extracellular vesicles, Exosomes, Direct isolation, Generic isolation, Analysis, Flow cytometry, Western Blot, Magnetic beads, CD9, CD63, CD81, Dynabeads™, Release, Electron microscopy, Functionality, Serum, Plasma, Urine

1 Introduction

The world of extracellular vesicles (EVs) is a fascinating and expanding field. In this book chapter the focus is on the smallest vesicles in this micro cosmos of vesicles, the exosomes, originating from the multivesicular bodies (MVBs) named due to its ultrastructural appearance in electron microscopy. In basic research the aim is to better understand the fundamental mechanisms for how exosomes are created, how they are loaded with cargo of proteins, nucleic acids, lipids and carbohydrates, how the organelles fuse with the plasma membrane releasing the messengers into the intercellular space and finally how they are interacting and transmitting information to target cells [1–6]. Simultaneously, the focus is in clinical settings as exosomes have been associated with several diseases, including malign transformation and tumor progression, infectious diseases, allergy, and autoimmunity. This opens up the

Winston Patrick Kuo and Shidong Jia (eds.), *Extracellular Vesicles: Methods and Protocols*, Methods in Molecular Biology, vol. 1660, DOI 10.1007/978-1-4939-7253-1_7, © Springer Science+Business Media LLC 2017

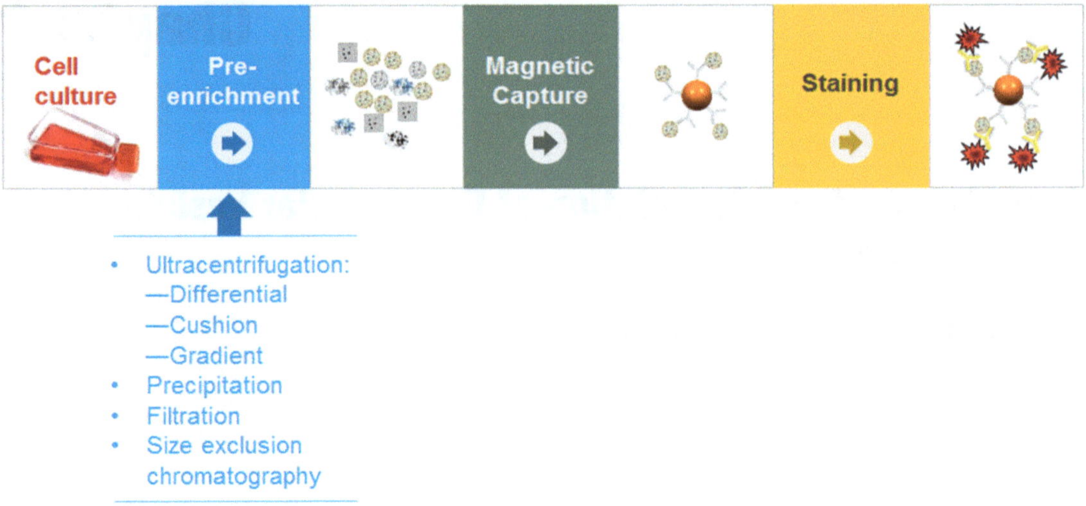

Fig. 1 The typical workflow, here exemplified by flow cytometry, is divided into three major steps. The first step is the pre-enrichment step which can be rather time consuming. In this step the cell culture medium is collected. A pre-enrichment step is done either by different forms of UC or precipitation. Alternative methods such as filtration or size exclusion are also in use. Specific capture using exosome surface markers as targets is performed in the second step. The third and final step is labeling of the captured exosomes

opportunity to use detection of tumor cell derived exosomes as diagnostic tools (liquid biopsies) and for monitoring the tumor status/progression and response to treatment [7].

The nature of these vesicles being in the nm size range, present in variable amounts and from numerous origins causes challenges in terms of isolation as well as analysis. At present, there is no consensus concerning standardization of vesicle isolation. This has led to a range of different isolation and enrichment methods such as ultracentrifugation (UC) in various forms, precipitation and filtration. Prior to detailed examination of the isolated vesicles, the origin needs to be verified. Equally important the level of contaminating vesicles should also be established. Currently, such work is in progress. Recommendations have been proposed by the International Society of Extracellular Vesicles (ISEV) defining verification strategies to ensure the presence of extracellular vesicle membranes and correspondingly the absence of contaminating vesicular membranes in the preparation [8]. Our previous work has contributed by providing detailed information on each step of the workflow (*see* Fig. 1) starting with pre-enriched samples followed by specific capture and finally addressing common methods for analysis [9].

In this book chapter the workflow has been simplified. The pre-enrichment step is omitted. Exosomes are isolated directly using magnetic beads. Direct capture is a useful tool to monitor exosome production during cell culture, analyze exosomes from limited sample volumes and in-depth characterization. The simplified strategy reduces risk for affecting exosome integrity and function as well

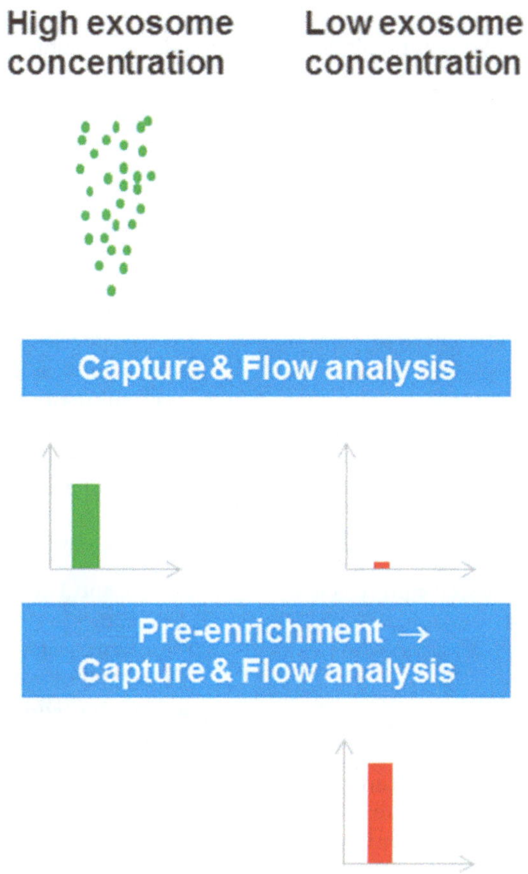

Fig. 2 When can direct isolation be performed? This depends on the content of exosomes in the starting material and is determined, e.g., by flow cytometry. If the signal is very low, indicating a very low exosome concentration in the sample a pre-enrichment step is needed

as making it more automation friendly and compatible with many downstream applications. When can direct isolation be performed? This will depend on a few factors such as exosome density in the sample, yield or purity, signal strength required for the assay and downstream application (Fig. 2). Three direct exosome isolation strategies are addressed: (1) "Specific and Direct," (2) "Semi Generic and Direct" and (3) "Generic and Direct." A selection guide is provided in (Table 1). "Specific and Direct" targets one exosomal marker during isolation. The "Semi Generic and Direct" combines multiple exosomal markers for isolation. The "Generic and Direct" isolation strategy utilizes generic, physical properties of the vesicles. A detailed description of exosome analysis is included covering flow cytometry, Western Blot and electron microscopy. Also included is a very fast release strategy after Generic and Direct

Table 1
Sample preparation and isolation protocol selection guide

Method	Sample	Protocol	Usage
Preparation of starting samples	Adherent cells (SW480)	Exosome enrichment from adherent cells (SW480) using cell culture bottles	Specific, Semispecific or Generic and direct isolation
	Adherent cells (SW480)	Exosome enrichment from adherent cells (SW480) using Integra CELLine AD 1000	
	Cells in suspension (Jurkat)	Exosome enrichment from adherent cells (Jurkat) using cell culture bottles	
	Cells in suspension (Jurkat)	Exosome enrichment from cells in suspension (Jurkat) using Integra CELLine CL 1000	
	Urine	Complex starting material	
	Plasma and serum	Complex starting material	
Exosome isolation with magnetic beads	Cell culture cells (adherent or suspension)	Specific and Direct Isolation of exosomes Semi Generic and Direct Isolation of exosome Generic and Direct Isolation of exosomes	Western Blot, qRT-PCR, Sequencing, Mass Spec
Release of magnetic bead-bound exosomes	Cell culture cells (adherent or suspension)	Release of exosomes after Generic and Direct Isolation	Functional studies, capture for flow analysis, TEM (negative stain)
Analysis of exosomes	Cell culture cells (adherent or suspension)	Flow analysis of bead bound exosomes Western Blot analysis of bead-bound exosomes Electron microscopy analysis of exosomes isolated with magnetic beads	n.a.
Function	Released exosomes (after Generic and Direct Isolation of exosomes)	Analysis of released exosomes by flow cytometry Analysis of released exosomes by electron microscopy	n.a.
Complex starting samples	Plasma and Serum	Semi Generic and Direct Isolation of exosomes from plasma and serum	Western Blot, qRT-PCR, Sequencing, Mass Spec
	Urine	Semi Generic and Direct Isolation of exosomes from Urine	Western Blot, qRT-PCR, Sequencing, Mass Spec

isolation providing intact, bead- and label-free exosomes to be used for any downstream application. Finally, a description of exosome isolation from more complex biological samples including urine serum and plasma is discussed.

2 Materials

2.1 Cell Culture

1. Jurkat cell line.
2. SW480 cell line.
3. RPMI 1640 Medium.
4. Heat-inactivated fetal calf serum (FCS).
5. Integra CELLine CL 1000 Culture System bottles.
6. Integra CELLine AD 1000 Culture System bottles.
7. Corning cell culture bottles 225 cm^2 with 0.2 μm ventilation lids.
8. 100 mM sodium pyruvate solution.

2.2 Immuno-magnetic Isolation

1. Exosome-Human CD9 Isolation (from cell culture) (Thermo Fisher Scientific).
2. Exosome-Human CD9 Flow Detection (from cell culture) (Thermo Fisher Scientific).
3. Exosome-Human CD63 Isolation (from cell culture) (Thermo Fisher Scientific).
4. Prototype magnetic bead for isolation of CD63 (Thermo Fisher Scientific).
5. Exosome-Human CD81 Isolation (from cell culture) (Thermo Fisher Scientific).
6. Exosome-Human CD81 Flow Detection (from cell culture) (Thermo Fisher Scientific).
7. Triple magnetic beads: Prototype magnetic bead for isolation of CD9, CD63, and CD81 (Thermo Fisher Scientific).
8. Mix magnetic beads: Prototype mix of three specific magnetic bead for isolation of CD9, CD63, and CD81, respectively (Thermo Fisher Scientific).
9. Generic magnetic beads: Prototype magnetic bead for targeting exosome membranes.
10. PBS: DPBS.
11. Isolation buffer: PBS with 0.1% BSA, filtered through a 0.2 μm filter.
12. 2 mL Sarstedt tubes.
13. Mixer.
14. Magnet.

2.3 Flow Cytometry Analysis

1. BD LSR Fortessa™ (BD Biosciences).
2. Anti-CD9 PE.
3. Anti-CD81 PE.

2.4 Electrophoresis

1. Lysis buffer: Radio Immunoprecipitation Assay (RIPA) buffer (3.75 mL 4 M NaCl, 1 mL NP40, 0.5 g SDS, 0.61 g Tris base, and H_2O up to 90 mL, adjust pH to 8 with 5 M HCl, H_2O up to 100 mL).
2. Protease inhibitor: add complete, EDTA-free 25× proteinase inhibitor solution to RIPA lysis buffer.
3. Sample Loading Buffer: 4× Bolt® LDS Sample Buffer.
4. Precast polyacrylamide gels of various polyacrylamide concentrations and gradient gels are commercially available (e.g., Bolt™ 4–12% Bis-Tris Plus Gels).
5. Power supply is required for electrophoresis and Western blot.
6. Electrophoresis running buffer: 20× Bolt® MES SDS Running Buffer.

2.5 Western Blot

1. Western Blot transfer buffer: Bolt® Transfer Buffer (20×).
2. Membrane for protein blotting: PVDF.
3. Membrane wash solution: TBS-T (2.42 g Tris–HCl, 0.56 g Tris base, add H_2O up to 900 mL, adjust pH to 7.60 with 5 M HCl, add CO_2 up to 1000 mL).
4. Blocking solution: dissolve 5% skim milk in TBS-T.
5. Antibodies: dilute primary antibody in blocking solution according to manufactures recommendations.
6. Substrate for antibody signal detection.

2.6 Generic Capture and Release Buffers

1. Loading buffer: 50 mM Na-phosphate pH 6.0, 1 M NaCl.
2. Binding and Washing buffer: 120 mM trietholamine pH 6.0, 10 mM NaCl.
3. Release buffer: 20 mM trietholamine pH 6.0, 2 M NaCl.

2.7 Transmission Electron Microscopy

1. 1% glutaraldehyde (GA) in 200 mM cacodylate buffer (pH 7.4).
2. 1.5% Potassium ferricyanide.
3. 1% osmium tetroxide.
4. 1.5% magnesium uranyl acetate.
5. 70% ethanol.
6. 90% ethanol.
7. 96% ethanol.
8. Absolute ethanol.
9. Epon.
10. Ultramicrotome for ultrathin sections.
11. Glow discharged copper grids.

12. Blocking solution: PBS with 0.5% BSA.
13. Protein A gold.
14. 0.3% uranyl acetate.

3 Methods

3.1 Sample Preparation

A protocol for direct isolation and analysis of exosomes derived from cell culture bottles is described. For some cell lines the production and release of exosome can be low. An alternative to the cell culture bottles is the Integra CELLine Culture System culture flasks, originally designed for hybridoma cultures. The Integra CELLine Culture System has been used to elevate exosome yield and concentration. A protocol for direct isolation and analysis of exosomes from Integra CELLine Culture System culture flasks is described.

Four specific cell culture systems are described to harvest exosome-enriched cell-conditioned medium, followed by a common protocol for further sample preparation.

Exosome enrichment from adherent cells (SW480) using cell culture bottles: Adherent cells are cultured to confluence in RPMI 1640 with 10% FCS (*see* **Notes 1** and **2**), 1 mM Sodium-pyruvate in cell culture bottles (225 cm^2) in 37 °C with 5% CO_2. The medium is removed and 50 mL of fresh medium is added to the confluent cells. After 3 days the cell-conditioned medium is harvested.

Exosome enrichment from cells in suspension (Jurkat) using cell culture bottles: Precultured Jurkat cells are seeded at a density of 0.4×10^6 cells/mL in fresh RPMI medium and grown for 3 days in 37 °C with 5% CO_2 before harvest of the cell-conditioned medium.

Exosome enrichment from adherent cells (SW480) using Integra CELLine AD 1000: Adherent cells are cultured to confluence in RPMI 1640 with 10% FCS and, 1 mM Sodium-pyruvate in Integra CELLine Culture System bottles in 37 °C with 5% CO_2. After 7 days the cell-conditioned medium is harvested.

Exosome enrichment from cells in suspension (Jurkat) using Integra CELLine CL 1000:

3×10^7 cells in 15 mL medium are cultured in RPMI with 10% FCS in Integra CELLine Culture System bottles. After 7 days the cell-conditioned medium is harvested.

After harvesting the cell-conditioned medium two centrifugation steps are performed to remove any cells and large particles (300 × *g* for 10 min at 2–8 °C and 2000 × *g* for 30 min 2–8 °C). The precleared conditioned medium is subjected to further centrifugation (10,000 × *g* for 40 min at 2–8 °C, fixed angle) or stored at −80 °C and upon thawing subjected to the 10,000 × *g* centrifugation (Beckman J2-21 M/E Centrifuge JA rotor).

3.2 Exosome Enrichment from Urine

There is currently no standard procedure for collection and treatment of urine samples from donors for downstream exosome harvest. A standardization of the sample collection process will have to be established [10]. Here, the urine samples have been pre-cleared by centrifugation (10,000 × *g* for 45 min at 2–8 °C, fixed angle) (Beckman J2-21 M/E Centrifuge JA rotor). The supernatant has been transferred to a new tube prior to isolation and analysis.

3.3 Exosome Enrichment from Plasma and Serum

There is currently no standard procedure for exosome sample collection from blood. A normalization of the process will have to be established. Here, human ACD plasma has been used. The plasma fraction was collected after centrifugation (1300 rpm for 15 min at 2–8 °C).

3.4 Isolation of Exosomes with Magnetic Beads

3.4.1 Specific and Direct Isolation of Exosomes

Specific and Direct isolation of exosomes using magnetic beads offers several advantages such as (1) upconcentration of the sample to ensure detection, (2) removal of potential protein aggregates or other nonexosomal vesicles which may obscure the results and (3) comparison of potential subpopulations of exosomes. The exosome isolation method described here is suitable for most downstream applications, including Western blot, mass spec, qRT-PCR, and sequencing. Preparation for flow cytometry and transmission electron microscopy (TEM) will be addressed separately since this requires a modified isolation step.

The protocol described uses 100 μL of sample volume and Dynabeads™ magnetic beads coated with antibodies (e.g., human CD9 or human CD81) using the Specific and Direct Isolation of Exosomes. The protocol can be scaled accordingly.

1. Add 300 μL PBS with 0.1% BSA in a Sarstedt tube.
2. Add 20 μL of magnetic beads (3×10^6 magnetic beads of a 1.3×10^8 beads/mL).
3. Place the tube on a magnet for 1 min.
4. Remove the buffer.
5. Add 100 μL exosome solution (*see* **Note 3**).
6. Incubation overnight (16–24 h) at 2–8 °C with mixing (*see* **Note 4**).
7. Place the tube on a magnet for 1 min.
8. Wash by removing the supernatant and adding 300 μL PBS with 0.1% BSA.
9. Place the tube on a magnet for 1 min, and repeat the washing step once.
10. The exosomes on the magnetic beads are now ready for downstream analysis (e.g., "Western Blot analysis of bead-bound exosomes").

3.4.2 Semi Generic and Direct Isolation of Exosomes

Now shifting focus from methods for direct isolation of exosomes by targeting individual exosomal markers (such as CD9, CD63 or CD81). In order to isolate several potential subpopulations of exosomes simultaneously in a semi generic and direct way, two isolation strategies can be considered. (1) Magnetic beads coated with antibodies targeting either CD9, CD63 or CD81 can be mixed together prior to isolation (Mix). (2) The second alternative is to couple antibodies targeting CD9, CD63 and CD81 on the same magnetic bead (Triple). The protocol describes both alternatives.

1. Add 300 μL PBS with 0.1% BSA in a Sarstedt tube.
2. Add 20 μL of Triple magnetic beads (3×10^6 magnetic beads of a 1.3×10^8 beads/mL)

   or

   Add 60 μL Mix magnetic beads (9×10^6 magnetic beads of a 1.3×10^8 beads/mL) (*see* **Note 5**).
3. Place the tube on a magnet for 2 min.
4. Remove the buffer.
5. Add 100 μL exosome solution (*see* **Note 3**).
6. Incubation overnight (16–24 h) at 2–8 °C with mixing (*see* **Note 4**).
7. Place the tube on a magnet for 2 min.
8. Wash by removing the supernatant and adding 300 μL PBS with 0.1% BSA.
9. Place the tube on a magnet for 2 min, and repeat the washing step once.
10. The exosomes on the magnetic beads are now ready for downstream analysis (e.g., "Western Blot analysis of bead-bound exosomes").

3.4.3 Generic and Direct Isolation of Exosomes

In order to isolate exosomes in a generic way, physical properties of the exosomes can be utilized. The following protocol covers exosome isolation with magnetic beads taking advantage of the negative charge of the exosomes [11–13].

Loading of beads before exosome isolation:

1. Add 10 μL Generic magnetic beads (5×10^8 beads) to a tube.
2. Add 500 μL Loading buffer to the magnetic bead.
3. Incubate for 2 min and repeat once.
4. Place the tube on a magnet.
5. Remove buffer.
6. Magnetic beads are ready for exosome isolation.

Isolation of exosomes:

1. Add 500 µL Binding and Washing buffer.
2. Incubate for 2 min and repeat once.
3. Remove buffer.
4. Add exosomes to the tube (*see* **Note 6**).
5. Add Binding and Washing buffer to 100 µL.
6. Incubation at RT for 10 min while mixing (*see* **Note** 7).
7. Place the tube on a magnet for 2 min.
8. Wash by removing the supernatant and adding 500 µL Binding and Washing buffer.
9. Place the tube on a magnet for 2 min, and repeat the washing step twice.

The magnetic bead-bound exosomes are now ready for release ("Release of exosomes after Generic and Direct Isolation") or downstream analysis such as Western Blot (e.g., "Western blot analysis of magnetic bead-bound exosomes"), mass spec, electron microscopy ("Electron microscopy analysis of magnetic bead-bound exosomes") or functional studies ("Analysis of released exosomes by flow cytometry").

3.5 Release of Magnetic Bead-Bound Exosomes

For some downstream applications exosomes must be released from the magnetic beads before use. Here we describe a fast and efficient release protocol to be performed after Generic and Direct Isolation of exosomes.

3.5.1 Release of Exosomes After Generic and Direct Isolation

1. Start with the bead bound exosomes from protocol "Generic and Direct Isolation of Exosomes" resuspended in 500 µL of Binding and Washing buffer.
2. Place the tube on a magnet for 2 min.
3. Remove the supernatant.
4. Add 75 µL Release Buffer to the bead pellet.
5. Incubate for 1 h at RT while shaking at 1200 rpm (*see* **Note** 7).
6. Place the tube on a magnet for 2 min.
7. Collect the supernatant with the released exosomes (*see* **Note 8**).

3.6 Analysis of Exosomes

This chapter covers analysis of exosomes by flow cytometry, Western blot and electron microscopy. Other analysis methods such as mass spec, qRT-PCR, sequencing and lipid or carbohydrate analysis are beyond the scope of this book chapter.

3.6.1 Flow Analysis of Magnetic Bead-Bound Exosomes

Exosomes are too small to be detected as individual events in an ordinary flow cytometer. Using magnetic bead-bound exosomes for the analysis, the magnetic beads provide a solid support for easy

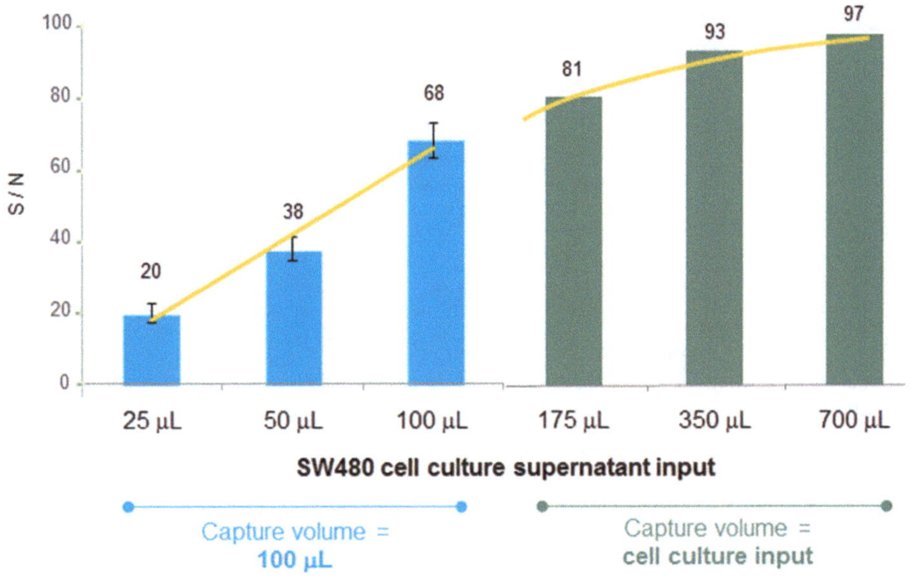

Fig. 3 Increase in S/N can be obtained in two ways. First, by keeping the capture volume constant and the number of beads constant (constant binding kinetics) the signal can be increased in a linear manner by increasing the input of exosomes. Second, the signal may be increased by increasing the sample volume but keeping the amount of beads constant. This will increase the signal in a nonlinear way as the binding kinetics will be reduced and the signal flattens out

handling during staining and standard scatter detection. For details regarding optimizing the conditions (*see* **Note 3**) and Fig. 3.

The protocol described uses 100 μL of sample volume and Dynabeads™ magnetic beads coated with antibodies (e.g., human CD9 or human CD81) using the "Specific and Direct Isolation of Exosomes" or "Semi Generic and Direct Isolation of Exosomes" (Triple). Mixing individual magnetic beads (Mix) is more challenging in terms of flow cytometry (*see* Figs. 4 and 5). The protocol can be scaled accordingly.

1. Add 300 μL PBS with 0.1% BSA in a Sarstedt tube.
2. Add 20 μL of magnetic beads (2×10^5 magnetic beads of a 1×10^7 beads/mL).
3. Place the tube on a magnet for 1 min (e.g., DynaMag™-2).
4. Remove the buffer.
5. Add 100 μL exosome solution (*see* **Note 3**).
6. Incubation overnight (16–24 h) at 2–8 °C with mixing (*see* **Note 4**).
7. Make a quick spin for a few seconds to remove magnetic beads from the lid (table top).
8. Place the tube on a magnet for 1 min.

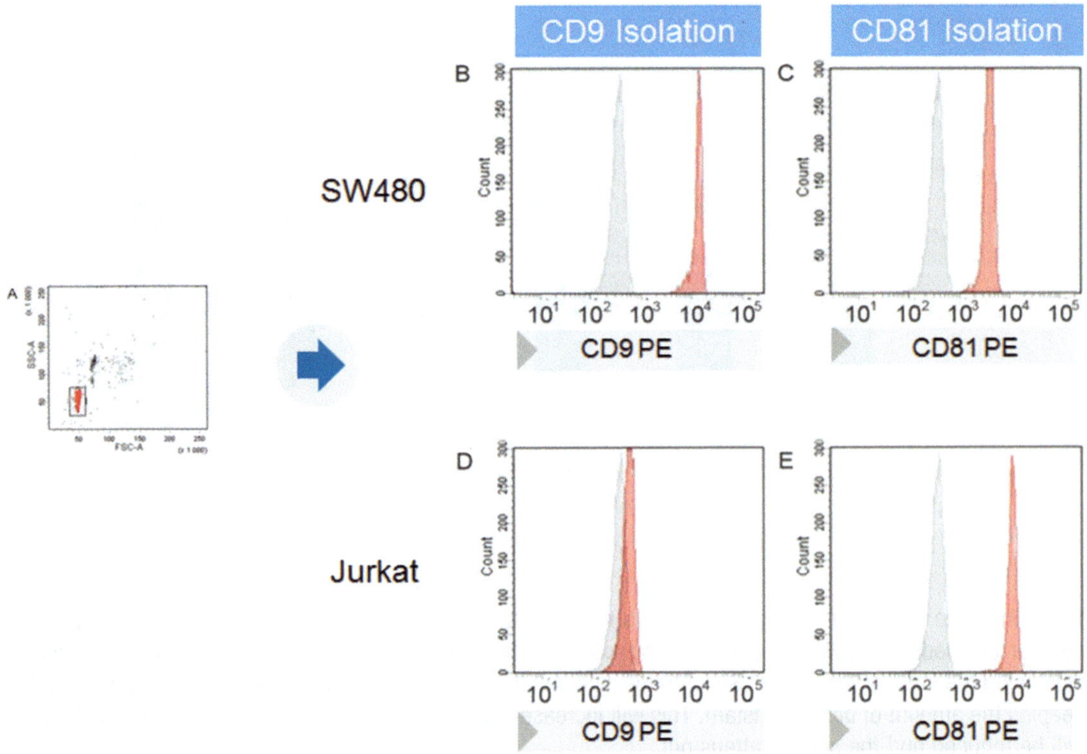

Fig. 4 SW480 (**b**, **c**) or Jurkat (**d**, **e**) derived exosomes were analyzed by flow cytometry ("Flow analysis of bead bound exosomes"). The exosomes were isolated by targeting CD9 (**b**, **d**) and CD81 (**c**, **e**) followed by staining. The scatter blot and gating profile is demonstrated in (**a**)

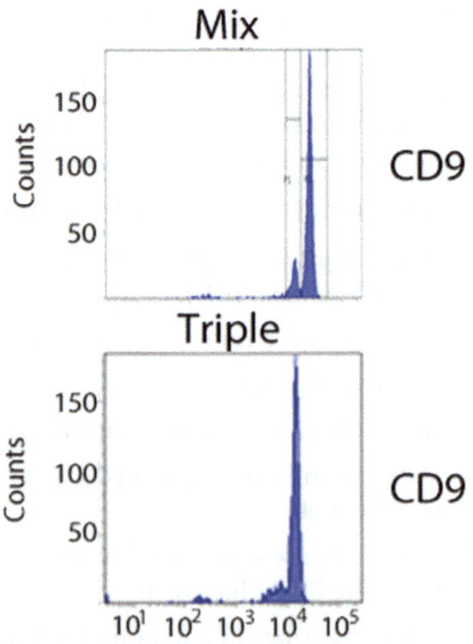

Fig. 5 CD9 labeled SW480 derived exosomes were analyzed by flow cytometry ("Flow analysis of bead bound exosomes") using a Semi Generic and Direct Isolation of Exosomes comparing Triple and Mix

9. Wash by removing the supernatant and adding 300 μL PBS with 0.1% BSA.
10. Place the tube on a magnet for 1 min, and repeat the washing step once.
11. Resuspend the beads in 150 μL PBS with 0.1% BSA.
12. Keep the sample at 4 °C until the samples are ready for flow staining.
13. Transfer 20 μL of staining antibodies to flow tubes (e.g., anti-CD9-PE and isotype control) (*see* **Note 9**).
14. Add 50 μL bead-bound exosomes to the tube. Mix gently by pipetting.
15. Incubate for approximately 45–60 min at RT on a sample shaker (approx. 1000 rpm (360 × *g*), protected from light).
16. Wash the bead-bound exosomes by adding 300 μL of Isolation Buffer. Gentle mixing by pipetting (do not vortex).
17. Place the tube on a magnet for 1 min and remove the supernatant.
18. Repeat the washing steps once and resuspend in 300 μL (or desired volume of Isolation Buffer appropriate for the flow instrument).

3.6.2 Western Blot Analysis of Magnetic Bead-Bound Exosomes

The protocol described uses 100 μL of sample volume and Dynabeads™ magnetic beads coated with antibodies (e.g., human CD9 or human CD81) using the "Specific and Direct Isolation of Exosomes" or the Triple/Mix beads for the "Semi Generic and Direct Isolation of Exosomes" (*see* Fig. 6). The protocol can be scaled accordingly. In terms of sensitivity *see* **Notes 3**, Figs. 5 and 7.

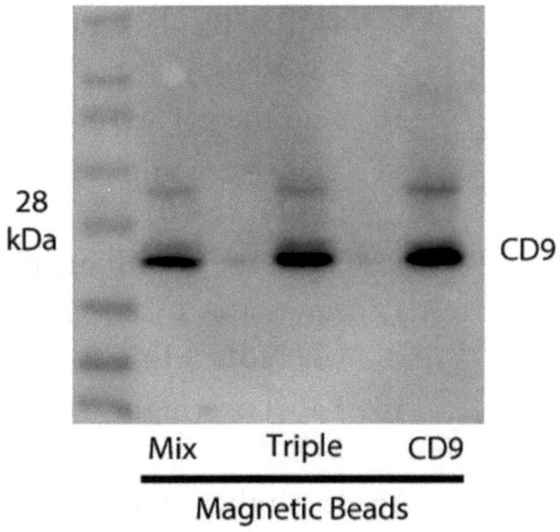

Fig. 6 CD9 labeled SW480 derived exosomes were isolated using the "Specific and Direct Isolation of exosomes" and "Semi Generic and Direct Isolation of exosomes" and analyzed by Western Blot ("Western Blot analysis of bead-bound exosomes")

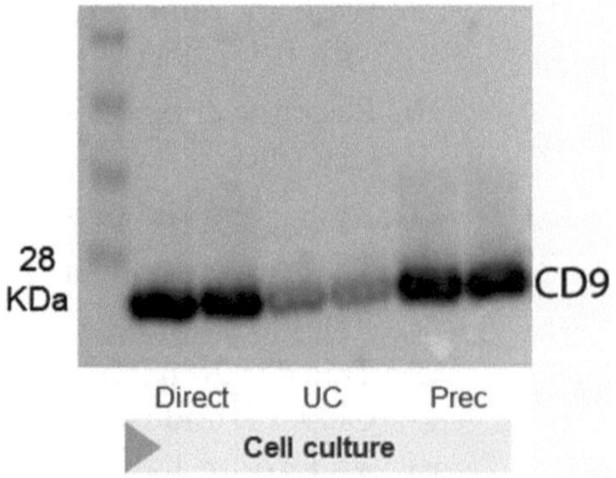

Fig. 7 CD9 labeled SW480 derived exosomes were isolated using the "Specific and Direct Isolation of exosomes" and analyzed by Western Blot ("Western Blot analysis of bead-bound exosomes"). The results were compared with equal amounts of exosomes pre-enriched using UC and Precipitation

1. Start with isolated exosomes on magnetic beads suspended in PBS with 0.1% BSA (cont. from Subheadings 3.4.1–3.4.3).
2. Place the tube on a magnet for 1 min.
3. Remove the supernatant.
4. Add lysis buffer to the bead–exosome complex (e.g., 6 μL 5 × RIPA + 1.25 μL Proteinase inhibitor + 24 μL H_2O).
5. Mix and sonicate for 20 s.
6. Incubate for 15 min (2–8 °C).
7. Add 10 μL of LDS Sample Buffer (4×).
8. Incubate for 10 min at 70 °C.
9. Load the sample in the well (40 μL per well for the Bolt™ 4–12% Bis-Tris Plus Gels, for other gels scale the volumes accordingly).
10. Run the gel at 165 V for 35 min at room temperature (RT).
11. Transfer samples to PVDF membranes by wet transfer (*see* **Note 10**).
12. Wash the PVDF membrane briefly in H_2O.
13. Block the PVDF membrane in blocking solution for 30 min (*see* **Note 11**).
14. Label overnight with primary antibody in blocking solution at 2–8 °C.
15. Wash three times in TBS-T for a total of 1 h.
16. Labeling with HRP conjugated secondary Ab for 1 h at RT (*see* **Note 12**).

17. Wash three times in TBS-T for a total of 1 h.
18. Signal detection by for example Super Signal West Dura Extended Duration substrate for 5 min at RT.
19. Visualized in imager.

3.6.3 Electron Microscopy Analysis of Magnetic Bead-Bound Exosomes

Electron microscopy can be used to examine the magnetic bead-bound exosomes at high resolution. The exosome isolation protocol is similar to the protocol for flow cytometry by aiming for a high density of exosomes per magnetic bead.

The protocol described uses 250 μL of sample volume and 50 μL of Dynabeads™ magnetic beads (1×10^7 beads/mL) coated with antibodies using the "Specific and Direct Isolation of exosomes" (e.g., human CD9 or human CD81) or the Triple beads using the "Semi Generic and Direct Isolation of Exosomes". For the "Generic and Direct Isolation of Exosomes" 200 μL of Dynabeads™ magnetic beads (1×10^7 beads/mL) is mixed with 200 μL sample. The protocol can be scaled accordingly.

1. Start with isolated exosomes on magnetic beads suspended in PBS with 0.1% BSA (cont. from Subheading 3.4.1–3.4.3).
2. Collect the sample by performing a short spin using a table centrifuge.
3. Add 500 μL of PBS with 0.1% BSA followed by mixing.
4. Apply the tube to a magnetic field and remove the supernatant.
5. Add sufficient amounts of 1% GA in 200 mM cacodylate buffer (pH 7.4).
6. Incubate for 1 h at RT (*see* **Note 13**).
7. Wash repeatedly in dH_2O.
8. Add cacodylate buffer with 1.5% potassium ferricyanide and 1% osmium tetroxide.
9. Incubate for 1 h (*see* **Note 14**).
10. Place the tube on a magnet and remove supernatant.
11. Add 1% tannic acid and incubate for 30 min at RT.
12. Place the tube on a magnet and remove supernatant.
13. Add 1.5% magnesium uranyl acetate and incubate for 30 min at RT.
14. Dehydration in.
    (a) 70% ethanol for 10 min.
    (b) 90% ethanol for 10 min.

(c) 96% ethanol for 10 min.

(d) Absolute ethanol for 4 × 15 min.

15. Add Epon and polymerize.
16. Ultrathin sections are prepared using an ultramicrotome.

3.7 Function

3.7.1 Analysis of Released Exosomes by Flow Cytometry

Exosomes can be subjected to detailed examination by flow cytometry upon release. The exosomes are first isolated according to the "Generic and Direct Isolation of Exosomes" protocol, then released before recapture and analyzed as described below (*see* Fig. 8).

1. Start with exosomes from "Release of exosomes after Generic amd Direct Isolation" protocol.
2. Add 200 μL PBS with 0.1% BSA in a Sarstedt tube.
3. Add 20 μL of magnetic beads (2×10^5 magnetic beads of a 1×10^7 beads/mL).
4. Place the tube on a magnet for 1 min.
5. Remove the buffer.
6. Add 50 μL released exosomes + 150 μL PBS w/BSA.
7. Incubation overnight (16–24 h) at 2–8 °C with mixing (*see* **Note 4**).
8. Make a quick spin for a few seconds to remove magnetic beads from the lid (table top).
9. Add 300 μL PBS w/0.1% BSA.
10. Place the tube on a magnet for 1 min.

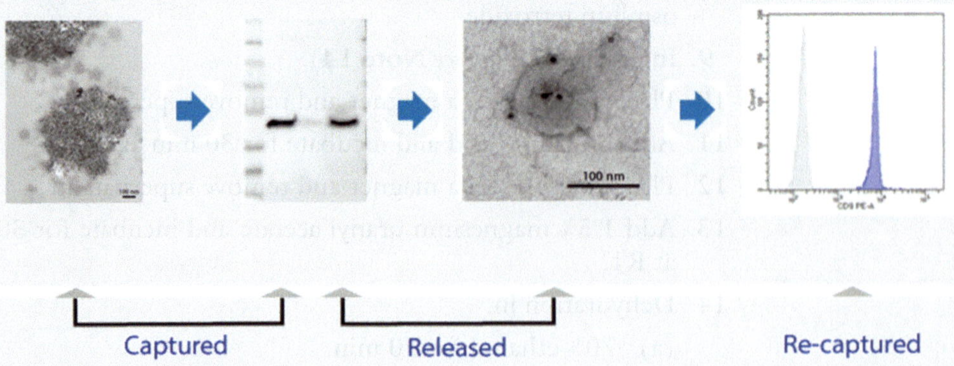

Fig. 8 Exosomes have here been isolated using the Generic and Direct Isolation of Exosomes and visualized by TEM and Western Blot. After isolation the exosomes was then released from the bead surface using the "Release of exosomes after Generic and Direct Isolation" protocol and processed for Western Blot and negative stain TEM ("Analysis of released exosomes by electron microscopy"). Finally, the released exosomes was recaptured and processed for flow cytometry

11. Remove the washing buffer.
12. Wash the magnetic bead-bound exosomes in 400 μL PBS w/0.1% BSA.
13. Place the tube on a magnet for 1 min.
14. Remove the washing buffer.
15. Resuspend in 350 μL buffer.
16. Add 20 μL of staining antibodies to flow tubes.
17. Add 100 μL magnetic bead-bound exosomes for each flow staining.
18. Incubate in a shaker at RT (approx. 1000 rpm (360 × *g*)) for 45 min.
19. Add 300 μL PBS w/0.1% BSA.
20. Place the tube on a magnet for 1 min.
21. Remove the washing buffer.
22. Wash the magnetic bead-bound exosomes in 400 μL PBS w/0.1% BSA.
23. Resuspend in 250 μL PBS w/0.1% BSA for flow analysis.

3.7.2 Analysis of Released Exosomes by Electron Microscopy

Exosomes released from magnetic beads after isolation can be subjected to detailed examination by performing immunolabeling followed by negative stain TEM. The protocol is quick and performed at RT on a clean surface.

The exosomes are first isolated according to the "Generic and Direct Isolation of Exosomes" protocol, and then released before immune-labeling/negative stain TEM as described below.

1. Start with exosomes from "Release of exosomes after Generic and Direct Isolation" protocol.
2. Incubate the grid on a drop of released exosomes (*see* **Note 15**).
3. Wash the grid on a drop of PBS (few minutes).
4. Block the grid on a blocking solution and incubate for 10 min (PBS with 0.5% BSA).
5. Incubate the grid on a drop of primary antibody and incubate for 15 min.
6. Wash the grid on four drops of PBS for a total of 10 min.
7. Incubate the grid on a drop of rabbit anti mouse secondary antibody for 15 min (if primary antibody is mouse origin).
8. Wash the grid on four drops of PBS for a total of 10 min.
9. Incubate the grid on a drop of protein A-gold for 15 min.
10. Wash grids on four drops of PBS for a total of 15 min.
11. Wash grids on four drops of water.
12. Move the grids to a drop of 0.3% UA and incubate for 15 min (*see* **Note 16**).

3.8 Complex Starting Materials

Here we describe the isolation of exosomes from biological samples, including urine, serum and plasma, all three being important exosomes sources in the context of liquid biopsies (Fig. 9).

3.8.1 Semi Generic and Direct Isolation of Exosomes from Plasma and Serum

Exosomes spiked into plasma or serum

1. Add 40 μL of cell culture derived exosomes to a tube.
2. Add 360 μL serum or plasma.

Isolation of exosomes:

1. Add 300 μL PBS w/0.1% BSA in a Sarstedt tube.
2. Add 20 μL of Triple magnetic beads (3×10^6 magnetic beads of a 1.3×10^8 beads/mL).
3. Place the tube on a magnet for 2 min.
4. Remove the buffer.
5. Add 100 μL plasma or serum with spiked exosomes.
6. Incubation overnight (16–24 h) at 2–8 °C with mixing (*see* **Note 4**).
7. Place the tube on a magnet for 2 min.
8. Wash by removing the supernatant and adding 300 μL PBS w/0.1% BSA.
9. Place the tube on a magnet for 2 min, and repeat the washing step once.
10. The exosomes on the magnetic beads are now ready for downstream analysis such as Western Blot ("Western Blot analysis of bead-bound exosomes").

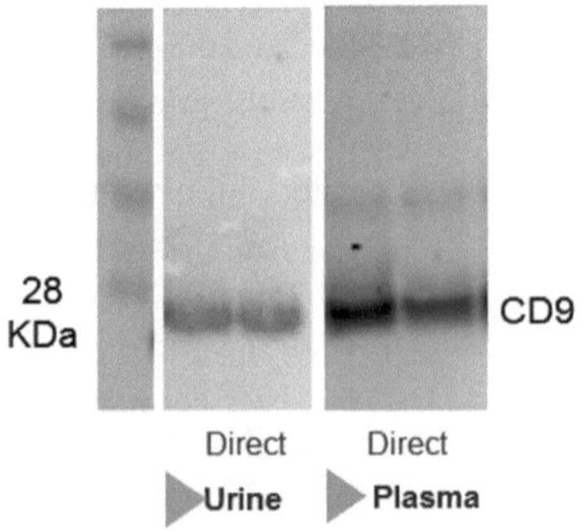

Fig. 9 CD9 labeled SW480 derived exosomes were isolated using the "Semi Generic and Direct Isolation of exosomes from Urine" and "Semi Generic and Direct Isolation of exosomes from plasma and serum" protocol and processed for Western Blot ("Western Blot analysis of bead-bound exosomes")

3.8.2 Generic and Direct Isolation of Exosomes from Plasma and Serum

In order to demonstrate exosome isolation from plasma or serum exosomes have been spiked into the biological material. The spiked exosomes is then subjected to isolation.

Exosome spiked into plasma or serum

1. Add 40 μL of cell culture derived exosomes to a tube.
2. Add 360 μL serum or plasma.

Loading of beads before exosome isolation:

1. Add 5 μL Generic magnetic beads (2.5×10^8 beads) to a tube.
2. Add 500 μL Loading buffer to the magnetic beads.
3. Incubate for 2 min and repeat once.
4. Place the tube on a magnet.
5. Remove buffer.
6. Magnetic beads are ready for exosome isolation.

Isolation of exosomes:

1. Add 500 μL Binding and Washing buffer.
2. Incubate for 2 min and repeat once.
3. Remove buffer.
4. Add 100 μL of exosome spiked plasma or serum to the tube.
5. Add 100 μL of Binding and Washing buffer.
6. Incubation at RT for 1 h while mixing.
7. Place the tube on a magnet for 2 min.
8. Wash by removing the supernatant and adding 500 μL Binding and Washing buffer.
9. Place the tube on a magnet for 2 min, and repeat the washing step twice.
10. The exosomes on the magnetic beads are now ready for release ("Release of Exosomes after Generic and Direct Isolation") or downstream analysis such as Western Blot ("Western Blot analysis of bead-bound exosomes"), Mass Spec, TEM ("Analysis of released exosomes by electron microscopy") or functional studies ("Analysis of released exosomes by flow cytometry").

3.8.3 Semi Generic and Direct Isolation of Exosomes from Urine

Isolation of exosomes:

1. Add 300 μL PBS w/0.1% BSA in a Sarstedt tube.
2. Add 20 μL of Triple magnetic beads (3×10^6 magnetic beads of a 1.3×10^8 beads/mL)

   For Specific isolation use for example *Exosome-Human CD9 Isolation (from cell culture)*.
3. Place the tube on a magnet for 2 min.

4. Remove the buffer.
5. Add 500 μL urine.
6. Incubation overnight (16–24 h) at 2–8 °C (*see* **Note 4**).
7. Place the tube on a magnet for 2 min.
8. Wash by removing the supernatant and adding 300 μL PBS w/ 0.1% BSA.
9. Place the tube on a magnet for 2 min, and repeat the washing step once.
10. The exosomes on the magnetic beads are now ready for downstream analysis such as Western Blot, Mass Spec ("Western blot analysis of bead-bound exosomes").

3.9 Analysis of Exosomes from Urine

3.9.1 Flow Analysis of Magnetic Bead-Bound Exosomes

1. Add 200 μL PBS w/0.1% BSA in a Sarstedt tube.
2. Add 20 μL of magnetic beads (2×10^5 magnetic beads of a 1×10^7 beads/mL).
3. Place the tube on a magnet for 1 min.
4. Remove the buffer.
5. Add 500 μL urine (*see* **Note 17**).
6. Incubation overnight (16–24 h) at 2–8 °C (*see* **Note 4**).
7. Make a quick spin for a few seconds to remove magnetic beads from the lid (table top).
8. Add 300 μL PBS w/0.1% BSA.
9. Place the tube on a magnet for 1 min.
10. Remove the washing buffer.
11. Wash bead–exosome complexes in 400 μL PBS w/0.1% BSA.
12. Place the tube on a magnet for 1 min.
13. Remove the washing buffer.
14. Resuspend in 350 μL buffer.
15. Add 20 μL of staining antibodies to flow tubes.
16. Add 100 μL bead bound exosomes for each flow staining.
17. Incubate in a shaker at RT (approx. 1000 rpm ($360 \times g$)) for 45 min.
18. Add 300 μL PBS w/0.1% BSA.
19. Place the tube on a magnet for 1 min.
20. Remove the washing buffer.
21. Wash the magnetic bead-bound exosomes in 400 μL PBS w/ 0.1% BSA.
22. Resuspend in 250 μL PBS w/0.1% BSA for flow analysis.

4 Notes

1. Cells are cultured with exosome-depleted FCS. An alternative is to grow the cell line in the absence of FCS if the cell line can handle that.
2. High cell viability is important to reduce the release of apoptotic bodies.
3. The number of magnetic beads used for capturing the exosomes should be low in order to capture as many exosomes per magnetic bead as possible. This will ensure the strongest possible signal. Increasing the bead concentration reduces the signal significantly. In terms of optimizing the signal there are two alternatives. (1) In the first alternative the capture volume and the number of beads are kept constant to ensure constant binding kinetics. By increasing the input of exosomes the signal will increase in a linear manner. (2) In the second alternative, the sample volume is increased but the number of magnetic beads is kept constant. The signal is increased in a nonlinear way as the binding kinetics will be reduced and eventually the signal will flatten. Increased capture time will give a slight increasing in signal. If maximum signal is not required the capture time may be shortened. More staining solution added to the isolated exosomes can adjust the signal.
4. Shorter incubation times can be applied but for optimal binding of exosomes to the magnetic beads incubations up to 21 h is recommended (reason for incubation at 2–8 °C).
5. Add the three different beads as if the other beads were not present—to maintain good binding kinetics.
6. More magnetic beads can also be added if complete exosome depletion is required. The protocol can be scaled up by increasing volumes and reagents proportionally. To achieve greater depletion of exosomes, increase the number of magnetic beads by 2–25 times/100 μL (final volume) of sample. In contrast to flow cytometry that uses relatively few magnetic beads in order to allow as many exosomes to dock onto each
7. Proper mixing is important. V-shaped tubes are not recommended. The isolation efficiency can also be increased by extending the incubation time to 1 h.
8. The released exosomes can be used for "any" downstream application, including further sub-fractionation based on specific surface markers. The released exosomes are suspended in buffer with 2 M NaCl. Optional: remove salt from released exosome solution by for example dialysis or spin columns.

9. Titration of the staining antibody is required for optimal signal to noise ratio (S/N). Start with the concentration recommended per 1 × 10^6 cells.
10. 20 V for 1 h at RT according to the Bolt instructions.
11. TBS-T and 5% skim milk for example.
12. Use TrueBlot® Ultra Ig HRP or similar for targets with the similar size (kDa) as heavy or light chain antibody (Rockland antibodies and assays).
13. Magnetic beads with exosomes subjected to fixation (GA) need more time on the magnet.
14. Lead citrate can be omitted if the contrast is too high.
15. Glow discharge carbon coated copper grids. This will improve adsorption by making the surface hydrophilic. Use 5–10 μL of released exosomes per grid. Cover the grids during incubation.
16. Contrast solutions: Aqueous uranyl acetate (1–3%), neutral phosphotungstic acid (1–3%; neutral phosphotungstic acid produces less contrast than the uranyl acetate), and ammonium molybdate (1%).
17. The signal may be increased by increasing the sample volume keeping the amount of magnetic beads constant.

Acknowledgment

Anette Kullmann for data and Lisbeth Larsen (ThermoFisher Scientific) for technical assistance; Antje Hoenen and Norbert Roos (University of Oslo, Norway) for technical assistance in electron microscopy.

References

1. Janowska-Wieczorek A, Wysoczynski M, Kijowski J, Marquez-Curtis L, Machalinski B, Ratajczak J, Ratajczak MZ (2005) Microvesicles derived from activated platelets induce metastasis and angiogenesis in lung cancer. Int J Cancer 113:752–760
2. Belting M, Wittrup A (2008) Nanotubes, exosomes, and nucleic acid-binding peptides provide novel mechanisms of intercellular communication in eukaryotic cells: implications in health and disease. J Cell Biol 183:1187–1191
3. Pegtel DM, van de Garde MD, Middeldorp JM (2011) Viral miRNAs exploiting the endosomal-exosomal pathway for intercellular cross-talk and immune evasion. Biochim Biophys Acta 1809(11-12):715–721
4. Thery C, Zitvogl L, Amigorena S (2002) Exosomes: composition, biogenesis and function. Nat Rev Immunol 2:11
5. Thery C, Ostrowski M, Segura E (2009) Membrane vesicles as conveyors of immune responses. Nat Rev Immunol 9(8):581–593
6. Valadi H, Ekstrom K, Bossios A, Sjostrand M, Lee JJ, Lotvall JO (2007) Exosome-mediated transfer of mRNAs and microRNAs is a novel mechanism of genetic exchange between cells. Nat Cell Biol 9(6):654–659
7. Al-Nedawi K, Meehan B, Micallef J, Lhotak V, May L, Guha A, Rak J (2008) Intercellular transfer of the oncogenic receptor EGFRvIII by microvesicles derived from tumour cells. Nat Cell Biol 10(5):619–624

8. Lotvall J, Hill AF, Hochberg F, Buzas EI, Di Vizio D, Gardiner C, Gho YS, Kurochkin IV, Mathivanan S, Quesenberry P, Sahoo S, Tahara H, Wauben MH, Witwer KW, Thery C (2014) Minimal experimental requirements for definition of extracellular vesicles and their functions: a position statement from the International Society for Extracellular Vesicles. J Extracell Vesicles 3:26913
9. Oksvold MP, Neurauter A, Pedersen KW (2015) Magnetic bead-based isolation of exosomes. In: Sioud M (ed) Methods in molecular biology: RNA interference challenges and therapeutic opportunities. Springer, New York, NY, p 18
10. Mitchell PJ, Welton J, Staffurth J, Court J, Mason MD, Tabi Z, Clayton A (2009) Can urinary exosomes act as treatment response markers in prostate cancer? J Transl Med 7:4
11. Akagi T, Hanamura N, Ichiki T (2015) Measuring of individual nanobioparticles on microfluidic chips by laser dark-field imaging. J Photopolym Sci Tech 28(5):727–730
12. Marimpietri D, Petretto A, Raffaghello L, Pezzolo A, Gagliani C, Tacchetti C, Mauri P, Meliolo G, Pistoia V (2013) Proteome profiling of neuroblastoma-derived exosomes reveal the expression of proteins potentially involved in tumor progression. PLoS One 8 (9):e75054
13. Sokolova V, Ludwig AK, Hornung S, Rotan O, Horn PA, Epple M, Giebel B (2011) Characterisation of exosomes derived from human cells by nanoparticle tracking analysis and scanning electron microscopy. Colloids Surf B Biointerfaces 87(1):146–150

Part II

Purification of Extracellular Vesicles

Chapter 8

Polymer-Based Purification of Extracellular Vesicles

Peter N. Brown and Hang Yin

Abstract

Lipid particles, including exosomes (extracellular vesicles [EVs]), are released from cells in vivo and in vitro. The contents of these EVs can be indicative of disease and therefore be utilized in diagnostic biophysical or biochemical assays. To make use of EVs in this way, methods for purification and quantification are required which vary depending on their particles origin and concentration. This chapter provides an overview of EV purification techniques and provides detailed instructions on the purification of EVs by polymer based precipitation and subsequent quantification. The subsequent quantification and characterization of these EVs also presents a challenge, as limited methods are capable of detecting EVs due to their small size.

Key words Exosome production, Extracellular vesicles, Exosome purification, Polymer based EV precipitation, Electron microscopy, Immunolabeling

1 Introduction

Extracellular vesicles (EVs) include microsomes and microvesicles. EVs differ from microvesicles by their mode of production and their small size (diameters around 30–100 nm) [1]. There is an increasing interest in EVs, as they have been shown to present a therapeutic and diagnostic potential [2]. This potential is a direct result of the ability to facilitate cell–cell communication by carrying nucleic acids, proteins, and lipids [3]. The isolation of EVs from complex biological solutions is essential for the subsequent analysis of its contents. Purification of EVs has been successfully achieved by at least four different principles: immunoaffinity [4], chromatography [5], ultracentrifugation and polymer-based precipitation [6].

Specific protein markers exist on the exosome membrane, e.g., tetraspanins, intergrins and adhesion molecules among many others [7]. Antibodies for these markers can be captured on Protein G coated Dynabeads™ [8]. These loaded Dynabeads™ can capture particles presenting the EVs specific protein markers, and specifically pull down EVs whilst leaving other cell debris and lipid vesicles in solution. The beads can be pelleted using a magnets leading to

Winston Patrick Kuo and Shidong Jia (eds.), *Extracellular Vesicles: Methods and Protocols*, Methods in Molecular Biology, vol. 1660, DOI 10.1007/978-1-4939-7253-1_8,

easy isolation [9]. Immunoaffinity has advantages over other methods due to its specificity for only targets presenting the exosome-specific markers. Further to purification, immunoaffinity can pull down EVs of different tissue origin. EVs present different protein markers based on their origin, e.g., EVs derived from immune cells carry CXCR4 [10]; similarly for EVs produced from diseased tissues, e.g., cancer cell derived EVs can carry TAG72, EGFRVIII, and CD163, among others [11]. Immunoaffinity is also a quick method to isolate particles [6]. However, this method is expensive can only be used for small sample volumes, making this method less useful for the isolation of low concentration samples, e.g., in urine and cell culture media. The exosome yield from immunoaffinity is also relatively small reducing the potential for analytical assays to low consumption methods, e.g., qPCR and electron microscopy (EM).

As previously discussed, one of the defining features of EVs is their specific range of diameters. Size exclusion chromatography (SEC) is a method which isolates particles based on hydrodynamic radius, which is simplified as size [12]. Large particles cannot enter the resin matrix and as a result, pass through the column in lower volumes; inversely, smaller particles are eluted after greater volumes [13]. This method of purification has also been scaled down for use in small scale spin columns reducing the initial cost and even allowing exosomal buffer exchange [5]. SEC is a very quick method and aids in the retention of intact EVs, especially when isolation is performed by gravity [14]. Though the method is limited to small volumes in each run; with the use of multiple runs or larger columns there is scope to scale up the system. As SEC isolation is based on hydrodynamic radius alone, there is a significant potential for contamination with other particles [15] reducing the reliability of the purification.

Ultracentrifugation is the most common and well established method of exosome purification [6]. This method consists of a two-stage process. The first is a medium–high speed spin (~10,000 × *g*) to remove cell debris. A subsequent high speed spin (~100,000 × *g*) for >16 h is used to pellet the exosome fraction which can then be isolated and resuspended [16, 17]. The use of a sucrose density gradient further increases the purity of the exosome sample [18]. Using ultracentrifugation has the benefit of being able to handle large volumes of sample for increased throughput. However, isolating EVs by this method is time consuming and usually results in a low yield of degraded product [19]. This method is also susceptible to variance based on viscosity of the biological solution from which there are isolated [20].

Recently, several commercially available isolations kits have been produced that use polymers to allow EVs to be pelleted with reduced force, producing a higher yield of useable particles, while still removing contaminant efficiently. Examples of these kits include ExoSpin™ [6], Invitrogen™ total exosome purification kit and ExoQuick™. Small volumes of precipitant are added to the sample, which is subsequently incubated for between 30 min and 12 h. This mixture can then be centrifuged at a relatively low speed (1500 × *g*) to pellet the exosome fraction. Polymer based precipitation of EVs from samples can be performed with different volumes and sources of biological fluids. Due to the speed, yield and throughput of these methodologies, this is an increasingly favorable method for exosome purification. However, the use of polymer based precipitation solutions can result in aggregates and coprecipitation of larger nonexosomal components [5]. The aggregation of vesicles results in errors during characterization as methodologies required to monitor 30–100 nm particles are highly influenced by slight variance in size or viscosity. Finally, precipitants used in purification can be difficult to remove and this contamination can skew downstream results [21]. These methods of exosome production and purification will be discussed in this chapter. Subsequent observation by EM can characterize purified EVs based on size; this can characterize the EVs immediately after purification. ExoQuick™ can result in large aggregated spots which may be remnants of the polymer used in purification whereas ultracentrifugation can result in damage causing empty dried "cup shaped" exosome particles of appropriate size (Fig. 1). Immunolabeling gives positive identification of the particles as EVs by the presence of the same protein markers used for immunoaffinity based purification. Immunolabeling will link antibodies to the protein markers in question and secondary antibodies are conjugated to 15 nm gold particles. These conjugated antibodies will confirm the particles as EVs (Fig. 2).

Thus, we have generated, purified and quantified EVs from different sources including cell media and blood sera. These purified EVs have been used successfully for qPCR, NanoSight analysis and electron microscopy. The methods for the purification of EVs from cell culture and subsequent analysis by electron microscopy are discussed in detail below.

2 Materials

All media and buffers should be made with ultrapure MilliQ water and stored at 4 °C. Buffers and media should be warmed to 37 °C and filtered prior to use to reduce the chance of contamination of cell cultures and exosome samples which could result in degraded particles.

2.1 Exosome Production

1. Cell Lines: Cancer cell lines have been shown to have increased production of EVs [22] (*see* **Note 1**). For ease of use we use MDA-MB-231 breast cancer cell lines as a model cancer system

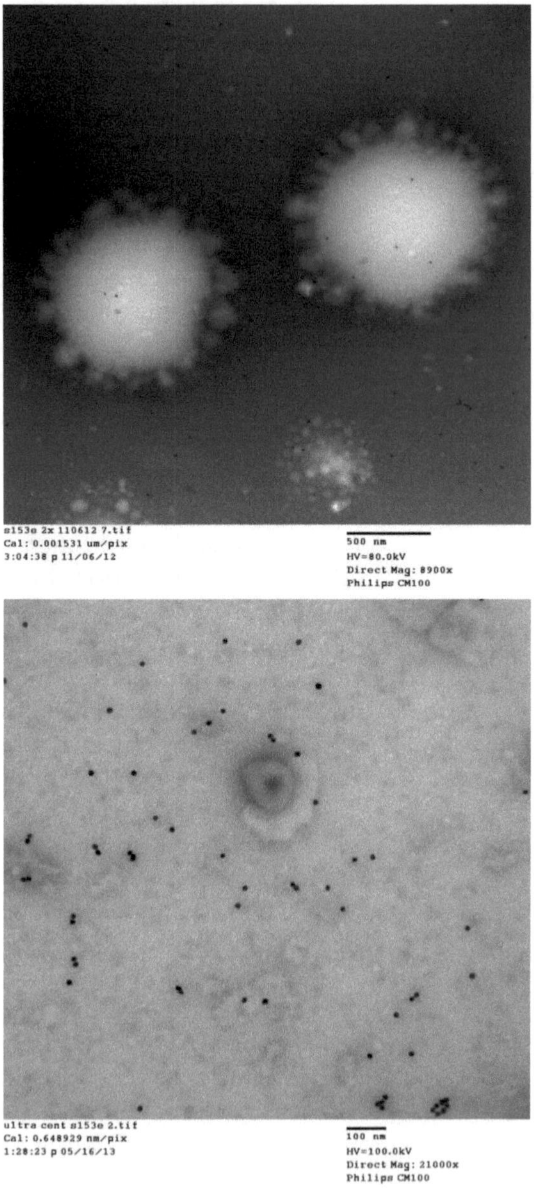

Fig. 1 Initial observation of EVs purified by ExoQuick™ or ultracentrifugation. These images show negative stained exosome samples purified by ExoQuick™ (*top*) or Ultracentrifugation (*bottom*) with 15 nm gold fiducial particles (*black spots*). The image of ExoQuick™ purified exosome sample imaged by TEM shows a large (diameter >500 nm) particle surrounded by smaller (diameter <100 nm) particles. The larger particle has the characteristics of glycerol or other viscous solution forming the core of a particle cluster. The image of the EVs purified by ultracentrifugation shows a characteristic 100 nm doughnut shape which is the remnants of a damaged dried exosome

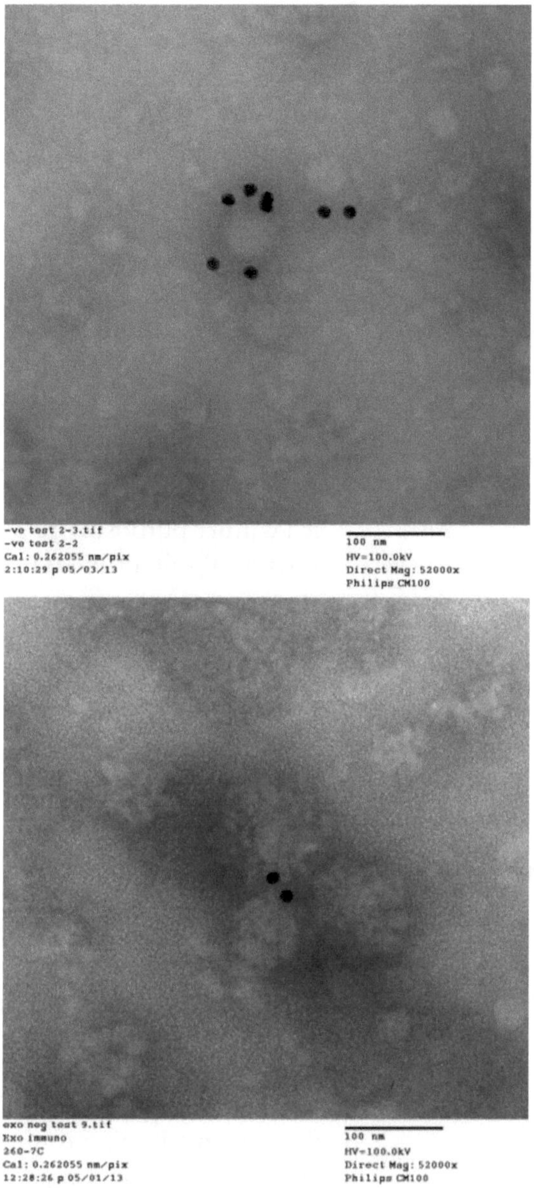

Fig. 2 Immunolabeled and negative stained EV samples at 52,000× magnification. These images show two observations of ExoQuick™ purified particles of >100 nm particles with conjugated 15 nm fiducial particles. These particles are shown to contain CD63 and have the size and shape characteristics of intact EVs

for the generation of EVs [22]. For exosome preparation and all work is carried out in a cell hood with all surfaces and utensils sprayed and wiped with 70% ethanol (*see* **Note 2**).

2. 100 mm polystyrene cell culture dishes with adsorbed poly-L-lysine (*see* **Note 3**)

3. Growth media: Cell culture is grown on 10 cm culture dishes in Dulbecco's modified Eagle's medium supplemented with 2 mM L-glutamine, 10% fetal bovine serum, 100 U/ml penicillin and 100 μg/ml streptomycin (*see* **Note 4**).
4. Exosome production media is unsupplemented DMEM (*see* **Note 5**).
5. PBS for cell culture and exosome storage: Weigh 8 g NaCl, 0.2 g KCl, 1.44 g Na_2HPO_4, 0.24 g KH_2PO_4, 0.133 g $CaCl_2 \cdot 2H_2O$ and 0.1 g $MgCl_2 \cdot 6H_2O$ and add to a 1 l glass beaker. Add 800 ml of milliQ water and completely dissolve the solid with stirring (*see* **Note 6**). Using HCl, pH the solution to pH 7.4. Transfer the solution to a 1 l measuring cylinder and make the solution up to 1 l with water. Using a bottle top filter, filter the buffer into a Duran bottle; this bottle can be sealed and used for storage (*see* **Note 7**). When used for cell culture purposes the buffer should be warmed to 37 °C for at least 30 min prior to use.
6. Normoxic to hypoxic incubator: We have found that cells grown in conditions of reduced O_2 return a greater exosome yield than those grown in atmospheric O_2 levels (*see* **Note 8**). Cells are grown in standard conditions of atmospheric gas +5% CO_2 at 37 °C. When cells are incubated for the generation of EVs we advise growing cells in a mix of 1% O_2, 5%CO_2 and 94% N_2 (*see* **Note 9**).

2.2 Exosome Purification

1. Precipitant: As stated above, multiple companies now sell solutions for the precipitation of EVs. In our work we use Exo-Quick-TC™ which is a bought solution with a proprietary recipe.
2. Refrigerated tabletop centrifuge (*see* **Note 10**).
3. 15 ml conical bottom tubes.
4. 1.5 ml microcentrifuge tubes.
5. Pipeteman autopipettor.
6. 10 ml disposable pipettes.

2.3 Immunolabeling for Electron Microscopy

1. Formvar and carbon coated nickel EM grids, 200 mesh (*see* **Note 11**).
2. Grid preparation humidity box.
3. Crossbar forceps for EM grids (*see* **Note 12**).
4. Whatman® filter paper.
5. Fixative solution: To a 15 ml conical bottom tube add 250 μl of 16% PFA, 400 μl 5% glutaraldehyde, 400 μl 0.5 M PIPES, 20 μl 0.1 M EGTA, 20 μl $MgSO_4$ and 910 μl MilliQ water (*see* **Note 13**).

6. Quenching solution: To a 15 ml conical bottom tube, add 2.5 ml ethanol, 2.5 ml PBS and 5 mg $NaBH_4$.
7. Blocking solution: 3% W/V nonfat milk in PBS (*see* **Note 14**).
8. Primary antibody of monoclonal anti-hCD63 (*see* **Note 15**).
9. Secondary antibody conjugated to 15 nm gold particles for visualization.
10. NanoVan® negative staining solution (*see* **Note 16**).
11. Philips CM100 Transmission electron microscope or similar.

3 Methods

Carry out all procedures at room temperature unless otherwise specified.

3.1 Exosome Generation

1. Grow MDA-MB-231 adherent cells in 10 cm to approximately 80% confluence with DMEM growth media in atmospheric gasses +5 %CO_2 at 37 °C (*see* **Note 17**).
2. Discard growth media (*see* **Note 18**) and wash cells with warmed PBS (*see* **Note 19**).
3. Repeat step three two more times.
4. Slowly add 8 ml of unsupplemented DMEM media to the cells.
5. Place the cells in a hypoxic incubator at 37 °C set with 1% O_2, 5%CO_2 and 94% N_2.
6. Allow the cells to incubate for 2 days to produce EVs for harvesting.
7. Gently remove the media from the cells being careful not to disturb the cell layer (*see* **Note 20**) and place the media in the 15 ml conical bottom tubes.

3.2 Exosome Purification

1. From this point on try to keep all samples at 4 °C whenever possible.
2. Centrifuge the falcon tubes with media at 4 °C, 3000 × *g* for 15 min to remove all cell debris and larger particulates.
3. Being sure not to disturb the pellet, place the supernatant into a clean 15 ml conical bottom tube (*see* **Note 21**).
4. Add 1/5th the volume of ExoQuick-TC™ as there is remaining media and mix (*see* **Note 22**).
5. Incubate the mix at 4 °C overnight (*see* **Note 23**).
6. The media ExoQuick-TC™ solution is then centrifuged at 4 °C, 1500 × *g* for 30 min to pellet the EVs from the solution.
7. Aspirate the supernatant and discard, being sure not to disturb the lighter lower phase (*see* **Note 24**).

8. Reseal the tube and centrifuged at 4 °C, 1500 × *g* for 15 min to repellet the EVs and drive any reaming contaminant media from the pellet (*see* **Note 25**).
9. Resuspend the pellet in 100 μl of PBS (*see* **Note 26**).
10. Dependent on the individual operators' uses of the EVs different volume aliquots may be dispensed into 1.5 ml microfuge tubes or similar style tube resistant to flash freezing.
11. Flash-freeze the samples in liquid nitrogen and store at −70 to −80 °C (*see* **Note 27**).

3.3 EM Immunolabeling

1. Thaw the frozen exosome sample slowly on ice (*see* **Note 28**).
2. Grids should be are formvar covered, carbons coated, and have a layer of adsorbed poly lysine. Take two of these nickel EM and glow discharge for 1 min under Argon gas.
3. Place EM forceps holding the two grids (*see* **Note 29**) into a humid chamber.
4. Add 10 μl of the exosome sample and 10 μl of PBS to each grid seal the chamber and incubate at room temperature for 1 h (*see* **Note 30**).
5. Blot excess EVs solution from the grid (*see* **Note 31**), add 10 μl of the fixative solution and incubate at room temperature for 15 min.
6. Blot excess fixative solution and add 10 μl of quenching solution.
7. Blot excess quenching solution. Wash the grids three times by adding 20 μl of PBS and incubating at room temperature for 15 min.
8. After the third wash, block the grids by adding 10 μl of blocking buffer and incubate the grids at room temperature for 1 h.
9. Blot excess blocking buffer from both grids; to the first grid add 10 μl of primary antibody in blocking buffer (this is the experimental grid); to the second grid add 10 μl of blocking buffer (this is the control grid) (*see* **Note 32**).
10. Seal the humid chamber and incubate the grids at room temperature for at least 16 h, ideally overnight.
11. After incubation repeat the wash step detailed in **step** 7.
12. Blot PBS and add 10 μl to fiducial conjugated secondary antibody to both grids and incubate at room temperature for 2 h.
13. After incubation repeat the wash step detailed in **step** 7.
14. Blot PBS and add 3 μl of NanoVan® negative staining solution.
15. Blot excess NanoVan® to dryness and leave to dry completely prior to imaging (*see* **Note 33**).

16. We visualize the grids on a Philips CM 100 Transmission electron microscope at a nominal magnification of 30,000×, however; any similar equipment would be acceptable.

4 Notes

1. We tested multiple cell lines and many showed similar exosome production in the same conditions. Cancer cell lines have been shown to produce increased quantities of EVs.
2. It is essential to sterilize all surfaces with 70% ethanol, not only is this good practices in cell culture applications but the generation of EVs is performed in unsupplemented media and any contamination will result in failure of the experiment.
3. The adsorbed poly-L-lysine layer is required for the cells to adhere to the surface of the plate and is essential for clean results.
4. Cell culture media should be mixed by agitations of the bottle and no foreign objects placed in the bottle. Once the media is made up it should be filter-sterilized into a storage bottle and stored at 4 °C.
5. Unsupplemented DMEM is used for the production of EVs and cells must be washed thoroughly as FBS contains biological EVs which would contaminate any samples.
6. When making buffers with large amounts of solid, especially in higher concentrations; it is possible to warm the glass beaker gently to aid the solid dissolving.
7. PBS can be made as a 10× stock solution and stored just as readily; if this is the desired option, be sure to use MilliQ water and filter the subsequent 1× buffers when made up.
8. There is a semantic argument about the O_2 levels in hypoxic or normoxic incubators. The general consensus seem to argue that 5–2% is normoxic to represent the O_2 levels in the blood therefore 1% O_2 is more representative of hypoxia. In solid tumors however it is common for O_2 levels to be reduced further; therefore, such levels may be more normal for cancerous cell lines.
9. At first we did not have access to a hypoxic incubator at 1% O_2. We used a desiccator with attached regulator to keep it under constant positive pressure with the gaseous mix. This was placed in a regular cell incubator to maintain a stable temperature.
10. The centrifuge should be able to take the 15 ml conical bottom tubes that the cell media will be collected in.

11. The grids should be selected with help from your EM technician for best results; however, we used nickel grids, as buffers used during immunolabeling would result in oxidation of copper grids. Nickel grids will not oxidize in the same way; however, it is then necessary to correct for objective lens astigmatism induced by the magnetic field from the grid. Gold grids can also be used and will not result in objective lens astigmatism but are significantly more expensive. When preparing sample, avoid touching the experimental surface of the grid to increase the usable area.
12. Crossbar forceps only release when pressed and will hold the grid gently yet firmly to make handling easier. When holding a grid, attach the tweezers to a small edge of the grid, minimizing disturbance to the main body of the grid.
13. Be sure to make the solution fresh each time as the efficacy of glutaraldehyde will significantly decrease over time
14. It is essential the milk powder be completely dissolved so that no particles remain in the buffer prior to use.
15. Antibody concentrations used will be at an approximate ratio of 1:2000; however, the exact concentration is batch dependent
16. NanoVan® is methylamine vanadate and contains the heavy metal Vanadium, this is safer than uranium based negative stains but must still be disposed of as heavy metal waste.
17. 30 min prior to cell line work all media and buffers should be prewarmed to 37 °C.
18. Gently tap the dish to dislodge dead and unhealthy cells so these can be discarded with the media
19. Tilt the culture dish at approximately 45° and slowly drip PBS at the top of the plate to minimize disturbance to the adhered cells and at no point should the cells be allowed to dry.
20. When removing media take media from the top and move down while aspirating, try to abandon a fine film of media at the cell layer to reduce large particles and any cell debris from being removed at the same time.
21. At this point, provided the media remains cool there should be little loss of sample; therefore, if a large amount of debris be present in the sample this initial centrifugation can be longer to improve purity in the final product.
22. The tubes should be mixed gently be inversion and left upright and undisturbed for the time of incubation
23. The material at the bottom of the tube is easily disturbed and great care to reduce any agitation of the tube should be taken. When removing the supernatant, it is best to discard the media

near the pellet and not to try and remove all the media. This improves the subsequent purity of the exosome solution.

24. The lower lighter phase of the tube is very soft and slightly viscous solution, it will leave an obvious track if you run a pipette tip through it, however; it will easily mix back into solution with any excessive agitation. Ideally all media should be removed at the stage to the best of the operators' ability
25. There should only be traces of media to be removed and if the volume of the remaining media is over 200 μl we would advise repeating this step.
26. We have recovered final pellet volumes of 50–200 μl and as such the final volume of the exosome pellet can vary substantially
27. We have stored sample in ultralow temperature freezers for over a year with no detectable degradation.
28. The slow thawing of samples reduces amount of particles and improves the quality of the remaining EVs.
29. One grid is the experimental grid and one grid is a control to determine if the secondary antibody shows any nonspecific interactions with the sample.
30. The hour long incubation allows time for the EVs to settle onto the grid we have found this to be the optimal amount of time for EVs to settle without resulting in a dense packing of EVs.
31. Tear the Whatman® filter paper to produce a rough edge. Touch the rough edge of this piece of filter paper to the liquid near the edge of the grid and try to disturb the surface of the grid as little as possible. It is important to avoid completely drying the surface.
32. The control grid will be used to monitor levels of nonspecific binding of the secondary antibody to the sample.
33. Grids can be stored for prolonged periods before observing; we have observed grids over a month old which have still shown clear immunostaining.

Acknowledgment

We thank the National Institutes of Health and Cancer Research UK for financial supports (NIH R01GM103843 and C596/A17096).

References

1. Kastelowitz N, Yin H (2014) EVs and microvesicles: identification and targeting by particle size and lipid chemical probes. ChemBioChem 15:923–928
2. Lässer C (2015) EVs in diagnostic and therapeutic applications: biomarker, vaccine and RNA interference delivery vehicle. Expert Opin Biol Ther 15:103–117
3. Roma-Rodrigues C, Fernandes AR, Baptista PV (2014) Exosome in tumour microenvironment: overview of the crosstalk between normal and cancer cells. Biomed Res Int 2014:10
4. Mathivanan S, Lim JWE, Tauro BJ, Ji H, Moritz RL, Simpson RJ (2010) Proteomics analysis of a33 immunoaffinity-purified EVs released from the human colon tumor cell line lim1215 reveals a tissue-specific protein signature. Mol Cell Proteomics 9:197–208
5. Taylor DD, Shah S (2015) Methods of isolating extracellular vesicles impact down-stream analyses of their cargoes. Methods 87:3–10
6. Lane RE, Korbie D, Anderson W, Vaidyanathan R, Trau M (2015) Analysis of exosome purification methods using a model liposome system and tunable-resistive pulse sensing. Sci Rep 5:7639
7. Lötvall J, Hill AF, Hochberg F, Buzás EI, Di Vizio D, Gardiner C, Gho YS, Kurochkin IV, Mathivanan S, Quesenberry P, Sahoo S, Tahara H, Wauben MH, Witwer KW, Théry C (2014) Minimal experimental requirements for definition of extracellular vesicles and their functions: a position statement from the international society for extracellular vesicles. J Extracell Vesicles 3:26913
8. Christianson HC, Svensson KJ, van Kuppevelt TH, Li J-P, Belting M (2013) Cancer cell EVs depend on cell-surface heparan sulfate proteoglycans for their internalization and functional activity. Proc Natl Acad Sci 110:17380–17385
9. Tauro BJ, Greening DW, Mathias RA, Mathivanan S, Ji H, Simpson RJ (2013) Two distinct populations of EVs are released from lim1863 colon carcinoma cell-derived organoids. Mol Cell Proteomics 12:587–598
10. Buschow SI, van Balkom BWM, Aalberts M, Heck AJR, Wauben M, Stoorvogel W (2010) MHC class II-associated proteins in b-cell EVs and potential functional implications for exosome biogenesis. Immunol Cell Biol 88:851–856
11. Jakobsen KR, Paulsen BS, Bæk R, Varming K, Sorensen BS, Jørgensen MM (2015) Exosomal proteins as potential diagnostic markers in advanced non-small cell lung carcinoma. J Extracell Vesicles 4:26659
12. Witwer KW, Buzás EI, Bemis LT, Bora A, Lässer C, Lötvall J, Hoen N-'t, EN PMG, Sivaraman S, Skog J, Théry C, Wauben MH, Hochberg F (2013) Standardization of sample collection, isolation and analysis methods in extracellular vesicle research. J Extracell Vesicles 2:20360
13. Böing AN, van der Pol E, Grootemaat AE, Coumans FAW, Sturk A, Nieuwland R (2014) Single-step isolation of extracellular vesicles by size-exclusion chromatography. J Extracell Vesicles 3:23430
14. György B, Módos K, Pállinger É, Pálóczi K, Pásztói M, Misják P, Deli MA, Sipos Á, Szalai A, Voszka I, Polgár A, Tóth K, Csete M, Nagy G, Gay S, Falus A, Kittel Á, Buzás EI (2011) Detection and isolation of cell-derived microparticles are compromised by protein complexes resulting from shared biophysical parameters. Blood 117:e39–e48
15. Hong P, Koza S, Bouvier ESP (2012) Size-exclusion chromatography for the analysis of protein biotherapeutics and their aggregates. J Liq Chromatogr Rel Technol 35:2923–2950
16. Théry C, Amigorena S, Raposo G, Clayton A (2001) Isolation and characterization of EVs from cell culture supernatants and biological fluids. In: Current protocols in cell biology. John Wiley & Sons, Inc., New York, NY
17. Lässer C, Eldh M, Lötvall J (2012) Isolation and characterization of rna-containing EVs. J Vis Exp 59:3037
18. Van Niel G, Mallegol J, Bevilacqua C, Candalh C, Brugière S, Tomaskovic-Crook E, Heath JK, Cerf-Bensussan N, Heyman M (2003) Intestinal epithelial EVs carry mhc class ii/peptides able to inform the immune system in mice. Gut 52:1690–1697
19. Jeppesen DK, Hvam ML, Primdahl-Bengtson B, Boysen AT, Whitehead B, Dyrskjøt L, Ørntoft TF, Howard KA, Ostenfeld MS (2014) Comparative analysis of discrete exosome fractions obtained by differential centrifugation. J Extracell Vesicles 3:25011
20. Momen-Heravi F, Balaj L, Alian S, Trachtenberg AJ, Hochberg FH, Skog J, Kuo WP

(2012) Impact of biofluid viscosity on size and sedimentation efficiency of the isolated microvesicles. Front Physiol 3:162

21. Taylor D, Zacharias W, Gercel-Taylor C (2011) Exosome isolation for proteomic analyses and RNA profiling. Humana Press, Serum/plasma proteomics

22. Parolini I, Federici C, Raggi C, Lugini L, Palleschi S, De Milito A, Coscia C, Iessi E, Logozzi M, Molinari A, Colone M, Tatti M, Sargiacomo M, Fais S (2009) Microenvironmental pH is a key factor for exosome traffic in tumor cells. J Biol Chem 284:34211–34222

Chapter 9

Size Exclusion Chromatography: A Simple and Reliable Method for Exosome Purification

Richard Lobb and Andreas Möller

Abstract

Exosomes were originally described 29 years ago as a mechanism for the removal of redundant molecules from reticulocytes, nothing but a process of removing cellular trash. It is now however, abundantly clear that exosomes have a more significant biological role. However, there is currently limited information pertaining to efficient isolation procedures that can be used to isolate exosomes from both cell culture media, and clinical samples such as plasma. Here, we present a reliable and efficient procedure that can be utilized for the isolation of exosomes from various starting materials.

Key words Exosomes, Extracellular vesicles, Isolation, Human plasma, Cell culture supernatant

1 Introduction

Exosomes are vesicles of endocytic origin with a size range of 30–100 nm. It is now clear that exosomes contribute to intercellular cross talk in an autocrine and paracrine fashion [1]. Exosomes contain lipids, proteins, mRNA, and miRNA and are released under physiological and pathological conditions [2]. Exosomes contain cell-type specific content, and it is for this reason that they represent a rich source of novel biomarkers. The isolation of exosomes from cell culture supernatants depends on a number of factors, including, but not limited to, the cell-type, conditioning time and media. However, one significant hurdle for exosome analysis is an efficient and standardized method for pure preparations [3].

Standard purification methods result in a large loss of exosomes, which can be improved using different methodologies. Various methods have been suggested for concentrating and purifying exosomes from cell culture supernatant or human biofluids. Most protocols rely on ultracentrifugation; however, there are some disadvantages that may detract from the utility of this method [4]. We have recently shown that repeated ultracentrifugation steps can lead to a reduced yield of exosomes in comparison to using

Winston Patrick Kuo and Shidong Jia (eds.), *Extracellular Vesicles: Methods and Protocols*, Methods in Molecular Biology, vol. 1660,
DOI 10.1007/978-1-4939-7253-1_9, © Springer Science+Business Media LLC 2017

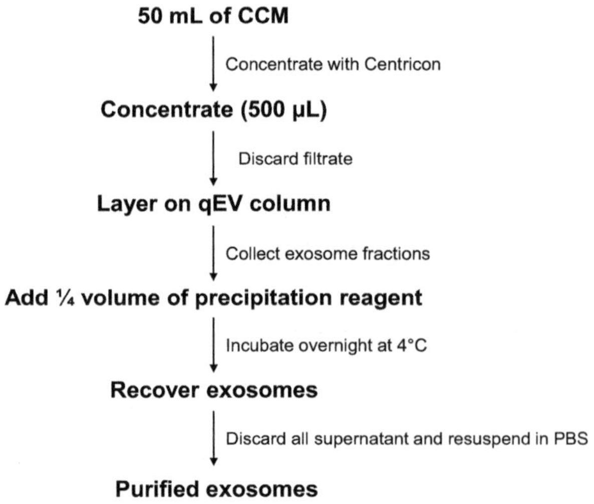

Fig. 1 Flow of exosome isolation. CCM—conditioned media

ultrafiltration [4]. Furthermore, we have shown that various isolation techniques can result in the co-isolation of contaminating nonexosomal proteins [4]. We find that purification with size exclusion chromatography (SEC), coupled to PEG-based retrieval of exosomes is reproducible, inexpensive, and importantly nondestructive (Fig. 1). To this end, we present an efficient isolation protocol for purifying exosomes based on their size from cell culture supernatant, which is also compatible with purifying exosomes from serum or plasma (Figs. 1, 2, and 3). This method does not depend on pelleting exosomes at high speed, and results in a highly purified exosome preparation (Fig. 3).

2 Materials

All solutions are prepared using ultrapure water and analytical grade reagents. Reagents are stored at room temperature, unless indicated otherwise.

1. Prepare 1× PBS pH 7.4 solution from stock 10× PBS pH 7.4. To prepare 10× PBS, Dissolve the following in 800 mL distilled H_2O: 80 g of NaCl, 2.0 g of KCl, 14.4 g of Na_2HPO_4 and 2.4 g of KH_2PO_4. Adjust volume to 1 L with water. Dilute 100 mL of 10× PBS with 900 mL of water.
2. Centricon Plus-70 Centrifugal Filter (Ultracel-PL Membrane, 100 kDa). qEV size exclusion columns (*see* **Note 1**).
3. Sodium azide: sodium azide 0.05% (w/v). Dissolve 0.5 g of sodium azide in 1 L of water.

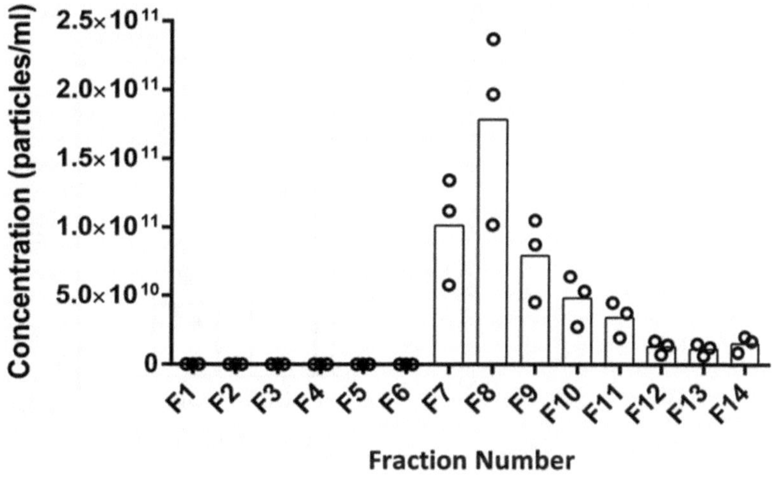

Fig. 2 Exosome containing fractions. 500 μL fractions were collected then analyzed using TRPS. Note the peak elution of vesicles at fraction 8

4. Sodium hydroxide: 0.5 M NaOH. Dissolve 20 g of NaOH in 800 mL of water, adjust volume to 1 L when NaOH is dissolved.
5. Litmus paper.
6. Polyethylene glycol (PEG) solution (precipitation reagent): PEG-8000 32% (w/v) (*see* **Note 2**), NaCl (1.2 M). To 320 g of PEG 8000, add 70.08 g of NaCl, add 800 mL PBS and warm in a water bath to dissolve. After PEG and NaCl have dissolved, adjust volume to 1 L with additional PBS.

3 Methods

All procedures are carried out at room temperature unless specified otherwise (*see* **Note 3**).

3.1 Preparing Columns and Centricons

1. Centricon devices are preserved with sodium azide and glycerol. To remove this, add 50 mL of PBS to a Centricon device and centrifuge for 10 min at 3500 × *g*.
2. Columns are stored in sodium azide (w/v). Equilibration of columns is dependent on the column volume; use at least 2 column volumes of PBS to equilibrate qEV columns before exosome purification.

3.2 Concentrating Conditioned Media

1. Collect and transfer 50 mL of (*see* **Note 4**) conditioned media to a 50 mL tube and centrifuge at 500 × *g* for 10 min at 4 °C to remove floating cells and large debris.

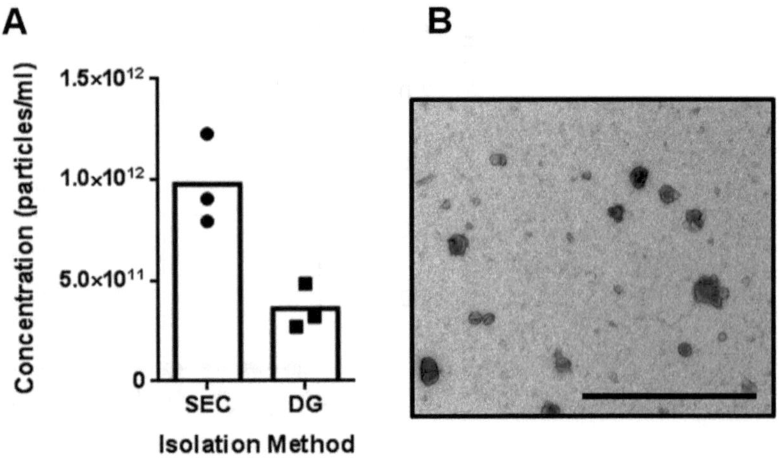

Fig. 3 (**a**) Size exclusion (SEC) provides a high yield of exosomes in comparison to density gradient (DG) purified exosomes. (**b**) Transmission electron microscopy image of exosomes purified with SEC columns, size bar = 500 μm

2. After centrifugation, take the supernatant and filter the samples through a 0.22 μm filter (Merck Millipore) in order to remove large particles.
3. Transfer the prepared conditioned media into a Centricon device (70 mL), and centrifuge at 3500 × *g* until the sample has been fully concentrated (300–500 μL) (*see* **Note 5**). The concentrated sample can then be recovered with a reverse spin at 1000 × *g* for 2 min.

3.3 Purifying Exosomes

1. Remove PBS from the top of an equilibrated qEV column and transfer concentrated media to the column. Allow the sample to run entirely into the column before adding additional PBS.
2. Overlay the column with 1 mL of PBS and run the sample through the column. *Always keep the column topped with PBS.* The void volume of the qEV columns are 3 ± 0.25 mL (Discard) (*see* **Note 6**) and exosomes will elute primarily in the following 1.5 ml (collect) (*see* **Note** 7) (Fig. 2). The columns can then be cleaned and reused (*see* **Note 8**).
3. Add 500 μL of precipitation reagent to 1.5 mL of exosome containing fractions (1/4 final volume), mix thoroughly. Incubate without agitation overnight at 4 °C.
4. After the overnight incubation, centrifuge the sample at 10,000 × *g* for 20 min.
5. Carefully remove all supernatant without disturbing the pellet.
6. Centrifuge again at 10,000 × *g* for 2 min and remove the remaining solution
7. Resuspend exosome pellet in 50 μL of PBS (*see* **Note 9**).

4 Notes

1. Size exclusion columns can be purchased, or prepared in the lab, *see* refs. 5, 6 for more information.
2. PEG concentrations can be manipulated to suit experimental setup and precipitate particles within a narrow size range.
3. All centrifugation steps are carried out at 4 °C.
4. We use serum-free media for exosome isolation. This method can be used with serum-containing media (exosome depleted); however, only small volumes of serum-containing media (20–50 mL) can be used due to the concentration of bovine proteins.
5. The length of time to concentrate the media depends on experimental setup (i.e., cell line used, conditioning time). We find that 15–20 min of centrifugation is generally sufficient.
6. The void volume is according to the manufacturer's specifications. This can vary and will have to be established prior to purification.
7. We collect 500 μL fractions and identify exosome containing fractions using tunable resistive pulse sensing (TRPS). Exosome containing fractions can also be identified with immunoblotting for exosome markers.
8. Columns can be cleaned and reused multiple times. We find a wash of 3 column volumes of PBS between samples is sufficient. If the column performance has been reduced from nonspecific adsorption, the column can be sanitized with NaOH. Wash the column with 1 column volume of NaOH, and then flush with 3 volumes of PBS. Check the pH of the column with litmus paper to determine if the column has been reequilibrated. After exosome collection, flush the column with 2 column volumes of sodium azide (w/v) and store at 4 °C.
9. The final volume of the exosome isolation should be seen as guide and can be modified for desired downstream purposes.

Acknowledgments

This work was supported by a National Health and Medical Research Council Australia project grant (NHMRC, APP1068510) and Cancer Council Queensland grant (APP1045620) to AM.

References

1. Thery C, Zitvogel L, Amigorena S (2002) Exosomes: composition, biogenesis and function. Nat Rev Immunol 2(8):569–579. doi:10.1038/nri855
2. Vlassov AV, Magdaleno S, Setterquist R, Conrad R (2012) Exosomes: current knowledge of their composition, biological functions, and diagnostic and therapeutic potentials. Biochim Biophys

Acta 1820(7):940–948. doi:10.1016/j.bbagen.2012.03.017

3. Witwer KW, Buzas EI, Bemis LT, Bora A, Lasser C, Lotvall J, Nolte-'t Hoen EN, Piper MG, Sivaraman S, Skog J, Thery C, Wauben MH, Hochberg F (2013) Standardization of sample collection, isolation and analysis methods in extracellular vesicle research. J Extracell Vesicles 2. doi:10.3402/jev.v2i0.20360
4. Lobb RJ, Becker M, Wen SW, Wong CS, Wiegmans AP, Leimgruber A, Moller A (2015) Optimized exosome isolation protocol for cell culture supernatant and human plasma. J Extracell Vesicles 4:27031. doi:10.3402/jev.v4.27031
5. Boing AN, van der Pol E, Grootemaat AE, Coumans FA, Sturk A, Nieuwland R (2014) Single-step isolation of extracellular vesicles by size-exclusion chromatography. J Extracell Vesicles 3. doi:10.3402/jev.v3.23430
6. de Menezes-Neto A, Saez MJ, Lozano-Ramos I, Segui-Barber J, Martin-Jaular L, Ullate JM, Fernandez-Becerra C, Borras FE, Del Portillo HA (2015) Size-exclusion chromatography as a stand-alone methodology identifies novel markers in mass spectrometry analyses of plasma-derived vesicles from healthy individuals. J Extracell Vesicles 4:27378. doi:10.3402/jev.v4.27378

Chapter 10

Purification Protocols for Extracellular Vesicles

Rebecca E. Lane, Darren Korbie, Matt Trau, and Michelle M. Hill

Abstract

This chapter provides a description of some of the standard methods used for the isolation of extracellular vesicles (EVs) from a variety of biological fluids, including cell culture media, urine, plasma and serum. The methods presented include ultracentrifugation, ultrafiltration, proprietary polymer-based reagents, size exclusion chromatography, density gradient separation, and immunoaffinity capture. Ultracentrifugation methods use high speed centrifugation to pellet vesicles, whilst polymer-based reagents are added to the sample to facilitate vesicle precipitation using lower speeds. Ultrafiltration involves the concentration of vesicles from a large volume of biological fluid using a centrifugal filter unit. Size exclusion chromatography and density gradient separation are both designed to allow the separation of vesicles from other nonvesicular debris. Immunoaffinity capture methods use antibody-coated beads to selectively isolate vesicles displaying a surface marker of interest. Ultimately, the choice of purification method for an individual experiment is influenced by time, cost, and equipment considerations, as well as the sample requirements for any downstream analyses.

Key words Extracellular vesicles, Exosomes, Microvesicles, Isolation

1 Introduction

Exosomes and microvesicles are submicron membranous particles shed from cells into circulation, and are collectively referred to as extracellular vesicles (EVs). Exosomes have garnered particular interest and are small (approximately 30–100 nm), relatively homogenous, and endosomally derived [1]. Microvesicles span a larger size range (100–1000 nm) and are directly shed from the plasma membrane [1]. EVs have been shown to participate in many normal and aberrant physiological processes [2]. For example, cancer cells have been shown to constitutively release exosomes into the circulation, and these promote several disease-associated processes including angiogenesis [3], establishment of the premetastatic niche [4] and mediation of organ-specific metastasis [5]. Additionally, there is substantial evidence that circulating EVs contain disease-associated mRNA, miRNA and protein species, endearing them as a potential biomarker source [6, 7]. As such,

Winston Patrick Kuo and Shidong Jia (eds.), *Extracellular Vesicles: Methods and Protocols*, Methods in Molecular Biology, vol. 1660, DOI 10.1007/978-1-4939-7253-1_10, © Springer Science+Business Media LLC 2017

there is considerable interest in studying the physical and molecular characteristics of EVs in order to utilize them for diagnostic or therapeutic purposes. Such analyses first require that extracellular vesicles be enriched and isolated from the surrounding biological material, which is a complex mixture of cells and cellular debris, protein, nucleic acids and lipids [8]. Depletion of these background contaminants is especially critical where nucleic acid or protein analysis is to be performed, such that low-abundance EV-associated species may be detected [8–10].

As exosomes have been the focus of much of the current EV research, EV isolation protocols are predominantly directed toward enrichment of these small ($\leq$100 nm) vesicles (*see* Subheading 3.1). These isolation protocols vary greatly in their cost and equipment requirements, processing time, type and volume of input material and the purity of the isolated material (Table 1). For example, ultrafiltration and proprietary precipitation-based reagents allow rapid enrichment of all EVs in a sample, however, the purity of this material is generally low due to the presence of coisolated protein contaminants [15]. Density gradient and immunoaffinity techniques are low throughput and require longer processing times; however, they produce more pure yields and may allow separation of EV subpopulations [14].

The following chapter will detail the standard protocols for six of the most commonly used techniques for isolation of vesicles from urine, cell culture media and plasma or serum samples. These standard protocols may be adapted for a variety of sample sources and input volumes. Ultimately, determination of the most appropriate protocol/s for an individual experiment will depend on the sample characteristics, as well as the nature and requirements of any downstream analyses.

2 Materials

All methods require standard 1× PBS prepared using deionized or ultrapure (e.g., MilliQ) water. It is recommended that PBS be filtered to 0.22 μm to eliminate contaminants prior to use.

2.1 Ultracentrifugation

1. Ultracentrifuge with compatible fixed angle or swinging bucket rotor.
   (a) Benchtop models: Beckman Coulter Optima MAX-XP, MAX-TL, or similar
   (b) Floor models: Beckman Coulter Optima L-series, XPN-series, or XE-series
2. Polycarbonate, polypropylene or similar tubes capable of withstanding forces $\geq$100,000 × g and that are compatible with selected rotor and sample input volume. Table 2 shows some example centrifuge, rotor and tube configurations for a variety of volumes. All listed components are available from Beckman Coulter (*see* **Note 1**).

Table 1
Overview of EV purification techniques, including the mechanism of separation of vesicles from other sample components, type and volume of input, approximate processing time, required equipment, sample purity, and other considerations

Method	Mechanism of separation	Input type	Processing time	Required equipment	Purity of isolates	Other considerations
Ultracentrifugation	Sedimentation rate	Raw sample, any volume	3–6 h	Ultracentrifuge and compatible tubes	Low	May induce vesicle disruption [11]
Ultrafiltration	Size	Raw sample, any volume	1–3 h	Benchtop centrifuge	Low	Retains EVs in original fluid, may need to perform buffer exchange
Polymer-based reagents	Sedimentation rate	Raw sample, any volume	2–18 h	Benchtop centrifuge	Low	Not recommended for use with mass spectrometry
Size-exclusion chromatography	Size	Concentrated EVs, ~500 μL volume	1–2 h	–	High	May separate vesicles from high density lipoprotein (HDL) particles [12]
Density gradient separation	Density	Concentrated EVs, ≤1 mL volume	20–24 h	Ultracentrifuge and compatible tubes	High	Coisolated vesicles and HDL particles [13]
Immunoaffinity isolation	Surface marker expression	Concentrated EVs, ~100 μL volume	18–20 h	Magnet separator, shaker	Very high	Allows targeting of specific subpopulations based on surface marker expression [14] but recovery may be low

Table 2
List of ultracentrifuge rotors and compatible tubes appropriate for use in extracellular vesicle isolation

Centrifuge type	Rotor	Rotor type	Rotor max capacity	Example compatible tube/bottle
Benchtop	TLA120.2	Fixed angle	10 × 2 mL	Thick-wall polycarbonate tube 1 mL (11 × 34 mm)
Benchtop	TLA100.3	Fixed angle	6 × 3.5 mL	Thick-wall polycarbonate tube 3.5 mL (13 × 57 mm)
Floor	Type 50.2 Ti	Fixed angle	12 × 39 mL	Thick-wall polycarbonate bottle 26.3 mL (25 × 89 mm)
Floor	Type 70 Ti	Fixed angle	8 × 39 mL	Thick-wall polycarbonate tube + cap (optional) 32 mL (25 × 89 mm)
Floor	SW32	Swinging bucket	6 × 38.5 mL	Thin-wall polypropylene tube 38.5 mL (25 × 89 mm)
Floor	SW40	Swinging bucket	6 × 14 mL	Thin-wall polypropylene tube 14 mL (14 × 95 mm)
Floor	SW41	Swinging bucket	6 × 13.2 mL	Thin-wall polypropylene tube 13.2 mL (14 × 89 mm)

2.2 Ultrafiltration

1. Filter columns with 10–100 kDa molecular weight cutoff (MWCO), with capacity appropriate to sample volume:
   (a) Amicon Ultra Columns (Merck Millipore, Germany), available in 0.5, 2, 4 and 15 mL capacities.
   (b) VivaSpin Centrifugal Concentrators (GE Healthcare Life Sciences, MA, USA) available in 0.5, 2, 4, 6 and 20 mL capacities.

2.3 Proprietary Polymer-Based Reagents

One of the following, compatible with the sample type of interest:

1. Invitrogen Total Exosome Isolation Kit (Life Technologies, USA) for cell culture media or for serum.
2. ExoSpin Exosome Purification Kit (Cell Guidance Systems, USA) for cell culture media/urine/saliva and other low protein biological fluids, or for blood sera/plasma.
3. ExoQuick Exosome Precipitation Solution (System Biosciences, USA) for tissue culture and urine, or for biofluids (serum and ascites).

2.4 Size Exclusion Chromatography (SEC)

1. qEV Size Exclusion Column (Izon Science, UK) (*see* **Note 2**).
2. PBS with a bacteriostatic agent for column storage, e.g., 20% ethanol and <0.05% w/v sodium azide.

2.5 Density Gradient Centrifugation

1. Optiprep Density Gradient Medium (60% w/v Iodixanol, Sigma-Aldrich, MO, USA).
2. Tris–Sucrose Buffer: 0.25 M sucrose, 10 mM Tris, pH 7.5.
3. Ultracentrifuge with swinging bucket rotor (*see* Subheading 2.1).
4. Centrifuge tubes capable of withstanding forces ≥100,000 × *g* (*see* Subheading 2.1).
5. Spectrophotometer and compatible cuvettes (if required).

2.6 Immunoaffinity Isolation

1. (a) Exosome-Streptavidin Isolation/Detection Reagent (Thermo Fisher, MA, USA) and biotinylated antibody against vesicle surface marker of interest (*see* **Note 3**); or
   (b) Exosome-Human CD9, CD81, CD63 or EpCAM Isolation Reagent (Thermo Fisher, MA, USA).
2. Magnetic Separator. DynaMag®-2 (Thermo Fisher, MA, USA).
3. Mixing/rocking device or sample shaker.
4. Isolation Buffer: PBS + 0.1% BSA, filtered to 0.22 μm.
5. Lysis buffer appropriate for downstream analysis.
   (a) For downstream protein analysis. RIPA buffer: 150 mM sodium chloride, 1.0% Triton X-100 or NP-40, 0.5% sodium deoxycholate, 0.1% sodium dodecyl sulfate (SDS), 50 mM Tris, pH 8.0, protease inhibitor cocktail (prepare as per individual manufacturer specifications).
   (b) For downstream RNA analysis: TRIzol reagent (*see* **Note 4**).

3 Methods

3.1 Ultracentrifugation

Ultracentrifugation is generally considered the 'gold standard' method for vesicle isolation, and is arguably the most commonly used method for this purpose [16]. This technique separates extracellular vesicles from other sample components based on their differential sedimentation rates, which are primarily affected by size and density [17]. The technique involves first pelleting cellular debris under slow (2000–5000 × *g*) centrifugation, then pelleting larger vesicles (apoptotic bodies and some larger microvesicles) at a faster (10,000–20,000 × *g*) centrifugation speed [18]. Small vesicles (exosomes and smaller microvesicles) can then be pelleted under very high speed (100,000 × *g*) centrifugation. It is worthwhile to note that due to the overlapping size profiles of exosomes (30–100 nm) and microvesicles (100–1000 nm), the separation of these two vesicle classes by differential centrifugation will invariably be incomplete, and the final pellet will contain a population of EVs secreted by different cellular mechanisms [17].

The following protocols have been adapted from those originally described by Théry et al. [19]. The standard ultracentrifugation protocol detailed below is suitable for cell culture media, urine and other low-protein fluids. For processing plasma and serum, there are minor differences to centrifugation conditions (*see* **Note 5**). At each step of the protocol, the alternate conditions for these high protein samples are noted inside square brackets [ ].

1. For plasma and serum samples, dilute sample 1:2 in PBS before beginning protocol.
2. Centrifuge sample at >1000 × *g* for 5–10 min at 4°°C [>1000 × *g* for 5–10 min] to remove residual cells. Discard pellet and transfer supernatant to new centrifuge tube (*see* **Note 6**).
3. Centrifuge sample at 2000–5000 × *g* at 4 °C for 10–20 min [2000 × *g* for 30 min] to remove dead cells and large debris. Discard pellet and transfer supernatant to new centrifuge tube.
4. Centrifuge sample at 10,000 × *g* for 30 min [12,000 × *g* for 45 min] to pellet small debris and larger vesicles. Transfer supernatant to ultracentrifuge tube/s (*see* **Note 7**). Retain pellet if analysis of larger vesicles is desired, else discard.
5. Centrifuge sample at 100,000 × *g* for 1–2 h [110,000 × *g* for 2 h]. Discard the supernatant, the pellet contains the small extracellular vesicles.
6. The pellet may be directly resuspended in a convenient volume of PBS (50–200 μL) for downstream use.
   (a) Optional wash step to remove contaminating proteins: resuspend pellet in 1 mL PBS and centrifuge for 1–2 h at 100,000 × *g* as per **step 5**. For further depletion of contaminants, this wash step may be performed twice (*see* **Note 8**). Resuspend the final pellet in a convenient volume of PBS (50–200 μL).
7. The isolated vesicles may be stored at −80 °C long term (up to 1 year).

3.2 Ultrafiltration

Ultrafiltration is commonly used as a first step to concentrate vesicles from a large volume of starting material (e.g., >100 mL cell culture media or urine) into a small, more manageable volume (typically 1–2 mL) which can then be subjected to further purification protocols (e.g., ultracentrifugation and size exclusion chromatography). Typically, the molecular weight cut off (MWCO) for ultrafiltration columns used for vesicle concentration is 50–100 kDa. Using liposomes as a model vesicle system as described in [20], we have demonstrated successful vesicle concentration by ultrafiltration using filter MWCOs of 10, 50 and 100 kDa respectively (Fig. 1).

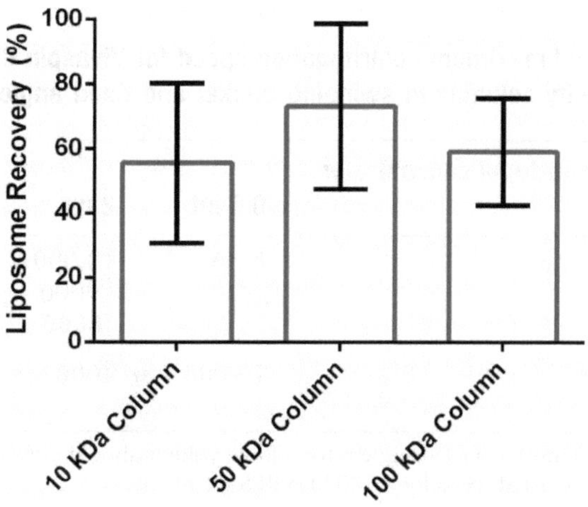

Fig. 1 Recovery (%) ± SD of liposomes (55:45 mol/mol 1,2-dioleoyl-*sn*-glycero-3-phosphocholine (DOPC): cholesterol, mean size = 100 nm, *FormuMax, CA, USA*) spiked into cell culture media (RPMI1640, *Sigma-Aldrich, MO, USA*) and concentrated using 10, 50, or 100 kDa molecular weight cutoff (MWCO) Amicon Ultra-2 filtration columns. Recovered particle concentrations measured by tunable resistive pulse sensing (TRPS) and compared to input concentrations

Table 3
Recommended maximum centrifugation speed for Amicon Ultra 0.5, 2, 4, and 15 mL columns in swinging bucket and fixed angle rotors, respectively

Amicon Ultra Column Capacity	0.5 mL	2 mL	4 mL[a]	15 mL[b]
Swinging bucket	N/A	4000 × *g*	4000 × *g*	4000 × *g*
Fixed angle	14,000 × *g*	7500 × *g*	7500 × *g*	5000 × *g*

[a]For 4 mL columns in a fixed angle rotor, filtrate volume should not exceed 3.5 mL
[b]For 15 mL columns in a fixed angle rotor, filtrate volume should not exceed 12 mL

1. Remove debris from sample as described the ultracentrifugation protocol (Subheading 3.1, **steps 2–4**). It is recommended to dilute plasma and serum at least 1:2 in PBS (or similar buffer) before filtration.
2. Assemble the ultrafiltration unit and place into/onto a flow-through collection tube.
3. Load sample into the assembled ultrafiltration column. If sample volume is larger than column capacity, sample can passed through the column in multiple sequential aliquots to concentrate (*see* **Notes 9** and **10**).
4. Centrifuge column to draw fluid through filter. Tables 3 and 4 give the centrifugation conditions required for each capacity

Table 4
Recommended maximum centrifugation speed for Vivaspin Centrifugal Concentrator 0.5, 2, 4, 6, and 20 mL capacity columns in swinging bucket and fixed angle rotors respectively

Vivaspin centrifugal concentrator capacity	0.5 mL	2 mL	4 mL	6 mL	20 mL[a]
Swinging bucket	N/A	12,000 × *g* [b]9000 × *g* [c]8000 × *g*	10,000 × *g* [b]7000 × *g*	10,000 × *g* [b]6000 × *g*	8000 × *g* [b]6000 × *g*
Fixed angle	15,000 × *g*	4000 × *g*	4000 × *g*	4000 × *g*	5000 × *g* [b]3000 × *g*

[a]For 20 mL columns in a fixed angle rotor, filtrate volume should not exceed 14 mL
[b]Indicates the adjusted speed for ≥100 kDa PES membranes
[c]Indicates the adjusted speed for CTA and HY membranes of all MWCO

column with either a fixed angle or swinging bucket rotor. Begin by centrifuging the column for 10 min, and adjust centrifugation time for subsequent runs as required to produce appropriate concentrate volume (*see* **Note 11**).

5. After each centrifugation step, discard column flow through. The vesicles will be retained in the concentrate above the filter.
6. Repeat **steps 3** and **4** until entire sample volume has been processed.
7. Recover concentrate either by directly pipetting from the filter unit (Vivaspin columns, 4 and 15 mL Amicon Ultra columns), or by inverting filter unit into a fresh collection tube and centrifuging at 1000 × *g* for 2 min (other Amicon Ultra columns).
8. The recovered concentrate can be subjected to further purification protocols as required (*see* **Note 12**). Concentrate may be stored long term (up to 1 year) at −80 °C.

3.3 Proprietary Polymer-Based Reagents

There are several proprietary polymer based reagents which are commercially available for use in EV isolation. The exact mechanism of these reagents has not been reported, however, they are designed to facilitate precipitation of vesicles during low-speed (≤20,000 × *g*) centrifugation. These reagents are said to simplify the process of vesicle isolation and avoid the need for specialized ultracentrifugation equipment. Despite these advantages, there is some evidence to suggest that these reagents may coprecipitate nonvesicular debris to a higher degree than other techniques [15, 21], and therefore may not be ideal where downstream proteomic or nucleic acid analysis is to be performed. The following section focuses on three of the most commonly used reagents: Invitrogen Total Exosome Isolation Kit (Life Technologies, CA, USA), ExoSpin Exosome Purification Kit

(Cell Guidance Systems, MO, USA), and ExoQuick Exosome Precipitation Solution (System Biosciences, CA, USA).

1. Remove debris from sample as described by the manufacturer protocol or as described in the ultracentrifugation protocol (Subheading 3.1, **steps 2–4**, *see* **Note 13**).
2. Add the appropriate volume of reagent to each sample (*see* Table 5).
3. Incubate sample at 4 °C for the time period recommended by the manufacturer (*see* Table 5 and **Note 14**). It is not necessary to agitate the sample during incubation.
4. Centrifuge sample at 4 °C at the speed and duration recommended by the manufacturer (*see* Table 4 and **Note 15**).
5. Remove and discard the supernatant. Resuspend the pellet in a convenient volume of PBS (50–200 μL).

   Steps 6–8 are for the ExoSpin kits only

6. Remove top and bottom caps from spin column. Equilibrate column by centrifuging at 50 × *g* for 15–30 s, applying 200 μL PBS to column and repeating centrifugation.
7. Apply EV sample to prepared column and centrifuge at 50 × *g* for 60 s. Discard the eluate.
8. Place column into collection tube, apply 200 μL PBS and centrifuge at 50 × *g* for 60 s. The eluate contains vesicles.
9. Isolated vesicles may be stored at −80 °C long term (up to 1 year).

3.4 Size Exclusion Columns

The following protocol is specifically for the qEV size exclusion column (Izon Science, UK). Size exclusion allows separation of sample components based on the differential rate of movement through a gel matrix of different sized components [22]. The smaller a component, the more it is able to penetrate into the gel matrix, the longer it is retained in the column and the later the elution time [22]. In the context of EV isolation, this means that the vesicles may be separated from smaller protein and nucleic acid contaminants [12]. The qEV column has been previously used for EV isolation, and is reported in the literature to produce yields depleted from contaminants and enriched in EV markers [9, 23].

1. Remove debris from sample as described in the ultracentrifugation protocol (*see* Subheading 3.1, **steps 2–4**). For samples with a low starting vesicle concentration (e.g., cell culture media and urine), concentrate vesicles to ≤1 mL volume using one of the previously described methods (*see* Subheadings 3.1–3.3).

Table 5
Reagent–sample ratio, incubation time, and centrifugation conditions for Invitrogen Total Exosome Isolation kit, ExoQuick Exosome Precipitation Solution, and ExoSpin Exosome Purification kit, respectively

Reagent	Compatible sample type/s	Reagent–sample ratio	Incubation time (*see* Note 14)	Centrifugation speed	Centrifugation time
Invitrogen Total Exosome Isolation Kit (for cell culture media)	Cell culture media	1:2	Overnight	10,000 × *g* (*see* Note 13)	1 h
Invitrogen Total Exosome Isolation Kit (for serum)	Serum, plasma	1:5	30 min	10,000 × *g*	10 min
ExoQuick-TC Exosome Precipitation Solution (for cell culture media and urine)	Cell culture media, urine	1:5	Overnight (minimum 12 h)	1500 × *g*	30 minutes
ExoQuick Exosome Precipitation Solution (for biofluids)	Serum, ascites, other biofluids	1:4	30 minutes (serum) Overnight (ascites)	1500 × *g*	30 min
ExoSpin Exosome Purification Kit (for cell culture media)	Cell culture media, urine, saliva	1:2	At least 1 h	16,000 × *g* (*see* Note 13)	1 h
ExoSpin Blood Exosome Purification Kit (for plasma/serum)	Plasma, serum	1:2	5 min–1 h	20,000 × *g*	30 min

2. To prepare column for use, secure in an upright position using a retort stand or similar. Ensure that there is sufficient space underneath the column to place collection tubes.
3. Remove the top cap of the column by pinching it inward and lifting it slowly and carefully. Do not remove the bottom cap until ready to start collecting flow through.
4. Equilibrate the column by passing through at least 10 mL of buffer (PBS or similar). Apply this to the top of the column in multiple 1–2 mL aliquots, ensuring that the top of the column never runs dry (*see* **Note 16**).
5. Measure and record the time taken for 5 mL of buffer to pass through the column. This will serve as a reference to indicate when column needs to be cleaned. Typically this time is around 5 min for a clean column.
6. When ready to process the sample, replace the bottom cap and carefully remove any buffer from the top of the column. Pipette sample in, and remove bottom cap when ready to begin collecting.
7. The first 3 mL to elute from the column is the void volume, and this will contain molecules which are too large to enter the gel matrix (>1 μm). This can be collected as six individual 0.5 mL fractions if desired, or as one 3 mL fraction.
8. The vesicles will begin to elute after the void volume has passed through. Collect at least 5 × 0.5 mL fractions and retain. Top up column with buffer as necessary during elution, but not until the sample has completely entered the column matrix. The vesicles should primarily be in the first 1–1.5 mL to elute after the void volume (*see* **Note 17**).
9. Wash column with 10 mL of buffer. Measure the time taken for 5 mL of buffer to pass through the column. If column is clean and not compromised, this should be comparable to the measurement taken before sample processing.
10. The presence of vesicles in each fraction can be determined by physical characterization techniques such as tunable resistive pulse sensing (TRPS, *see* ref. 24 for technique details) or nanoparticle tracking analysis (NTA, *see* ref. 25 for technique details). Sample purity and protein contamination may be assessed by performing SDS PAGE with a Coomassie stain and/or a Western blot for vesicle markers (e.g., CD63, CD81 and Flotillin).
11. The vesicle-containing fractions may be pooled for further downstream analysis. Samples may be stored at −80 °C for up to 1 year.

3.5 Density Gradient Centrifugation

Density gradient centrifugation is a method which separates vesicles from other contaminants based on their buoyant density, which for exosomes is estimated to be 1.13–1.19 g/mL [1]. A discontinuous gradient is formed by layering solutions of different densities, and the sample overlaid on top. The column is centrifuged at high speed (100,000 × *g*) overnight (~16–18 h) to separate out sample components. This method has been found to produce purer yields than ultracentrifugation alone, and can be used to 'clean up' preparations of crude vesicle concentrates produced by ultrafiltration, ultracentrifugation or other methods [21, 26]. It has been noted, however, that this method cannot separate abundant high density lipoprotein (HDL) particles from extracellular vesicles in plasma and serum samples, and may therefore not be an ideal method for this sample type [12]. Further, this method is associated with a loss of vesicle yield during processing [8].

1. Prepare 40, 20, 10 and 5% iodixanol solutions by diluting the 60% w/v Optiprep stock solution with Tris–Sucrose buffer. Table 6 shows the volumes of stock solution and buffer required to produce 12 mL of each concentration.
2. Layer 3 mL of each of the 40, 20 and 10% solutions and 2.5 mL of the 5% solution in a 13–15 mL ultracentrifuge tube (*see* **Note 18**).
3. Overlay 500 μL of the sample on top of the gradient. This should be a preprocessed concentrated sample, produced by ultracentrifugation (*see* Subheading 3.1), ultrafiltration (*see* Subheading 3.2) or a polymer-based reagent (*see* Subheading 3.3) and reconstituted to approximately 500 μL with PBS or similar buffer.
4. Prepare a second blank gradient as per **step 2**, and overlay 500 μL of Tris–Sucrose buffer in place of the sample.
5. Centrifuge both sample and blank gradients at 100,000 × *g* for 16–18 h, in a swinging bucket rotor (*see* Table 2).

Table 6
Preparation of 40, 20, 10, and 5% iodixanol solutions (12 mL each) from Optiprep (60% w/v iodixanol) stock and Tris–sucrose buffer

Iodixanol concentration (w/v %)	Optiprep stock (mL)	Tris–sucrose buffer (mL)
40	8	4
20	4	8
10	2	10
5	1	11

6. Following centrifugation, collect 12 × 1 mL fractions from top to bottom (i.e., with increasing density) from sample and blank gradients. Ensure that each collected fraction is clearly labeled.
7. Using the fractions from the blank gradient, calculate the density of each fraction by diluting the solution 1:10,000 and measuring the absorbance of the solution at 244 nm with a spectrophotometer (*see* **Note 19**).
8. Centrifuge the fractions collected from the sample gradient at 100,000 × *g* and 4 °C for 1–2 h (*see* Subheading 3.1 for ultracentrifugation details). Resuspend each in 50–100 μL PBS.
9. The presence of vesicles in each fraction can be determined by physical characterization techniques such as tunable resistive pulse sensing (TRPS, *see* ref. 24 for technique details) or nanoparticle tracking analysis (NTA, *see* ref. 25 for technique details). Sample purity and protein contamination may be assessed by performing SDS PAGE with a Coomassie stain and/or a Western blot for vesicle markers (e.g., CD63, CD81 and flotillin).
10. Vesicle-containing fractions may be pooled for further downstream analysis. Samples may be stored at −80 °C for up to 1 year.

3.6 Immunoaffinity Isolation

Immunoaffinity isolation allows the separation of EV subpopulations based on the expression of surface markers. Preconcentrated EV samples are incubated with magnetic beads coated with an antibody against the target marker, allowing the specific pull-down of EVs expressing this marker on their surface. There are commercially available beads precoated with a general exosome surface marker (e.g., CD81, CD9, and CD63), and these are designed to allow separation of exosomes from other coisolated EVs. In addition, there are several reports of the purification of EVs expressing cell-specific target markers, allowing the isolation of a subpopulation of EVs derived from a particular cell type [14, 27]. The advantage of this method is that it produces pure and homogenous EV yields [28]. This method by design, however, excludes the capture of potentially biologically relevant subpopulations which do not express the particular surface marker of interest.

Steps 1–8 are for the Exosome-Streptavidin Isolation reagent only

1. Resuspend streptavidin-coated beads, either by vortexing for 30 s or by placing tube on a mixer/shaker for at least 10 min.
2. Transfer 1 mL (1×10^7) of beads to a new tube (*see* **Notes 20** and **21**).

3. Place the tube on a magnet separator for at least 1 min. Aspirate the supernatant and discard, and then remove tube from magnet.
4. To wash beads, resuspend in 1 mL isolation buffer, return tube to magnet and aspirate supernatant.
5. Remove tube from magnet, resuspend beads in 1 mL isolation buffer, and add an appropriate amount (typically ~4 μg) of biotinylated antibody. Mix well, and then incubate for 30–60 min at room temperature under gentle agitation. The recommended speed for a shaker is 650 rpm.
6. Place tube on a magnet separator for at least 1 min. Aspirate the supernatant and discard.
7. To wash beads, remove tube from the magnet, resuspend beads in 1 mL isolation buffer. Place tube on magnet for at least 1 min, aspirate the supernatant and discard. Repeat twice, for a total of three washes.
8. Remove tube from the magnet; resuspend antibody-coupled beads in 1 mL isolation buffer.

Steps 9–16 apply to all reagents

9. (a) For Exosome-Streptavidin or Human CD63 Isolation reagent: Resuspend bead mixture by vortexing for 30 s or by placing on a mixer for at least 10 min. Transfer 100 μL of this mixture to a new round or flat-bottomed tube.
   (b) For Exosome-Human CD9/CD81/EpCam Isolation reagent: Resuspend beads as per **step 9a**. Transfer 40 μL of this mixture to a new round or flat-bottomed tube.
10. Reconstitute concentrated EV sample to a total volume of 100 μL using isolation buffer. The protein content of the sample should be approximately 25 μg, as determined by Bradford assay or similar technique.
11. Add at least 500 μL of isolation buffer to the bead mixture and mix well. Place bead mixture on a magnet separator for at least 1 min, aspirate the supernatant and discard.
12. Remove the bead tube from the magnet, and add 100 μL of reconstituted EV sample. Mix sample well, and incubate overnight (at least 18 h) at 2–8 °C with gentle agitation (shaker at 650 rpm).
13. Centrifuge the sample tube for 3–5 s. For Exosome-Human CD81/CD9/EpCAM reagent, add 1 mL of isolation buffer and mix gently by pipetting. For Exosome-Streptavidin or Human CD63 Isolation reagent, add 300 μL isolation buffer and mix gently by pipetting. Do not vortex.

14. Place the sample on a magnet for at least 1 min, aspirate supernatant and discard, taking extra care not to dislodge any beads from the side of the tube.
15. Remove tube from the magnet. For Exosome-Streptavidin or Human-CD63 reagent, add 400 μL isolation buffer and mix gently as above (*see* **step 13**). For Exosome-Human CD81/CD9 or EpCAM reagent, add 500 μL isolation buffer and mix gently as above (*see* **step 13**). Place tube on a magnet for at least 1 min, aspirate supernatant and discard, again taking extra care not to dislodge beads.
16. Elute bound EVs from beads using a buffer appropriate for downstream analysis (*see* **Note 22**):
    (a) For protein extraction: Add 15–20 μL RIPA buffer, mix well and incubate at 2–8 °C (or on ice) for 15 min to complete vesicle lysis. Magnetic beads can be retrieved by placing the sample on the magnet and withdrawing the supernatant. Add appropriate sample and/or loading buffers to the supernatant and proceed to gel electrophoresis.
    (b) For RNA extraction: Add 1 mL TRIzol reagent and vortex to homogenize sample. Magnetic beads can be retrieved by placing the sample on the magnet, if desired. Proceed with RNA extraction as per manufacturer instruction.

4 Notes

1. The list of rotors and tubes given in Table 1 is by no means exhaustive; there are numerous other rotor and tube configurations which have successfully been used for EV isolation. The Beckman Coulter website contains detailed and up-to-date information on centrifuge, rotor and tube compatibilities and it is recommended to check equipment details here before beginning any ultracentrifuge protocols.
2. The qEV column is a commercially available column developed for extracellular vesicle isolation (Izon Science, UK). Other size exclusion columns which have been used in EV experiments include the Hi-Prep Sephacryl S-400 HR column (GE Healthcare Life Sciences, MA, USA) [11] and Sepharose CL-2B (Sigma-Aldrich, MO, USA) packed in a syringe to a volume of ~10 mL [8, 29]. Details of the protocols for these columns can be found in the respective publications referenced above.
3. Isolation of intact exosomes has also been performed using antibodies coupled to Protein G Dynabeads® (Thermo Fisher, MA, USA). For a description of the associated protocol, please *see* ref. 14.

4. There are numerous commercially available RNA extraction kits which could be used in place of a TRIzol-based method. Substitute TRIzol with the appropriate lysis buffer as per the kit instructions, and proceed as per manufacturer protocol. Eldh et al. [30] provide a comparison of different techniques for exosomal RNA extraction.
5. Other biofluids may require further modifications to the standard protocol. Ultracentrifugation-based protocols have previously been reported for ascites [31], human breast milk [32] and saliva [33].
6. Performing the short (<30 min) centrifugation steps at 4 °C is desirable but not necessary. In our experience, exosomes may be centrifuged at room temperature without substantial yield loss. However, for downstream applications which require intact functional exosomes (e.g., coculture assays) or where RNA or proteomic analysis is to be performed, refrigeration during all centrifuge runs would be recommended.
7. **Steps 1–3** (Subheading 3.1) of the ultracentrifugation protocol may be replaced by filtration to approximately 0.2 μm. We have observed that this substitution does not lead to a substantial reduction in yield, provided that the sample volume is ≥2 mL and the sample is not viscous.
8. Each repeat ultracentrifugation step is likely to result in some yield loss [17]. It is therefore important to consider both yield and purity requirements for isolated vesicles before deciding on the final protocol.
9. When centrifuging sample through filter in multiple aliquots, keep remaining sample on ice or at 4 °C to avoid yield loss during processing.
10. The manufacturer recommendations suggest that filters should not be centrifuged more than 4–5 times as this may cause filter clogging and decrease filtration efficiency. For very large starting sample volumes, it may therefore be necessary to split the sample across multiple filter units and pool concentrates.
11. We recommend beginning with a 10 min centrifugation speed and scaling duration up or down as necessary, as it is difficult to predict the exact behavior of individual samples within each column. We have observed that samples with a high concentration of vesicles require a longer centrifugation times to achieve the same concentration factor as those with a lower amount of vesicles. Centrifugation times may therefore need to be adjusted for different individual samples of the same type (e.g., cell culture media from two different cell lines, urine from two different individuals). Additionally, the centrifugation time may need to be increased over subsequent

centrifugation steps to produce the same concentrate volume, as filter efficiency is decreased with each centrifugation.

12. Ultrafiltration columns may also be used for buffer exchange of concentrated vesicle samples. For example, vesicles can be exchanged from phenol red-containing cell culture media to a phenol red-free buffer which will not interfere with downstream fluorescence assays. This process is similar to that described in Subheading 3.2. The major difference is that following each centrifugation step, the sample is reconstituted in the new buffer and this is repeated until the original buffer has been removed.
13. Typically, manufacturer provided protocols for proprietary reagents will include several steps for clearing samples of contaminating debris. These are generally similar to what is described in **steps 2–4** (Subheading 3.1) of the ultracentrifugation protocol, and either method may be used.
14. Extending the incubation time from that recommended by the manufacturer may slightly improve yields. Alternately, we have found that it is possible to reduce overnight incubation steps to 3–4 h with only minor yield loss, however, this is not recommended for downstream applications where maximizing vesicle, protein or nucleic acid input is critical (e.g., mass spectrometry and RNA sequencing).
15. We and others [15] have noted that increasing the centrifugation speed and/or time from that recommended by the manufacturer may improve yields.

    The standard ExoQuick-TC protocol calls for a 30 min centrifugation at 1500 × *g*. Alvarez and colleagues report an adapted method using a higher ratio of reagent to sample (from 1:5 to 1:3.33), and increasing centrifugation speed to 10,000 × *g*, which resulted in improved vesicle yield [15]. There is also a possibility that more nonvesicular debris may be coprecipitated during a longer or faster centrifugation step, and this should be considered for applications where sample purity is critical.
16. When using the column for the first time or after storage, be aware that the initial 5–6 mL to elute will contain 0.05% sodium azide and must be disposed of in accordance with local regulations.
17. The exact time for the vesicles to elute may vary. According to the manufacturer, peak elution is typically at 4 mL, but can occur anywhere between 3 and 4.5 mL.
18. The volumes given are for 13–15 mL ultracentrifuge tubes. These volumes may be scaled up or down accordingly for use in different sized tubes.

19. Iodixanol has a molar extinction coefficient of 320 L/g/cm at this wavelength [34].
20. The protocol described here to conjugate 4 μg biotinylated antibody to 1 mL (1×10^7) beads, as per manufacturer's recommendation. These volumes may be adjusted accordingly, however, it is recommended to keep the concentration of beads constant at 1×10^7 per mL to maximize the efficiency of the coupling reaction.
21. Other streptavidin coated Dynabeads may be used in place of the Exosome-Streptavidin Isolation reagent. The stock concentration of beads and their antibody binding capacity varies between individual bead types. The input of beads and antibody to the reaction will need to be scaled appropriately to account for this. We recommend starting at a bead concentration of 1 mg/mL during the reaction and adjusting as necessary.
22. The amount of material recovered after immunoaffinity isolation depends on a number of factors, including efficiency of antibody coupling to beads, composition of input sample and the method of elution. As such, achieving sufficient yields for downstream protein and RNA analysis is likely to require optimization of conditions for each individual reaction.

References

1. van der Pol E, Boing AN, Harrison P et al (2012) Classification, functions, and clinical relevance of extracellular vesicles. Pharmacol Rev 64(3):676–705. doi:10.1124/pr.112.005983
2. Anderson HC, Mulhall D, Garimella R (2010) Role of extracellular membrane vesicles in the pathogenesis of various diseases, including cancer, renal diseases, atherosclerosis, and arthritis. Lab Investig 90(11):1549–1557. doi:10.1038/labinvest.2010.152
3. Park JE, Tan HS, Datta A et al (2010) Hypoxic tumor cell modulates its microenvironment to enhance angiogenic and metastatic potential by secretion of proteins and exosomes. Mol Cell Proteomics 9(6):1085–1099. doi:10.1074/mcp.M900381-MCP200
4. Costa-Silva B, Aiello NM, Ocean AJ et al (2015) Pancreatic cancer exosomes initiate pre-metastatic niche formation in the liver. Nat Cell Biol 17(6):816–826. doi:10.1038/ncb3169
5. Hoshino A, Costa-Silva B, Shen TL et al (2015) Tumour exosome integrins determine organotropic metastasis. Nature 527 (7578):329–335. doi:10.1038/nature15756
6. Skog J, Wurdinger T, van Rijn S et al (2008) Glioblastoma microvesicles transport RNA and proteins that promote tumour growth and provide diagnostic biomarkers. Nat Cell Biol 10 (12):1470–1476. doi:10.1038/ncb1800
7. Keller S, Konig AK, Marme F et al (2009) Systemic presence and tumor-growth promoting effect of ovarian carcinoma released exosomes. Cancer Lett 278(1):73–81. doi:10.1016/j.canlet.2008.12.028
8. Muller L, Hong CS, Stolz DB et al (2014) Isolation of biologically-active exosomes from human plasma. J Immunol Methods 411:55–65. doi:10.1016/j.jim.2014.06.007
9. Lobb RJ, Becker M, Wen S et al (2015) Optimized exosome isolation protocol for cell culture supernatant and human plasma. J Extracell Vesicles 4:27031. doi:10.3402/jev.v4.27031
10. Rood IM, Deegens JK, Merchant ML et al (2010) Comparison of three methods for isolation of urinary microvesicles to identify biomarkers of nephrotic syndrome. Kidney Int 78(8):810–816. doi:10.1038/ki.2010.262
11. Nordin JZ, Lee Y, Vader P et al (2015) Ultrafiltration with size-exclusion liquid chromatography for high yield isolation of extracellular

vesicles preserving intact biophysical and functional properties. Nanomedicine 11 (4):879–883. doi:10.1016/j.nano.2015.01.003

12. Boing AN, van der Pol E, Grootemaat AE et al (2014) Single-step isolation of extracellular vesicles by size-exclusion chromatography. J Extracell Vesicles 3. doi:10.3402/jev.v3.23430
13. Yuana Y, Levels J, Grootemaat A et al (2014) Co-isolation of extracellular vesicles and high-density lipoproteins using density gradient ultracentrifugation. J Extracell Vesicles 3. doi:10.3402/jev.v3.23262
14. Mathivanan S, Lim JW, Tauro BJ et al (2010) Proteomics analysis of A33 immunoaffinity-purified exosomes released from the human colon tumor cell line LIM1215 reveals a tissue-specific protein signature. Mol Cell Proteomics 9(2):197–208. doi:10.1074/mcp.M900152-MCP200
15. Alvarez ML, Khosroheidari M, Kanchi Ravi R et al (2012) Comparison of protein, microRNA, and mRNA yields using different methods of urinary exosome isolation for the discovery of kidney disease biomarkers. Kidney Int 82(9):1024–1032. doi:10.1038/ki.2012.256
16. Momen-Heravi F, Balaj L, Alian S et al (2013) Current methods for the isolation of extracellular vesicles. Biol Chem 394(10):1253–1262. doi:10.1515/hsz-2013-0141
17. Livshts MA, Khomyakova E, Evtushenko EG et al (2015) Isolation of exosomes by differential centrifugation: theoretical analysis of a commonly used protocol. Sci Rep 5:17319. doi:10.1038/srep17319
18. Mause SF, Weber C (2010) Microparticles: protagonists of a novel communication network for intercellular information exchange. Circ Res 107(9):1047–1057. doi:10.1161/CIRCRESAHA.110.226456
19. Théry C, Amigorena S, Raposo G et al. (2006) Isolation and characterization of exosomes from cell culture supernatants and biological fluids. Curr Protoc Cell Biol. Chapter 3: Unit 3.22. doi:10.1002/0471143030.cb0322s30
20. Lane RE, Korbie D, Anderson W et al (2015) Analysis of exosome purification methods using a model liposome system and tunable-resistive pulse sensing. Sci Rep 5:7639. doi:10.1038/srep07639
21. Van Deun J, Mestdagh P, Sormunen R et al (2014) The impact of disparate isolation methods for extracellular vesicles on downstream RNA profiling. J Extracell Vesicles 3. doi:10.3402/jev.v3.24858
22. Wu CS (2003) Handbook of size exclusion chromatography and related techniques: revised and expanded. CRC Press, Boca Raton, FL
23. Clark DJ, Fondrie WE, Liao Z et al (2015) Redefining the breast cancer exosome proteome by tandem mass tag quantitative proteomics and multivariate cluster analysis. Anal Chem 87(20):10462–10469. doi:10.1021/acs.analchem.5b02586
24. Anderson W, Lane R, Korbie D et al (2015) Observations of tunable resistive pulse sensing for exosome analysis: improving system sensitivity and stability. Langmuir 31 (23):6577–6587. doi:10.1021/acs.langmuir.5b01402
25. Dragovic RA, Gardiner C, Brooks AS et al (2011) Sizing and phenotyping of cellular vesicles using nanoparticle tracking analysis. Nanomedicine 7(6):780–788. doi:10.1016/j.nano.2011.04.003
26. Kalra H, Adda CG, Liem M et al (2013) Comparative proteomics evaluation of plasma exosome isolation techniques and assessment of the stability of exosomes in normal human blood plasma. Proteomics 13 (22):3354–3364. doi:10.1002/pmic.201300282
27. Torregrosa Paredes P, Gutzeit C, Johansson S et al (2014) Differences in exosome populations in human breast milk in relation to allergic sensitization and lifestyle. Allergy 69 (4):463–471. doi:10.1111/all.12357
28. Tauro BJ, Greening DW, Mathias RA et al (2012) Comparison of ultracentrifugation, density gradient separation, and immunoaffinity capture methods for isolating human colon cancer cell line LIM1863-derived exosomes. Methods 56(2):293–304. doi:10.1016/j.ymeth.2012.01.002
29. de Menezes-Neto A, Saez MJ, Lozano-Ramos I et al (2015) Size-exclusion chromatography as a stand-alone methodology identifies novel markers in mass spectrometry analyses of plasma-derived vesicles from healthy individuals. J Extracell Vesicles 4:27378. doi:10.3402/jev.v4.27378
30. Eldh M, Lotvall J, Malmhall C et al (2012) Importance of RNA isolation methods for analysis of exosomal RNA: evaluation of different methods. Mol Immunol 50(4):278–286. doi:10.1016/j.molimm.2012.02.001
31. Runz S, Keller S, Rupp C et al (2007) Malignant ascites-derived exosomes of ovarian carcinoma patients contain CD24 and EpCAM. Gynecol Oncol 107(3):563–571. doi:10.1016/j.ygyno.2007.08.064

32. Admyre C, Johansson SM, Qazi KR et al (2007) Exosomes with immune modulatory features are present in human breast milk. J Immunol 179(3):1969–1978. doi:10.4049/jimmunol.179.3.1969
33. Sharma S, Gillespie BM, Palanisamy V et al (2011) Quantitative nanostructural and single-molecule force spectroscopy biomolecular analysis of human-saliva-derived exosomes. Langmuir 27(23):14394–14400. doi:10.1021/la2038763
34. Schroder M, Schafer R, Friedl P (1997) Spectrophotometric determination of iodixanol in subcellular fractions of mammalian cells. Anal Biochem 244(1):174–176. doi:10.1006/abio.1996.9861

Part III

Characterization of Extracellular Vesicles

Chapter 11

Characterization of Extracellular Vesicles by Surface Plasmon Resonance

Hyungsoon Im, Katherine Yang, Hakho Lee, and Cesar M. Castro

Abstract

Surface plasmon resonance (SPR) enables real-time, label-free detection of ligand binding to target receptors immobilized on a sensing surface. SPR has emerged as a promising technique for extracellular vesicle (EV) characterization with its label-free detection scheme, and exquisite sensitivity. Among the various system configurations, nanohole-based SPR sensors are of particular interest because of their simple optical setup, tunability, and scalability. Here, we describe the characterization of circulating EVs or exosomes from human clinical samples using a nanohole-based SPR sensor, named nPLEX (nano-plasmonic exosome) sensing.

Key words Surface plasmon resonance, Periodic nanoholes, Label-free detection, Protein analysis, Clinical specimens

1 Introduction

The nPLEX (nano-plasmonic extracellular vesicle) sensor comprises periodic nanohole arrays patterned in an optically opaque gold film (100–200 nm) (Fig. 1) [1]. Nanohole-based SPR sensors have been widely studied and demonstrated to confer unique advantages over prism-based conventional SPR systems [2–4]. While conventional SPR systems operate in a total internal reflection mode, named the Krestchman configuration [5], nPLEX operates in a collinear transmission mode, which allows for simple optical setup and system miniaturization (Fig. 2). In addition, the nanohole geometry is readily scalable to high-density sensor arrays for high-throughput parallel detection [6, 7].

nPLEX sensing with nanohole structures has exceptionally high sensitivity that is about 100 times better than ELISA and 10,000 times better than Western blotting analysis (Fig. 3). In addition, the label-free detection scheme allows for fast EV characterization.

Winston Patrick Kuo and Shidong Jia (eds.), *Extracellular Vesicles: Methods and Protocols*, Methods in Molecular Biology, vol. 1660, DOI 10.1007/978-1-4939-7253-1_11,

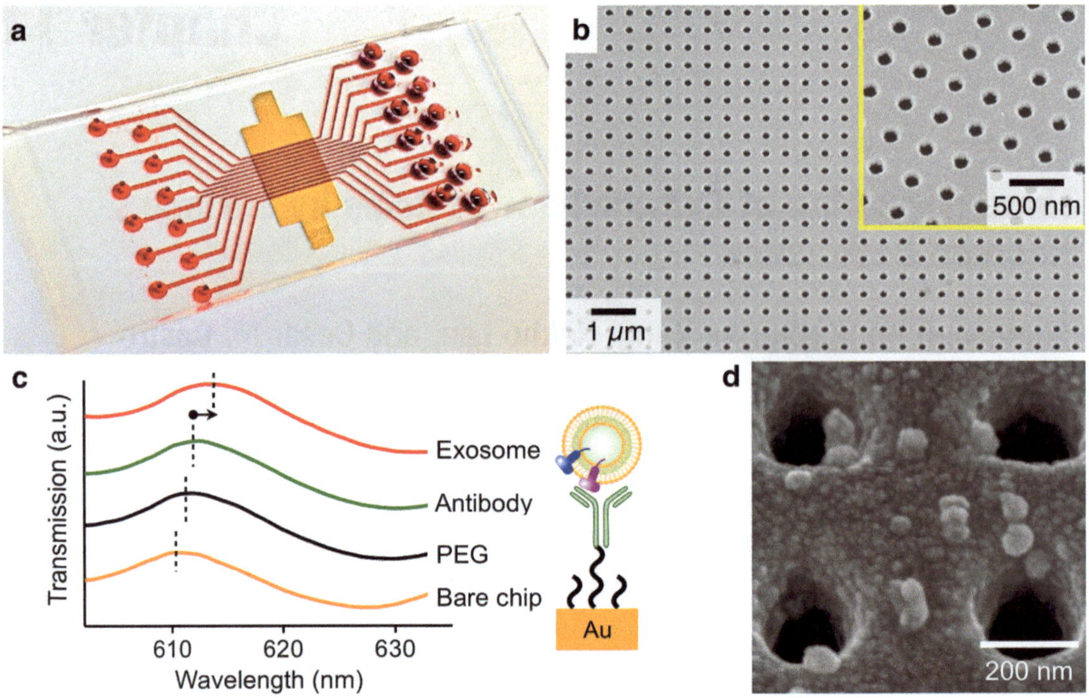

Fig. 1 Nanoplasmonic extracellular vesicle (nPLEX) system. (**a**) A photograph of nPLEX chip integrated with a multichannel microfluidic cell. (**b**) A scanning electron micrograph of nanoholes in the nPLEX sensor. The *inset* shows a zoomed-in image. (**c**) A representative schematic of changes in transmission spectra. (**d**) A scanning electron micrograph of exosomes captured on the nanohole surface (reproduced from ref. 1 with permission from Nature Publishing Group)

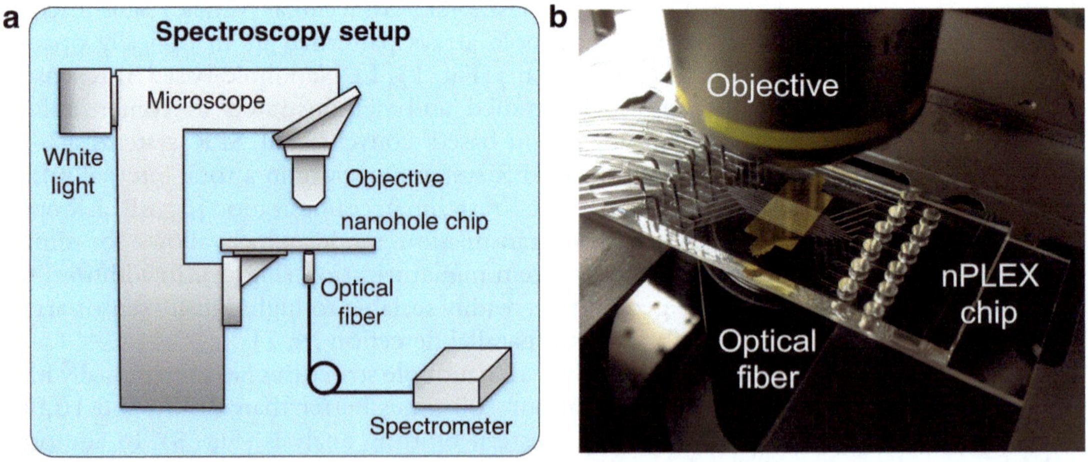

Fig. 2 A spectrometer setup was constructed on a conventional upright microscope in a schematic (**a**) and photograph. (**b**) A white light source illuminated the nPLEX sensor through a microscope objective, and the transmitted light was collected by an optical fiber placed underneath the nPLEX sensor. The collected light was then analyzed by a miniaturized spectrometer (Reproduced from ref. 1 with permission from Nature Publishing Group)

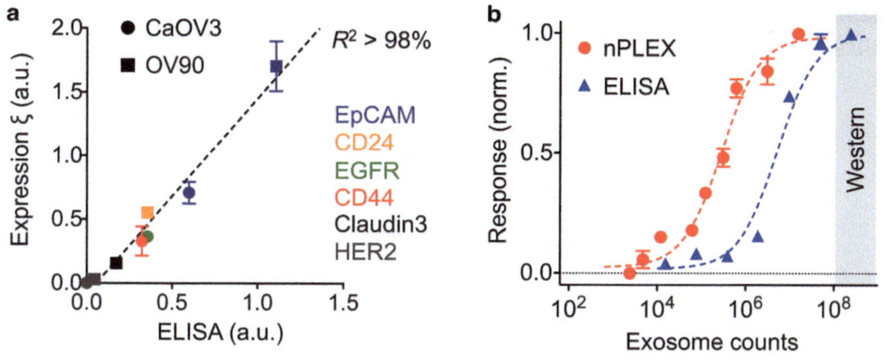

Fig. 3 (**a**) Correlation between nPLEX and ELISA measurements. (**b**) Comparison of the detection sensitivity of nPLEX and ELISA (Reproduced from ref. 1 with permission from Nature Publishing Group)

For example, the first nPLEX prototype has been used to screen 12 exosomal protein markers in less than 1 h. With its high sensitivity, nPLEX has enabled reliable detection of cancer-derived exosomes directly from clinical specimens (i.e., ascites) based on high EpCAM and CD24 expression [1].

In this chapter, we detail the nPLEX sensing method, including EV isolation from clinical specimen, chip fabrication and assay procedures.

2 Materials

These materials are necessary to isolate EVs from blood samples and process them using the nPLEX system. The same materials could be used to process other clinical sample (e.g., ascites, urine and cerebrospinal fluid) and cell culture supernatant. All reagents are to be stored as recommended by suppliers. Proper antibody storage temperature is especially important to minimize degradation.

2.1 Extracellular Vesicle Isolation from Whole Blood

1. Purple-top EDTA tubes for blood collection.
2. Ultracentrifuge tubes (Beckman 344058 or other appropriate tube for ultracentrifuge rotor).
3. Ultracentrifuge supporting up to 100,000 × *g* centrifugation speed.
4. 60 ml syringe (BD 309654).
5. 0.22 μm PVDF syringe filters (Millipore Millex®-GV 0.22 μm PVDF, SLGV033RS).
6. Sterile PBS (Corning Cellgro 21-040-CV).
7. Microcentrifuge tubes.
8. Nanoparticle tracking analysis (NTA), dynamic light scattering (DLS) or tools for EV size and concentration measurements.

2.2 Reagents for nPLEX Assays

1. Antibody (BSA-free); pre-biotinylated form could be used if available.
2. Zeba Spin Desalting Columns, 0.5 ml (Thermo Scientific, 89882 or 89883).
3. EZ-Link Sulfo-NHS-LC-Biotin (Thermo Scientific, 21335 or 21327).
4. PBS (Thermo Scientific 28372 or 28348).
5. Nanodrop or other spectrophotometer to check antibody concentration.
6. Thiol-polyethylene (PEG, MW = 1 and 0.2 kDa).
7. 1 mg/ml neutravidin.
8. 2 and 0.2% bovine serum albumin (BSA).

2.3 Fabrication of nPLEX Chip and Microfludic Flow Cell

1. Standard microscope glass slides.
2. Acetone, methanol and isopropyl alcohol.
3. Thermal or electron-beam evaporator.
4. Metal sources (gold and Ti/Cr for an adhesion layer).
5. SU-8 photoresist and SU-8 developer (MicroChem).
6. Chromium sodalime mask for microfluidic channels.
7. Spin coater and aligner for photolithography.
8. PDMS and curing agent.

2.4 nPLEX Measurement System

1. Upright microscope with a mount holder under a microscope stage.
2. Spectrometer with a working range around 500–900 nm.
3. Optical fiber and holder.
4. Software (e.g., Excel, Numbers and Matlab) that can analyze the spectrometer data.

3 Methods

All procedures are performed at room temperature unless otherwise specified.

3.1 Extracellular Vesicle Isolation from Whole Blood

1. Centrifuge tubes containing blood at 400 × *g* for 10 min (4 °C).
2. Transfer the plasma to a 15 ml tube without disturbing the buffy coat (*see* **Note 1**).
3. Centrifuge tubes at 1100 × *g* for 10 min (4 °C).
4. Transfer the plasma supernatant to a 15 ml tube and store at −80 °C or continue processing immediately (*see* **step 5**).
5. Thaw plasma in warm water.

6. Transfer to an ultracentrifuge or 50 ml tube. Dilute to 30–35 ml with PBS.
7. Centrifuge at 14,000 × *g* for 20 min.
8. Filter through a 0.22 μm PVDF syringe filter attached to a 60 ml syringe into a new ultracentrifuge tube [1] (*see* **Note 2**).
9. Centrifuge 70 min at 100,000 × *g* (4 °C).
10. Remove most of the supernatant with a 25 ml serological pipette and P1000 pipette, leaving ~100–500 μl supernatant remaining on the exosome pellet.
11. Dilute the EVs in 30–35 ml PBS (fill the ultracentrifuge tube).
12. Centrifuge 70 min at 100,000 × *g* (4 °C).
13. Remove most of the supernatant with a 25 ml serological pipette and P1000 pipette. Resuspend the EV pellet in an appropriate volume of PBS (*see* **Notes 3** and **4**) and transfer the sample to a microcentrifuge tube.
14. Aliquot an appropriate volume of EVs for NanoSight (or other) analysis and quality control.
15. Store EV samples at −80 °C until ready to use.

3.2 Antibody Biotinylation for nPLEX

1. Remove sodium azide from the antibody (if present) using a Zeba column prepared according to the manufacturer's instructions.
2. Apply antibody to the prepared column (if antibody volume is less than 70 μl, use a stacker for optimal recovery) and centrifuge at 1500 × *g* for 2 min.
3. Check the antibody concentration using a NanoDrop (or other spectrophotometer).
4. Calculate a 20-fold molar excess of Sulfo-NHS-LC-Biotin using the following calculations (for an IgG):
   (a) ml IgG × (mg IgG/ml IgG) × (mmol IgG/150,000 mg IgG) × (20 mmol Biotin/1 mmol IgG) = mmol Biotin
   (b) mmol Biotin × (1,000,000 μl/l) × (l/10 mmol) = μl Biotin Reagent
5. Prepare Sulfo-NHS-LC-biotin by making a 10 mM stock or adding 180 μl ultrapure water to a no-weigh vial.
6. Add the appropriate volume of Biotin reagent to the antibody solution. Mix gently by pipetting up and down.
7. Incubate at room temperature for 30 min.
8. During the incubation, prepare another Zeba column to remove the excess biotin.
9. Apply the antibody to the Zeba column and centrifuge at 1500 × *g* for 2 min.

10. Check the antibody concentration using a nanodrop (or other spectrophotometer).

3.3 nPLEX Chip Fabrication

Various fabrication methods are developed to pattern periodic nanohole structures, including focused ion-beam milling [8, 9], electron-beam lithography [10], nanosphere lithography [11, 12], nano-imprint lithography [13], interference lithography [10, 14, 15] and template stripping [16, 17]. This section focuses on a method based on ion-beam milling. Please refer to the aforementioned literature for alternative methods.

1. Clean glass slides using acetone, methanol and isopropyl alcohol for 5 min each (*see* **Note 5**). Alternatively, piranha cleaning using a mixture of sulfuric acid and hydrogen peroxide can be used (*see* **Note 6**).
2. Following washing with deionized (DI) water, dry the slides with nitrogen stream and leave them on a hot plate at 180 °C for 5 min (*see* **Note 7**).
3. Deposit a 200 nm gold film with a 5 nm-thick adhesion layer (Ti or Cr) via metal evaporation (*see* **Note 8**).
4. Use focused ion-beam (FIB) to pattern periodic nanoholes in the gold film (*see* **Note 9**). Typical hole size and hole-to-hole distances (i.e., periodicity) are 200 and 450 nm, respectively (*see* Fig. 1b). Smaller holes or longer periodicity can also be used (*see* **Note 10**).
5. Inspect nanohole patterns using scanning electron microscopy (*see* **Note 11**).

3.4 Microfluidic Flow Cell Fabrication

1. Prepare a chrome mask for patterning SU-8 template using optical lithography (*see* **Note 12**).
2. Remote native oxide on a silicon wafer using a diluted hydrofluoric acid solution (*see* **Note 13**).
3. Spin-coat SU-8 resist on a silicon wafer and bake the wafer on a hot plate or in an oven (*see* **Note 14**).
4. Expose the SU-8 through the chrome mask. This will transfer patterns in the mask to the SU-8 (*see* **Note 15**).
5. After post-exposure baking, develop the exposed resist using a SU-8 developer.
6. Rinse the developed wafer using isopropyl alcohol and dry the wafer with nitrogen stream (*see* **Note 16**).
7. Put the silicon wafer with a few drops of silane solution in a vacuum desiccator and leave for 30 min. The silane solution will be vaporized and immobilized on the silicon surface. PDMS is easily peeled off from the wafer with the surface treatment.

8. Mix PDMS with a curing agent in a 10:1 mass ratio (e.g., 30 g of PDMS and 3 g of curing agent). Leave the PDMS mixture in a vacuum desiccator for 30 min to degas the mixture.
9. Pour the degassed PDMS onto the silicon wafer and bake at 60 °C overnight (*see* **Note 17**).
10. Cut the PDMS using a razor blade and peel off the PDMS piece out of the silicon wafer. The SU-8 patterned wafer can be reused multiple times.
11. For permanent bonding of the flow cell on the nPLEX chip, treat the PDMS and nPLEX chip with oxygen plasma and leave on a hot plate at 60 °C for 10 min after bonding to each other (*see* Fig. 1a).

3.5 nPLEX Assay

1. Incubate the chip surface with thiolated PEG overnight.
2. Prepare 50 μg/ml neutravidin with a 2 mg/ml BSA in PBS solution and incubate the chip with the neutravidin solution for 30 min. The neutravidin acts as a linker to capture biotinylated antibodies on the peglyated gold surface.
3. Immobilize biotinylated antibodies (10 μg/ml) on desired nanohole arrays for 30 min.
4. Introduce isolated EVs on the antibody coated surface. Measure transmission spectra before and after EV binding and calculate spectral shifts upon EV binding (*see* **Note 18** and Fig. 1c).

4 Notes

1. Serum can also be used for EV isolation (*see* ref. 18)
2. Multiple filters may be necessary to process one sample, as the 0.22 μm filters clog easily.
3. A pellet of EVs may be visible at this step, depending on the volume and type of the sample.
4. It may be useful to resuspend the sample in a consistent volume, so that all samples processed for a given study are in an equal final volume.
5. Wash the slides with acetone first and followed by methanol and isopropyl alcohol. Doing sonication, gentle shaking, or washing with squeeze bottles would be suggested.
6. Be careful handling the solution. Upon adding hydrogen peroxide to sulfuric acids, the solution is heated quickly. Wash the slides with deionized water multiple times.
7. Drying on a hot plate will ensure the glass substrates are solvent-free. The temperature and drying time can vary.

8. Because the adhesion of gold to a glass surface is poor, you need to use an adhesion layer. Chromium and titanium are common materials for the adhesion layer. A thinner gold (e.g., 100 nm) or an adhesion layer (e.g., 2 nm) can be also used. Less than 10 nm of adhesion layer is desired as Cr or Ti dampens the surface plasmons.
9. Follow manufacture instructions to determine optimal beam voltage and current. Usually higher voltages reduce the patterning time, but also sacrifice the patterning resolution.
10. While the periodicity is a major parameter to determine the resonance wavelength [19], the hole size, lattice arrangement, and film thickness also affect the resonance wavelength [20]. Computer simulation (finite-difference time-domain, FDTD) is useful to optimize the dimension parameters.
11. Measuring the transmission spectra is another way to confirm the quality of nanohole arrays.
12. A film mask can be used for large patterns over 500 μm. A CAD program (e.g., AutoCAD) is often used to draw the mask design.
13. Native oxide on a silicon wafer results in poor adhesion of SU-8 on the silicon wafer. Silicon wafers are intrinsically hydrophobic, but become hydrophilic as the native oxide grows. HMDS can also be used to improve the SU-8 adhesion.
14. Refer to manufacturer instructions for spin-coat time, baking condition, exposure dose and development time. Those parameters vary depending on the thickness of SU-8 layers (i.e., channel thickness of flow cell in the microfludic chip).
15. SU-8 is a negative photoresist; the area exposed to UV light (365 nm) is hardened and remains after development.
16. If SU-8 is underdeveloped, white residue will be seen when washed with isopropyl alcohol. In this case, develop the patterns a bit longer and check with isopropyl alcohol again.
17. Shorter baking time is required when baked at higher temperature.
18. In each step, wash out the reagents with a PBS solution after incubation and perform spectral measurements.

Acknowledgment

This work was supported in part by US NIH Grants R01-HL113156 (to H.L.), K12CA087723-11 A1 (to C.M.C.) and 1K99CA201248-01 (to H.I.), the Massachusetts General Hospital

Physician Scientist Development Award (to C.M.C.), and the Department of Defense Ovarian Cancer Research Program Award W81XWH-14-1-0279 (to H.L.)

References

1. Im H, Shao H, Park YI, Peterson VM, Castro CM, Weissleder R, Lee H (2014) Label-free detection and molecular profiling of EVss with a nano-plasmonic sensor. Nat Biotechnol 32:490–495
2. Brolo AG (2012) Plasmonics for future biosensors. Nat Photonics 6:709–713
3. Homola J (2008) Surface plasmon resonance sensors for detection of chemical and biological species. Chem Rev 108:462–493
4. Rupert DL, Lasser C, Eldh M, Block S, Zhdanov VP, Lotvall JO, Bally M, Hook F (2014) Determination of EV concentration in solution using surface plasmon resonance spectroscopy. Anal Chem 86:5929–5936
5. Homola J (2003) Present and future of surface plasmon resonance biosensors. Anal Bioanal Chem 377:528–539
6. Im H, Lesuffleur A, Lindquist NC, Oh SH (2009) Plasmonic nanoholes in a multichannel microarray format for parallel kinetic assays and differential sensing. Anal Chem 81:2854–2859
7. Lindquist NC, Lesuffleur A, Im H, Oh SH (2009) Sub-micron resolution surface plasmon resonance imaging enabled by nanohole arrays with surrounding Bragg mirrors for enhanced sensitivity and isolation. Lab Chip 9:382–387
8. Brolo AG, Gordon R, Leathem B, Kavanagh KL (2004) Surface plasmon sensor based on the enhanced light transmission through arrays of nanoholes in gold films. Langmuir 20:4813–4815
9. Lesuffleur A, Im H, Lindquist NC, Oh S-H (2007) Periodic nanohole arrays with shape-enhanced plasmon resonance as real-time biosensors. Appl Phys Lett 90:243110
10. Yanik AA, Cetin AE, Huang M, Artar A, Mousavi SH, Khanikaev A, Connor JH, Shvets G, Altug H (2011) Seeing protein monolayers with naked eye through plasmonic Fano resonances. Proc Natl Acad Sci U S A 108:11784–11789
11. Dahlin AB, Jonsson MP, Hook F (2008) Specific self-assembly of single lipid vesicles in nanoplasmonic apertures in gold. Adv Mater 20:1436–1442
12. Lee SH, Bantz KC, Lindquist NC, Oh S-H, Haynes CL (2009) Self-assembled plasmonic nanohole arrays. Langmuir 25:13685–13693
13. Wittenberg NJ, Im H, Xu X, Wootla B, Watzlawik J, Warrington AE, Rodriguez M, Oh S-H (2012) High-affinity binding of remyelinating natural autoantibodies to myelin-mimicking lipid bilayers revealed by nanohole surface plasmon resonance. Anal Chem 84:6031–6039
14. Henzie J, Lee MH, Odom TW (2007) Multiscale patterning of plasmonic metamaterials. Nat Nanotechnol 2:549–554
15. Menezes JW, Ferreira J, Santos MJL, Cescato L, Brolo AG (2010) Large-area fabrication of periodic arrays of nanoholes in metal films and their application in biosensing and plasmonic-enhanced photovoltaics. Adv Funct Mater 20:3918–3924
16. Im H, Lee SH, Wittenberg NJ, Johnson TW, Lindquist NC, Nagpal P, Norris DJ, Oh S-H (2011) Template-stripped smooth Ag nanohole arrays with silica shells for surface plasmon resonance biosensing. ACS Nano 5:6244–6253
17. Im H, Sutherland JN, Maynard JA, Oh S-H (2012) Nanohole-based surface plasmon resonance instruments with improved spectral resolution quantify a broad range of antibody-ligand binding kinetics. Anal Chem 84:1941–1947
18. Thery C, Amigorena S, Raposo G, Clayton A (2006) Isolation and characterization of exosomes from cell culture supernatants and biological fluids. Curr Protoc Cell Biol Chapter 3:Unit 3.22
19. Ebbesen TW, Lezec HJ, Ghaemi HF, Thio T, Wolff PA (1998) Extraordinary optical transmission through sub-wavelength hole arrays. Nature 391:667–669
20. Genet C, Ebbesen TW (2007) Light in tiny holes. Nature 445:39–46

Chapter 12

Extracellular Vesicle Isolation and Analysis by Western Blotting

Emma J. K. Kowal, Dmitry Ter-Ovanesyan, Aviv Regev, and George M. Church

Abstract

Extracellular vesicles (EVs) are released by mammalian cells and are thought to be important mediators of intercellular communication. There are many methods for isolating EVs from cell culture media, but the most popular methods involve purification based on ultracentrifugation. Here, we provide a detailed protocol for isolating EVs by differential ultracentrifugation and analyzing EV proteins (such as the tetraspanins CD9, CD63 and CD81) by western blotting.

Key words Exosomes, Extracellular vesicles, Exosome isolation, Extracellular vesicle isolation, Ultracentrifugation, Immunoblotting, Western blot, Tetraspanins, CD63, CD81, CD9

1 Introduction

Extracellular vesicles (EVs) are involved in intercellular communication. All mammalian cell types secrete heterogeneous vesicles of a variety of sizes [1]. Here, we focus on small EVs, which are often called exosomes. Although the exact definition of exosomes varies in different papers and remains debated [2], we define an exosome as any EV between 30 and ~200 nanometers (nm). The upper limit is imposed by using a 0.22 μm filter.

There are many different protocols for isolating EVs, which yield vesicles of varying purity (all protocols, to some degree co-isolate soluble proteins) [3]. We describe an ultracentrifugation-based protocol, which remains one of the most commonly used protocols for isolating extracellular vesicles from cell culture media [4]. There are many variations on this protocol, and our protocol includes a 0.22 μm filtration step. This cutoff provides a specific size

Emma J.K. Kowal and Dmitry Ter-Ovanesyan contributed equally to this work.

Winston Patrick Kuo and Shidong Jia (eds.), *Extracellular Vesicles: Methods and Protocols*, Methods in Molecular Biology, vol. 1660, DOI 10.1007/978-1-4939-7253-1_12, © Springer Science+Business Media LLC 2017

range for the isolated extracellular vesicles, which enhances reproducibility in EV analysis across cell types or conditions.

EVs can be characterized by a wide variety of techniques, including profiling the proteins they contain. Tetraspanins are a group of proteins that contain four transmembrane domains and certain characteristic features. The tetraspanins CD9, CD63 and CD81 are transmembrane proteins that are commonly found in extracellular vesicles across cell types [5]. In addition to our EV isolation protocol, we describe a detailed protocol for analysis of these characteristic EV markers by western blotting.

2 Materials

Store all materials at room temperature unless otherwise stated.

2.1 Cell Culture

1. Cells and cultureware.
2. Fetal bovine serum (FBS)-depleted media or defined media without FBS (*see* **Note 1**).

2.2 EV Isolation

1. PBS without Ca^{2+}/Mg^{2+}.
2. HEPES buffer (optional) (*see* **Note 2**).
3. 50 mL Falcon tubes (Fisher Scientific).
4. 0.22 μm Steriflip filter tubes (Fisher Scientific).
5. Ultracentrifuge and rotor.
6. Polyallomer ultracentrifuge tubes (Beckman Coulter).

2.3 Western Blot

1. Sample Buffer: Bolt 4× LDS Sample Buffer (Thermo Fisher Scientific). Store at 4 °C.
2. Bolt 10× Reducing Buffer (optional) (Thermo Fisher Scientific) (For proteins that require reducing conditions, not tetraspanins; *see* **Note 3**).
3. RIPA buffer (optional) (*see* **Note 4**).
4. A660 or BCA protein quantification assay (optional) (*see* **Note 5**).
5. Running buffer: 100 mL 20× MES SDS running buffer (Thermo Fisher Scientific), 1900 mL deionized water.
6. 4–12% Bis-Tris Gels (Thermo Fisher Scientific). Store at 4 °C.
7. Gel tank (XCell SureLock® Mini, Thermo Fisher Scientific) and electrophoresis equipment.
8. MagicMark XP Western protein standard (Thermo Fisher Scientific). Store at −20 °C.
9. SeeBluePlus2 protein ladder (Thermo Fisher Scientific). Store at 4 °C.

10. XCell II Blot Module and sponges (Thermo Fisher Scientific) (*see* **Note 6**).
11. Methanol.
12. Transfer buffer: 100 mL Bolt 20× transfer buffer (Thermo Fisher Scientific), 400 mL methanol, 1500 mL deionized water.
13. PVDF or nitrocellulose membranes (Thermo Fisher Scientific).
14. Milk powder.
15. Tween 20.
16. PBST: PBS with 0.1% vol/vol Tween 20. Store at 4 °C.
17. Cold room with a tilting rocker (not orbital).
18. Plastic containers to hold membranes, such as PerfectWestern™ containers.
19. Flat tweezers for handling membranes.
20. Antibodies to proteins of interest.
21. HRP-conjugated secondary antibody for visualization (Bethyl Laboratories).
22. HRP substrate, such as SpectraQuant™-HRP CL Chemiluminescent detection reagent (BridgePath Scientific).

3 Methods

3.1 EV Isolation

1. Culture cells under standard conditions to 50–70% confluency.

Day 1

For suspension cells:

1. Spin down desired total number of cells (*see* **Note 7**) in six 50 mL Falcon tubes at 300 × *g* for 5 min.
2. Aspirate media and resuspend each cell pellet in 40 mL FBS-depleted or defined media without FBS (*see* **Note 1**). Transfer contents of each Falcon tube to T75 flask and return to incubator.

For adherent cells:

1. Aspirate media from twelve 15 cm plates.
2. Add 20 mL FBS-depleted or defined media without FBS per plate (*see* **Note 1**). Return cells to incubator.

Day 2

1. After 24 h, take off all media and divide among 50 mL falcon tubes.
2. Spin at 300 × *g* for 10 min at RT (to pellet the cells).

3. Transfer supernatant to new 50 mL tubes leaving cell pellet behind. If cell protein is to be analyzed alongside EVs, one cell pellet can at this step be resuspended in the desired lysis buffer (*see* **Notes 4** and **5**).
4. Spin at 2000 × *g* for 10 min at RT (to pellet any dead cells).
5. Transfer supernatant to new 50 mL tubes leaving cell pellet behind.
6. Spin supernatant at 16,500 × *g* for 20 min at 4 °C (to pellet large EVs).
7. Transfer supernatant to new 50 mL tubes, leaving pellet behind.
8. Pass supernatant through Steriflip 0.22 μm filter.
9. Transfer supernatant to polyallomer ultracentrifuge tubes. Centrifuge at 120,000 × *g* (26,500 RPM with SW32Ti rotor) for 70 min at 4 °C.
10. Remove most of supernatant, leaving ~2 cm of media above pellet. Add 5 mL PBS to each tube. Vortex on medium speed for a few seconds. Fill to top of each tube with PBS.
11. Again, centrifuge at 120,000 × *g* for 70 min at 4 °C.
12. Aspirate all of supernatant with Pasteur pipet without touching bottom of tube where pellet is located (*see* **Note 8**).
13. Resuspend pellet either in PBS or directly in the desired lysis buffer for western blot (*see* **Notes 4** and **5**).

3.2 Western Blot

14. Add 100 μL 1× Sample Buffer to each pellet, or add 4× sample buffer to a final concentration of 1× (i.e., add 25 μL 4× sample buffer to 75 μL sample) in a pre-isolated EV sample. Vortex on high speed to mix. If in ultracentrifuge tubes pipet up and down to further disrupt pellet, then transfer to Eppendorf tubes.
15. Incubate at 70 °C for 10 min.
16. Make 2 L of 1× MES SDS running buffer (100 mL buffer into 1900 mL deionized water).
17. Make 5% milk buffer: 2.5 g dried milk into 50 mL PBST (PBS + 0.1% v/v Tween 20). Tumble at 4° for an hour (*see* **Note 9**).
18. Prepare 4–12% bis-tris gel in gel tank. Add 1× MES running buffer to top of gel. Do not forget to rinse wells.
19. Load wells. For ladder use 3 μL MagicMark + 6 μL SeeBlue-Plus2, in separate lanes if possible (*see* **Note 10**).
20. Run 40 min at 150 V, 22 min at 200 V, or until blue dye reaches gel foot.

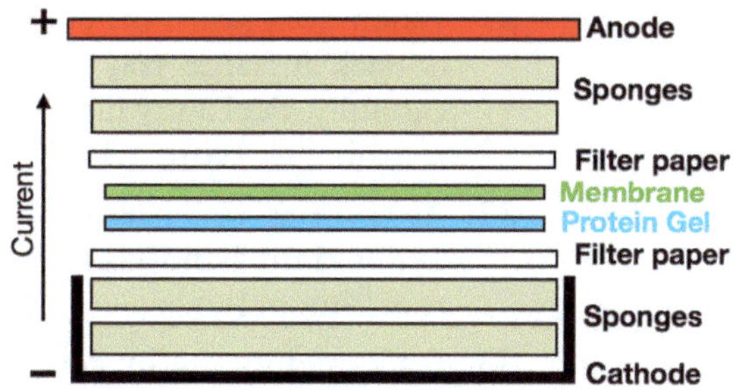

Fig. 1 Schematic of membrane transfer sandwich. This should be constructed from the bottom (cathode) up, using as many sponges as necessary to put even pressure across the surface of the membrane with no air bubbles, and all internal components except gel pre-soaked in transfer buffer

21. While gel is running, prepare for transfer (see XCell surelock manufacturer instructions [6] for more detail).
22. Make 2 L transfer buffer (100 mL Bolt 20× transfer buffer, 400 mL methanol, 1500 mL water).
23. If using PVDF membranes, place one membrane in empty tip box lid with a few milliliters methanol to activate it. Rinse several times with transfer buffer, dumping excess into large sandwich making tray, then rock gently in hand for several minutes. For nitrocellulose membranes, simply soak in transfer buffer for a minute.
24. Soak sponges in transfer buffer, squeezing out bubbles as much as possible (*see* **Note 11**). Briefly immerse filter papers in transfer buffer as well.
25. Build sandwich up from the bottom in the following order: anode, sponges, filter paper, gel, transfer membrane, filter paper and sponges (*see* Fig. 1, **Note 12**).
26. When sandwich is ready for gel, take gel out of tank and rinse it off. Crack open plastic casing. Cut off wells and foot so that remainder is completely flat and lay carefully on filter paper.
27. Squeeze sandwich together in holder and insert into gel tank. If reusing the same tank make sure to pour out gel running buffer and rinse with deionized water.
28. Use fresh transfer buffer to fill in sandwich from the top. Open and close clamp several times to let the buffer soak down through.
29. Fill the rest of the gel box with deionized water, which will serve as a heat sink.
30. Put on lid and run 1.5–2 h at 30 V, tapping firmly on occasion to remove bubbles (*see* **Note 11**).

31. When done, turn off current, pull out sandwich in holder, put it back in large tray (minus transfer buffer) then unpack it carefully. Peel away filter paper very slowly to check for protein transfer (*see* **Note 13**).
32. As soon as you peel off membrane, take a blade and cut off upper right hand corner to mark "top" face (face which was touching gel).
33. Place membrane in PerfectWestern box containing 5–10 mL milk buffer (as much as necessary to cover membrane completely). If using PVDF membranes, ensure that the membrane does not dry out at any step.
34. To block, place membrane in milk buffer on rocker in the cold room and let rock at least half an hour. Conduct all further steps in the cold room if possible.
35. After at least half an hour, pour off block and add 10 mL primary antibody diluted 1:1000 in milk buffer (*see* **Note 14**). Leave overnight rocking in the cold room.

Day 3:

1. Pour off primary, take PBST, pour in, swish, pour off, 2×, then do three washes in PBST of ~10 min each, rocking in the cold room.
2. Add 10 mL secondary antibody diluted 1:2000 in milk buffer (*see* **Note 14**). Leave rocking in the cold room for at least 1 h.
3. Pour off secondary, take PBST, pour in, swish, pour off, 2×, then do three washes in PBST of ~10 min each, rocking in the cold room.
4. Bring the membrane (in fresh PBST), equal volumes of each component of HRP substrate (reagent A and B; *see* Subheading 2.3, **item 21**) and an empty falcon tube to the imaging stage.
5. Mix reagent A and B together immediately before use. Pour PBST off membrane and pour A/B mix on. Let sit for a minute then image, using tweezers to handle membrane (*see* **Note 15**).

4 Notes

1. Since fetal bovine serum contains bovine EVs, it is important for downstream analysis that media from which EVs will be isolated is either FBS-free or has been depleted of vesicles by overnight ultracentrifugation at 120,000 × *g*. A convenient formulation is to make media with 2× FBS and ultracentrifuge it overnight, then remove and keep the supernatant, diluting it 1:1 in the base media to bring it to 1×. Some cells will still not like this media and so we advise collecting EVs for 24 h.

2. For storage of EVs at −80 °C we recommend the addition of HEPES buffer to a final concentration of 20 mM to stabilize pH over freeze–thaw cycles (to PBS or other buffers).
3. Protein gel electrophoresis can be either denaturing or non-denaturing ("native," i.e., retaining the original folded structure) and either reducing (where Cys-Cys disulfide bonds are specifically broken) or nonreducing. Though reducing can help to solubilize a concentrated or complex sample, tetraspanins such as CD63, CD81 and CD9 require nonreducing electrophoresis for western blotting, as the epitope recognized by antibodies to these proteins usually relies on several disulfide bonds to fold properly and be recognized.
4. Transmembrane proteins, particularly those with four or more membrane-spanning regions, can be difficult to extract from lysates. We have had success extracting tetraspanins with LDS sample buffer alone (Subheading 2.3, **item 1**) but other proteins may require some optimization of lysis buffer for efficient extraction. RIPA buffer is one of the harsher common buffers and is well suited for this purpose. When extracting membrane proteins from cells, it is often helpful to centrifuge the lysate at high speed (>12,000 × *g*) for 10 min and take the supernatant, leaving behind the membrane and insoluble material which can interfere with electrophoresis.
5. Many common protein quantification assays (such as A660 and BCA) rely on a colorimetric readout, and are thus incompatible with the bromophenol blue-containing LDS sample buffer. This protocol does not explicitly describe how to quantify protein in a lysate, but note that if you do wish to quantify the protein in your samples, you should lyse the cells or vesicles in RIPA (Subheading 2.3, **item 2**) or another clear buffer, quantify, and then add 4× LDS sample buffer to 1× concentration prior to immunoblotting.
6. The reagents listed as subheading 2.3, **items 10–13** are required for a traditional wet transfer of proteins to a membrane. These can be substituted with other materials of your choice for dry or semidry transfer. For example, we have found the iBlot dry blotting system from Thermo Fisher is convenient and effective, though not all labs may have the required equipment.
7. The total number of cells per isolation should be determined by the total volume of media from which you are able to isolate EVs. The limiting factor will likely be the volume capacity of your ultracentrifuge tubes (e.g., the SW32Ti rotor can hold six tubes with a volume of ~38 mL each, so the max volume per isolation is 228 mL). Start with a few extra milliliters of media per flask to account for some loss throughout the

centrifugation steps and culture the number of cells necessary to achieve 50–70% confluence in this volume.

8. The pellet at this stage will most likely not be visible. It is possible to remove all but 20–30 μL of the supernatant by tilting the tube to pool the liquid on one side and carefully avoiding touching the center of the tube bottom. We have also found that it is helpful to remove all but ~2 cm of supernatant and wait 30 s before aspirating the final few milliliters, as otherwise some liquid clings to the sides of the tube and makes the final residual volume >50 μL.

9. The proteins in the milk buffer associate with proteins in the membrane and block nonspecific antibody interactions. There are many formulations of blocking solution available but we have found milk to be cheap and effective. It is important to make this buffer fresh (it should be a few days old at most and stored at 4 °C with rotation).

10. MagicMark XP is a protein standard ladder containing IgG binding sites (you will see it on the final western blot, not in the gel), while SeeBlue is a prestained protein standard ladder which you should see in the gel and membrane but not in the final blot. These can be mixed if necessary but will run better in separate wells. SeeBlue is useful for evaluating how far the gel has run and if the transfer was successful (*see* **Note 13**) as well as for horizontally cutting the membrane in order to blot for proteins of different molecular weights, e.g., CD63 and CD81 (*see* Fig. 2).

11. Air bubbles anywhere in the sandwich can prevent successful transfer of proteins to the membrane in that spot, so it's important to squeeze the sandwich tightly and firmly tap the XCell mini tank periodically (as many times as is convenient) while transfer occurs.

12. Use as many sponges as necessary to form a tight sandwich. Generally at least three sponges on either side of the gel and membrane (six total) will suffice, but the tighter the better. *See* Fig. 1 for schematic.

13. Carefully peel away the top corner of the membrane closest to where the SeeBlue ladder was run and check for the location of the colored bands. If the transfer worked, some or all of them should now be on the membrane instead of the gel. Specifically, check that the SeeBlue bands in the molecular weight range of your protein of interest (for example, the 28 kDa band is close to the size of CD81) are on the membrane. If they are still on the gel, you can carefully reconstruct the sandwich (ensure that the gel and membrane do not shift relative to one another) and run it slightly longer. Keep in mind that running the transfer for too long will cause the lower molecular weight bands to

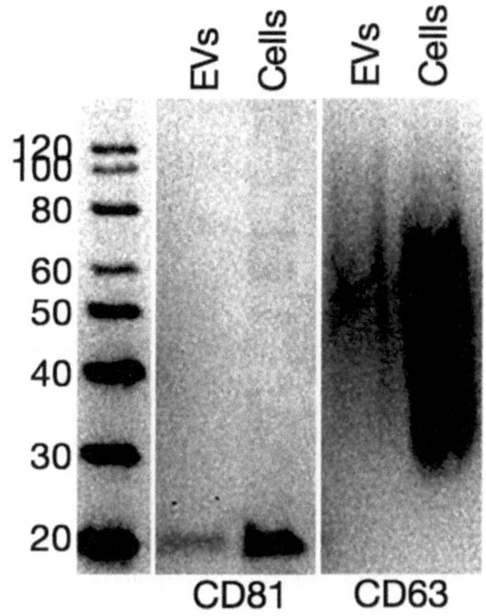

Fig. 2 Western blot showing CD81 and CD63 blotting on K562 EV lysate (7 μg) and cell lysate (86 μg), resuspended in RIPA and quantified before addition of LDS Sample Buffer. Note that CD63 appears as a smear in the range of 30–60 kDa; this is due to heavy glycosylation and is expected [7]. Each membrane was blocked for 30 min at 4 °C in milk buffer, then incubated with primary antibody diluted in 10 mL milk buffer for 12 h (1:1000 dilution mouse anti-human CD81, clone M38, or 1:1000 dilution mouse anti-human CD63, clone TS63), washed three times for 10 min each in PBST, then incubated for 2 h with 10 mL secondary antibody (Rabbit anti-mouse HRP, 1:2000 dilution in milk buffer), washed three times for 10 min each in PBST, then imaged on a Bio-Rad ChemiDoc MP system with SpectraQuant'™-HRP CL Chemiluminescent detection reagent

pass through the membrane onto the filter paper, at which point they cannot be recovered.

14. As different antibodies have different affinities for their targets, it is often necessary to experimentally determine the optimal antibody dilutions for immunoblotting. Generally these fall within 1:100 and 1:5000 and are lower (i.e., more dilute) for the secondary antibody. We recommend starting with a higher dilution (more concentrated) to ensure a strong signal and diluting further as necessary to eliminate background or conserve reagents.
15. If using Image Lab software to visualize the blot, it can be set to "signal accumulation mode" to determine optimal exposure time by monitoring blot over the course of imaging.

Acknowledgment

This work was supported by US National Institutes of Health National Human Genome Research Institute grant P50 HG005550.

References

1. Tkach M, Thery C (2016) Communication by extracellular vesicles: where we are and where we need to go. Cell 164(6):1226–1232
2. Gould SJ, Raposo G (2013) As we wait: coping with an imperfect nomenclature for extracellular vesicles. J Extracell Vesicles 2. doi:10.3402/jev.v2i0.20389
3. Witwer KW, Buzas EI, Bemis LT, Bora A, Lasser C, Lotvall J, Nolte-'t Hoen EN, Piper MG, Sivaraman S, Skog J, Thery C, Wauben MH, Hochberg F (2013) Standardization of sample collection, isolation and analysis methods in extracellular vesicle research. J Extracell Vesicles 2. doi:10.3402/jev.v2i0.20360
4. Thery C, Amigorena S, Raposo G, Clayton A (2006) Isolation and characterization of exosomes from cell culture supernatants and biological fluids. Curr Protoc Cell Biol Chapter 3:Unit 3.22
5. Kowal J, Arras G, Colombo M, Jouve M, Morath JP, Primdal-Bengtson B, Dingli F, Loew D, Tkach M, Thery C (2016) Proteomic comparison defines novel markers to characterize heterogeneous populations of extracellular vesicle subtypes. Proc Natl Acad Sci U S A 113(8):E968–E977
6. XCell SureLock® Mini-Cell. https://tools.thermofisher.com/content/sfs/manuals/surelock_man.pdf. Accessed 27 Apr 2016

7 Ageberg M, Lindmark A (2003) Characterisation of the biosynthesis and processing of the neutrophil granule membrane protein CD63 in myeloid cells. Clin Lab Haematol 25(5):297–306

Chapter 13

Analysis of Extracellular Vesicles Using Fluorescence Nanoparticle Tracking Analysis

Pauline Carnell-Morris, Dionne Tannetta, Agnieszka Siupa, Patrick Hole, and Rebecca Dragovic

Abstract

Fluorescence nanoparticle tracking analysis (fl-NTA) allows for accurate sizing, counting, and phenotyping of extracellular vesicles (EV). Here, we present two protocols for the analysis of EVs using fl-NTA, highlighting the potential pitfalls and challenges. The first protocol utilizes CellMask Orange™ (CMO) as a general membrane marker to label EVs derived from plasma. The second protocol describes the use of a Qdot-conjugated antibody to identify syncytiotrophoblast (STB)-derived EVs. "Standard" preparations of STB-derived EVs enriched for either microvesicles (STBMV) or exosomes (STBEX), containing a known amount of EV positive for the STB specific antigen placental alkaline phosphatase (PLAP), were also used to optimize fl-NTA camera settings.

Key words Extracellular vesicles, Exosomes, Microvesicles, Fluorescence nanoparticle tracking analysis, Quantum dots

1 Introduction

Robust characterization of the cell-derived extracellular vesicles (EV), exosomes (~50–150 nm) and microvesicles (~100–1000 nm), is integral to understanding their role in both physiological and pathological processes. This requires techniques that are able to accurately size and phenotype membranous vesicles ranging from ~50 to 1000 nm in diameter. Presently, methods have to be adapted to try to meet the technical difficulties associated with detecting and phenotyping nanosized vesicles. Methodologies employed to analyze both the size and phenotype of EV in biological samples include flow cytometry, electron microscopy (EM) and cryo-EM (as reviewed in [1–3]). Of these, flow cytometry is the most widely used as it is able to quickly and accurately phenotype large numbers of individual EV by detecting bound fluorescently labeled antibodies. However, most flow cytometers are designed to detect cells and thus

Winston Patrick Kuo and Shidong Jia (eds.), *Extracellular Vesicles: Methods and Protocols*, Methods in Molecular Biology, vol. 1660, DOI 10.1007/978-1-4939-7253-1_13, © Springer Science+Business Media LLC 2017

have poor sizing capabilities and detection limits of ~300 nm at best—the size range of large microvesicles and apoptotic bodies.

Many researchers are now using the alternative method for EV sizing of Nanoparticle Tracking Analysis (NTA). NTA determines the size of vesicles from their Brownian motion, by tracking the movement of individual vesicles in real time, allowing accurate sizing and counting of EV in biological fluids down to ~50 nm in diameter [4]. The range of NanoSight instruments (Malvern Instruments Ltd) equipped with Nanoparticle Tracking Analysis (NTA) software also has fluorescence capability to enable detection of fluorescently labeled EV (fl-NTA). This gives users the capability to both size and phenotype EV well below the size range of most flow cytometers. This chapter describes two protocols for the analysis of EV using fl-NTA. The first protocol describes the use of the general cell membrane dye; CellMask Orange™ (CMO), to label plasma EV and their detection by fl-NTA. The second protocol is an antibody based method, for labeling placenta-derived EV, with a Qdot-conjugated antibody specific for the syncytiotrophoblast marker placental alkaline phosphatase (PLAP), and then quantifying PLAP positive EV using fl-NTA [5]. The aim of this chapter is, by using these specific examples, to highlight general concepts and considerations when developing protocols for EV characterization using fl-NTA.

2 Materials

All reagents should be analytical grade and unless stated prepare all solutions using deionized water (18 MΩ cm at 25 °C). We recommend that all EV dilutions be made using Dulbecco's Phosphate Buffered Saline (PBS) manufactured by Sigma-Aldrich (D8537) as this is particle free (*see* **Note 1**). All samples must be taken up in a 1 mL syringe prior to performing a NTA measurement.

2.1 NanoSight Instrumentation

Select the NanoSight instrument of choice (*see* **Note 2**). For the detection of CMO labeled EVs we recommend using the NS300 or NS500 instrument equipped with a sCMOS camera, 532 nm (green) laser, and 565 nm long-pass filter. For the detection of Qdot labeled EVs we recommend using the NS300 or NS500 instrument equipped with a sCMOS camera, 405 nm (violet) laser and 430 nm long-pass filter. We recommend using the NanoSight syringe pump module for the detection of EVs in both scatter and fluorescence modes. Flow-controlled detection of EVs improves the repeatability of concentration measurements and when in fluorescence mode it mitigates the effect of any photobleaching exhibited by the sample.

2.2 NanoSight Instrument Performance Check

1. Silica microspheres of a known size and concentration: select microspheres that are of a similar size to the EVs being measured [6]. We recommend using silica microspheres (colloidal) (Polysciences).
2. Particle free deionized water: using a 10 mL syringe, first filter the deionized water through a 0.2 μm filter (polyethersulfone; PES membrane) and then again using a 0.02 μm filter (Whatman™ Anotop™ 25 syringe filter).

2.3 Preparation of CellMask Orange™ (CMO) Labeled EVs

1. Preparation of EVs (*see* **Note 3**).
2. CellMask Orange™ plasma membrane stain; 5 mg/mL solution in DMSO (Thermo Fisher Scientific; Life Technologies-Molecular Probes).

2.4 Preparation of EV Standards for Immunobead Depletion and Antibody Q-Dot Labeling

1. Preparation of EVs (*see* **Note 4**).
2. Antibody of choice and matching isotype control antibody (*see* **Note 5**).
3. Pan mouse IgG Dynabeads (Thermo Fisher Scientific; Life Technologies-Molecular Probes).
4. Isolation buffer (PBS supplemented with 0.1% BSA and 2 mM EDTA, pH 7.4).
5. FcR blocking reagent. Filter using a Nanosep 0.2 μm centrifugal device (Pall Life Sciences). Store at 4 °C.
6. Magnetic separator for 1.5 mL microcentrifuge tubes.
7. Loading buffer (3×): 0.24 M TRIS pH 6.8, 6% sodium dodecyl sulfate (SDS), 30% glycerol, 16% β-mercaptoethanol, and 0.06% bromophenol blue (*see* **Note 6**).

2.5 Preparation of Qdot-Conjugated Antibodies

1. SiteClick™ Qdot Antibody Conjugation Kit (Thermo Fisher Scientific; Life Technologies-Molecular Probes) (*see* **Note** 7).
2. 1% Sodium azide.
3. Nanosep 0.2 μm centrifugal device (Pall Life Sciences).

3 Methods

3.1 NTA: Instrument Performance Check

The following procedure should be carried out at the beginning of each day to ensure optimal instrument performance.

1. Prime the fluidics with particle free deionized water, making sure there are no bubbles in the sample chamber and tubing. Start the camera and assess the level of background particles by examining a static live image. No more than three particles should be visible.

2. Using a preparation of silica microspheres of known size and concentration, prepare at least 1 mL by performing a dilution in particle free water to give an expected concentration of 5–6 × 10^8 particles/mL. Take up the sample using a 1 mL syringe, place into the syringe pump module and perform a NTA scatter measurement. *NTA scatter measurement*: Infuse the sample into the chamber, focus the image, set the camera level and measure (for details regarding infusion rate, camera level, focus position, video capture, detection threshold and processing settings; *see* **Note 8**). Check the accuracy of the size and concentration measurements (both measurements should be within 10% of the expected size and concentration values).

3.2 Labeling of EVs Using CMO

1. Prime the fluidics of the NanoSight instrument with PBS, making sure there are no bubbles in the sample chamber and tubing. Start the camera and assess the level of background particles by examining a static live image. No more than three particles should be visible.
2. Perform a NTA scatter measurement under flow conditions (*see* **Notes 8** and **9**) of the EV stock to determine the concentration of EVs. Dilute the EV stock in PBS to give an appropriate concentration (*see* **Note 10**). Make a note of the user-selectable parameters (i.e., camera level, focus position, infusion rate, number and length of videos captured, detection threshold and processing settings) as the same settings should be used for subsequent measurements. From the concentration data obtained and the dilution factor used, calculate the concentration of EVs in the stock solution.
3. Prepare an EV solution of 2 × 10^{11} EV/mL by diluting the EV stock with PBS, allowing 10 μL for each EV incubation (*see* **Note 11**).
4. Freshly prepare a 2.5 μg/mL solution of CMO by diluting the stock CMO with PBS, allowing 20 μL for each EV incubation (*see* **Notes 12** and **13**). Keep in the dark at room temperature.
5. Place 10 μL of the 2.5 μg/mL CMO solution prepared in **step 4** in a tube and add 10 μL of EVs (2 × 10^9 EV) prepared in **step 3** (*see* **Note 14**). Mix by gently pipetting up and down. Incubate in the dark for 15 min at room temperature (*see* **Note 15**).
6. Transfer 10 μL of the CMO labeled EVs (1 × 10^9 EV) to a 2 mL tube. Add 990 μL of PBS and mix by gently pipetting up and down. Add a further 1 mL of PBS and gently mix (CMO labeled EV: 5 × 10^8 EV/mL).
7. *fl-NTA measurement*: Using the same infusion rate and camera level determined in **step 2**, load the CMO labeled EV sample into the chamber. Insert the 565 nm long-pass filter. Slowly adjust the focus a few steps in the positive and negative

direction to ensure the EVs are in sharp focus (*see* **Notes 16** and **17**). Also, the contrast of the image can be enhanced by adjusting the Low and High Threshold Advanced Camera settings (*see* **Note 18**). Examine the image and make sure the EVs are not photo bleaching (*see* **Note 19**). Measure the sample, ensuring that the number and length of videos captured, detection threshold and processing settings are the same as those used in **step 2**. Make a note of all the user-selectable parameters (i.e., camera level, focus position, flow rate, number and length of videos captured, detection threshold, and processing settings) as these same settings will be used when performing **steps 10** and **11**.

8. With the CMO labeled EV sample still infusing (do not adjust the flow rate) perform a NTA scatter measurement. Remove the 565 nm long-pass filter and refocus the image (use the same focus position from **step 2**). The fluorescence intensity signal of the labeled EVs is likely to be similar to the scatter intensity signal; therefore, the same camera level used for the fl-NTA measurement should be appropriate for the NTA scatter measurement. Ensure that the particle brightness appears similar, adjusting the Low and High Threshold Advanced Camera settings if required. Measure the sample, ensuring that the number and length of videos captured, detection threshold and processing settings are the same as those measured in fluorescence mode (**step** 7).
9. Set up a dye-control incubation (CMO control sample) by mixing 10 μL of the 2.5 μg/mL CMO solution (prepared in **step 4**) in a tube and add 10 μL of PBS. Mix by gently pipetting up and down. Incubate in the dark for 15 min at room temperature and finally dilute the sample as described in **step 6**.
10. Infuse the CMO control sample into the chamber. Perform a fl-NTA measurement (described in **step** 7, using the same user-selectable parameters), followed by a NTA scatter measurement (described in **step 8**, using the same user-selectable parameters) (*see* **Note 20**) (Fig. 1).
11. Finally, check the EVs display no autofluorescence. Pipette 5 μL of EVs (prepared in **step 3**) into a 2 mL tube and top up with 1995 μL of PBS (5×10^8 EV/mL). Perform a fl-NTA measurement (described in **step** 7, using the same user-selectable parameters) (*see* **Note 21**). No EVs should be visible, as nonlabeled EVs should not exhibit a fluorescence signal.
12. Subtract any CMO control values from the CMO labeled EV data for both scatter and fluorescence modes.
13. Examine the NTA size and concentration data for the unlabeled EV sample vs. CMO labeled EV sample both measured in scatter mode. These should be very similar (Fig. 1). Compare

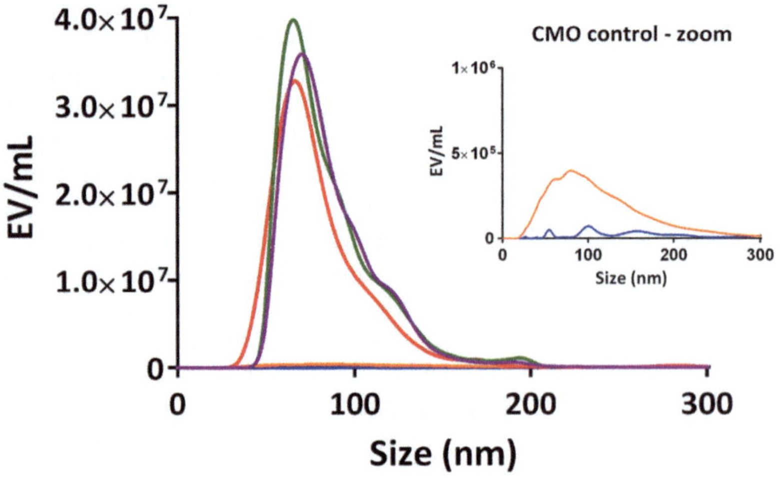

Fig. 1 Representative NTA profile of plasma EVs. Scatter mode: unlabeled EVs (*purple line*), CMO labeled EVs (*green line*), CMO control (*orange line*). Fluorescence mode: CMO labeled EVs (*red line*) and CMO control (*blue line*). Plasma EVs were measured using camera level 16 with the following camera thresholds used for fluorescence mode measurements; Low 130, High 715. Concentration of CMO labeled EVs: Scatter mode; 1.8×10^9 EV/mL, Fluorescence mode; 1.5×10^9 EV/mL. The CMO labeling efficiency of plasma EVs is 83%

the NTA size and concentration data for the CMO labeled EV sample in scatter and fluorescence modes. Similar size and concentration values should be obtained if the labeling process has been successful (Fig. 1).

14. The labeling efficiency can be calculated using the following equation:

$$\text{Labeling efficiency } (\%) = 100 \times \left(\frac{\text{Fluorescence mode: Concentration of EV}}{\text{Scatter mode: Concentration of EV}}\right)$$

3.3 fl-NTA of Qdot-Conjugated-Antibody-Labeled EVs

EV brightness, which is set by camera level and/or detection threshold, determines the sample depth in which EV are counted (i.e., brighter EV will be visible to the camera from lower depths in the sample). This becomes an issue when analyzing EV with a different brightness in scatter and fluorescence mode. Such a scenario is more likely to occur with antibody labeling, where binding is determined by antigen expression, but tends not to occur with membrane dyes, such as CMO, that are incorporated into the EV membrane in much higher quantities. As a consequence, appropriate camera levels, that make the visible EV approximately the same brightness in scatter and fluorescence mode, need to be determined to ensure EV are measured in the same sample volume. For antibody labeling this typically means a higher camera level for fluorescence measurements than is used in scatter mode.

To determine appropriate NTA scatter and fluorescence mode camera levels for the analysis of EV labeled with a fluorescently tagged antibody, we suggest the use of a "standard" that contains a known amount of antigen positive EV. This standard can then be analyzed by fl-NTA and camera levels adjusted until comparable results are obtained. This procedure would need to be carried out for specific antigens using a "standard" appropriate for the EV to be subsequently analyzed. Here we describe the use of immunobead depletion to generate placenta-derived EV standards containing known quantities of placental alkaline phosphatase (PLAP—a syncytiotrophoblast marker) positive EV, and their use to set appropriate camera levels for fl-NTA measurement of PLAP positive EV in placenta-derived microvesicle (STBMV) and exosome (STBEX) enriched samples.

3.3.1 Preparation of EV Standards: Immunobead Depletion

1. Using a representative preparation of EVs (*see* **Note 22**), perform a NTA scatter measurement to determine the EV stock concentration (as described in Subheading 3.2, **step 2**). In this example, we used pools ($n = 4$) of STBMV and STBEX. Make a note of the user-selectable parameters (i.e., camera level, focus position, infusion rate, number and length of videos captured, detection threshold and processing settings). From the concentration data obtained and the dilution factor used, calculate the concentration of EVs in the stock solution.
2. Wash the Pan mouse IgG Dynabeads (*see* **Note 23**). Coat with the antibody of choice and respective isotype control antibody according to the manufacturer's instructions, ensuring that the final resuspension of antibody coated Dynabeads is in PBS. Store at 4 °C until use.
3. A saturating concentration of antibody-coated beads needs to be added to a fixed number of EVs in order to remove all positive antigen signal from any unbound EV portion. Preparing multiple tubes, dilute the EVs in PBS, bearing in mind that the final concentration in each tube should be 5×10^9 EV/mL (*see* **Note 24**) and in the final step the volume will be made up to 1 mL with PBS. Pipette the EVs into each tube and incubate with 10 μL of FcR blocking reagent for 10 min at 4 °C. To determine a saturating concentration of antibody-coated beads, add these (or isotype-coated beads) in increasing amounts to the respective tubes containing EVs + FcR blocking reagent in a total volume of 1 mL. Incubate the tubes overnight at 4 °C using a rotating laboratory mixer (*see* **Note 25**). Be sure to include a tube that only has EV [5×10^9 EV/mL] + FcR blocking reagent (i.e., no Dynabeads). This tube will be used as the control tube for the total number of starting

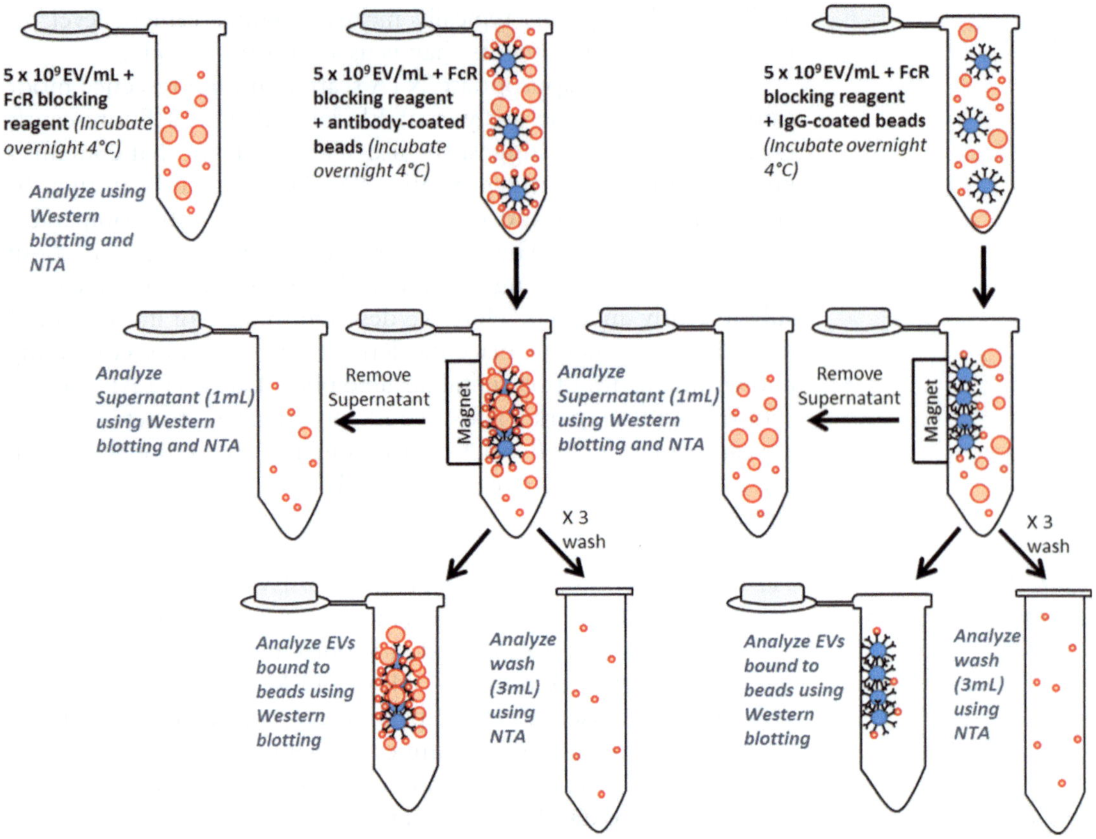

Fig. 2 Schematic diagram illustrating the samples and steps involved in the quantification of antigen positive EV using immunobead depletion, NTA analysis and Western blotting, to produce a "standard" EV preparation for establishing correct fl-NTA camera level settings

EVs. A schematic diagram illustrating the samples and steps involved using this method is shown in Fig. 2.

4. With the exception of the tube that contains EV + FcR blocking reagent only, place all tubes in a magnet for 1 min and gently pipette the 1 mL of supernatant (unbound EV) into a clean tube. Store the supernatants at 4 °C.

5. Wash the beads by removing the tubes from the magnet and add 1 mL of PBS. Gently pipette up and down a few times and place the tube in a magnet for 1 min. Pipette the 1 mL of supernatant into a clean 5 mL tube. Repeat this step twice more, each time collecting the 1 mL of supernatant and placing this into the same 5 mL tube (a total of 3 mL should be collected). Store the wash supernatants at 4 °C.

6. To solubilize EVs bound to the beads directly add 30 μL of 1× reducing buffer (*see* **Notes 6** and **26**). Centrifuge at 11,500 × *g* for 1 min and place the tube in a magnet for 1 min. Pipette the 30 μL 1× reducing buffer (avoiding the beads) into a clean

tube and set aside for SDS-PAGE and Western blotting (*see* **Note 27**).

7. Pipette 20 μL of the unbound EV supernatant (collected in **step 4**) and mix with 10 μL of 3× loading buffer. Pipette 20 μL from the tube which contains EV [5×10^9 EV/mL] + FcR blocking reagent (i.e., no Dynabeads) and mix with 10 μL 3× loading buffer (this tube serves as the control and assumes that no EV were bound). Set tubes aside for SDS-PAGE and western blotting (*see* **Notes 26** and **27**).
8. To determine the saturating dose of antibody-coated beads that effectively depletes all positive antigen signal, perform SDS-PAGE and Western blotting (*see* **Note 28**) on the bound EV fractions (collected in **step 6**) (Fig. 3a STBMV; Fig. 3b

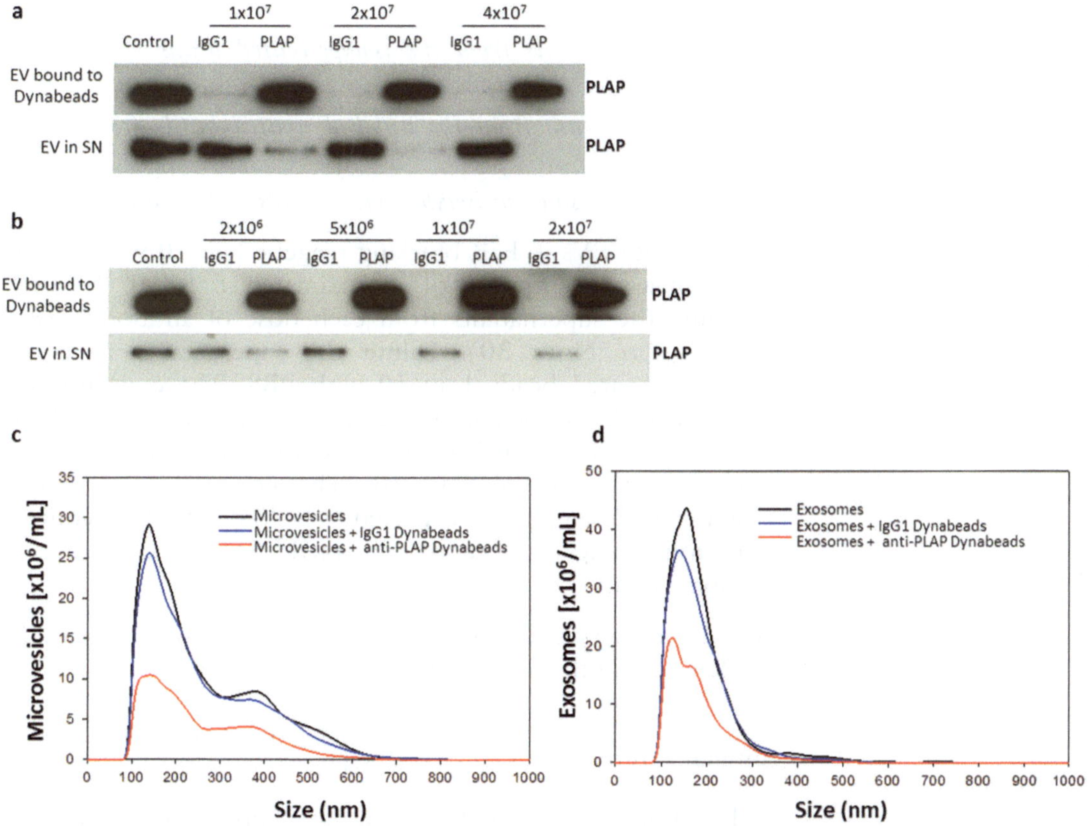

Fig. 3 Immunobead depletion and NTA measurements of syncytiotrophoblast (placental) microvesicle (STBMV) and exosome (STBEX) enriched pools. Representative immunoblot images of placental alkaline phosphatase (PLAP) in the (**a**) STBMV pool and (**b**) STBEX pool showing bound EV (PLAP positive) and those remaining in the supernatant (SN; PLAP negative) with increasing doses of IgG1 Dynabeads or anti-PLAP Dynabeads. NTA profiles of (**c**) STBMV pool and (**d**) STBEX pool alone (*black line*) or post incubation with a saturating concentration [STBMV: 4×10^7; STBEX: 1×10^7] of IgG1 Dynabeads (*blue line*) or anti-PLAP Dynabeads (*red line*). 59.8 ± 5.2% of the STBMV pool and 51.6 ± 3.4% of the STBEX pool were PLAP positive. Reproduced from [5]

STBEX) and unbound EV fractions (collected in **step 7**) (EV in SN: Fig. 3a STBMV and Fig. 3b STBEX). Load an equivalent amount of starting EV (total: 5×10^9 EVs) onto the gel when analyzing the bead bound EV fractions.

9. The efficiency of the immunobead depletion method can be determined using NTA, by measuring the EVs in the supernatant and comparing with the starting EV preparation (*see* **Note 29**) (Fig. 3c, d). This data should also corroborate with that of the Western blots. To determine the number of total EVs [total] and unbound EVs [antigen negative], perform a NTA scatter measurement (as outlined in Subheading 3.1, **step 2**, also *see* **Note 8**) on the following samples:
   (a) *EVs + FcR blocking reagent (prepared in* ***step 3****).*
   (b) *Supernatants from the antibody-coated beads (collected in* ***step 4****).*
   (c) *Supernatants from the isotype-coated beads (collected in* ***step 4****).*
   (d) *3 mL wash from the antibody-coated beads (collected in* ***step 5****).*
   (e) *3 mL wash from isotype-coated beads (collected in* ***step 5****).*

   Dilute EVs + FcR blocking reagent 1 in 10 with PBS to obtain a concentration of approximately 5×10^8 EV/mL. Dilute the supernatants from each dose of antibody-coated beads (*see* **Note 30**). Dilute the supernatants from the isotype-coated beads 1 in 10 with PBS. The concentration obtained should be approximately 5×10^8 EV/mL (*see* **Note 31**). Measure the 3 mL wash samples from the antibody-coated beads and isotype-coated beads. The measured concentration in the wash samples are expected to be low. Add these concentration values to those obtained from the respective supernatants.

10. The percentage of bound EVs (antigen positive) can be calculated using the following equation:

$$\text{Antigen Positive EVs (\%)} = 100 \times \left(\frac{\text{Total} - \text{Antigen negative}}{\text{Total}}\right)$$

This value can be used to determine appropriate scatter and fluorescence mode camera levels when measuring Qdot-labeled EVs in this same standard.

3.3.2 Antibody Qdot Conjugation

Select the antibody and corresponding isotype control antibody to be conjugated (*see* **Note 32**). Following the manufacturer's instructions, conjugate the antibodies using a SiteClick™ Qdot 605 Antibody Conjugation Kit (*see* **Note 7**). Once the antibody conjugation is complete, we recommend adding sodium azide (0.02% w/v) to the antibody. Filter the antibodies through a Nanosep 0.2 μm centrifugal device (1500 × *g* 2 min) and store at 4 °C in the dark until use (*see* **Note 33**).

3.3.3 Labeling the EV Standard with Qdot-Conjugated Antibodies

1. Using the EV standard, first test a single concentration of antibody-Qdot (*see* **Note 34**).
2. Dilute the EV standard in PBS, preparing a concentration of 1×10^{10} EV/mL. Pipette 100 μL (1×10^{9}) of EVs and 10 μL FcR blocking reagent into a tube and incubate at 4 °C for 10 min.
3. Pipette the chosen concentration of antibody-Qdot into the tube and incubate for 20 min at room temperature in the dark. Top up the tube with PBS to a total volume of 1 mL.

3.3.4 NTA of the EV Standard Labeled with Qdot-Conjugated Antibodies

1. Perform a NTA scatter measurement (as outlined in Subheading 3.1, **step 2**, also *see* **Note 8**). Make a note of the user-selectable parameters (i.e., camera level, focus position, infusion rate, number and length of videos captured, detection threshold, and processing settings). The concentration of EVs should be toward the upper limit of the instrument, measuring ~1×10^{9} EV/mL. This is to maximize the number of fluorescently labeled EVs that are captured later.
2. With the sample still infusing, insert a 430 nm long-pass filter and perform a fl-NTA measurement. Slowly adjust the focus a few steps in the positive and negative direction to ensure the EVs are in sharp focus (*see* **Note 16**). Set the camera to a level in which the EVs appear to be of a similar brightness to that measured in scatter mode. This will typically be a higher camera setting and we recommend starting with a camera level that is one level higher than that used in scatter mode. If no fluorescent EVs appear, further increase the camera level. If no fluorescent EVs are apparent at these higher camera levels, then do not continue any further as this suggests that a higher concentration of antibody-Qdot is required. Repeat the above procedure using a higher concentration of antibody-Qdot until a population of fluorescent EVs are apparent. Measure the sample, ensuring that the number and length of videos captured, detection threshold and processing settings are the same as those used in **step 1**.
3. Next, using the EV standard we recommend refining the concentration of antibody-Qdot by carrying out a titration. All NTA scatter measurements should be carried out using a single camera level, whereas multiple camera levels can be tested in fluorescence mode in order to establish which camera level is the most suitable and will give comparable results of percentage positive EVs to that obtained using immunobead depletion.

3.3.5 Titration of Qdot-Conjugated Antibodies Using the EV Standard

Using the EV standard, test 3–4 different antibody-Qdot concentrations.

1. Dilute the EVs in PBS, preparing a concentration of 1×10^{10} EV/mL. Prepare the following tubes:
   (a) *Positively labeled EVs: EVs + FcR blocking reagent + antibody-Qdot.*
   (b) *Negative Control: EVs + FcR blocking reagent + IgG-Qdot.*
   (c) *Unlabeled EVs: EVs + FcR blocking reagent.*
   (d) *Antibody-Qdot background control: FcR blocking reagent + antibody-Qdot.*
   (e) *IgG-Qdot background control: FcR blocking reagent + IgG-Qdot.*
2. Pipette 100 μL (1×10^9) of EVs into tubes 1–3. Pipette 100 μL of PBS into tubes 4 and 5. Pipette 10 μL FcR blocking reagent into all tubes and incubate at 4 °C for 10 min. Pipette the antibody-Qdot or IgG-Qdot into the respective tubes and incubate for 20 min at room temperature in the dark. Top up each tube with PBS to a total volume of 1 mL.
3. Repeat **steps 1** and **2** for each antibody concentration tested (*see* **Note 35**).
4. Load the first tube containing EV + FcR blocking reagent + antibody-Qdot into the chamber and perform a NTA scatter measurement (as outlined in Subheading 3.1, **step 2**, also *see* **Note 8**). The concentration of EVs should measure ~1×10^9 EV/mL. Make a note of the user-selectable parameters (i.e., camera level, focus position, infusion rate, number and length of videos captured, detection threshold, and processing settings) as these same settings will be selected each time a scatter measurement is performed.
5. With the sample still infusing, insert a 430 nm long-pass filter and perform a fl-NTA measurement (Subheading 3.3.4, **step 2**) (*see* **Note 36**). The EVs need to be of a similar brightness to that measured in scatter mode, therefore we recommend starting with a camera level which is one higher than that used when performing the scatter measurement. We recommend testing multiple camera levels (*see* **Note 37**). Take note of the following user-selectable parameters (camera level(s), focus position and infusion rate). Measure the sample, ensuring that the number and length of videos captured, detection threshold and processing settings are the same as those used for the NTA scatter measurement (**step 4**) (*see* **Note 38**).
6. Load tube 2 (EV + FcR blocking reagent + IgG-Qdot) into the chamber and perform a NTA scatter measurement (same

user-selectable parameters as determined in **step 4**). The EV concentration should measure ~1×10^9 EV/mL and the size profile should be similar to that in tube 1 (EV + FcR blocking reagent + antibody-Qdot).

7. With the contents of tube 2 (EV + FcR blocking reagent + IgG-Qdot) still infusing into the chamber, perform a fl-NTA measurement (same user-selectable parameters as determined in **step 5**). The concentration of EVs should be very low, measuring $<0.5 \times 10^8$ EV/mL (*see* **Note 39**). Repeat the fl-NTA measurement for each camera level tested, ensuring that the concentration of EVs remains $<0.5 \times 10^8$ EV/mL.
8. Load tube 3 (EV + FcR blocking reagent) into the chamber and perform a NTA scatter measurement (same user-selectable parameters determined in **step 4**). The concentration of EVs should measure ~1×10^9 EV/mL and the size profile should be similar to that in tube 1 (EV + FcR blocking reagent + antibody-Qdot) and tube 2 (EV + FcR blocking reagent + IgG-Qdot) (*see* **Note 40**). With the sample still infusing into the chamber, perform a fl-NTA measurement (same user-selectable parameters determined in **step 5**). Repeat the fl-NTA measurement for each camera level tested. Ensure that no EVs are visible.
9. Load tube 4 (FcR blocking reagent + antibody-Qdot) into the chamber and perform a NTA scatter measurement (same user-selectable parameters determined in **step 4**), followed by a fl-NTA measurement (same user-selectable parameters determined in **step 5**). Repeat the fl-NTA measurement for each camera level tested. Repeat the NTA scatter and fl-NTA measurements (same user-selectable parameters determined in **steps 4** and **5** respectively), loading tube 5 (FcR blocking reagent + IgG-Qdot) into the chamber. The concentration of EVs measured in scatter and fluorescence in tubes 4 and 5 should measure <0.5 $\times 10^8$ EV/mL (*see* **Note 41**).
10. Repeat all of the above using 3–4 antibody-Qdot concentrations.
11. For each antibody concentration tested, use the scatter and fluorescence measurements to calculate the percentage of positive EVs for the antigen of interest. Complete this analysis for each camera level used for the fluorescence measurements. This will provide information regarding which concentration of antibody-Qdot to use for future experiments and also determine which camera level is the most suitable (i.e., gives comparable results for the percentage of antigen positive EVs to those achieved using immunobead depletion).

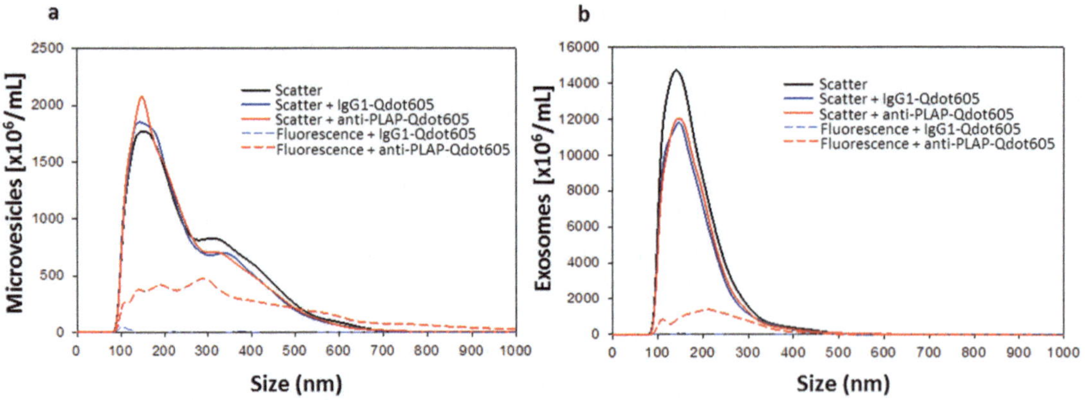

Fig. 4 NTA of the (**a**) STBMV pool and (**b**) STBEX pool measured in scatter mode: alone (*black line*), labeled with IgG1-Qdot605 control (*blue solid line*) or labeled with anti-PLAP-Qdot605 antibody (*red solid line*) and fluorescence mode: labeled with IgG1-Qdot605 control (*blue dash*) or anti-PLAP-Qdot605 antibody (*red dash*). 48.5 ± 3.0% of the STBMV pool and 17.9 ± 3.3% of the STBEX pool were PLAP positive. Reproduced from [5]

$$\text{Scatter} \times 10^8\ \text{EV/mL} = (\text{EVs} + \text{Qdot antibody}) - (\text{Qdot antibody background particles})$$

$$\text{Fluorescence} \times 10^8\ \text{EV/mL} = (\text{EVs} + \text{Qdot antibody}) - (\text{Qdot antibody background particles})$$

$$\text{Positive EVs (\%)} = 100 \times \left(\frac{\text{Fluorescence} \times 10^8\ \text{EV/mL}}{\text{Scatter} \times 10^8\ \text{EV/mL}}\right)$$

Following the above methodology, STBMV (Fig. 4a) and STBEX (Fig. 4b) were labeled with anti-PLAP-Qdot605 and IgG-Qdot605 and measured using fl-NTA [5]. The percentage of PLAP positive STBMV and STBEX were calculated using the above equation. Our data showed that fl-NTA worked well for STBMV, giving results for the percentage of PLAP positive STBMV (48.5 ± 3.0%) comparable to those obtained for immunobead depletion (59.8 ± 5.2%). However, immunobead capture is more sensitive than fl-NTA for the detection of PLAP positive STBEX (immunobead capture 51.6 ± 3.4%; fl-NTA 17.9 ± 3.3%). STBEX may have low surface expression of PLAP, therefore limiting the number of Qdots per EV and their subsequent detection by fl-NTA [5].

4 Notes

1. If using an alternative product or buffer, or if the solution is not found to be particle free, polyethersulfone (PES) membrane syringe filters can be used to obtain a particle-free solution. We

recommend using a range of pore sizes in decreasing size, finishing with a 20 nm pore to obtain a particle-free solution.

2. fl-NTA can be carried out using a variety of NanoSight instruments, including LM10, NS300, and NS500. The NS300 and NS500 instruments are optimized for fluorescence detection and have been used for the procedures outlined in this chapter.
3. We have experience using EVs isolated from human healthy-volunteer plasma and urine; however, this labeling protocol can be used on other types of EVs. The isolation process used filtration and ultracentrifugation to generate a sample with a low level of polydispersity and the majority of EVs being 70–120 nm (hydrodynamic diameter).
4. We have experience using syncytiotrophoblast (placental) EVs (STBEVs) enriched for microvesicles (STBMV; 100–600 nm) or exosomes (STBEX; 100–300 nm) using the technique described by [5]. This protocol can be used to isolate EVs from other fluids such as culture medium, plasma and urine.
5. We recommend using mouse monoclonal antibodies. We have experience using NDOG2, an in-house antibody to PLAP [5]. Appropriate mouse Isotype control IgGs are commercially available (e.g., Cambridge Biosciences and R&D Systems).
6. To make 1× Reducing buffer, dilute Reducing buffer (×3) 1 in 3 with deionized water.
7. SiteClick™ Qdot Antibody labeling kits are available in a range of wavelengths (525, 565, 585, 625, 655, 705 and 800 nm emission). However, we only have experience using the Qdot 605 antibody labeling kit.
8. Set the following parameters when performing a NTA measurement:

   Camera level: Select a suitable camera level whereby the particles/EVs can be visualized clearly. For example, when analyzing 100 nm silica microspheres we recommend using a camera level of 12. If analyzing silica microspheres or EVs <100 nm a higher camera level should be used.

   Infusion rate: Load at least 0.5–0.75 mL of sample into the chamber (do not exceed an infusion rate of >1000) and ensure that there are no bubbles in the sample chamber and tubing. Adjust the flow speed to ensure that a single silica microsphere or EV takes approximately 5–10 s to move across the screen.

   Image focus: Adjust the focus to ensure that each silica microsphere or EV is a single spherical point.

   Video capture: For the analysis of silica microspheres or homogeneous EV populations we recommend capturing at least

5×30 s video recordings. If the EV preparation is heterogeneous, we recommend capturing at least 5×60 s video recordings. Ensure that >200 completed tracks are analyzed per video recording.

Detection Threshold and processing settings: Each video file should be processed using an appropriate detection threshold, which will vary depending on the version of NTA software used. If using NTA version 2.3 or below we recommend using a detection threshold of 10. If using NTA version 3.0 or above, we recommend using a detection threshold of 5. Always ensure that all particle centers attain a red cross while blue crosses (indicative of potential centers of noise) are kept to below 10 per frame. Check settings are correct by viewing additional frames of the video. We recommend keeping all other post acquisition settings (max jump mode, blur, min track length) as auto.

PBS Flush: In between EV samples, flush PBS into the sample chamber and ensure all residual EVs are removed before moving onto the next sample.

9. For EVs in the expected size range of 70–120 nm, a suitable camera level is typically 14, 15, or 16.
10. NTA has a working concentration measurement ranging from 1×10^6 EV/mL to ~3×10^9 EV/mL when using a syringe pump. For optimum measurements, approximately 20–60 EVs in the field of view is ideal, and can be used as a guide for obtaining the appropriate dilution prior to running the measurement. When the number of EVs viewed exceeds approximately 60 in the field of view a more dilute sample should be prepared and measured.
11. An EV stock of 2×10^{11} EV/mL is approximately 200× concentrated stock. If the initial EV isolation produces a stock less concentrated than this and the sample cannot be centrifuged to generate a more concentrated sample, we recommend that the amount of CMO is decreased proportionately (also *see* **Note 13** below).
12. Dilute the supplied 5 mg/mL CMO stock ×1000 by adding 1 μl of CMO stock to 999 μL of PBS followed by a further 2× dilution in PBS.
13. The appropriate amount of CMO to add to EVs will depend on a number of factors including the source of the EVs, the size of the EVs, and the isolation technique used to prepare the EV stock.
14. In our experience, best results are obtained when the biological sample is added to the dye solution and when the incubation is

carried out in a small volume with the interacting partners at a high concentration.

15. For the EVs used in this study, incubating for longer than 15 min did not achieve a higher labeling efficiency. The optimum incubation time may vary with different EV samples (*see* **Note 13**).
16. The insertion of the filter changes the optical path of the instrument which in turn changes the focal path. Therefore the focus will have to be adjusted slightly for a sharp image.
17. To visualize the labeled EVs it may be necessary to alter the camera level. However, if the camera level is changed the unlabeled EV sample will need to be remeasured using this same camera level.
18. The Advanced Camera histogram settings can be used to improve the overall contrast of the image. Dimmer EVs can then be viewed as brighter EVs on a darker background.
19. If photo bleaching of EVs is an issue, try to readjust the infusion rate. Speed up the flow of EVs so that they are still visible over 50% of the screen prior to any photo bleaching. Ensure that the flow rate is not too fast that it compromises accurate tracking of the EVs.
20. The same user-selectable parameters used for measuring the labeled EV sample must also be used to measure the CMO control in scatter and fluorescence mode, respectively. This allows for any contribution of background particles and fluorescent particles from the CMO dye to be accounted for (and subtracted from) the labeled EV sample. In our experience, the contribution of CMO-derived background particles measured in scatter and fluorescence mode were negligible and did not impact size or concentration measurements.
21. The same user-selectable parameters used for measuring both the labeled EV sample and CMO control in fluorescence mode should be used.
22. To give a representative sample, we recommend preparing a pool of individual EV preparations (using a minimum of $n = 3$).
23. Wash the Dynabeads in the recommended Isolation buffer, according to the manufacturer's instructions. After coating the prewashed Dynabeads with the antibody of interest and respective isotype control (according to the manufacturer's instructions), carry out repeated washes in Isolation buffer to remove excess antibodies. The final wash should be done in PBS and the antibody-coated Dynabeads/isotype-coated Dynabeads should be resuspended in a given volume of PBS. Do not resuspend the coated Dynabeads in Isolation buffer as

this contains particulate matter that will interfere with downstream NTA readings. We also recommend that freshly coated Dynabeads are used in all experiments.

24. In our experience, 5×10^9 EV/mL was an optimum concentration that ensured sufficient numbers for downstream analyses. Although it may be possible to lower the concentration of EVs, this may impact downstream NTA and Western blotting analysis.
25. The dose of saturating beads will be variable, depending on: (a) the type of EV preparation and (b) the antigen of interest. We recommend to always use a fixed concentration of EV [5×10^9 EV/mL] and to increase the amount of beads in a dose dependent manner. Start with a concentration of beads at 2×10^6 (2500 EV per bead) and increase up to 4×10^7 beads (125 EV per bead).
26. For SDS-PAGE gels with a well capacity of below 30 μL, scale down volumes appropriately.
27. Analyze samples immediately or store at −20 °C for subsequent analysis.
28. SDS-PAGE and Western blotting can be performed using the method as described in [7]. If using the same primary antibody used to coat the Pan mouse IgG Dynabeads, it is important to note that this will result in the detection of both heavy (~50 kDa) and light chains (~25 kDa) of IgG. If the antigen of interest has a similar molecular weight to either the heavy chain or light chain then this may result in the antigen being masked. To overcome this problem, we recommend using a primary antibody which is raised in a different species.
29. We recommend performing NTA measurements within 24–48 h of sample collection. We do not recommend samples be frozen for later analysis as freezing may affect EV size and concentration.
30. Sample dilution will vary depending on how many EVs are positive for the antigen of interest. We recommend diluting the sample to a concentration of $2\text{–}10 \times 10^8$ EV/mL.
31. The concentration obtained should be within ~10% of that measured in the EVs + FcR blocking reagent sample (sample 1). Values below this range could indicate nonspecific binding. This may be evident in the Western blot, showing a band in the lane of the saturating dose of isotype-beads. Further titration of the antibody-coated beads/isotype-coated beads may therefore be required.
32. We recommend using the same antibodies used for the immunobead depletion method.

33. We recommend using flow cytometry to test if the Qdot-conjugated antibodies are functional. Use a preparation of cells positive for the marker of interest.
34. The initial concentration of antibody will need to be determined by the user and this will vary for individual antibodies.
35. Perform each antibody titration separately, to avoid long delays between staining and NTA sample measurement.
36. In our experience, there was no significant photo bleaching of EVs labeled with the Qdot-conjugated antibody. Therefore, no readjustment to the infusion rate was required between NTA scatter and fluorescence measurements. If photo bleaching of EVs is an issue, *see* **Note 19**.
37. We recommend testing several camera levels. However in our experience, using camera levels of $\geq$14 will result in increased numbers of blue trajectory tracks, indicative of unbound Qdots. Tracking of unbound Qdots will be evident in the size and concentration data; therefore, we recommend avoiding such camera levels.
38. The NTA size profiles of the EVs measured in scatter and fluorescence modes should be similar. If the NTA size profile of EVs measured in fluorescence is weighted to only track the larger EVs, this may indicate a higher concentration of antibody is required or a higher camera level should be used. Using high camera levels may result in the tracking of unbound Qdots (single or aggregates) which could mistakenly be identified as positively labeled EVs. In scatter mode, ensure the unlabeled EVs and EVs labeled with antibody-Qdot display very similar size profiles. If there are a large number of very small EVs/particles (<100 nm) present in the EV sample labeled with antibody-Qdot, but not the unlabeled EV sample, this suggests that unbound Qdots are being tracked. Furthermore, ensure the NTA concentration measurement in fluorescence never exceeds that measured in scatter, as this would be indicative that the camera level used is too high.
39. A concentration which exceeds 0.5×10^8 EV/mL is indicative of nonspecific binding and therefore a lower antibody concentration should be tested.
40. In our experience, bound Qdots do not affect EV physical characteristics. Analysis of EVs + FcR blocking reagent, EVs + blocking reagent + antibody-Qdot and EVs + FcR blocking reagent + IgG1-Qdot will show overlapping size profiles.
41. It is important to analyze the level of background particles that are solely from the antibody-Qdot/IgG-Qdot. To eliminate single or aggregated Qdots from the analysis we recommend

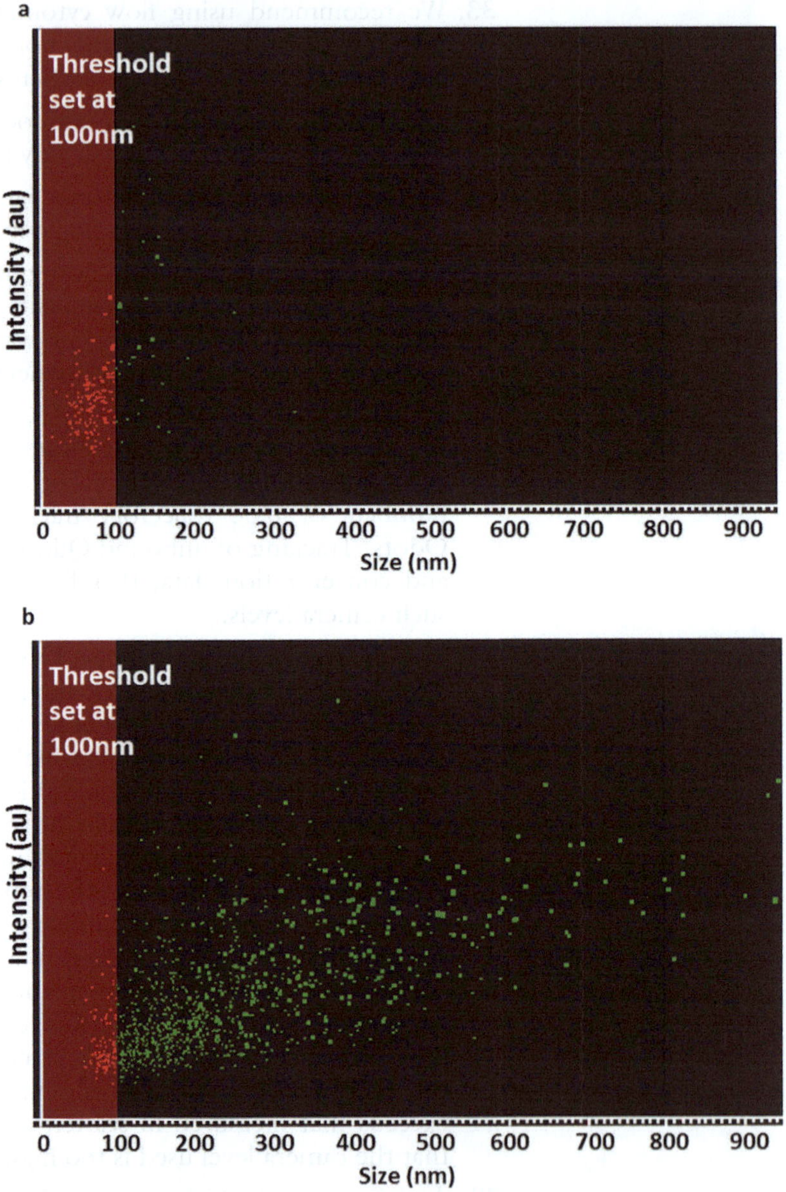

Fig. 5 NTA scatter mode intensity plots for (**a**) Qdot-conjugated antibody alone and (**b**) Qdot-conjugated-antibody-labeled EV, showing the use of a threshold set at 100 nm to remove interference by unbound Qdot-conjugated antibody

using the scatter plot ((size (nm) vs. intensity (a.u.)) and placing an exclusion box from 0 to 100 nm (Fig. 5a, b). With the exclusion box in place, the total concentration of EV coming from the antibody-Qdot/IgG-Qdot should measure $<0.5 \times 10^8$ EV/mL. If the exclusion box is used, then this must be applied when analyzing all samples (i.e., those containing EVs). In our experience using an exclusion box from 0 to

100 nm excludes <10% of the total EVs being analyzed, however this will be sample and software build number dependent and the user may need to adapt the size of the exclusion box accordingly. For software versions NTA 3.0 and above, a smaller exclusion region can be applied, e.g., 0–60 nm.

Acknowledgment

This work was supported by MRC programme grant MR/J003360/1.

References

1. van der Pol E, Coumans F, Varga Z et al (2013) Innovation in detection of microparticles and exosomes. J Thromb Haemost 11(Suppl 1):36–45. doi:10.1111/jth.12254
2. van der Pol E, Hoekstra AG, Sturk A et al (2010) Optical and non-optical methods for detection and characterization of microparticles and exosomes. J Thromb Haemost 8 (12):2596–2607. doi:10.1111/j.1538-7836.2010.04074.x
3. Witwer KW, Buzas EI, Bemis LT et al (2013) Standardization of sample collection, isolation and analysis methods in extracellular vesicle research. J Extracell Vesicles 2. doi:10.3402/jev.v2i0.20360
4. Dragovic RA, Gardiner C, Brooks AS et al (2011) Sizing and phenotyping of cellular vesicles using Nanoparticle Tracking Analysis. Nanomedicine 7(6):780–788. doi:10.1016/j.nano.2011.04.003
5. Dragovic RA, Collett GP, Hole P et al (2015) Isolation of syncytiotrophoblast microvesicles and exosomes and their characterisation by multicolour flow cytometry and fluorescence Nanoparticle Tracking Analysis. Methods 87:64–74. doi:10.1016/j.ymeth.2015.03.028. http://creativecommons.org/licenses/by/4.0/
6. Gardiner C, Ferreira YJ, Dragovic RA et al (2013) Extracellular vesicle sizing and enumeration by nanoparticle tracking analysis. J Extracell Vesicles 2. doi:10.3402/jev.v2i0.19671
7. Tannetta DS, Dragovic RA, Gardiner C et al (2013) Characterisation of syncytiotrophoblast vesicles in normal pregnancy and pre-eclampsia: expression of Flt-1 and endoglin. PLoS One 8(2): e56754. doi:10.1371/journal.pone.0056754

Chapter 14

Characterization of Extracellular Vesicles by Flow Cytometry

Virginia Camacho, Vasilis Toxavidis, and John C. Tigges

Abstract

Here, we describe a comprehensive methodology for the setup and standardization of EV analysis using nanoscale flow cytometry. Controls of different size ranges, fluorescent intensities, and materials can be used to set up distribution curves that are then used for instrument optimization and as a reference guide. Using these controls, flow cytometry instruments can be primed for the detection, analysis, and sorting of specific EV populations. This allows for cross platform comparison and the ability to monitor both quality control (QC) and quality assurance (QA). The method here describes the use of nanoparticles to optimize a flow cytometer for small particle detection. It also outlines the procedures necessary to recover EVs for downstream applications.

Key words Extracellular vesicles, Small particle detection, Nanoscale, Flow cytometry methods, Nanoparticle sorting, Submicron

1 Introduction

Over the past decade, there has been a rapid growth in studies of secreted membrane vesicles, collectively called extracellular vesicles (EVs) [1]. The release of EVs has been reported in the pathologies of cancer [2–5], neurological, hematological [5], cardiovascular [6], autoimmune and rheumatologic diseases [7] and viral infections such as malaria [8]. The study of EVs is gaining increasing interest within both the medical and scientific communities due to the diagnostic and therapeutic possibilities [9, 10]. However, the identification and classification of EVs is problematic. Although advances in various fields, including microscopy, have addressed some of the preliminary hindrances, flow cytometry remains the preferred approach for the characterization of submicron cell-derived particles.

Flow cytometry is an advantageous tool for the analysis and characterization of EVs because of its robust statistical power and its multiparametric capabilities. The overarching goal of this

Winston Patrick Kuo and Shidong Jia (eds.), *Extracellular Vesicles: Methods and Protocols*, Methods in Molecular Biology, vol. 1660, DOI 10.1007/978-1-4939-7253-1_14,

technique is to accurately represent the size distribution and scatter profiles of these vesicles; one method of achieving this is by the use of nanoscale flow cytometry. As with conventional flow cytometry, the suspended vesicles pass through a chamber and are hydrodynamically focused. As vesicles pass through, the laser light is refracted and scattered in all directions. Their scatter properties are measured by detectors; concurrently fluorescent dyes can be used to tag various properties of interest. This principle holds for nanoscale flow cytometry; however, given the unique physical properties of extracellular vesicles, there are additional variables to consider when characterizing them. These hurdles are generally surmounted through specialized instrumentation, and by combining an array of verification techniques.

Characterization of EVs is greatly impeded by several factors: their size (below 100 nm), their overlapping size distribution, particle polydispersity, and an overall low refractive index [11, 12]. Because conventional flow cytometers are not equipped with necessary optical capacities to reach this lower threshold, instruments that are used for nanoscale flow cytometry typically have specific adaptations. In light of these enhancements, there are a number of modifications to the conventional form of analysis that should be considered.

The primary hurdle in analyzing particles at the submicron level has been to accurately represent their size distribution and light scatter profiles. Instrumentation thresholds were originally designed using whole blood as the standard, thereby excluding cellular measurements below 3 μm. Therefore, the most challenging elements have been the availability of sufficient forward scatter (FSC) resolution and precision optics to delineate target populations from background signal and instrument noise. Recently, flow cytometric technology has been developed to distinguish populations spanning the <200 nm to 1 μm range. This is accomplished by increasing the instruments sensitivity via optical enhancements. These particular flow cytometers have higher laser powers, more precise collection optics, and high sensitivity photomultiplier tubes or avalanche photodiodes. In addition, careful consideration has been given to the fluidics. Increased pressure setting control, leading to tighter hydrodynamic focusing, and filtration of the sheath fluid have decreased instrument dark noise. This allows for smaller particles to be seen where instrument noise had overwhelmed the signal. However, the accuracy of these measurements and the validity of the results are questioned. Therefore, hardware upgrades that will not only improve accuracy, but will allow for validation of the procedure by recovery of the EVs, are utilized (*see* **Note 1**) [3, 13].

The particular instrumentation and methodology has been termed nanoscale flow cytometry. The primary hurdle with this technique is always to verify that the vesicles being analyzed are indeed the population of interest. Complementary technologies such as NTA, TRPPS, imaging and microscopy are used in verification of EV distribution [1, 14–17].

Prior to flow cytometric analysis, the isolation techniques and preparation of sample is of utmost importance. In the current climate there are a variety of protocols, and lack of agreement with regard to standards for isolating EVs. It is therefore imperative to establish a reproducible methodology by which to prepare the vesicles. The ISEV has put out general guidelines of minimal general requirements for proceeding with EV research [1].

One of the tools for assessing the size distribution of EVs is the bead sizing curve [14, 18]. This curve is generated prior to EV analysis. Photon Correlation Spectroscopy (PCS) controls and Dragon Green Beads are two types of bead standards that can be employed for this process; they are discussed within this methodology. It is worth noting that within the field there is a lack of suitable controls by which to generate a cohesive sizing curve. The size parameters, in congruence with the refractive indices of EVs (approximately 1.38) [19], make it difficult to find suitable controls that can mimic both the size distribution and light scatter properties of vesicles. Currently there is no consensus on a bead standard which comes close to approximating the profile of the biological. The variety in the materials used to generate bead controls adds an extra variable to the standardization process. Therefore, ApogeeMix beads are analyzed to confirm an instrument's ability to differentiate between fluorescent and nonfluorescent particles. Furthermore, the particles have differing refractive indices for evaluation of scatter resolution.

In testing out various bead candidates, there are notable differences in how different materials affect the position of the beads on scatter plots (*see* **Note 2**). Given the versatile nature of EVs and biological discrepancies, it is useful to have other sources of comparison to supplement the beads used for the sizing curves. One favorable standard by which to assess size and scatter distribution of EVs are liposomes. Given the phospholipid membranes, liposomes present a more attractive candidate than beads for mimicking the biological behavior of other membrane-bound vesicles (*see* **Note 3**).

With respect to instrumentation, the notes described here are specific to the flow cytometers discussed. The adaptations performed were done to maximize optical efficiency of the instrument and to enhance the resolution of the vesicles.

2 Materials

2.1 Dragon Green Polystyrene Beads: Bangs Laboratories, Inc. 9025 Technology Drive Fishers, Indiana 46038-2886 USA

Many imaging applications rely on fluorescent microspheres for the detection of binding events or signal enhancement. Addressable bead populations may be created with different intensities of fluorescence for the development of multiplexed flow cytometric assays, and small fluorescent spheres can function as reporters for ELISA-type assays. Fluorescent microspheres are also useful for fluid tracing, cell tracking and phagocytosis studies. For this protocol, internally labeled Dragon Green fluorescent microspheres were used. Fluorescent microspheres are internally dyed using a solvent swelling/dye - entrapment technique. Internal dyeing produces very bright and stable particles with typically narrow fluorescence CVs. Internally dyed beads are also used extensively in imaging applications, as they offer a greater resistance to photobleaching. Dragon Green is an excellent spectral surrogate for fluorescein (488 nm/530 nm), and is suitable for use with fluorescein filter sets (530/30, 525BP or equivalent). By using beads of differing sizes and fluorescent intensities, one can optimize the flow cytometer for nanoscale flow cytometry.

A stock solution of filtered PBS with 0.1% Tween 20 is prepared. 192, 520, and 780 nm beads are diluted with the PBS/ 0.1% Tween 20 solution, to a final concentration of 1.29×10^7 beads/mL. Prior to dilution, the stock solutions of Dragon Green Beads were sonicated to eliminate clumps.

2.2 The ApogeeMix Product (Cat #1493): Apogee Flow Systems Unit 7, Grovelands, Boundary Way, Hemel Hempstead, HP2 7TE, UK

A mixture of nonfluorescent silica beads and fluorescent latex beads with sizes from 110 nm to 1300 nm can be used to prepare flow cytometers for the analysis of small biological particles. They offer an easy means to assess the sensitivity and resolution of the flow cytometer's light scatter and fluorescence optics and the silica beads offer a means to calibrate a flow cytometers light scatter optics at a refractive index of approximately 1.43. The ApogeeMix (Cat#1493) contains 25 mL of an aqueous mixture of plastic spheres with diameters 180, 240, 300, 590, 880 and 1300 nm with refractive index $\eta = 1.43$, and 110 nm and 500 nm green fluorescent (blue laser) beads with refractive index $\eta = 1.59$ (latex). The product is intended to be used to assess a flow cytometer's light scatter and fluorescence performance (both sensitivity and resolution).

2.3 Photon Correlation Spectroscopy Latex Beads 6602336: Beckman Coulter, Inc. Life Science Division Headquarters 5350 Lakeview Parkway S Drive Indianapolis IN 46268

Utilized in a wide variety of applications including electron microscopy and cell counter calibration, antibody mediated agglutination diagnostics, phagocytosis experiments and many others. Photon Correlation Spectroscopy (PCS) controls are uniform latex particles intended for quality control analysis of Dynamic Light Scatter instrumentation.

2.4 0.1% Tween 20 in PBS

Nonionic detergent, surfactant, widely used in biochemical applications. As nanobeads can be sticky, 0.1% Tween 20 is added to PBS to prevent the beads from clumping together. Sterile Phosphate Buffered Saline: PBS is a balanced salt solution used for a variety of cell culture applications, such as washing cells before dissociation, transporting cells or tissue, diluting cells for counting, and preparing reagents. PBS is formulated without calcium and magnesium for rinsing chelators from the culture before cell dissociation.

2.5 Specialized Instrumentation

1. MoFlo Astrios EQ

   The beam splitter is removed from the FSC assembly. This allows the maximum amount of laser light to be detected through the FSC1 PMT detector (*see* **Note 4**). For small particle detection the P1 mask was chosen, the larger opening allows for maximum resolution and dynamic range (*see* **Note 5**). Finally, 561 nm SSC was assigned as the Trigger parameter and FSCA Log versus SSCA Log was plotted for population determination (*see* **Note 6**).

2. MoFlo XDP with NanoView module

   The new optical system design utilizes a custom aspheric imaging lens that has been optimized to collect the scattered light from the core stream and image it onto a 200 μm pinhole (*see* **Note 7**). The collection angles in the forward scatter direction extend up to 18° and are directed to the FSC PMT. For this study, the largest available blocker bar was used (±6.3° to ±12.6° angles blocked as rotated from horizontal to vertical) and set to an approximate angle of 45°. 488 SSC was used as the trigger parameter and FSCA Log versus SSCA Log was plotted for population determination.

3. CytoFLEX

   Beckman Coulter's CytoFLEX is equipped with custom fluidics and the ability to use Violet (405 nm) Side Scatter (VSSC) as a trigger parameter. The CytoFLEX has the ability to both trigger off and analyze by VSSC. This is an important feature when considering Mie theory and its effects on small particle detection (*see* **Note 8**). The Fiber Array Photo Diode

(FAPD) provides better optical resolution especially with small particle measurements due to low-noise detection with high quantum efficiency and minimum light loss ensuring high signal to noise ratio. VSSC was used as the trigger parameter and VSSCA Log versus SSCA Log was plotted for population determination.

3 Methods

The method described is for determination of a flow cytometry instrument's ability to evaluate submicron particles. As instruments differ between models and manufacturers, specific details in regard to modifications and setup may not be covered.

QC was performed according to manufacturer's recommendations. All instrumentation and protocols were configured for small particle detection. While instrumentation differs, the protocol remained consistent throughout. However, due to the MoFlo XDP with NanoView and MoFlo Astrios EQ's alignment system, micrometers on the FSC attachments were adjusted to maximize FSC signaling. The CytoFLEX required a bandpass filter change (405 nm placed to detectable APD); no alignment or manual adjustments were performed.

3.1 Dragon Green Beads: Evaluation of Size and Fluorescence

Appropriate Dot Plots and Histograms are displayed in instrument software (*see* **Note 9**).

The DGBs are acquired for instrument optimization:

1. 780 nm Dragon Green Beads.
2. 520 nm Dragon Green Beads.
3. 192 nm Dragon Green Beads.
4. 192 nm, 520 nm, and 780 nm Dragon Green Beads Mixed.

780 nm Dragon Green Beads were acquired to set the Scatter properties for differentiation between beads and low-end noise. In addition, scatter properties were adjusted to maximize resolution and dynamic range. The largest size is chosen first for ease of particle identification and to prove instrument ability to analyze below 1 μm.

192 nm Dragon Green Beads were acquired to test the ability of the instrument to differentiate between the particle and noise. As particle size decreases, instrument Noise populations will begin to overwhelm the Dragon Green Bead's signal.

520 nm Dragon Beads were acquired for accuracy of separation of 192 nm and 780 nm beads. This allows for visualization of dynamic range of the instrument.

Gates were drawn to encompass the three distinct populations.

Mixed Dragon Beads were acquired to ensure proper gating and maximum separation of bead populations for the determination of a relative sizing distribution matrix (*see* **Note 10**).

The fluorescent characteristics of the Dragon Green Beads, allow for the verification of size distribution based on fluorescent intensity. This serves as a form of secondary assessment and can be used for data assurance. In addition, a sample of the stock PBS solution is acquired for quantification of the background contribution of the PBS.

All three populations were easily separated from background and from one another. Instrument has been optimized and template and settings are saved (Fig. 1).

3.2 Photon Correlation Spectroscopy Beads (See Note 11)

The PCS controls are acquired for instrument optimization:

1. 500 nm PCS control Beads.
2. 300 nm PCS control Beads.
3. 200 nm PCS control Beads.
4. MIXTURE: 200 nm, 300 nm, 500 nm PCS.

Using the Dragon Green Beads saved protocols on each instrument, the PCS controls were acquired according to the Dragon Beads protocol listed above. Gates were drawn to encompass the three distinct populations.

Mixed PCS controls were acquired to ensure proper gating and maximum separation of bead populations (Fig. 2) for the determination of a relative sizing distribution matrix (*see* **Note 12**). Any changes to template should be saved as a new protocol.

3.3 ApogeeMix Sensitivity and Resolution Performance

Prior to dilution, the stock solutions of ApogeeMix beads were sonicated to eliminate clumps.

100 μL of ApogeeMix, as described in Table 1, is added to 1 mL of filtered PBS.

Using the nanoscale flow cytometry size distribution saved protocols on each instrument, the ApogeeMix was acquired. Previous gates were used from the three previous PCS control and DGB size and fluorescent distribution curves (*see* **Note 13**).

The use of ApogeeMix verifies the instrument's ability to differentiate samples based on refractive indices. If particle size distribution curves are based solely on controls of higher refractive index (approximately 1.5–1.6), this may lead to a gross misrepresentation of actual EV size [19]. Analyzing samples of similar refractive index (*see* **Note 14**), allows for more accurate representation of EV size. This is essential in the characterization of EVs due to their translucent membrane and characterization of exosome populations ranging from 80 nm to 150 nm (Fig. 3).

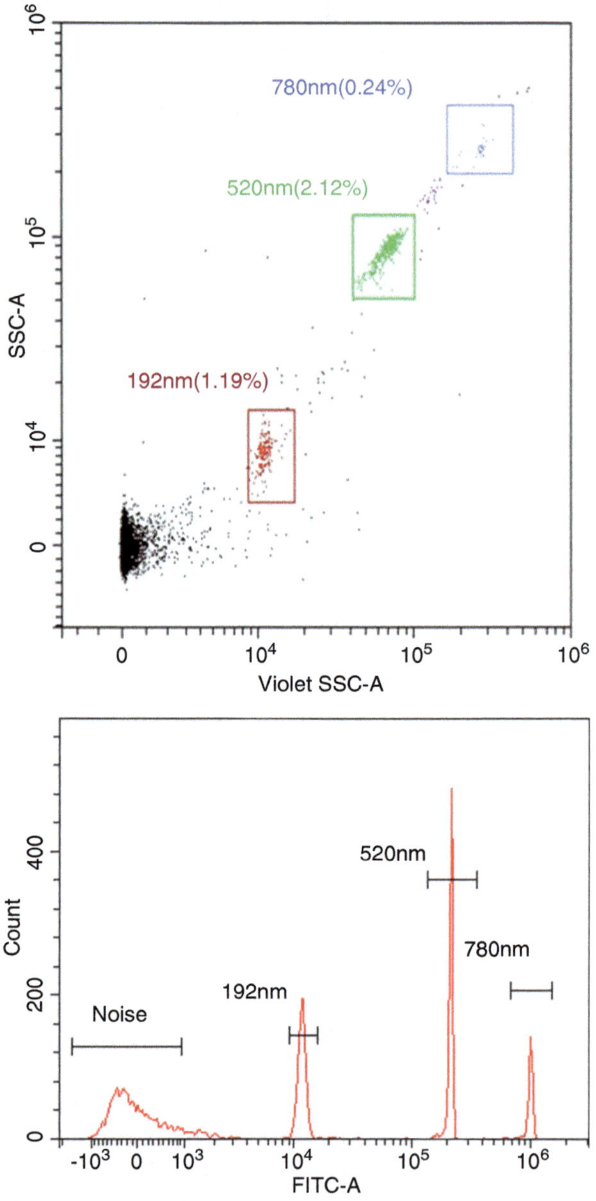

Fig. 1 Dragon Green Beads acquired on the CytoFLEX flow cytometer for determination of a relative size and fluorescence profile

3.4 100 nm Photon Correlation Spectroscopy Bead: Lower Detection Limit (See Note 15)

100 nm PCS control will be acquired to determine if the particles can easily be separated from noise without sacrificing dynamic range.

Adjustments in the scaling, threshold and instrument alignment may be required to maximize separation from noise (Fig. 4).

Using the final protocol with saved voltages, gains, threshold, and templates, EVs can analyzed and recovered for further evaluation (*see* **Note 16**).

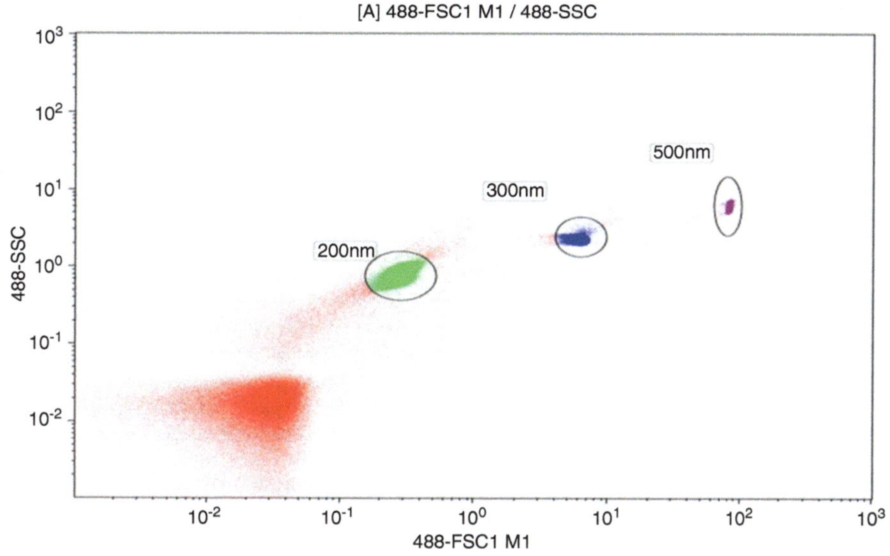

Fig. 2 PCS controls acquired on the MoFlo Astrios EQ to determine relative sizing curve of nonfluorescently labeled EVs

Table 1
Size, concentration, and fluorescence qualities of particles in Apogee Calibration Bead Mix. Information provided by Apogee Flow Systems

Particle size (nm)	Approximate number per microliter	Fluorescence from 488 nm excitation
110	8000	Green
180	23,000	None
240	10,000	None
300	9000	None
500	3600	Green
590	2700	None
880	3900	None
1300	3400	None

For future experiments, the saved protocol is reloaded and adjustments are made to the alignment. Adjustments are made to guide populations to original gates made. This allows for consistency in analysis between differing time points (*see* **Note 17**).

3.5 Acquisition of EVs

Keeping determined settings from the above described protocols, evaluate biological EV samples for acquisition, analysis, and sorting.

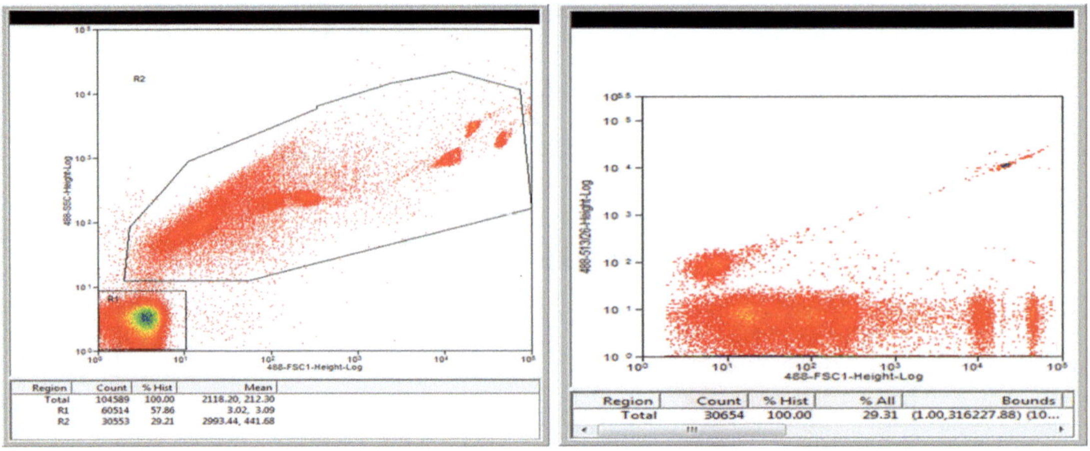

Fig. 3 Determination of relative size and fluorescence of Apogee Calibration Bead Mix. In addition, due to the known refractive index, the beads can be used for comparison to biological particles

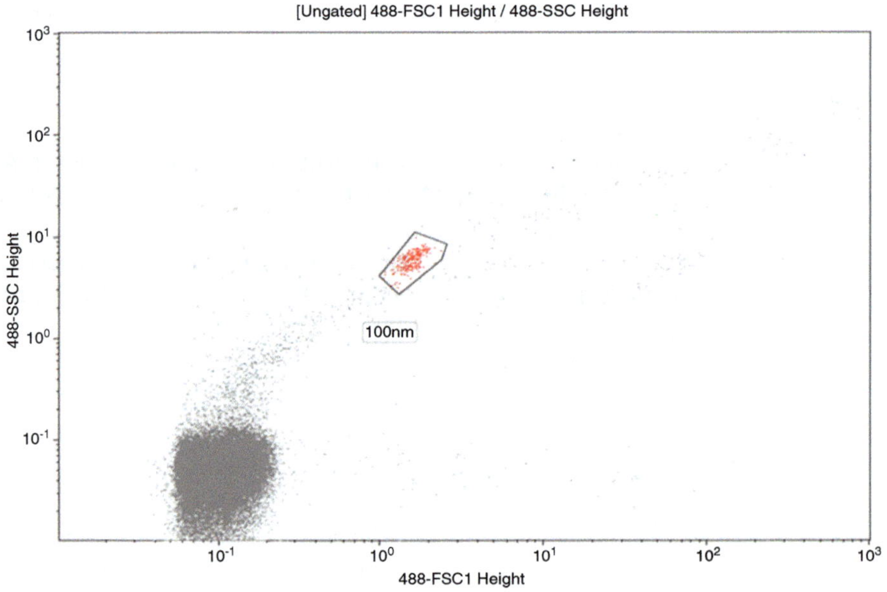

Fig. 4 100 nm PCS nonfluorescent beads to evaluate the ability to differentiate particles from noise

Samples from plasma, urine, serum and other biological fluids can be evaluated (*see* **Note 18**). Samples are diluted at a ratio of 1:1000 (*see* **Note 19**).

Samples are acquired based on total number of events or a specific amount of time (*see* **Note 20**). 500 μL of a 500 nm bead at a fixed concentration is spiked into plasma samples as an internal control. Events, including background noise events, are acquired to a total of 500 k to 1 M. Time is set to a total of 180 s after a 30 s equilibration. Samples are acquired at a rate of 1–5 k events per second.

Gates to identify 500 nm bead control and sub 500 nm EVs are drawn. In addition, fluorescent antibodies and/or proteins can be identified from populations of interest (*see* **Note 21**).

3.6 Sorting of EVs

With regard to EVs, the sorting scheme is similar to conventional strategies [20]. The selected populations are separated via electronic charge into the appropriate collection medium. For the methods discussed here the MoFlo AstriosEq and the MoFlo XDP with NanoView module were the instruments employed. A 70 μm nozzle was used, and pressure was kept at 60 psi.

For determination of the Amplitude and Frequency on the MoFlo AstriosEQ and MoFlo XDP, a camera is used to scan the drops formed by the strobe (*see* **Note 22**). A breakoff, or the last attached drop, is found and used to calculate drop delay.

Intellisort II is used for calculation of the drop delay, time from interrogation to sort breakoff point. Intellisort II is a beadless drop-delay technology. This alleviates the requirement to remove sterile samples during drop-delay recalculation due to nozzle blockages and ensures uninterrupted sort sterility. Intellisort II maintains droplet breakoff for uninterrupted sorting (*see* **Note 23**).

Populations of interest, either size or fluorescent based, are gated and sorted into appropriate collection media (Fig. 5). Sorted samples can be utilized for downstream applications (*see* **Note 24**) such as dark-field and fluorescent microscopy (Fig. 6).

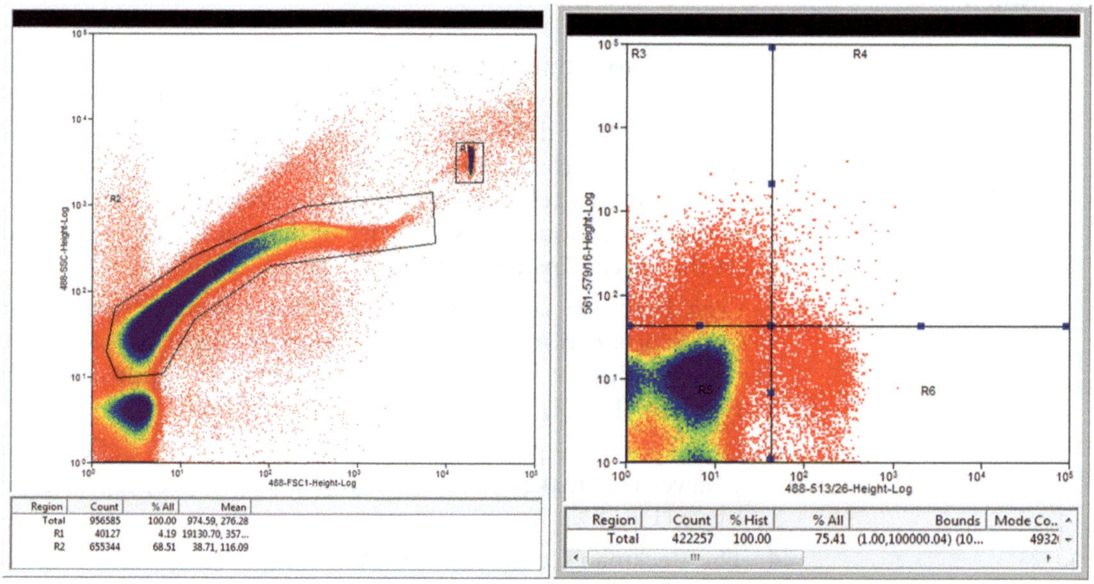

Fig. 5 Plasma derived EVs, spiked with 500 nm PCS beads for QA, were analyzed and sorted based on size and/or fluorescent properties

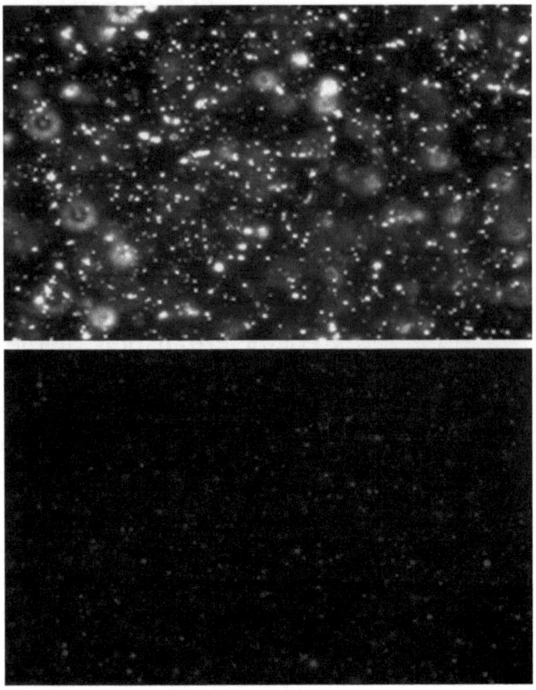

Fig. 6 Darkfield and Fluorescent Microscopy images of sorted plasma EVs. Images provided by Dr. Ionita Ghiran

4 Notes

1. In order to reduce the level of user bias into the instrument setup, it is highly advised to generate a comprehensive setup protocol which delineates basic standards. Parameters to consider for this protocol include threshold levels, pressure differentials, and the relative amount for instrument noise to include in the acquisition. The sensitivity of detection will be dictated on the specific instrument used.
2. Even beads of same reported size can appear differently or scatter proportionally different amounts of light based on the refractive index of the particular material (approximately 1.4–1.6) (Fig. 7).
3. Homogeneous populations of liposomes are difficult to achieve; this makes it difficult to generate a proper standard curve (Fig. 8).
4. The Beckman Coulter MoFlo AstriosEq is equipped with two FSC PMT pathways separated by a beam splitter (60/40 split). FSC1 is a direct laser beam pathway (60), with the FSC2 being directed at an angle from the beam splitter (40). Upon removal

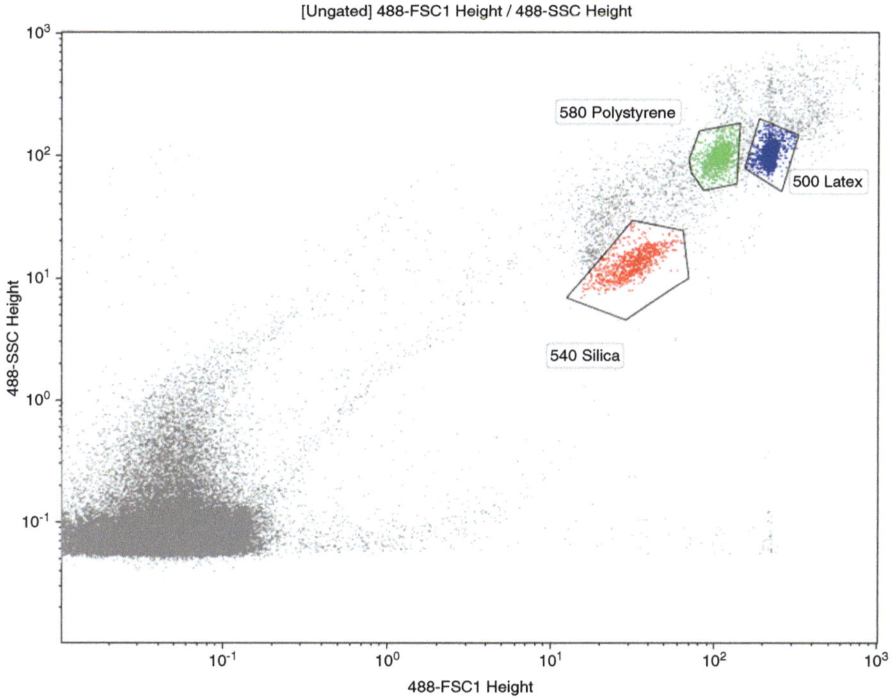

Fig. 7 Reference beads of similar size show inconsistencies in size profile. This is due to the difference in refractive indices of the bead materials. Silica beads appear smaller than latex and polystyrene due to their lower RI

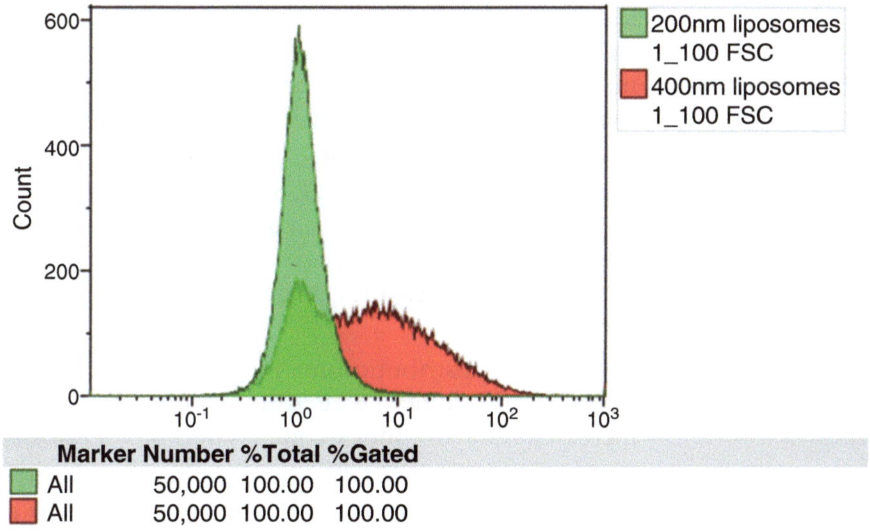

Fig. 8 200 nm and 400 nm liposome fractions. Due to the inclusion of smaller liposome populations in the 400 nm sample, mixed liposome sizing curves are difficult to achieve

of the beam splitter, 100% of laser light is directed to the FSC1 pathway.

5. Seven different and unique masks are provided to optimize particle identification and focus laser light to the PMTs. After testing of all the masks, P1 mask was chosen.
6. Ability to trigger from any of the scatter parameters associated with the six collimated laser lines. The 561 nm SSC was chosen due to its location as the center of the seven pinholes. Testing of the remaining five SSC parameters was performed and similar results were recorded.
7. The pinhole serves to align the system and remove the stray laser light that has not been generated by the particle of interest and greatly reduces the background light that is received at the FSC PMT.
8. Mie theory predicts that the scattering cross section of a particle, and thus its scattering intensity is dependent on the wavelength of light, the angle of collection , and the size, shape, and refractive index of the particle.

   All other factors being equal, using a shorter illumination wavelength will result in an increase in scattering cross section, and thus more scattered light.
9. A single bivariate Dot Plot is set on FSC vs. SSC or VSSC vs. SSC is plotted. In addition, a single univariate Histogram of a FITC (525 nm) channel is plotted.
10. Larger particles may mask the existence of their smaller counterparts. Therefore, mixed populations verify the ability to separate and distinguish multiple populations. Voltage, gain, and threshold settings can be adjusted to maximize performance. However, it is strongly suggested that single bead populations be acquired again for quality assurance purposes.
11. Results are expected to be similar to Dragon Green Beads based on the similarity in size and refractive index. However, PCS controls do not contain a fluorescent marker. Therefore, measurements will be based solely on Scatter detection. This template will be used for analysis of EVs that are not fluorescently labeled.
12. Relative size distribution matrices are produced based on the entire bead profiles. Size distributions are relative due to increased refractive index compared to biological.
13. Fluorescent Threshold/Trigger can be used to identify fluorescently labeled EVs. Both DGBs and ApogeeMix are excited using a FITC trigger due to their Green fluorescent properties.
14. EVs have a measured refractive index of approximately 1.38. ApogeeMix beads unlabeled bead controls have a refractive

index of 1.42. This reduced refractive indices difference allows for more accurate determination of size of EVs.

15. Due to exosomes measuring from 80 nm to 150 nm on average, 100 nm PCS controls are acquired to assess ability to differentiate exosome populations from noise.
16. The general time scale for setting up various calibration controls, albeit beads or liposomes is approximately 30 min. This excludes general instrument warm-up time, and the otherwise general alignment and daily maintenance and calibration involved; which can add another 30–45 min.
17. It is suggested that if either major adjustments are needed or bead lots change, then the entire method is repeated to ensure optimal performance. Persistent changes in performance may be indicative of an optical issue. Due to the need for increased optical sensitivity, laser powers and optical alignment need to be consistently monitored.
18. Studies represented in this particular protocol are from plasma samples that were processed according to Diurnal Variations of Circulating Extracellular Vesicles Measured by Nano Flow Cytometry [14].
19. Serial dilutions of the plasma were run to test for the evaluation of coincidence events, due to multiple particles being analyzed simultaneously. This is a common issue when dealing with nanoparticles as the size of the laser beam exceeds the size of the particle. Therefore, multiple particles are at the interrogation point at the same time. In the analysis of EVs and nanoparticles, the phenomenon is known as swarming [21].
20. Total number of events or set acquisition time limit is chosen based on particular study.
21. Fluorescent labeling of EVs is problematic and beyond the scope of this methodology [22].
22. Amplitude and frequency for a 70 μm nozzle size at 60 PSI are approximately 15 V and 95 kHz respectively. These values are reasonable initial settings for manual adjustments.
23. Drop delay noise can enter into the system presenting as a uniform population. Adjustments to the frequency and/or micrometers will allow for removal of the drop delay noise. In addition, air in the system will cause fluctuations in the calculation of drop delay.
24. Due to drop volume , sorted EVs are heavily diluted. Either centrifugation or vacuum aspiration may be necessary to obtain proper concentrations for other applications.

References

1. Lötvall J, Hill AF, Hochberg F, Buzas EI, Di Vizio D, Gardiner C, Gho YS, Kurochkin IV, Mathivanan S, Quesenberry P, Sahoo S, Tahara H, Wauben MH, Witwer KW, Thery C (2014) Minimal experimental requirements for definition of extracellular vesicles and their functions: a position statement from the International Society for Extracellular Vesicles. J Extracell Vesicles 3:26913
2. Rabinowits G, Gercel-Taylor C, Day JM, Taylor DD, Kloecker GH (2009) Exosomal microRNA: a diagnostic marker for lung cancer. Clin Lung Cancer 10(1):42–46
3. Taylor DD, Gercel-Taylor C (2011) Exosomes/microvesicles: mediators of cancer-associated immunosuppressive microenvironments. Semin Immunopathol 33(5):441–454
4. Taylor DD, Gercel-Taylor C (2005) Tumour-derived exosomes and their role in cancer-associated T-cell signalling defects. Br J Cancer 92(2):305–311
5. Toth B, Liebhardt S, Steinig K, Ditsch N, Rank A, Bauerfeind I, Spannagl M, Friese K, Reininger AJ (2008) Platelet-derived microparticles and coagulation activation in breast cancer patients. J Thromb Haemost 100(4):663–669
6. Danielson KM, Das S (2014) Extracellular vesicles in heart disease: excitement for the future? Exosomes Microvesicles 2:1
7. Sellam J, Proulle V, Jüngel A, Ittah M, Miceli Richard C, Gottenberg JE, Toti F, Benessiano J, Gay S, Freyssinet JM, Mariette X (2009) Increased levels of circulating microparticles in primary Sjögren's syndrome, systemic lupus erythematosus and rheumatoid arthritis and relation with disease activity. Arthritis Res Ther 11(5):1–11
8. Mantel PY, Hoang AN, Goldowitz I, Potashnikova D, Hamza B, Vorobjev I, Ghiran I, Toner M, Irimia D, Ivanov AR, Barteneva N, Marti M (2013) Malaria-infected erythrocyte-derived microvesicles mediate cellular communication within the parasite population and with the host immune system. Cell Host Microbe 13(5):521–534
9. Cicero A, Stahl P, Raposo G (2015) Extracellular vesicles shuffling intercellular messages: for good or for bad. Curr Opin Cell Biol 35:69–77
10. De Toro J, Herschlik L, Waldner C, Mongini C (2015) Emerging roles of Exosomes in normal and pathological conditions: new insights for diagnosis and therapeutic applications. Front Immunol 6:203
11. Orozco AF, Lewis DE (2010) Flow cytometric analysis of circulating microparticles in plasma. Cytometry A 77(6):502–514
12. van der Pol E, Hoekstra AG, Sturk A, Otto C, van Leeuwen TG, Nieuwland R (2010) Optical and non-optical methods for detection and characterization of microparticles and exosomes. J Thromb Haemost 8(12):2596–2607
13. Stoffel CL, Kathy RF, Rowlen KL (2005) Design and characterization of a compact dual channel virus counter. Cytometry A 65 (2):140–147
14. Danielson KM, Estanislau J, Tigges J, Toxavidis V, Camacho V, Felton EJ, Khoory J, Kreimer S, Ivanov AR, Mantel PY, Jones J, Akuthota P, Das S, Ghiran I (2016) Diurnal variations of circulating extracellular vesicles measured by Nano flow cytometry. PloS One 11(1):e0144678
15. Headland SE, Jones HR, D'Sa AS, Perretti M, Norling LV (2014) Cutting-edge analysis of extracellular microparticles using ImageStreamX imaging flow cytometry. Sci Rep 4:5237
16. Momen-Heravi F, Balaj L, Alian S, Tigges J, Toxavidis V, Ericsson M, Distel RJ, Ivanov AR, Skog J, Kuo WP (2012) Alternative methods for characterization of extracellular vesicles. Front Physiol 3:354
17. van der Pol E, Coumans F, Varga Z, Krumrey M, Nieuwland R (2013) Innovation in detection of microparticles and exosomes. J Thromb Haemost 11(1):36–45
18. van der Pol E, Coumans FA, Grootemaat AE, Gardiner C, Sargent IL, Harrison P, Sturk A, van Leeuwen TG, Nieuwland R (2014) Particle size distribution of exosomes and microvesicles determined by transmission electron microscopy, flow cytometry, nanoparticle tracking analysis, and resistive pulse sensing. J Thromb Haemost 12(7):1182–1192
19. Green RE, Sosik HM, Olson RJ, Durand MD (2003) Flow cytometric determination of size and complex refractive index for marine particles: comparison with independent and bulk estimates. Appl Opt 42(3):526–541
20. Orfao A, Ruiz-Argüelles A (1996) General concepts about cell sorting techniques. Clin Biochem 29(1):5–9
21. van der Pol E, van Gemert MJ, Sturk A, Nieuwland R, van Leeuwen TG (2012) Single vs. swarm detection of microparticles and exosomes by flow cytometry. J Thromb Haemost 10(5):919–930
22. Choi H, Lee DS (2016) Illuminating the physiology of extracellular vesicles. Stem Cell Res Ther 7(1):55

Chapter 15

Characterization of Extracellular Vesicles by Size-Exclusion High-Performance Liquid Chromatography (HPLC)

Tao Huang and Jiang He

Abstract

Extracellular vesicles (EVs) have recently attracted substantial attention due to the potential diagnostic and therapeutic relevance. Although a variety of techniques have been used to isolate and analyze EVs, it is still far away from satisfaction. Size-exclusion chromatography (SEC), which separates subjects by size, has been widely applied in protein purification and analysis. The purpose of this chapter is to show the applications of size-exclusion high-performance liquid chromatography (HPLC) as methods for EV characterization of impurities or contaminants of small size, and thus for quality assay for the purity of the samples of EVs.

Key words Size-exclusion chromatography, High-performance liquid chromatography (HPLC), Extracellular vesicles, Exosomes

1 Introduction

Extracellular vesicles (EVs), which include exosomes, are membrane enclosed vesicles that are secreted by all cells, circulated in blood and are readily accessible in most body fluids [1–7]. Carrying abundant biomolecules like proteins, RNAs, DNAs, metabolites and lipids, these vesicles are important messengers and mediators in intercellular communications, and play pivotal role in tumor progression and metastasis [8]. In addition to the efforts to understand the biology of EVs in different disease, extensive attention has been recently also paid to the studies of these nanoparticles as biomarker and the engineering to be used as therapeutic vehicles for drug and gene delivery. All these research activities using different approaches including proteomics [9–11], transcriptomics [10, 11], lipidomics [10, 12] and vesicle engineering [13, 14] demand samples of high purity.

However, current isolation and characterization methods are far to be satisfied to assure high purity EVs for clinic use. Various methods have been used for vesicle isolation, including ultracentrifugation [15], size exclusion (filtration or chromatography)

Winston Patrick Kuo and Shidong Jia (eds.), *Extracellular Vesicles: Methods and Protocols*, Methods in Molecular Biology, vol. 1660, DOI 10.1007/978-1-4939-7253-1_15, © Springer Science+Business Media LLC 2017

[15–19], immunoaffinity isolation [15, 20], precipitation (Exo-Quick or "salting-out") [21, 22] or the combinations of above techniques. Each of these methods has advantages and disadvantages. There is a lack of clear consensus for an optimal method of isolation as well as characterization of the pure samples of these EVs [23]. A few techniques have been commonly used to characterize isolated vesicles, such as transmission electron microscopy (TEM) or Cryo-EM [24], nanoparticle tracking analysis (NTA, Nano-Sight) [25] and western blot for protein marker confirmation [15]. Recently, Webber et al. proposed to use the ratio of EVs counts to protein concentration as an indirect means to check the purity of EVs preparation [26]. All these approaches are based on the presence of vesicular particles. On the other hand for impurity detection, while the presence of protein biomarkers that are not expressed by vesicles [26] was proposed for quality assay, the selection of these markers is challenging and may not be accurate due to the limited understanding of how proteins are packaged into EVs.

To assure high quality EVs, not only the desired population of EVs must be confirmed to be present, but contaminants and impurities must also be demonstrated to be absent. Unfortunately, there is no robust and reproducible method available to rule out the presence of contaminants. We hereby describe a simple SE-HPLC analysis to detect water-soluble impurities of small size in the samples of exosomes or extracellular vesicles. SE-HPLC is widely used and accessible in most institutions and could be used along with other current techniques to characterize EVs [27]. It provides a convenient and robust means to quantify the EV products for further biology and function studies.

2 Materials

Chemicals and reagents are purchased from commercial sources and used without further processing unless indicated otherwise.

1. A Beckman Coulter Optima LE-80K ultracentrifuge with an SW28 rotor for ultracentrifugation.
2. An Agilent 1260 Infinity high-performance liquid chromatography (HPLC) system equipped with an autosampler and a variable wavelength UV detector.
3. A size-exclusion column (Superose 12 10/300 GL from GE Healthcare) (9–13 μm) for HPLC.
4. DPBS buffer (pH 7.4) (Hyclone Laboratories, Inc.) for HPLC mobile phase and the ultracentrifugation.
5. The Gel Filtration HMW Calibration Kits (GE Healthcare) which contains five proteins from 43 kD to 669 kD and Blue Dextran 2000.

6. 0.45 μm and 0.22 μm polyvinylidene fluoride (PVDF) syringe filters (Sigma-aldrich).
7. Dulbecco's Modified Eagle's Medium (DMEM, Thermo Fisher Scientific).
8. Fetal bovine serum, FBS (research grade and sterile triple 100 nm filtered, Thermo Fisher Scientific).
9. The EV-depleted fetal bovine serum (FBS) used for cell culture are processed in-house from the commercial available FBS as shown below.

3 Methods

3.1 Preparation of Vesicle-Depleted Fetal Bovine Serum (FBS)

1. The commercial FBS is subject to ultracentrifugation at 120,000 × *g* overnight (15 h).
2. The upper liquid of 80% volume is reserved for cell culture while the remaining bottom liquid was used for EV preparation from FBS itself (*see* **Note 1**).
3. The EV-depleted FBS is frozen at −20 °C for storage.

3.2 Preprocessed Cell Culture Supernatant for Isolation of EVs

1. Cells such as MDA-MB-231, MSTO-211H, and endothelial cells are maintained at 37 °C at 5% CO_2 in Dulbecco's Modified Eagle's Medium supplemented with 4% EV-depleted FBS (*see* **Note 2**).
2. When cell confluence reached 90%, the supernatant is collected.
3. The supernatant is preprocessed by subsequent centrifugations at 400 × *g* for 5 min and 3000 × *g* for 30 min to remove cell debris.
4. The preprocessed supernatant is used immediately or kept at −80 °C for future use.
5. Before ultracentrifugation, the preprocessed supernatant is sequentially filtered through 0.45 μm and 0.22 μm PVDF filters (*see* **Note 3**).

3.3 EV Isolation from FBS or Preprocessed Cell Culture Supernatant

1. The preprocessed supernatant or FBS is ultracentrifuged using SW28 flying rotor at 120,000 × *g* for 90 min to pellet EVs (*see* **Note 4**).
2. The EV pellet is resuspended, washed with DPBS and ultracentrifuged at 120,000 × *g* for 90 min, and the washing step is repeated multiple times, during which SE-HPLC can be used to monitor the removal of contaminants.
3. The final EV pellet is resuspended in DPBS and aliquots are used immediately for various characterizations or stored at −80 °C.

3.4 Size-Exclusion HPLC Analysis

1. A superpose 12 10/300 GL column with exclusion limit of 2,000,000 Da is selected for impurity analysis (*see* **Note 5**, Table 1).
2. The column is calibrated with high molecular weight (HMW) kit, which contains five proteins from 43 kD to 669 kD and Blue Dextran 2000 (Fig. 1).
3. DPBS buffer (pH 7.4) is used as mobile phase (*see* **Note 6**).
4. The flow rate is set for 1 mL/min (*see* **Note 7**).
5. 50–100 μL of an aliquot is injected via the autosampler (*see* **Note 8**).
6. UV absorbance is detected at a wavelength of 254 nm (*see* **Notes 9** and **10**).

Table 1
Comparison of HPLC size-exclusion columns

Manufacturer	Product	Pore diameter (nm)	Particle (mm)	Exclusion limit Protein M	Optimal range	pH range
Bio-Rad (silica-based diol)	Bio-Sil SEC 125	12.5	5	100 kD	5–100 kD	2.0–8.0
	Bio-Sil SEC 250	25	5	300 kD	10–300 kD	
	Bio-Sil SEC 400	40	5	1000 kD	20–1000 kD	
MICRA (silica-based diol)	SynChropak GPC Peptide	5	5	35 kD	1–35 kD	2.0–8.0
	SynChropak GPC100	10	5	500 kD	5–160 kD	
	SynChropak GPC300	30	5	2000 kD	10–500 kD	
	SynChropak GPC500	50	7	5000 kD	40–1000 kD	
	SynChropak GPC1000	100	7	10,000 kD	40–10,000 kD	
	SynChropak GPC4000	400	10	>10,000kD		
GE Healthcare (semirigid agarose)	Superose 6		13	40,000 kD	5–5000 kD	1.0–14
	Superose 12		10	2000 kD	1–300 kD	3–12
	Superdex 75		11–15	100 kD	3–70 kD	
	Superdex 200		11–15	1300 kD	10–600 kD	
Agilent Technologies (silica-based diol, stabilized)	ZORBAX GF-250	15	5	400 kD	4–400 kD	2.5–8.5
	ZORBAX GF-450	30	6	1000 kD	10–1000 kD	

Data from manufacturer's catalogs

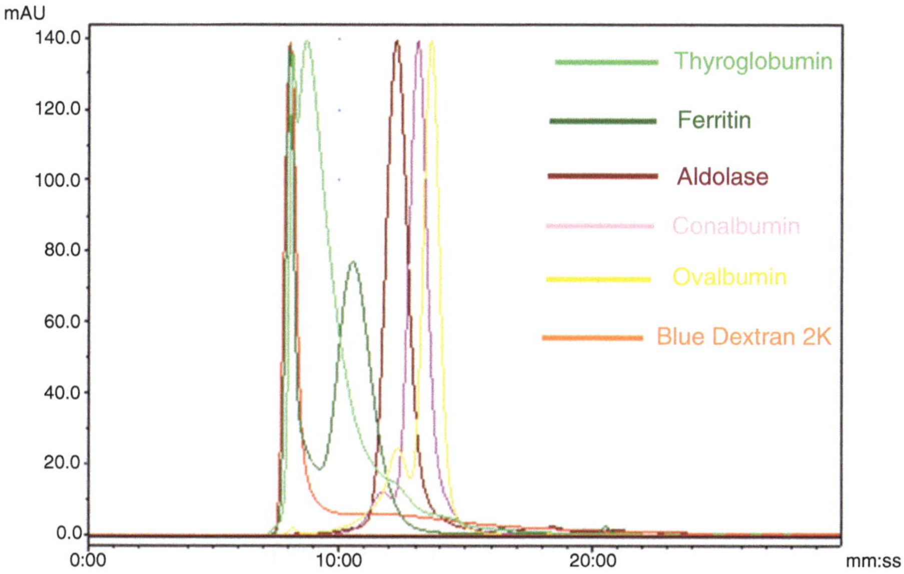

Fig. 1 The column calibration with HMW kit provided by the column manufacturer. Elution profiles of Calibration Kit standards: Thyroglobumin (669 kD, 8 min 43 s and 8 min 4 s), Ferritin (440 kD, 10 min 34 s and 8 min 4 s), Aldolase (158 kD, 12 min 17 s), Conalbumin (78 kD, 13 min 6 s), Ovalbumin (44 kD, 13 min 35 s), and Blue Dextran 2000 (2000 kD, 8 min 4 s) (Courtesy of Journal of Circulation Biomarkers)

4 Notes

1. FBS is used as nutrients during cell culture. To avoid irrelevant vesicle contamination for further biological analysis and studies, it is critical to deplete those vesicles from FBS before adding it into cell media.
2. A variety of cell lines could be cultured to “produce” EVs. In the authors’ lab, MDA-MB-231 cancer cells, MSTO-211H cancer cells, and endothelial cells have been used for study.
3. Filtration further reduced cell debris or large vesicle particles although particles larger than 220 nm could still squeezed through the filter. Filtration through 0.45 μm first makes it easier to filter further through 0.22 μm. Otherwise, the filter will get clogged.
4. The ultracentrifugation speed and time can vary depending on the equipment and rotor type [15, 23].
5. The selection of size-exclusion column is critically important. Packing materials, resolving limit, pressure sustainability, and so forth should be taken into consideration. The high exclusion limit of 2×10^6 Daltons was used in this study with the aim to give a wide detection range and better resolution of small impurities for the extracellular vesicles with size less than around 200 nm. Other range of exclusion limit may give

different results depending on the vesicles and the sources to prepare these vesicles. From our experience, size-exclusion columns with higher exclusion limit may be able to resolve large particles but will give less desirable resolution of small size molecules. (Figs. 2 and 3).

Size-exclusion columns for HPLC are usually packed with materials based on silica which has been covered with a carbohydrate bonded phase. The particle size for SE-HPLC column packing is usually much smaller than that for low pressure SEC, which provides better resolution at higher pressure. Parameters of some commercially available columns for SE-HPLC are listed in the Table 1, from which an appropriate column can be selected for each application.

6. Here we used DPBS buffer as mobile phase since the EV samples was isolated through multiple DPBS wash steps and kept in DPBS. Generally, it is critical to choose suitable mobile phase in size-exclusion chromatography to ensure the mobile phase can solubilize the sample and maintain low viscosity. Also the mobile phase could influence the interactions between the stationary phase and the sample materials. If there is mobile phase recommended by column providers, it is wise to follow

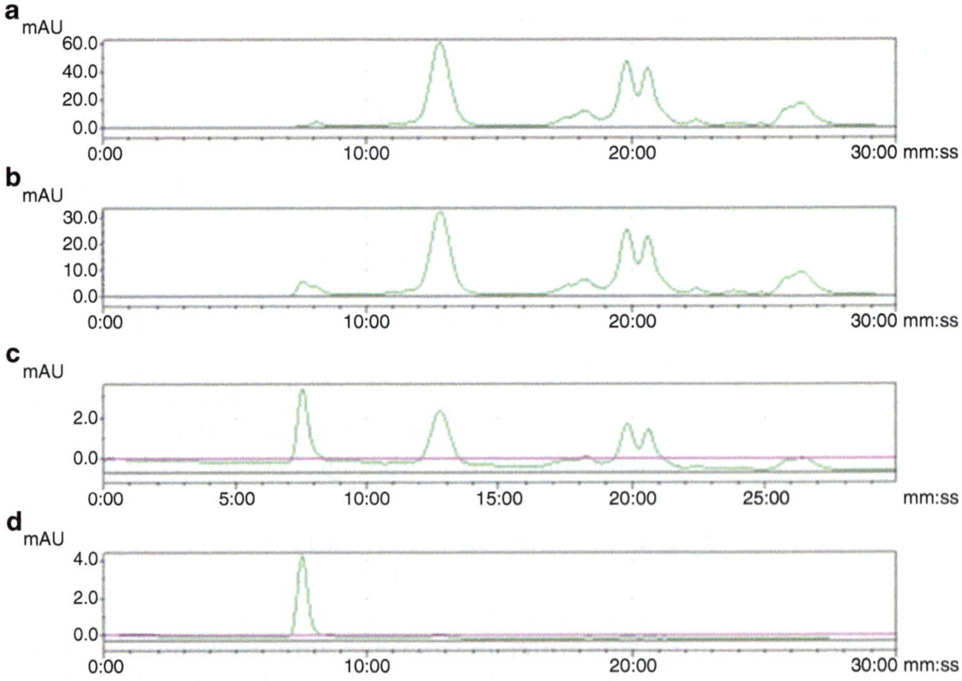

Fig. 2 Representative chromatograms of extracellular vesicles from MSTO cell culture supernatant. Aliquots after once (**a**), twice (**b**), three times (**c**), and four times (**d**) ultra-centrifugations were analyzed. The peak with a retention time of 7.5 min stands for the exosome particles while the rest peaks are impurities of small size. (Courtesy of Journal of Circulation Biomarker)

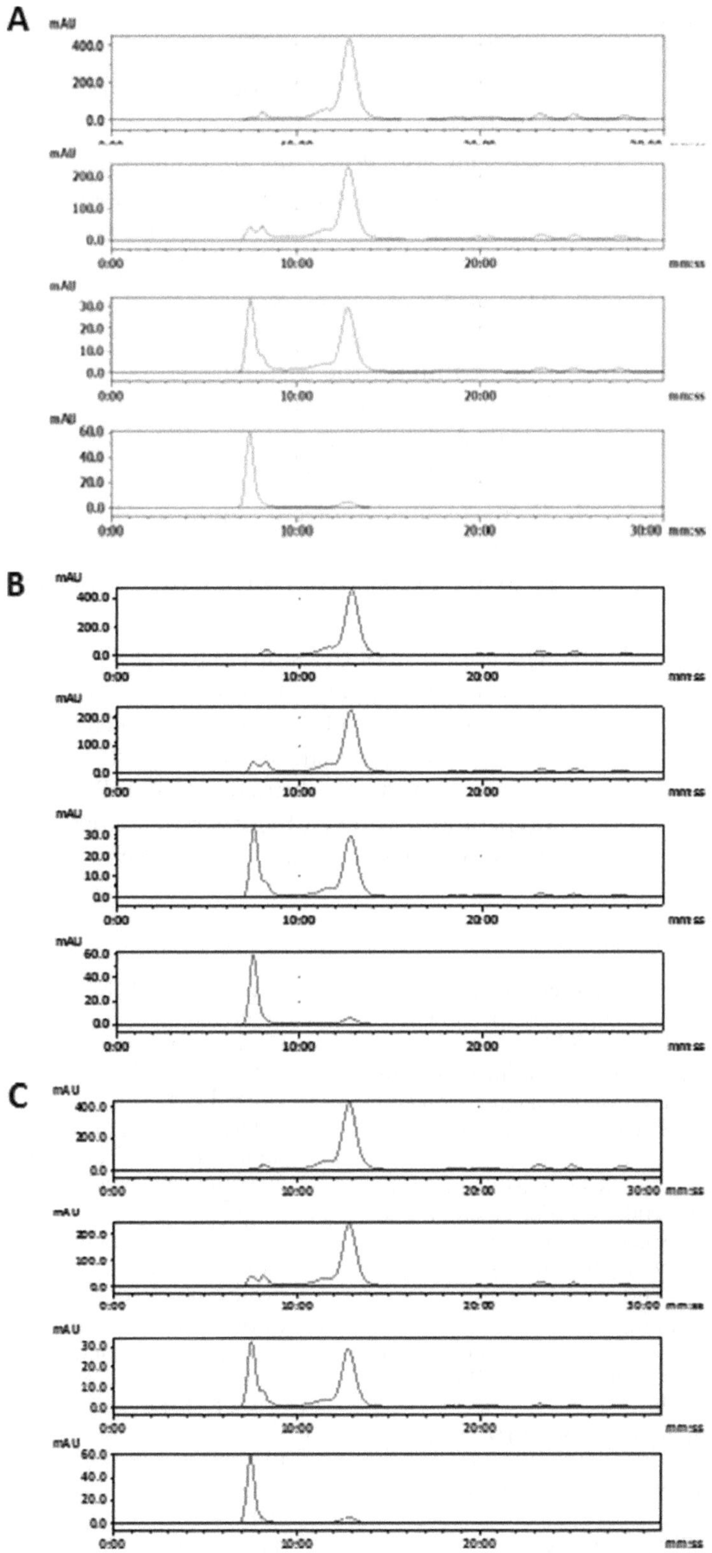

Fig. 3 SE-HPLC chromatogram of extracellular vesicles purified from FBS: (**a**), (**b**) and (**c**) represent HPLC results from three batches of EV isolations demonstrating high reproducibility. Four chromatograms from top to bottom in (**a**) or (**b**) or (**c**) are representing aliquots after once, twice, three times, and four times ultracentrifugations respectively. The peak with a retention time of 7.5 min stands for the exosome particles while the rest peaks are impurities of small size. (Courtesy of Journal of Circulation Biomarkers)

because it should be optimized for the specific packing materials.

7. Generally, low flow rates are desired to get optimal resolution. For commercially available columns, flow rate range and maximal are recommended. Take this column as an example, 0.5–1 mL/min is recommended with a maximum of 1.5 mL/min.
8. The loading capacity for each analysis is related to multiple factors such as column size and the sample characters. As a rule of thumb, the loading volume should be no more than 2% of the column volume otherwise band spreading may occur. According to this empirical rule, the maximal loading volume for this column is 500 μL.
9. To make small molecule impurity detectable, the most commonly used 254 nm was selected where RNAs, DNAs and proteins are all detectable. Other wavelength could also be used in case by case.
10. The intention of this study is to test if SE-HPLC can be used as a means for quality analysis of soluble proteins and other impurities of small size. For a detailed size measurement by HPLC, more evaluations are needed. It would be best to use the simple and straightforward SE-HPLC to qualify the vesicle preparation followed by other characterizations such as TEM, NTA, western blot and flow cytometry.

References

1. Thery C, Zitvogel L, Amigorena S (2002) Exosomes: composition, biogenesis and function. Nat Rev Immunol 2:569–579
2. Thind A, Wilson C (2016) Exosomal miRNAs as cancer biomarkers and therapeutic targets. J Extracell Vesicles 19(5):31292. doi:10.3402/jev.v5.31292
3. Conde-Vancells J, Rodriguez-Suarez E, Embade N et al (2008) Characterization and comprehensive proteome profiling of exosomes secreted by hepatocytes. J Proteome Res 7(12):5157–5166
4. Van der Pol E, Boing AN, Harrison P et al (2012) Classification, functions, and clinical relevance of extracellular vesicles. Pharmacol Rev 64(3):676–705
5. Wang F, Nazarali AJ, Ji S (2016) Circular RNAs as potential biomarkers for cancer diagnosis and therapy. Am J Cancer Res 6 (6):1167–1176
6. Pisitkun T, Shen RF, Knepper MA (2004) Identification and proteomic profiling of exosomes in human urine. Proc Natl Acad Sci 101 (36):13368–133373
7. Srivastava A, Babu A, Filant J et al (2016) Exploitation of Exosomes as Nanocarriers for Gene-, chemo-, and immune-therapy of cancer. J Biomed Nanotechnol 12(6): 1159–1173
8. Kahlert C, Kalluri R (2013) Exosomes in tumor microenvironment influence cancer progression and metastasis. J Mol Med 91 (4):431–437
9. Simpson RJ, Jensen SS, Lim JW (2008) Proteomic profiling of exosomes: current perspectives. Proteomics 8(19):4083–4099
10. Choi DS, Kim DK, Kim YK et al (2013) Proteomics, transcriptomics and lipidomics of exosomes and ectosomes. Proteomics 13 (10–11):1554–1571
11. Xiao D, Ohlendorf J, Chen Y et al (2012) Identifying mRNA, microRNA and protein profiling of melanoma exosomes. PLoS One 7 (10):e46874
12. Subra C, Laulagnier K, Perret B et al (2007) Exosome lipidomics unravels lipid sorting at the level of multivesicular bodies. Biochimie 89(2):205–212

13. Johnsen KB, Gudbergsson JM, Skov MN et al (2014) A comprehensive overview of exosomes as drug delivery vehicles-endogenous nanocarriers for targeted cancer therapy. Biochim Biophys Acta 1846(1):75–87
14. Van Dommelen SM, Vader P, Lakhal S et al (2012) Microvesicles and exosomes: opportunities for cell-derived membrane vesicles in drug delivery. J Control Release 161 (2):635–644
15. Thery C, Amigorena S, Raposo G, et al. (2006) Isolation and characterization of exosomes from cell culture supernatants and biological fluids. Curr Protoc Cell Biol Chapter 3: Unit 3.22
16. Muller L et al (2014) Isolation of biologically-active exosomes from human plasma. J Immunol Methods 411:55–65. doi:10.1016/j.jim.2014.06.007
17. Tan SS, Yin Y, Lee T et al (2013) Therapeutic MSC exosomes are derived from lipid raft microdomains in the plasma membrane. J Extracell Vesicles 2:22614
18. Arroyo JD, Chevilleta JR, Kroha EM et al (2011) Argonaute2 complexes carry a population of circulating microRNAs independent of vesicles in human plasma. Proc Natl Acad Sci 108(12):5003–5008
19. Boing AN, van der Pol E, Grootemaat AE et al (2014) Single-step isolation of extracellular vesicles by size-exclusion chromatography. J Extracell Vesicles 3:23430
20. Clayton A, Court J, Navabi H et al (2001) Analysis of antigen presenting cell derived exosomes, based on immune-magnetic isolation and flow cytometry. J Immunol Methods 247:163–174
21. http://www.systembio.com/downloads/ExoQuick_ProdSheet.pdf
22. Brownlee Z, Lynn KD, Thorpe PE et al (2014) A novel "salting-out" procedure for the isolation of tumor-derived exosomes. J Immunol Methods 407:120–126
23. Witwer KW, Buzas EI, Bemis LT et al (2013) Standardization of sample collection, isolation and analysis methods in extracellular vesicle research. J Extracell Vesicles 2:20360. doi:10.3402/jev.v2i0.20360
24. Yuan Y, Koning RI, Kuil ME et al (2013) Cryo-electron microscopy of extracellular vesicles in fresh plasma. J Extracell Vesicles 2:21494. doi:10.3402/jev.v2i0.21494
25. Gardiner C, Ferreira YJ, Dragovic RA et al (2013) Extracellular vesicle sizing and enumeration by nanoparticle tracking analysis. J Extracell Vesicles 2:19671. doi:10.3402/jev.v2i0.19671
26. Webber J, Clayton A (2013) How pure are your vesicles? J Extracell Vesicles 2:19861. doi:10.3402/jev.v2i0.19861
27. Huang T, Banizs AB, Shi W et al (2015) Size-exclusion HPLC detection of small- size impurities as a complementary means for quality analysis of extracellular vesicles. J Circ Biomark. doi:10.5772/61148

Chapter 16

Multi-Surface Antigen Staining of Larger Extracellular Vesicles

Veronika Lukacs-Kornek, Henrike Julich-Haertel, Sabine Katharina Urban, and Miroslaw Kornek

Abstract

Larger extracellular vesicles, microparticles (MPs) or microvesicles (MVs), especially their acquisition and characterization by flow cytometry (FACS), is increasingly in focus of clinical/translational research efforts. Several laboratories have shown that MPs/MVs might be suitable for the diagnosis and predicting prognosis in various diseases including cancer. However, FACS staining of larger extracellular vesicles (EVs) can be difficult and results potentially in false positive and inconsistent data interpretation. Despite that FACS equipment is well maintained and the operators have ample experience, a reliable and for larger EVs optimized staining protocol is missing. Here, we aim to close that gap and provide a working multi-antibody staining protocol for larger EVs isolated from human serum samples. We describe in detail the needed steps as currently done in our laboratory. Staining is demonstrated exemplarily for multi-antibody mix including CD147, a potential cancer marker if applied in combination with other MP/MV surface markers.

Key words Larger extracellular vesicles, Microparticles, Microvesicles, Surface antigen staining, Antibody, Flow cytometry

1 Introduction

The quantification of larger extracellular vesicles (>100 nm in diameter), known as microparticles (MPs) or microvesicles (MVs), by flow cytometry (FACS), due to its robust statistical power and possibility of measuring various MP/MV subpopulations and other MP/MV associated parameters simultaneously, is raising constantly more and more attention [1, 2]. It became apparent that extracellular vesicles (EVs) are beneficial for disease surveillance especially tumor diagnosis and tumor treatment monitoring as discussed and shown by others and us [1, 3–6]. Recently, we have shown that the so-called tumor-associated microparticles (taMPs) could most likely aid in detecting cancer neoplasia and that

Winston Patrick Kuo and Shidong Jia (eds.), *Extracellular Vesicles: Methods and Protocols*, Methods in Molecular Biology, vol. 1660, DOI 10.1007/978-1-4939-7253-1_16, © Springer Science+Business Media LLC 2017

other MP populations associated with inflammatory events could distinguish between chronic liver diseases [2, 4, 6].

Nowadays, flow cytometers are commonly available at larger medical clinics and research institutes with access to human cancer patients. However, some of those are lacking the expertise to establish a robust, reliable MP/MV multi-antibody staining protocol for the quantitative and qualitative detection of MPs/MVs by FACS. MPs/MVs are shed from the plasma membrane of the parental cell. If the surface antigens for the identification of the parental cell are known, the same markers (antibodies) should be applicable for MPs/MVs originating from the associated parental cell [2, 7].

Since it is known that MPs/MVs have the tendency to take up proteins unspecifically, including IgGs, as some of their parental cells do, it appears almost inevitable that to a certain degree false positive events could be measured and interpreted as MPs/MVs positive events. In order to avoid this, the use of Fc-Block and BSA is recommended and a careful evaluation of the acquired data is required [2, 4, 6, 8]. To set up a new staining for MPs/MVs, the below-described MP staining protocol for the used singular or multi-antibody mix needs to be established, which should follow certain basic rules to evade false positive events during the quantification of MPs/MVs via flow cytometry.

One of the cornerstones is the right use of isotype controls. Here we describe a comprehensive strategy for the setup of the needed amount or dilution of the used antibodies as part of a multi-antibody mix. As part of good laboratory practice, every antibody should be titrated against the matching isotype as part of MPs/MVs multi-antibody mix, even if a successful dilution for a singular use was found for cells or MPs/MVs previously. However, our strategy guide will allow the interested reader to set up a surface antigen staining of MPs/MVs ensuring both Quality Control (QC) and Quality Assurance (QA) for cross-platform comparison as discussed in a previous chapter (*see* Chapter 14).

Our methodology will describe how to exemplarily set up CD147 (EMMPRIN) MP/MV surface staining as part of a multi-antibody mix as used currently in our laboratory and partly published [4, 6]. However, each staining protocol is constantly evolving. This protocol represents current knowledge and practice as done in our laboratory. We will also outline the procedures for troubleshooting (*see* associated Notes). The International Society for Extracellular Vesicles (ISEV) has put out general guidelines for proceeding with EV research that we followed regarding the isolations of these larger EVs [9]. With respect to instrumentation, here our MP/MV surface staining guide/protocol and notes are optimized for our flow cytometer but will apply for other FACS instruments too.

2 Materials

All solutions except antibody containing solutions as provided by the vendor were centrifuged or filtered prior to their use to remove contaminations such as possible protein aggregates or particles with similar size as larger EVs.

2.1 Solutions

1. PBS, was prior filtered using the Steriflip™ Vacuum-driven Filtration System (0.22 μm Millipore Express® Plus Membrane from Millipore™, Merck KGaA, Darmstadt, Germany (Cat# SCGP00525).
2. As Fc block, Hu FcR Binding Inhibitor Purified, eBioscience Inc., San Diego, CA, USA (Cat# 14-9161-73); was utilized and filtered before use with Steriflip™ Vacuum-driven Filtration System (0.22 μm Millipore Express®Plus Membrane from Millipore™, Merck KGaA, Darmstadt, Germany (Cat# SCGP00525).
3. BSA, Miltenyi Biotec GmbH, Bergisch Gladbach, Germany (Order# 130-091-376). BSA stock, before use, is centrifuged at 20,000 × *g* for 60 min at 4 °C.
4. Annexin V Binding Buffer (20× stock solution), Miltenyi Biotec GmbH, Bergisch Gladbach, Germany (Order# 130-092-820). Annexin V Binding Buffer stock, was filtered before use with Steriflip™ Vacuum-driven Filtration System (0.22 μm Millipore Express®Plus Membrane from Millipore™, Merck KGaA, Darmstadt, Germany (Cat# SCGP00525).

2.2 Antibodies

1. Annexin V-FITC, working dilution is 1:20 for a total staining volume of 100 μL (*see* **Note 1**), Miltenyi Biotec GmbH, Bergisch Gladbach, Germany (Order# 130-092-052).
2. CD147-APC-Vio770, our recommended working dilution optimized for our multi-antibody mix is 1:2000 for a total staining volume of 100 μL (*see* **Note 2**), Miltenyi Biotec GmbH, Bergisch Gladbach, Germany (Order# 130-104-493, REA clone REA282).
3. REA Control APC-Vio770, our recommended working dilution matched to the corresponding antibody, here CD147-APC-Vio770, is 1: 720 for a total staining volume of 100 μL (*see* **Note 2**), Miltenyi Biotec, Bergisch Gladbach, Germany (Order# 130-104-618, REA clone REA293).

2.3 Software

1. Cytometric data was analyzed with FlowJo X software for MAC OSX, Tree Star, Inc., Ashland, Oregon, USA.

2.4 Hardware

1. The MACSQuant® Analyzer 10, Miltenyi Biotec GmbH, Bergisch Gladbach, Germany (Order# 130-096-343).

3 Methods

The method described is optimized for our used multi-antibody mix with five antibodies in total. We lay out our protocol in detail and even more crucial how to find a good working antibody dilution. Saying this, each antibody was titrated as part of the multi-antibody mix against the corresponding isotype including the other antibodies (*see* **Note 3**). Prepare and store all antibody mixes on ice and in the dark prior to use (unless indicated otherwise). Since human fluid, here human serum, is used please follow waste disposal regulations when disposing those materials or leftovers. We do not use EDTA, because it interferes with the AnnexinV binding buffer and reduces AnnexinV binding to larger EVs. Our methodology is not focused on MP/MV isolation, our protocol starts at that point when the final centrifugation was done, the supernatant removed and the pure MP/MV pellet is ready to be suspended in ultrapure PBS.

1. MPs/MVs were freshly isolated and the final MP/MV pellet resuspended in 300 μL ultrapure PBS and stored on ice (*see* **Note 4**).
2. Preparation of the *Fc-Block Mix*: mix gently 11.5 μL of ultrapure PBS, 2.5 μL of Hu FcR Binding Inhibitor Purified and 1 μL of BSA for each staining, spin down shortly and store on ice (*see* **Note 5**).
3. Add 15 μL of *Fc-Block Mix* to 50 μL resuspended MP/MV sample, mix by snipping or gently pipetting up and down and incubate at 4 °C for 60 min (*see* **Note 6**).
4. Preparation of the *AnnexinV Mix*: mix 5 μL 1:2 diluted 20× Annexin V Binding Buffer and 5 μL Annexin V-FITC for each staining (*see* **Notes 5** and 7). Resuspend gently by snipping or pipetting up and down and store on ice in the dark prior to use.
5. Add 10 μL of the *AnnexinV Mix* to each sample that has been previously incubated with the *Fc-Block Mix*. If needed, spin down shortly. Incubate at 4 °C for 30 min in the dark (*see* **Notes 8** and **9**). Resuspend by snipping or gently pipetting up and down.
6. Preparation of the *Multi-Antibody Mix*: To 25 μL ultrapure PBS we add 1 μL from each antibody that has been previously 1:10 diluted (*see* **Notes 5**, **10**, and **11**). Resuspend by gently pipetting up and down. Store *Multi-Antibody Mix* prior to use on ice and in the dark.
7. Add 25 μL of the *Multi-Antibody Mix* to each sample that has received previously the *Multi-Antibody-Mix*, the *Fc-Block Mix*, and the *AnnexinV Mix*. Resuspend gently by snipping or pipetting up and down. If needed spin down shortly. Incubate at

44 °C for 20 min in the dark (*see* **Note 12**). In sum the total staining volume is 100 μL for each sample.

8. Add 100 μL of ultrapure PBS to each sample that was incubated with the *Multi-Antibody Mix* for a total volume of 200 μL (*see* **Notes 13** and **14**). Resuspend gently by snipping or pipetting up and down.

4 Notes

1. The working dilution of the indicated antibody was set up accordingly to **Note 11** (*see* **Note 11**).
2. The working dilution of the isotype was matched to the recommended dilution of the corresponding antibody ensuring that the total amount of the antibody was equal to the total amount of the isotype (*see* **Note 11**). Be aware that the antibody concentration as compared to the corresponding isotype can be different, frequently observed if the manufacturer is not the same.
3. This is crucial, since we frequently observed that certain antibodies, when part of a mix, must be used differently regarding their dilution. This might be because of self-aggregation or because their fluorochromes interfere with each other.
4. We recommend preparing always-new MPs/MVs from serum, since we observed that with each freeze–thaw cycle MPs/MVs will be partially degraded and therefore previous data could likely not be reproduced.
5. We recommend mixing a master-mix depending on the amount of samples that should be measured by FACS.
6. The unstained control and the Annexin only control will also be incubated with the Fc-Block Mix as the other samples. Additionally we recommend running an Fc-Block Mix only control without MPs/MVs to check for possible contaminations in our 0.22 μm filtered Fc-Block stock.
7. We do not observe any major differences between Annexin V Binding buffers provided by various vendors.
8. We do not recommend the incubation on ice.
9. One sample should serve as the unstained control for the used Annexin V-FITC. This means that one AnnexinV Mix will be provided without Annexin V-FITC. Accordingly, 5 μL Annexin V-FITC will be exchanged with 5 μL ultrapure PBS. This control is important to set the Annexin Gate later on (*see* Fig. 1a).
10. Optional: spin the antibodies prior to use at 2000 × *g* for 10 min to remove IgG aggregates.

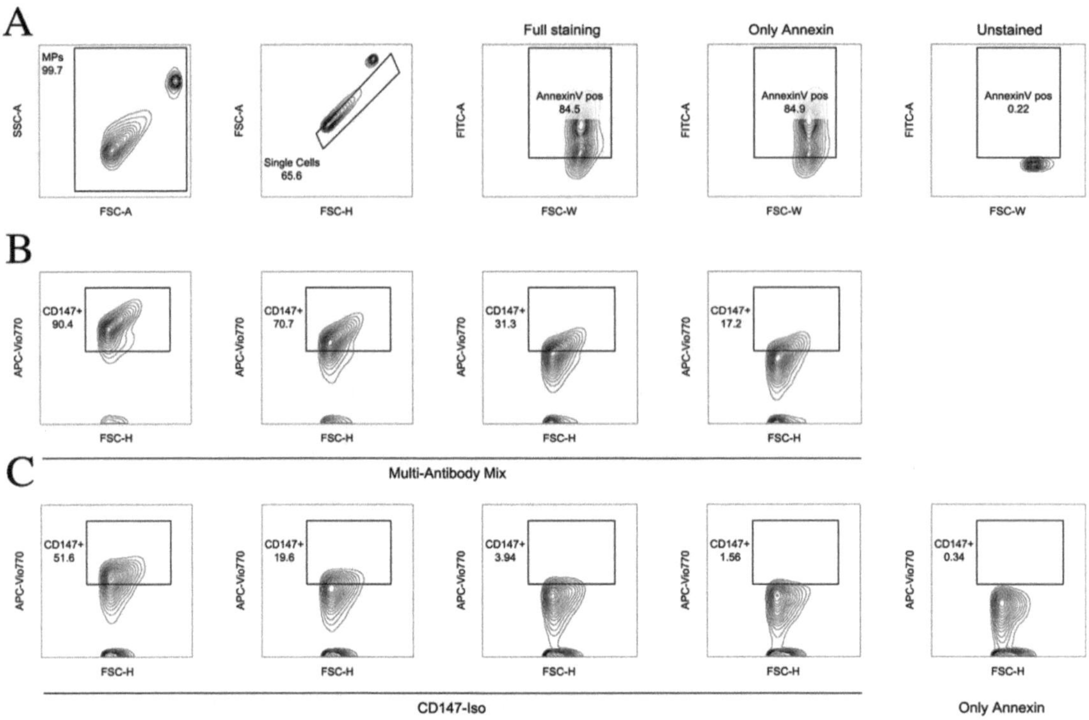

Fig. 1 Overview of the FlowJo FACS analysis of the CD147-APC-Vio770 titration against matched isotype control. Human tumor samples were measured on MACSQuant® Analyzer 10, Miltenyi Biotec GmbH, Bergisch Gladbach, Germany. Given are the names and percentages of the indicated gates within each contour FACS plot. (**a**) Gating strategies are depicted for *Annexin gate*, *Only Annexin* and *Unstained* (MPs plus Fc-Block Mix) control as indicated. (*Single cells gate* is set based on FSC-A vs. FSC-H as for doublet exclusion. Here it is used for the exclusion of larger protein and EV aggregates). (**b**) Results of the titration with various CD147-APC-Vio770 dilutions are depicted: from left to the right 1:200, 1:500, 1:1000, and 1:2000. (**c**) Matched isotype (*CD147-Iso*) controls are shown including *Only Annexin* control demonstrating the background in the channel of detection for CD147 (APC-Vio770 on MACSQuant® Analyzer 10)

11. How to find the right working dilution of each used antibody? Exemplarily we show a titration for CD147-APC-Vio770 against its matching isotype directly conjugated with the same fluorochrome. We do not recommend using secondary antibodies; best results will be achieved with directly conjugated antibodies. We recommend using the same manufacturer for the antibody and isotype.

    Usually we use 4 different dilutions. Accordingly to the concentration of CD147-APC-Vio770 the dilution of the corresponding isotype was adjusted ensuring equal total amounts of used antibody and isotype IgGs per dilution group. We usually perform triplicate measurements using three different pooled serum tumor samples, ensuring that individual operator errors will not affect the outcome and most likely result in a meaningful titration as seen in Fig. 1a, b (*see* Fig. 1a, b). Compared to the Annexin only control, it

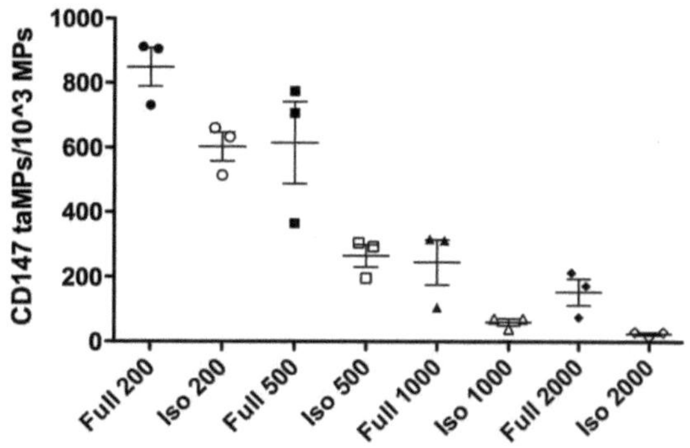

Fig. 2 Graphical summary of the CD147-APC-Vio770 antibody titration data after FlowJo analysis. Plotted are the calculated numbers of indicated AnnexinV$^+$CD147$^+$ double positive taMPs per 1000 AnnexinV MPs as part of the used multi antibody staining mix. 3 pooled human tumor serum samples were used for each indicated dilution. The difference between CD147-APC-Vio770 used 1:2000 (Full 2000) and matched isotype (Iso 2000) was significant ($p < 0.05$, two tailed Student's *t*-test)

shows that the highest CD147-APC-Vio770 dilution including the matched isotype dilution results in least false positive events that could be interpreted as AnnexinV$^+$CD147$^+$ tumor-associated microparticles (taMPs). FACS plot analysis was additionally visualized as a graph underlining our interpretation and summarizing our titration effort graphically (*see* Fig. 2). In this example, in this particular setting with other antibodies we recommend to use the 1:2000 dilution. That means using 1 μL from the CD147-APC-Vio770 antibody that has been previously 1:20 diluted for our particular Multi-Antibody Mix. The titration should be done for each antibody as part of the used Multi-Antibody Mix.

12. Besides the unstained control we recommend to include an Only Annexin control. This sample is only incubated with the Fc-Block Mix and AnnexinV-Mix, no Multi-Antibody Mix, allowing us the correct position of the Annexin gate as part of the FlowJo FACS data analysis (*see* Fig. 1a).
13. We usually measure 150 μL from the total volume of 200 μL.
14. It is not part of this chapter to provide the FACS machine instrument or individual gate settings, in general we refer to chapter (*see* Chapter 14) that provided an excellent introduction how the right instrument setting should be set for various FACS machines.

Acknowledgment

This work was supported by a Deutsche Krebshilfe grant (111184) to Miroslaw Kornek; and by the Alexander von Humboldt Foundation, Sofja Kovalevskaja Award to Veronika Lukacs-Kornek.

References

1. Julich H, Willms A, Lukacs-Kornek V, Kornek M (2014) Extracellular vesicle profiling and their use as potential disease specific biomarker. Front Immunol 5:413. doi:10.3389/fimmu.2014.00413
2. Kornek M, Lynch M, Mehta SH, Lai M, Exley M, Afdhal NH, Schuppan D (2012) Circulating microparticles as disease-specific biomarkers of severity of inflammation in patients with hepatitis C or nonalcoholic steatohepatitis. Gastroenterology 143(2):448–458. doi:10.1053/j.gastro.2012.04.031
3. Melo SA, Luecke LB, Kahlert C, Fernandez AF, Gammon ST, Kaye J, LeBleu VS, Mittendorf EA, Weitz J, Rahbari N, Reissfelder C, Pilarsky C, Fraga MF, Piwnica-Worms D, Kalluri R (2015) Glypican-1 identifies cancer exosomes and detects early pancreatic cancer. Nature 523 (7559):177–182. doi:10.1038/nature14581
4. Willms A, Muller C, Julich H, Klein N, Schwab R, Gusgen C, Richardsen I, Schaaf S, Krawczyk M, Krawczyk M, Lammert F, Schuppan D, Lukacs-Kornek V, Kornek M (2016) Tumour-associated circulating microparticles: a novel liquid biopsy tool for screening and therapy monitoring of colorectal carcinoma and other epithelial neoplasia. Oncotarget 7(21):30875–30867. doi:10.18632/oncotarget.9018
5. D'Souza-Schorey C, Clancy JW (2012) Tumor-derived microvesicles: shedding light on novel microenvironment modulators and prospective cancer biomarkers. Genes Dev 26 (12):1287–1299. doi:10.1101/gad.192351.112
6. Julich-Haertel H, Urban SK, Krawczyk M, Willms A, Jankowski K, Patkowski W, Kruk B, Krasnodębski M, Ligocka J, Schwab R, Richardsen I, Schaaf S, Klein A, Gehlert S, Sänger H, Casper M, Banales JM, Schuppan D, Milkiewicz P, Lammert F, Krawczyk M, Lukacs-Kornek V, Kornek M (2017) Cancer-associated circulating large extracellular vesicles in cholangiocarcinoma and hepatocellular carcinoma. J Hepatol. doi:10.1016/j.jhep.2017.02.024. pii: S0168-8278(17)30123-X
7. Boilard E, Nigrovic PA, Larabee K, Watts GF, Coblyn JS, Weinblatt ME, Massarotti EM, Remold-O'Donnell E, Farndale RW, Ware J, Lee DM (2010) Platelets amplify inflammation in arthritis via collagen-dependent microparticle production. Science 327 (5965):580–583
8. Headland SE, Jones HR, D'Sa AS, Perretti M, Norling LV (2014) Cutting-edge analysis of extracellular microparticles using ImageStream (X) imaging flow cytometry. Sci Rep 4:5237. doi:10.1038/srep05237
9. Lotvall J, Hill AF, Hochberg F, Buzas EI, Di Vizio D, Gardiner C, Gho YS, Kurochkin IV, Mathivanan S, Quesenberry P, Sahoo S, Tahara H, Wauben MH, Witwer KW, Thery C (2014) Minimal experimental requirements for definition of extracellular vesicles and their functions: a position statement from the International Society for Extracellular Vesicles. J Extracell Vesicles 3:26913. doi:10.3402/jev.v3.26913

Chapter 17

Microcapillary Chip-Based Extracellular Vesicle Profiling System

Takanori Akagi and Takanori Ichiki

Abstract

A microcapillary chip-based particle electrophoresis system developed for characterizing extracellular vesicles (EVs) is described. So far, it is technologically difficult to analyze or identify a heterogeneous population of particles ranging from several tens to one hundred nanometers, and hence, there is a growing demand for a new analytical method of nanoparticles among researchers working on extracellular vesicles. The analytical platform presented in this chapter allows detection of individual nanoparticles or nanovesicles of less than 50 nm in diameter and enables the characterization of nanoparticles based on multiple indexes such as concentration, diameter, zeta potential, and surface antigenicity. This platform will provide a useful and easy-to-use solution for obtaining both quantitative and qualitative information on EV samples used in research and development of exosome biology and medicine.

Key words Exosomes, Microcapillary chip, Dark-field microscopy, Immonoelectrophoresis, Particle electrophoresis, Multiplex assay, Zeta potential, Brownian motion analysis, Subpopulation

1 Introduction

Extracellular vesicles (EVs) including exosomes and microvesicles have attracted considerable attention in the fields of cell biology and medicine. For a better understanding of EVs and further exploration of their applications, the development of analytical methods for biological nanovesicles is required. However, it is technologically difficult to analyze or identify a heterogeneous population of particles ranging from several tens to one hundred nanometers, and hence, there is a growing demand for a new analytical method of nanoparticles among researchers working on extracellular vesicles.

To understand biological functions of EVs, various analytical methods are used at present. For characterization of molecular composition of exosomes, the omics techniques such as microarray, sequencer, and mass spectrometry can be used [1–3]. Furthermore, for evaluation of physical properties of exosomes, nondestructive

Winston Patrick Kuo and Shidong Jia (eds.), *Extracellular Vesicles: Methods and Protocols*, Methods in Molecular Biology, vol. 1660,
DOI 10.1007/978-1-4939-7253-1_17, © Springer Science+Business Media LLC 2017

physical techniques such as electron microscopy, atomic force microscopy, optical single particle tracking [4], resistive pulse sensing [5], dynamic light scattering, flow cytometry and particle electrophoresis are available. However, when studying the physiological and pathological roles of exosomes, it is important to remind the presence of EV subtypes with different biogenesis pathways [6]. Namely, surface protein profiling is strongly desired in characterizing EVs. Although recent improvements in high-resolution fluorescent flow cytometry have enabled the detection and analysis of fluorescence-labeled vesicles with a diameter of up to 100 nm, the lower detection limit in size is far from sufficient for exosome research and the requirement for an experienced operator and expensive apparatus will limit their use [7]. On the other hand, surface proteins of biological particles can also be evaluated by measuring their zeta potential. The electrophoresis system that uses a microcapillary chip as an electrophoresis chamber is recognized as the most advanced and reliable method for obtaining the zeta potential of heterogeneous particles like cells or EVs [8]. Recently, an on-chip microcapillary electrophoresis (μCE) system has been developed and applied to the characterization of EVs [9, 10]. This system has enabled the sensitive imaging of individual EVs in a dark field by detecting laser light scattered from EVs. Since fluorescence intensity depends on the number of fluorescent molecules bound on the EV surface, low signal to noise ratio (S/N) and photobleaching are serious problems. In contrast, zeta potential does not depend on size but surface charge density of EVs.

In this chapter, overview of evaluation method of the diameter, zeta potential, and protein expression of individual EVs using a microchip-based nanoparticle profiling platform is described.

2 Materials

2.1 Analytical Instrumentation

Schematic and photograph of the microfluidic chip-based profiling system for extracellular vesicles are shown in Figs. 1 and 2. Briefly, the system comprises a μCE chip, a pair of platinum electrodes, a DC power supply [Matsusada Precision HVL-1.1P(A)], a 404 nm laser source (Coherent Obis laser system, 50 mW), a microscope (Nikon, Ti-U), and a CMOS camera (Andor Neo sCMOS). Details of a μCE chip are described in the Subheading 2.2. EVs in microchannels are observed by a laser dark-field microscopy method to detect the motion of small EVs. The laser beam is shaped within 100 μm waist height using cylindrical lens to prevent the background signal generated by scattering at the top and bottom surfaces of the microchannel. Additionally this system is equipped with custom-developed software for adjusting the position of laser beam and objective lens, applying voltage, particle tracking velocimetry and calculating the diameter and zeta potential of EVs.

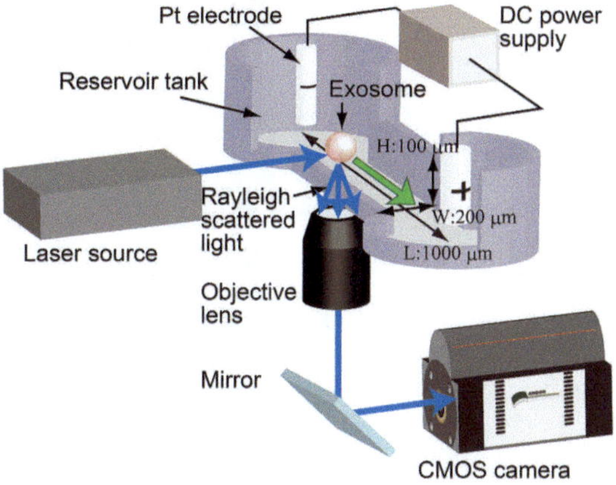

Fig. 1 Schematic of the analytical platform. This platform comprises a microcapillary electrophoresis (μCE) chip and a laser dark-field microscope with a CMOS camera. Reprinted with permission from [10]

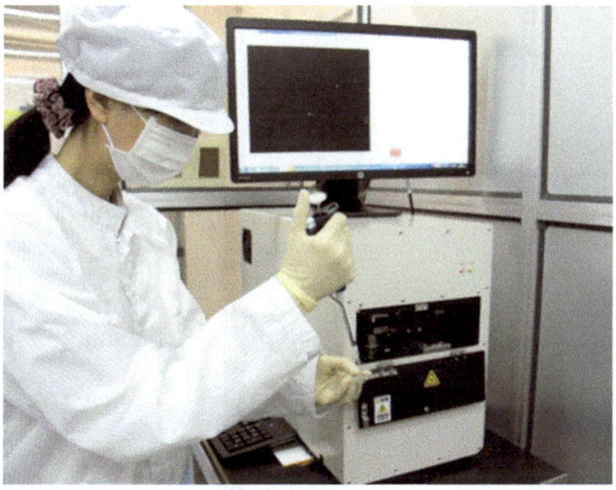

Fig. 2 Photograph of the analytical instruments

2.2 Analysis Chip

Figure 3 shows photographs of the analysis chip. Poly(dimethylsiloxane) (PDMS) microchannels fabricated by a soft lithography method [11] was bonded to a glass substrate. The chip is designed to have four microchannels (width: 200 μm; height: 200 μm; length: 10 mm) with the rectangular cross-section and open reservoirs on both ends. The inner surface of microchannels is coated with anti-biofouling polymers such as a phospholipid copolymer containing 2-methacryloyloxyethyl phosphorylcholine (MPC) and 3-methacryloxyethyl triethoxysilane (METESi) to suppress nonspecific adsorption of proteins and EVs.

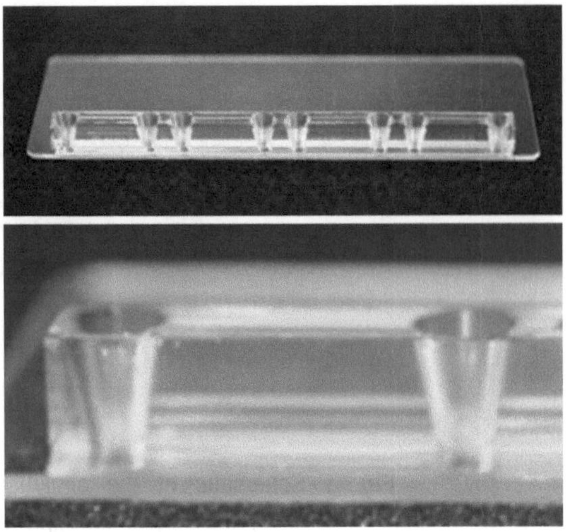

Fig. 3 Photographs of the analysis chip

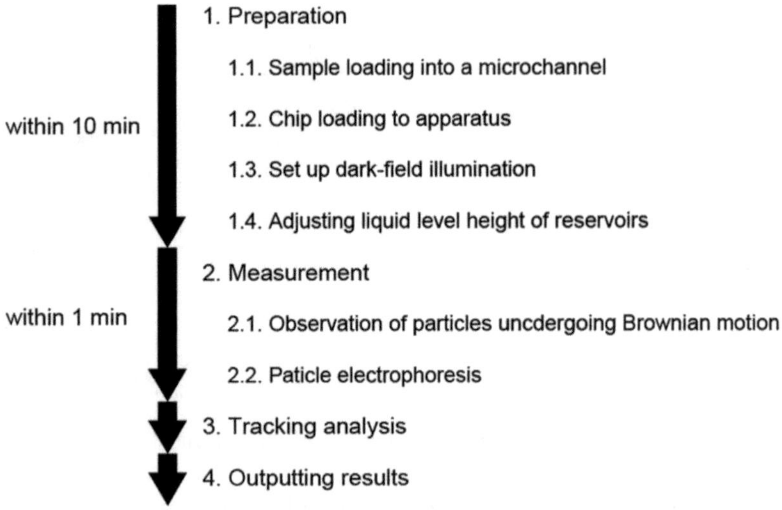

Fig. 4 The procedure of EV evaluation on a microfluidic chip, which consists of sample loading, chip placement onto the instrument, and water level adjustment between reservoirs

3 Methods

3.1 Preparation of Microfluidic Chips

As shown in Fig. 4, the procedures to perform EV evaluations on microfluidic chips consist of sample loading, chip placement onto the instrument, and water level adjustment between reservoirs (*see* **Note 1**). EV samples are loaded as follows:

1. Chip is placed with its microchannels on the near side of an operator.

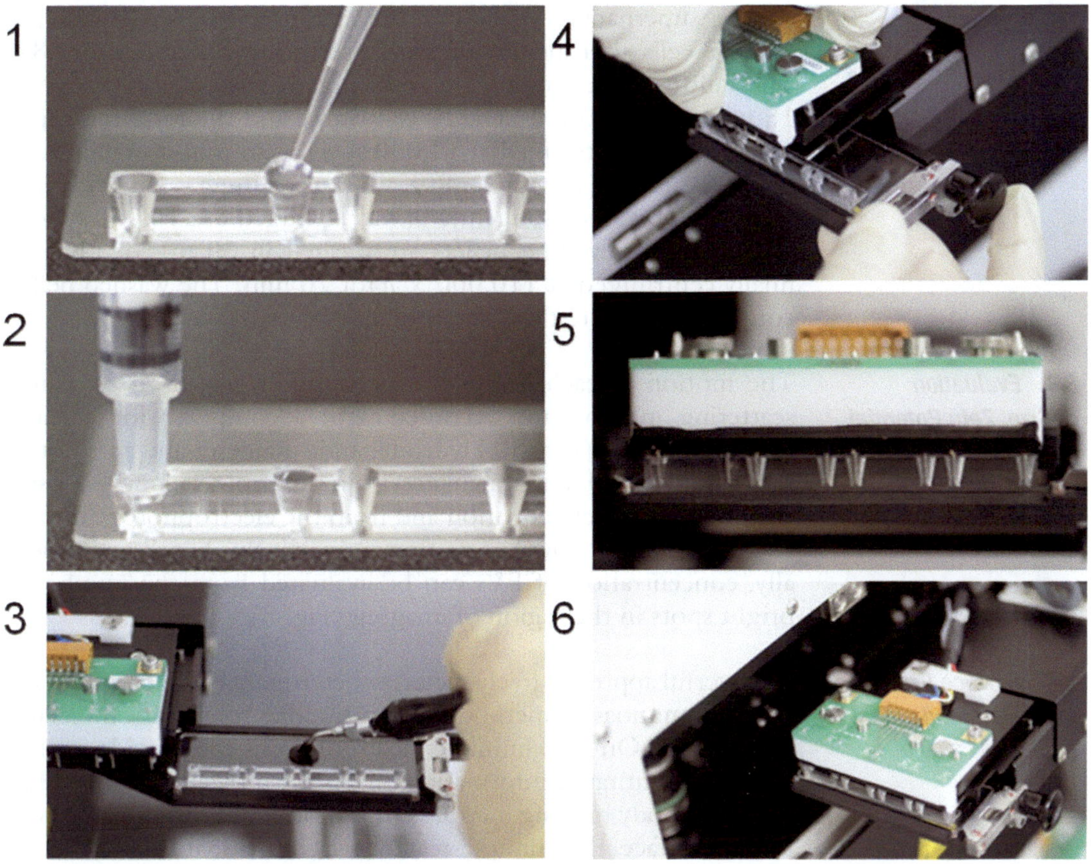

Fig. 5 Photographs of the procedure of EV evaluation on a chip

2. 250 μL of EV sample is pipetted into the right reservoir (Fig. 5 (1)).
3. The sample is introduced into a microchannel by suctioning from left reservoir with a syringe (Fig. 5(2)).
4. The difference in the water level between reservoirs is preadjusted by adding buffer solution.
5. The chip is placed on the holder (Fig. 5(3)).
6. Put on an attachment on the chip (Fig. 5(4)).
7. Platinum electrodes are introduced into the reservoirs (Fig. 5 (5)).
8. The chip is loaded into the instrument (Fig. 5(6)).

After the chip is loaded inside the instrument, again, the water level of reservoirs is carefully adjusted to stop the hydrodynamic flow perfectly.

3.2 Preparation of EV Samples

The typical procedures for preparing EV samples used in the demonstration experiment are as follows. EV samples are collected from culture supernatants of human breast cancer MDA-MB-231 and Sk-Br-3 cells after incubation with a serum-free medium for 48 h. The culture supernatants are centrifuged twice at 2000 × *g* for 15 min using a Tomy LC-220 centrifuge and the resulting supernatant was recentrifuged at 12,000 × *g* for 35 min using an Beckman Coulter Optima L-90K ultracentrifuge. The clarified supernatant was subsequently ultracentrifuged at 110,000 × *g* for 70 min. The resulting pellets are washed in 30 mL of PBS and finally centrifuged at 110,000 × *g* for 70 min. The washed pellets containing EVs are then suspended in PBS.

3.3 Evaluation of Size, Zeta Potential, and Concentration of EVs

The motion of each EV in microchannel is visualized by a light scattering method and recorded with a computer for particle-tracking analysis (Fig. 6). Hydrodynamic diameter and zeta potential of individual EV are calculated from the Brownian motion using the Stokes–Einstein equation and from the electrophoretic motion using the Smoluchowski equation, respectively (Fig. 7). Additionally, concentration of EVs can be evaluated from the number of bright spots in the region of imaging (Fig. 8).

3.4 Surface Protein Analysis of Individual EVs

A powerful approach to characterize heterogeneous biological samples is immunoassay method using antibodies for specific molecular recognition. On-chip immune-electrophoresis can provide qualitative and quantitative information on surface molecules of EVs [12]. Since antibody binding increases the amount of positive charges on the EV surface, the immune-reactivity of individual EVs is reflected in their electrophoretic mobility, and hence, their zeta potential as shown in Fig. 9.

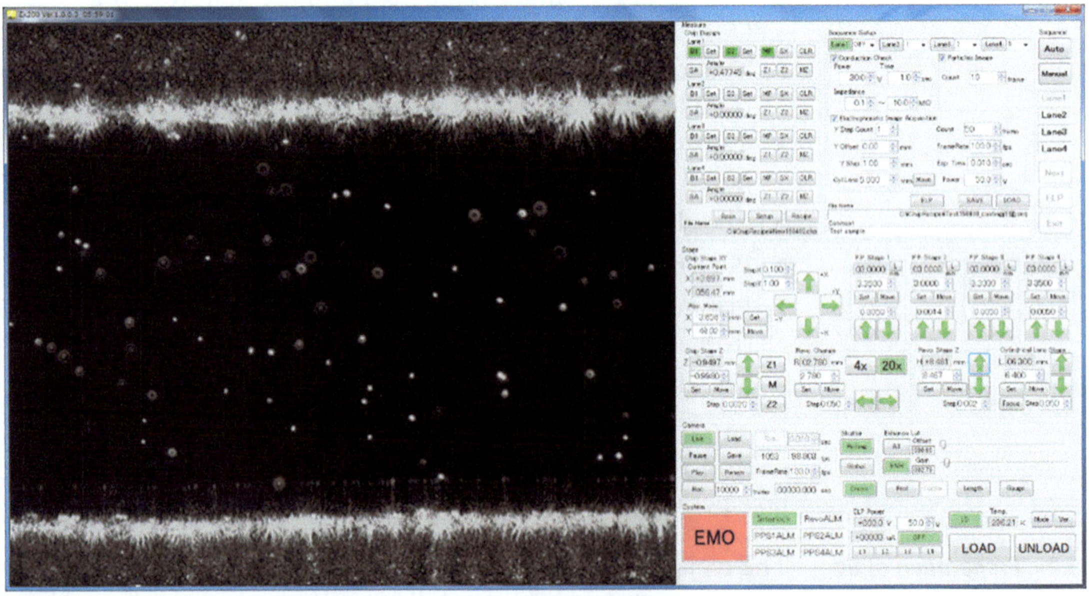

Fig. 6 The control panel of the system with a dark-field image of observed particles

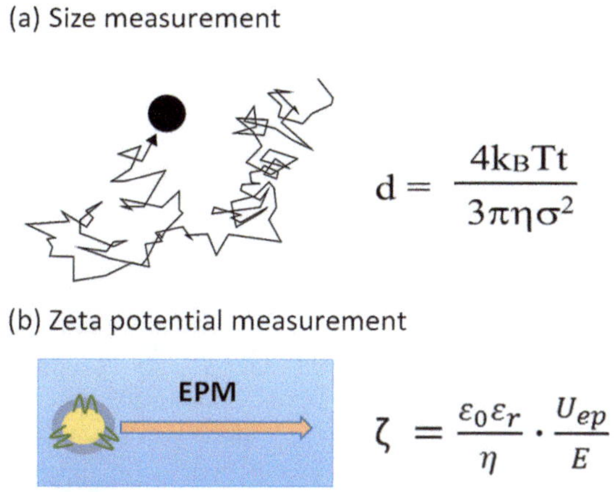

Fig. 7 Evaluation of hydrodynamic diameter and zeta potential of individual EVs using tracking analysis. (**a**) Hydrodynamic diameter was calculated from the Brownian motion using the Stokes–Einstein equation. (**b**) Zeta potential was calculated from the electrophoretic motion using the Smoluchowski equation

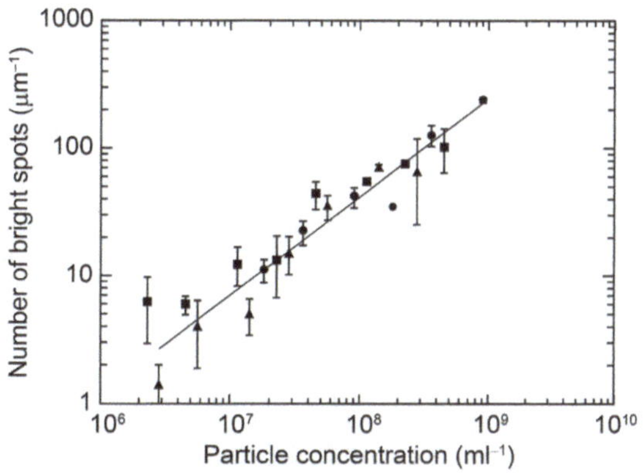

Fig. 8 Relationship between the concentration of size standard polystyrene particles and the number density of bright spot in the region of imaging. *Closed circle*: 100 nm, *Closed square*: 200 nm, *Closed triangle*: 400 nm

3.5 Multiparametic Characterization; Evaluation of Both Diameter and Surface Protein Expression of Individual Exosomes

When on-chip immunoelectrophoresis is performed immediately after the Brownian motion analysis, two-dimensional histograms of the diameter and zeta potential of each EV and exosome of Sk-Br-3; cells can be obtained as shown in Fig. 10. Here, exosomes were purified from EVs by OptiPrep density-gradient ultracentrifugation. There was no significant difference in zeta potential between EVs treated with anti-CD9 antibody and EVs treated with IgG, which was used as an isotype control antibody to estimate the nonspecific adsorption of anti-CD9Ab. In contrast, the zeta potential of exosomes treated with anti-CD9 antibody was different from that of exosomes treated with IgG ($p < 0.05$, t-test).

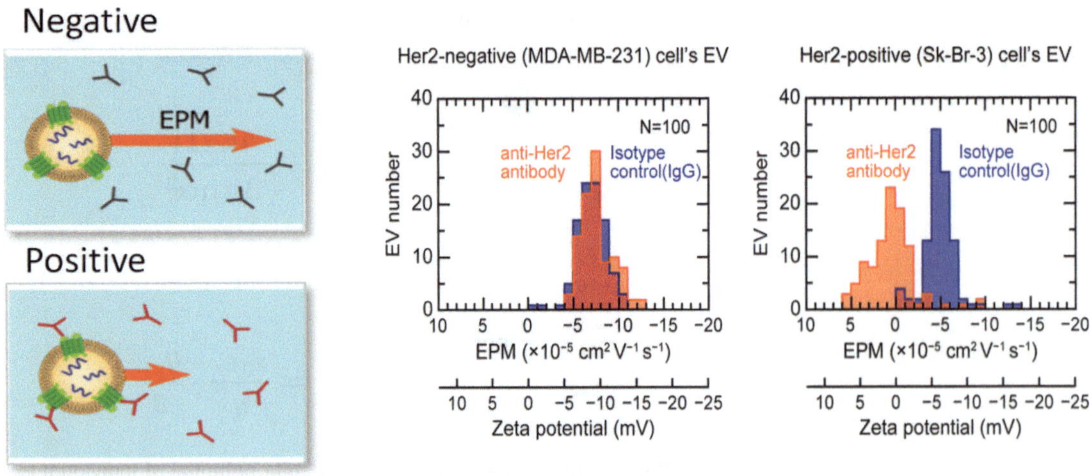

Fig. 9 Schematic of immunoelectrophoresis of EVs on μCE chip (*left*). Because antibody binding increases the number of positive charges on the EV surface, the immunoreactivity of individual EVs is reflected in their electrophoretic mobility, that is, their zeta potential. Zeta potential distributions of EV secreted from Her2-negative (MDA-MB-231) and Her2-positive (Sk-Br-3) human breast cancer cells (*right*)

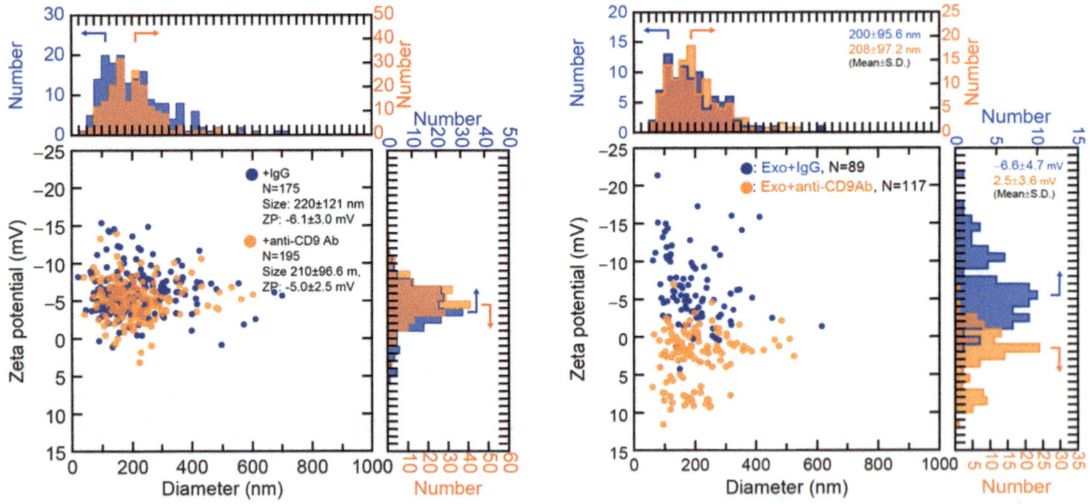

Fig. 10 Simultaneous evaluation of diameter and surface protein expression of individual EVs (*left*) and exosomes (*right*) of human breast cancer Sk-Br-3 cells

4 Notes

Removal of bubbles and dusts

1. In this system, bubbles and dusts in microchannels are detrimental for the performance and longevity of a microfluidic device. Before using reagents, filtration with a submicron-pore filter is recommended. In manual pipetting of EV samples into

reservoirs, it is important to touch the tip of the pipet to the bottom of the reservoir, not to the side wall of the reservoir, to avoid generating bubbles in the reservoir. During introducing EV samples into a microchannel, it is important to perform the suction slowly with a syringe to avoid generating bubbles in the microchannel.

References

1. Kim D-K, Lee J, Simpson RJ, Lötvall J, Gho YS (2015) EVpedia: a community web resource for prokaryotic and eukaryotic extracellular vesicles research. Semin Cell Dev Biol 40:4–7
2. Kreimer S, Belov AM, Ghiran I, Murthy SK, Frank DA, Ivanov AR (2015) Mass-spectrometry-based molecular characterization of extracellular vesicles: lipidomics and proteomics. J Proteome Res 14:2367–2384
3. Hill AF, Pegtel DM, Lambertz U, Leonardi T, O'driscoll L, Pluchino S, Ter-Ovanesyan D, Nolte EN (2013) ISEV position paper: extracellular vesicle RNA analysis and bioinformatics. J Extracell Vesicles 2. doi:10.3402/jev.v2i0.22859
4. Dragovic RA, Gardiner C, Brooks AS, Tannetta DS, Ferguson DJ, Hole P, Carr B, Redman CW, Harris AL, Dobson PJ (2011) Sizing and phenotyping of cellular vesicles using nanoparticle tracking analysis. Nanomedicine 7:780–788
5. Balaj L, Tigges J, Toxavidis V, Ivanov AR, Skog J, Momen Heravi F, Alian S, Ericsson M, Distel R, Kuo WP (2012) Alternative methods for characterization of extracellular vesicles. Front Physiol 3:354
6. Kowal J, Arras G, Colombo M, Jouve M, Morath JP, Primdal-Bengtson B, Dingli F, Loew D, Tkach M, Théry C (2016) Proteomic comparison defines novel markers to characterize heterogeneous populations of extracellular vesicle subtypes. Proc Natl Acad Sci 113:E968–E977
7. van der Vlist EJ, Nolte-'t Hoen EN, Stoorvogel W, Arkesteijn GJ, Wauben MH (2012) Fluorescent labeling of nano-sized vesicles released by cells and subsequent quantitative and qualitative analysis by high-resolution flow cytometry. Nat Protoc 7:1311–1326
8. Mehrishi JN, Bauer J (2002) Electrophoresis of cells and the biological relevance of surface charge. Electrophoresis 23:1984–1994
9. Kato K, Kobayashi M, Hanamura N, Akagi T, Kosaka N, Ochiya T, Ichiki T (2013) Electrokinetic evaluation of individual exosomes by on-chip microcapillary electrophoresis with laser dark-field microscopy. Jpn J Appl Phys 52:06GK10
10. Akagi T, Kato K, Hanamura N, Kobayashi M, Ichiki T (2014) Evaluation of desialylation effect on zeta potential of extracellular vesicles secreted from human prostate cancer cells by on-chip microcapillary electrophoresis. Jpn J Appl Phys 53:06JL01
11. Whitesides GM (2006) The origins and the future of microfluidics. Nature 442:368–373
12. Akagi T, Kato K, Kobayashi M, Kosaka N, Ochiya T, Ichiki T (2015) On-chip immuno-electrophoresis of extracellular vesicles released from human breast cancer cells. PLoS One 10:e0123603

Part IV

Imaging of Extracellular Vesicles

Chapter 18

Detection and Characterization of Extracellular Vesicles by Transmission and Cryo-Transmission Electron Microscopy

Petr Cizmar and Yuana Yuana

Abstract

Transmission electron microscopy (TEM) and transmission scanning electron Microscopy (TSEM), which denotes application of a scanning electron microscope (SEM) in the transmission mode, have been used to detect and characterize particles down to an imaging resolution of ~1 nm. In the field of EVs, TEM also has been valued for its capability to detect and characterize single EV. Furthermore, employing immunogold labeling in TEM could give information regarding biochemical properties of EV surface proteins. Significant shortcomings in TEM such as dehydration, chemical fixation, and/or staining of the biological specimens are eluded by the use of cryo-TEM. In cryo-TEM imaging, samples are directly applied onto an EM grid, vitrified and visualized, thus allowing for characterization of EVs near its native state. In this chapter, we describe a step-by-step guide for preparing EVs on the grid before TEM and cryo-TEM imaging. Finally, we provide a guide to an automated image-processing analysis to provide the size distribution of EVs.

Key words Immunogold labeling, Negative staining, Lactadherin, CD9, CD63, Exosomes, Image processing

1 Introduction

The term "extracellular vesicles" (EVs) accommodates all types of vesicles released or secreted from prokaryotic and eukaryotic cells [1]. Cells secrete EVs either for maintaining their physiological balance or as a response of the extracellular/environmental stimuli [2]. In human body fluids EVs can be isolated from for example blood, urine, saliva, and synovial and seminal fluids [3]. Research has shown that EVs inherit proteins, lipids, and genetic materials from their parental cells and also harbor certain unique proteins or genetic materials resulted from the stimuli [4]. Recent clinical applications of EVs are as surrogate markers, costimulatory, and drug carriers [5]. These vesicles are heterogeneous varied from size ranging from 30 to 1000 nm. Unfortunately, the size

Winston Patrick Kuo and Shidong Jia (eds.), *Extracellular Vesicles: Methods and Protocols*, Methods in Molecular Biology, vol. 1660,
DOI 10.1007/978-1-4939-7253-1_18, © Springer Science+Business Media LLC 2017

determination indicated here mainly depends on the method used [6]. Transmission Electron Microscopy (TEM) and Transmission Scanning Electron Microscopy (TSEM), which denotes application of a Scanning Electron Microscope (SEM) in the transmission mode, have been used to detect and characterize particles down to an imaging resolution of ~1 nm [6]. In the field of EVs, TEM also has been valued for its capability to detect and characterize single EV [7]. Furthermore, employing immunogold labeling in TEM could give information regarding biochemical properties of EV surface proteins [8]. Since TEM is performed in a vacuum environment and EV sample requires fixation and dehydration, these processes may affect their size and change their morphology. EVs chemically fixed and contrasted with uranyl acetate with or without methylcellulose prior to TEM have cup shape appearance, whereas those analyzed by cryo transmission electron microscopy (cryo-TEM) show rounded shape EVs [9, 10] Therefore, cryo-TEM where samples are analyzed in frozen conditions at temperatures below −100 °C without being stained or fixed has been recently applied for EV analysis [10, 11]. EV also could be stained using immunogold labeled antibodies to phenotype surface protein expression of EV and analyzed with cryo-TEM [12]. Comparisons between TEM and cryo-TEM with their advantages and disadvantages are summarized shortly in Table 1. Although TEM and cryo-TEM have been used extensively for detection of EVs, this method only provides semiquantitative information on EVs. The measurement time also is in the order of hours. In addition, limited and non-uniform adhesion of vesicles on the surface may affect the particle size distribution [6]. Therefore, other particle/vesicle detection methods such as nanoparticle-tracking analysis, dynamic light scattering or resistive pulse sensing may need to be done in

Table 1
Advantages and disadvantages using TEM with negative staining and cryo-TEM

Transmission electron microscopy (negative staining)	Cryo transmission electron microscopy
Advantages	
• Simple sample preparation on grid • Good signal-to-noise ratio • Radiation resistant	• No artifacts due to fixation/dehydration • Preservation of native conformation • No stain is needed for visualization
Disadvantages	
• Structural artifacts due to staining and dehydration • Imaging under a non-native environment	• Technical demands in preparing, handling, and storing frozen samples prior to imaging • Low signal-to-noise ratio due to ice thickness • Freezing artifacts • Radiation sensitivity

parallel to electron microscopy analysis to provide size distribution and concentration of a large number of vesicles [7].

As the method used for isolation of EV could affect numbers and composition of EVs, it is important to choose suitable isolation method which can ensure reliable and comparable measurements of EVs. A common high-speed centrifugation technique induces the formation of EV aggregates composed of a mixture of EVs of various phenotypes and morphologies [13, 14]. Thus, high-speed centrifugation technique should not be used for EV isolation. Other non-centrifugation techniques, such as size-exclusion chromatography and ultrafiltration may be more suited than the centrifugation techniques for EV isolation, especially to purify EVs from non-EVs without affecting the numbers and composition of EVs [15–17].

In this chapter, we describe a step-by-step guide for preparing EVs on the grid before TEM and cryo-TEM imaging. Furthermore, we provide a guide to an automated image-processing analysis to process large amounts of images for obtaining good statistical analysis. This automatic processing is implemented as a computer program and used to detect particle (vesicle) and determine its diameter size [13].

2 Materials

General materials needed for preparing the EM grids are indicated in Fig. 1.

All prepared solutions should be filtered through 0.2 μm filter before use. Incubation of EVs with antibodies is performed at room temperature (RT) and dark, unless stated otherwise.

1. Phosphate Buffer Saline (PBS). Weigh 90 g sodium chloride (1540 mmol/L), 2.2 g di-sodium hydrogen phosphate dehydrate (12.4 mmol/L), and 0.32 g sodium dihydrogen phosphate dehydrate (2.05 mmol/L). Dissolve in 1 L ultrapure water. This PBS is ten times concentrated. Stored as a PBS stock solution not longer than 2 months at 2–8 °C. Before use, dilute 1 part of PBS stock solution with 9 parts of ultrapure water and mix well. Check the pH. This should be around 7.5 and does not need to be adjusted.
2. Sodium Citrate (3.2%). Weigh 1.6 g sodium (tri)citrate in 50 mL ultrapure water. Stored for ±3 months at 2–8 °C.
3. PBS with 0.32% sodium citrate. Mix 1 part of 3.2% sodium citrate with 9 parts of PBS solution (1 time concentration). Check the pH, it should be around 7.4, otherwise, adjust the solution by adding Hydrochloric acid (HCl) 0.1 M.
4. Bovine serum albumin (BSA)-c (0.1%) for immunogold labeling. Pipet 200 μL BSA-c (Aurion, Wageningen, The

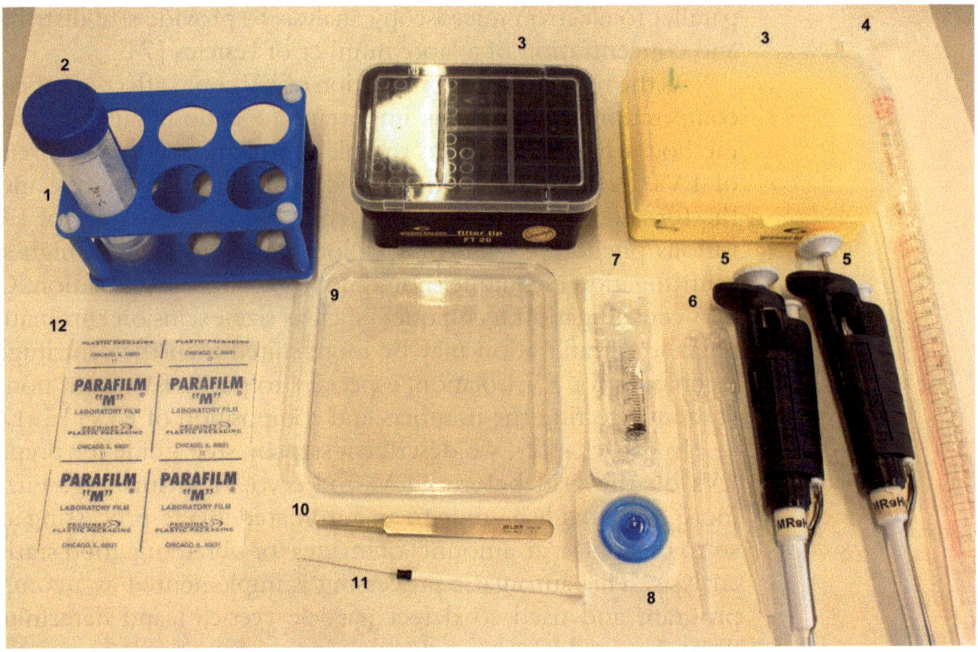

Fig. 1 As depicted in the clockwise direction, some materials needed for preparing EM grids are tube rack with disposable sterile tube (*1, 2*), disposable micropipette tips (*3*), disposable pipette (*4*), micropipettes (*5*), disposable transfer pipettes (*6*), sterile syringe (*7*), 0.2 μm sterile syringe filter (*8*), square petri dish (*9*), tweezers (*10*), nichrome loop (*11*) and Parafilm (*12*)

Netherlands) in a 50-mL tube Fill up with PBS (1 time concentration) till 20 mL. This solution should be freshly made and can be stored up to 1 day at 4 °C.

5. Glutaraldehyde (2%) In the fume hood, pipet 400 μL of 25% glutaraldehyde in a 15-mL tube. Fill up with PBS (1 time concentration) till 5 mL. This solution should be freshly made.
6. Uranyl acetate (2%). Use gloves while making this solution. In the fume hood, weigh 4 g of uranyl acetate powder and put it into the volumetric flask. Pour just enough hot ultrapure water to make the powder dissolved as the liquid has to be at RT for the flask to be accurate. Swirl the flask so that the powder dissolves. Once the contents of the flask are at RT, add ultrapure water to the flask to make 100 mL of solution. Pour the solution from the volumetric flask into a glass bottle. Label the stock bottle (4% uranyl acetate) and store dark at 4 °C. Make sure to dispose of all contaminated waste properly as uranyl acetate is radioactive. Pipet 200 μL of 4% uranyl acetate into a tube. Add 200 μL ultrapure water. Use this 2% uranyl acetate (weight/volume) for negative staining or store dark at 4 °C (shelf life: 3 months).
7. Methyl cellulose (2%). Heat 196 mL distilled water to a temperature of 90 °C. Add 4 g methyl cellulose 25 centipoises

while stirring. Rapidly cool the solution on ice while stirring. Seal with Parafilm and stir overnight at 4 °C. Stop stirring and store the solution for 2 days at 4 °C. Add distilled water till 200 mL. Centrifuge for 1.5 h at 100,000 × *g*. Collect the supernatant in tubes with a screw cap. Store at 4 °C in the dark (shelf life: 6–9 months).

8. Methyl cellulose–uranyl acetate. Use gloves while making this solution. Add 1 mL of 4% uranyl acetate solution into 9 mL of 2% methyl cellulose solution. Mix gently. Store at 4 °C in the dark (shelf life: 3 months).
9. Paraformaldehyde (1% weight/volume) (Electron Microscopy Sciences, Hatfield, PA, USA).
10. Carbon coated copper grids and nickel grids (200 mesh, Electron Microscopy Sciences).
11. EM grid with lacey carbon (300 mesh, TedPella).
12. Blocking solution which contains normal goat serum (Aurion).
13. Vitrobot Mark IV (FEI Company, The Netherlands) or a Leica EM GP (Leica, Germany).
14. Tecnai 12 electron microscope (FEI Company, Eindhoven, The Netherlands).

3 Methods

3.1 Negative Staining

1. Fix EVs at RT for 30 min by 0.1% paraformaldehyde in PBS. Use gloves and fume hood while working.
2. Glow-discharged the 200-mesh formvar and carbon coated copper grid.
3. Cut a piece of Parafilm big enough to put all the samples on. Stretch and lay this Parafilm on a square disposable petri dish. Mark the Parafilm with waterproof marker to specify the samples.
4. Pipet 10 μL of each sample on the Parafilm.
5. Put the shiny side of the grid (this side is coated) on top of the sample and incubate for 7 min.
6. Spot 3 small drops of ultrapure water and place the grid on top of it. Incubate the grid 2 min for each wash (*see* **Notes 1** and **2**) (Fig. 2).
7. Pipet 10 μL of 2% uranyl acetate on the Parafilm and incubate the grid for 7 min in this solution.
8. Excess solution was removed by blotting the grid at 45 degree angle once from the side of the grid with filter paper (*see* **Note 3**).

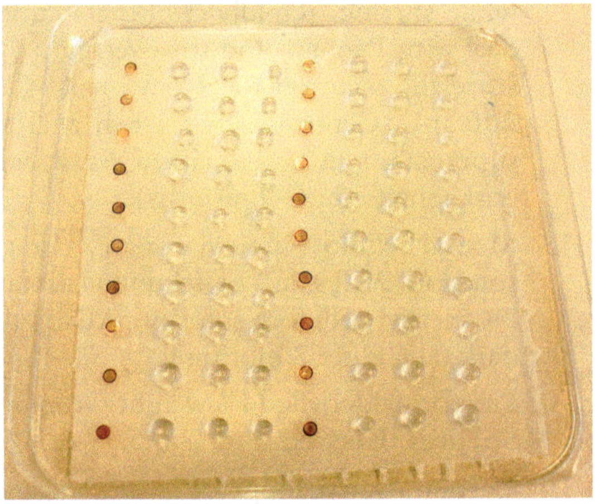

Fig. 2 Grids are placed on top of the samples and incubated for 7 min. Then, grids are transferred to drops of ultrapure water for washing

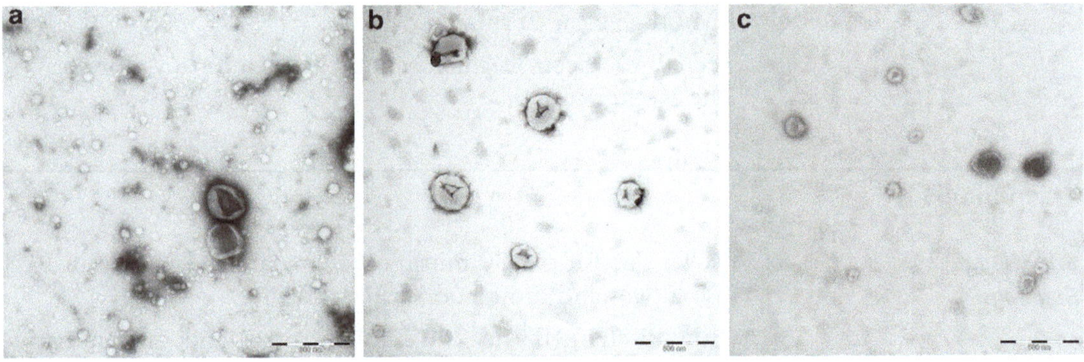

Fig. 3 EVs were isolated from plasma (**a**), erythrocyte concentrate (**b**), and platelet concentrate (**c**) using Sepharose CL-2B size-exclusion chromatography. EVs were visualized using uranyl acetate negative staining and imaged by TEM

9. Store the grid in the petri dish with the shiny side up or grid box at RT and dark until imaging.
10. The grid is imaged using a Tecnai 12 electron microscope operated at 80 kV (Fig. 3).

3.2 Immunogold Labeling

1. Use 3 of 200 mesh formvar and carbon coated nickel grid to perform single or double labeling. One grid will be used for the antibody labeling. The other grids will be for the antibody control (e.g., isotype control) and for the control of gold-coupled secondary antibody. The latest will be done by replacing the primary antibody with PBS.
2. Glow-discharged all the grids just before use to make the grid surface hydrophilic (*see* **Note 4**).

Table 2
Antibodies, lactadherin, and immunogold reagents for immunogold staining of EVs

Antibody/control	Clone	Dilution	Company
Mouse anti-human CD9	M-L13	1:50	Pharmingen
Mouse anti-human CD63	MEM-259	1:10	AbCam
Lactadherin FITC		1:20	Haematologic Technologies Inc.
Mouse IgG1 isotype control	X40	1:50	BD Biosciences
Goat anti-mouse 6 nm		1:50	Aurion
Mouse anti-FITC 10 nm		1:20	Aurion

3. Cut a piece of Parafilm big enough to put all the samples on. Stretch and lay this Parafilm on a square disposable petri dish. Mark the Parafilm with waterproof marker to specify the samples.
4. Pipet 10 μL of each EV sample on the Parafilm (*see* **Note 5**).
5. Put the shiny side of the grid (this side is coated) on top of the sample and incubate for 7 min.
6. Spot 3 small drops of filtered PBS on the Parafilm. Incubate the grid 2 min for each wash.
7. Pipet 10 μL blocking solution which contains normal goat serum. Use this solution undiluted (*see* **Note 6**). Place the grid for 30 min on top of this blocking solution to reduce background (nonspecific) reactions.
8. In the meantime, make the dilution of primary antibody and isotype control in 0.1% BSA-c (*see* Table 2).
9. Spot 3 small drops of 0.1% BSA-c on the Parafilm.
10. Wash the grid three times by placing the grid on top of these 0.1% BSA-c drops and incubate the grid 2 min for each wash.
11. Pipet 10 μL of mouse anti-human CD63, mouse anti-human CD9, lactadherin FITC, mouse IgG1 isotype control, or PBS on the Parafilm for single staining. For double staining with lactadherin FITC, make a mixture of 5 μL antibody and 5 μL lactadherin FITC on the Parafilm.
12. Incubate the grid with this antibody/isotype control/PBS solution for 1 h (*see* **Note 7**). Cover the grid so that the drop would not be evaporated.
13. Spot 6 small drops of 0.1% BSA-c on the Parafilm and wash the grid 6 times by placing the grid on top of these drops and incubate the grid 2 min for each wash.
14. Dilute goat anti-mouse 6-nm gold particles (*see* Table 2) in 0.1% BSA-c and pipet 10 μL on the Parafilm.

15. Incubate the grid with this gold coupled-antibody for 1 h. Put the lid of the petri dish on and place it in a dark place or just cover it with aluminum foil during incubation.
16. Spot 6 small drops of 0.1% BSA-c on the Parafilm and wash the grid six times by placing the grid on top of these drops. Incubate the grid 2 min for each wash.
17. If lactadherin FITC staining is performed, dilute mouse-anti FITC 10 nm gold particles in 0.1% BSA-c (*see* Table 2) and pipet 10 μL on the Parafilm. Repeat **step 16** and continue with **step 18**.
18. Spot 3 small drops of PBS and wash the grid three times by placing the grid on top of PBS drops. Incubate the grid 2 min for each wash.
19. Spot a drop of 2% glutaraldehyde on the Parafilm and place the grid on top of this solution followed by incubation for 5 min.
20. Spot 6 small drops of ultrapure water and place the grid on top of it. Incubate the grid 2 min for each wash.
21. Pipet 10 μL of 0.4% uranyl acetate in 2% methyl cellulose on the Parafilm and incubate the grid for 7 min on top of this solution on ice (*see* **Note 8**).
22. Excess solution was removed by blotting the grid at 45 degree angle once from the side of the grid with filter paper.
23. Store the grid in the petri dish with the shiny side up or grid box at RT and dark until imaging.
24. The grid is imaged using a Tecnai 12 electron microscope operated at 80 kV (Fig. 4).

3.3 Cryo-Electron Microscopy

1. Glow-discharge 300 mesh EM grid with lacey carbon.
2. Pipet 10 μL of each EV sample on grid.
3. Blot the excess sample once between 1 and 2 s with filter paper.
4. Plunged the grid into liquid ethane that was kept in equilibrium with solid ethane.
5. **Steps 3** and **4** are done using a Vitrobot Mark IV or a Leica EM GP at room temperature and 100% humidity.
6. Store the grid after vitrification until use.
7. The grid is imaged using a Tecnai 12 electron microscope operated at 120 kV. Images were recorded at 13,000× magnification (Fig. 5).

3.4 TEM-/TSEM-Image Processing

The image processing procedure below describes the simplest working algorithm and in general it can be implemented using MATLAB, Python or other programming languages (*see* **Note 9**). We also provide a video tutorial explaining the analysis of an image which can be found at http://vesclas.cizmar.org.

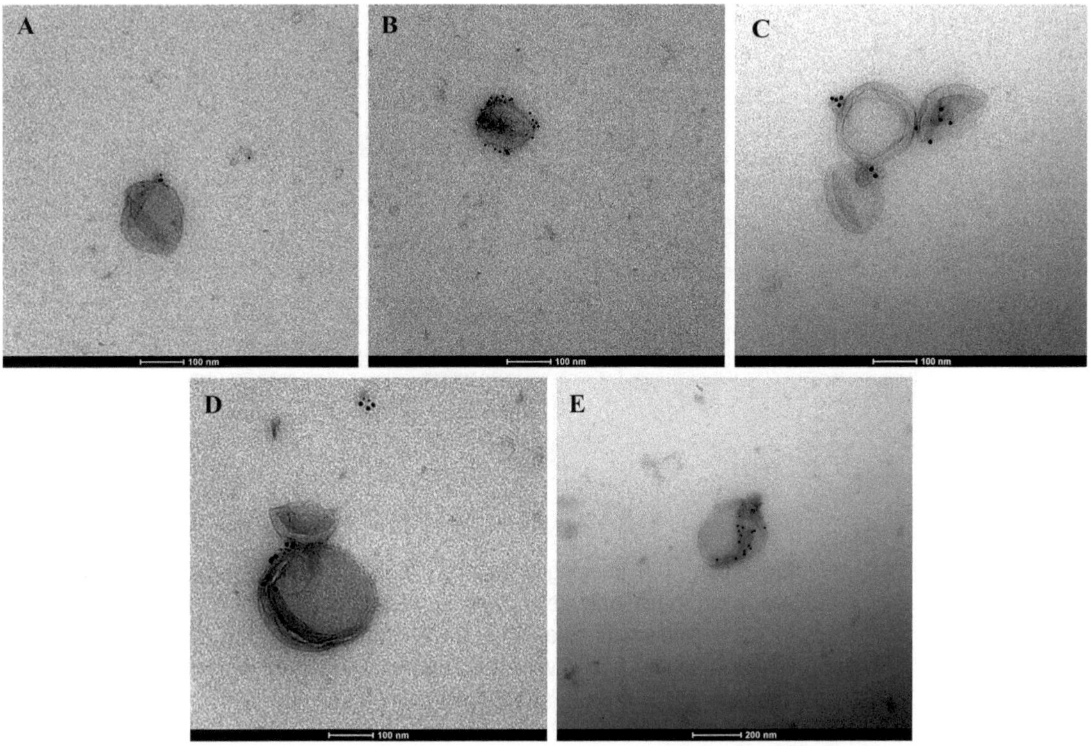

Fig. 4 EVs were isolated from cell culture supernatant and stained with antibody CD63 (**a**), antibody CD9 (**b**), and lactadherin, a marker of a negatively charged phosphatidylserine (**c**). Antibodies CD63 and CD9 are coupled with 6 nm gold particle. Lactadherin is coupled with 10 nm gold particle. Double labeling of antibody CD63 and lactadherin (**d**) or antibody CD9 and lactadherin (**e**) are shown. These images are taken by using magnification 105,000× (a-d, scale bars 100 nm) and 60,000× (e, scale bar 200 nm)

1. Crop information bar from the image, if present. This helps to avoid confusion of the detection algorithm.
2. Even out the background with removal of the lowest frequencies from the image using for example Discrete Cosine Transform.
3. Blur the image with a two-dimensional Gaussian function with low sigma to remove noise.
4. Set a starting point S in the image.
5. Set a direction angle θ.
6. Starting from the point S, search for the next local maximum of image intensity in the direction of θ. (Fig. 6).
7. When found, continue searching for the following local minimum in the same direction. Else go to point 8. If the following local minimum is found, it is a border point candidate.
8. Repeat from point 5 until all directions have been processed (whole 360 degrees).

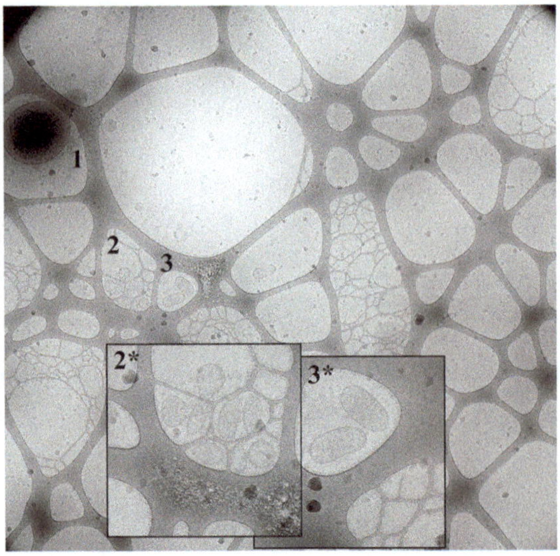

Fig. 5 Cyro-TEM imaging of isolated plasma EVs. Different types and sizes of EVs are found in plasma: electron dense vesicle with a size >200 nm (*1*), electron lucent vesicles with clear lipid bilayer and mostly with a size <200 nm (*2, 3*). Bigger magnification of vesicles *2* and *3* are depicted in *2** and *3**. The big scan image is 4.8 × 4.8 μm and the small scan images are 600 × 600 nm (*2*, 3**)

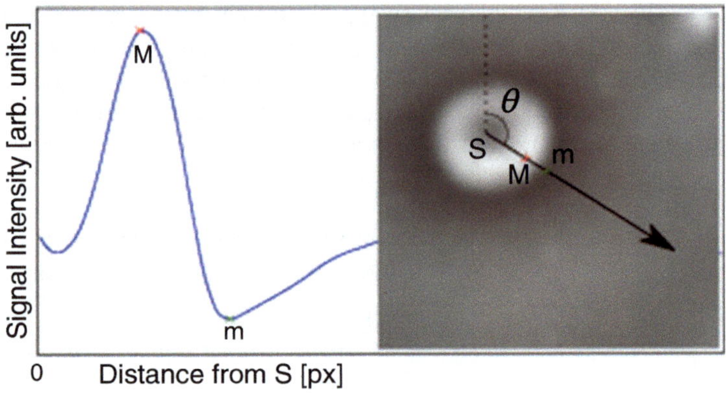

Fig. 6 Search for border points in TSEM image of a donut-shaped vesicle. Right: TSEM image of a donut-shaped vesicle, left: graph of the intensity profile in the direction of θ. *S* denotes the starting point, θ is the search-direction angle, *M* is the detected local maximum and *m* the local minimum and the detected border point

9. Check how many border points have been found. If too few (less then a prior set threshold), this object is either not a donut-shaped vesicle or it cannot be detected.
10. Fit a circle on the saved border-point candidates.
11. If the radius is smaller than the smallest allowed radius or it is bigger than the biggest allowed radius, this is not a valid donut-shaped vesicle.

12. Calculate the mean square deviation of all border-point candidates from the fitted circular shape. If the sum is too high (higher than a prior set threshold), this is not a valid donut-shaped vesicle.
13. For all border-point candidates, check if the deviation of the border point is higher than three times of the mean deviation. If so, delete the candidate point.
14. Repeat from point 10, until no more candidates shall be deleted.
15. Save the position of the center and the radius of the last fitted circle. This is a detected donut-shaped vesicle and its measured parameters. Save them.
16. Repeat from point 4, until the entire image is not scanned through (usually in a raster pattern).

4 Notes

1. During the incubation and washing steps, the shiny side (coated side) of the grid is facing the solution/suspension.
2. A precise volume of solution is not needed in the washing step. Thus, a plastic transfer pipet can be used for spotting the solution.
3. Whatman No. 1 filter paper is used for blotting the excess of solution.
4. Nonmagnetic tweezers are used for handling the nickel grid or use nichrome loop for handling the grid during immunogold labeling.
5. EVs should not be fixed when the primary antibodies are not suited for fixation.
6. Blocking solution used is from the species that the secondary antibody was generated in, as opposed to the species of the primary antibody.
7. At this step, the grids also could be stored dark overnight at 4 °C before proceeding to the wash steps.
8. To keep the grids cold, use a glass dish covered with Parafilm on ice.
9. This working procedure finds and measures donut-shaped vesicle structures only. Example of such implementation can be downloaded from http://vesclas.cizmar.org.

Acknowledgment

Both authors thank the technical supports given by A.E. Grootemaat from Department of Cell Biology and Histology, Academic Medical Centre, University of Amsterdam, Dr. R.I. Koning from the Molecular Cell Biology Department at the Leiden University Medical Center, and Dr. G. Posthuma from Cell Microscopy Center, Department of Cell Biology of University Medical Center Utrecht, The Netherlands.

References

1. van der Pol E, Boing AN, Gool EL, Nieuwland R (2016) Recent developments in the nomenclature, presence, isolation, detection and clinical impact of extracellular vesicles. J Thromb Haemost 14(1):48–56
2. Yuana Y, Sturk A, Nieuwland R (2013) Extracellular vesicles in physiological and pathological conditions. Blood Rev 27(1):31–39
3. van Der Pol E, Boing AN, Harrison P, Sturk A, Nieuwland R (2012) Classification, functions, and clinical relevance of extracellular vesicles. Pharmacol Rev 64(3):676–705
4. Yanez-Mo M, Siljander PR, Andreu Z et al (2015) Biological properties of extracellular vesicles and their physiological functions. J Extracell Vesicles 4:27066
5. Lener T, Gimona M, Aigner L et al (2015) Applying extracellular vesicles based therapeutics in clinical trials - an ISEV position paper. J Extracell Vesicles 4:30087
6. van der Pol E, Coumans FA, Grootemaat AE, Gardiner C, Sargent IL, Harrison P, Sturk A, van Leeuwen TG, Nieuwland R (2014) Particle size distribution of exosomes and microvesicles determined by transmission electron microscopy, flow cytometry, nanoparticle tracking analysis, and resistive pulse sensing. J Thromb Haemost 12(7):1182–1192
7. Lotvall J, Hill AF, Hochberg F, Buzas EI, Di Vizio D, Gardiner C, Gho YS, Kurochkin IV, Mathivanan S, Quesenberry P, Sahoo S, Tahara H, Wauben MH, Witwer KW, Thery C (2014) Minimal experimental requirements for definition of extracellular vesicles and their functions: a position statement from the International Society for Extracellular Vesicles. J Extracell Vesicles 3:26913
8. Crescitelli R, Lasser C, Szabo TG, Kittel A, Eldh M, Dianzani I, Buzas EI, Lotvall J (2013) Distinct RNA profiles in subpopulations of extracellular vesicles: apoptotic bodies, microvesicles and exosomes. J Extracell Vesicles 2:20677
9. Raposo G, Stoorvogel W (2013) Extracellular vesicles: exosomes, microvesicles, and friends. J Cell Biol 200(4):373–383
10. Yuana Y, Koning RI, Kuil ME, Rensen PC, Koster AJ, Bertina RM, Osanto S (2013) Cryo-electron microscopy of extracellular vesicles in fresh plasma. J Extracell Vesicles 2:21494
11. Hoog JL, Lotvall J (2015) Diversity of extracellular vesicles in human ejaculates revealed by cryo-electron microscopy. J Extracell Vesicles 4:28680
12. Arraud N, Linares R, Tan S, Gounou C, Pasquet JM, Mornet S, Brisson AR (2014) Extracellular vesicles from blood plasma: determination of their morphology, size, phenotype and concentration. J Thromb Haemost 12(5):614–627
13. Yuana Y, Boing AN, Grootemaat AE, van der Pol E, Hau CM, Cizmar P, Buhr E, Sturk A, Nieuwland R (2015) Handling and storage of human body fluids for analysis of extracellular vesicles. J Extracell Vesicles 4:29260
14. Linares R, Tan S, Gounou C, Arraud N, Brisson AR (2015) High-speed centrifugation induces aggregation of extracellular vesicles. J Extracell Vesicles 4:29509
15. Boing AN, van der Pol E, Grootemaat AE, Coumans FA, Sturk A, Nieuwland R (2014) Single-step isolation of extracellular vesicles by sizeexclusion chromatography. J Extracell Vesicles 3:23430
16. Nordin JZ, Lee Y, Vader P, Mager I, Johansson HJ, Heusermann W, Wiklander OP, Hallbrink M, Seow Y, Bultema JJ, Gilthorpe J, Davies T, Fairchild PJ, Gabrielsson S, Meisner-Kober NC, Lehtio J, Smith CI, Wood MJ, Andaloussi SE (2015) Ultrafiltration with size-exclusion liquid chromatography for high yield isolation of extracellular vesicles preserving intact biophysical and functional properties. Nanomedicine 11(4):879–883
17. Lozano-Ramos I, Bancu I, Oliveira-Tercero A, Armengol MP, Menezes-Neto A, Del Portillo HA, Lauzurica-Valdemoros R, Borras FE (2015) Size-exclusion chromatography-based enrichment of extracellular vesicles from urine samples. J Extracell Vesicles 4:27369

Chapter 19

Imaging of Isolated Extracellular Vesicles Using Fluorescence Microscopy

Dmitry Ter-Ovanesyan, Emma J. K. Kowal, Aviv Regev, George M. Church, and Emanuele Cocucci

Abstract

High-resolution fluorescence microscopy approaches enable the study of single objects or biological complexes. Single object studies have the general advantage of uncovering heterogeneity that may be hidden during the ensemble averaging which is common in any bulk conventional biochemical analysis. The implementation of single object analysis in the study of extracellular vesicles (EVs) may therefore be used to characterize specific properties of vesicle subsets which would be otherwise undetectable. We present a protocol for staining isolated EVs with a fluorescent lipid dye and attaching them onto a glass slide in preparation for imaging with total internal reflection fluorescence microscopy (TIRF-M) or other high-resolution microscopy techniques.

Key words Exosomes, Extracellular vesicles, EVs, Imaging, Microscopy, TIRF

1 Introduction

Over the last decade, interest in extracellular vesicles (EVs) has significantly grown due to their role in normal physiology, as well as pathological processes. EVs are thought to signal with recipient cells by interaction with their surface ligands or by transferring their contents (such as RNA or protein) into the recipient cell's cytoplasm [1]. Furthermore, as EVs are secreted into all biological fluids, they also hold great promise as novel biomarkers [2].

One of the major challenges in understanding the role of EVs in intercellular communication is characterizing the composition of individual EVs. When EVs are isolated in bulk, information regarding EV heterogeneity is lost. Existing EV isolation techni-

Dmitry Ter-Ovanesyan and Emma J.K. Kowal contributed equally to this work.

Winston Patrick Kuo and Shidong Jia (eds.), *Extracellular Vesicles: Methods and Protocols*, Methods in Molecular Biology, vol. 1660, DOI 10.1007/978-1-4939-7253-1_19, © Springer Science+Business Media LLC 2017

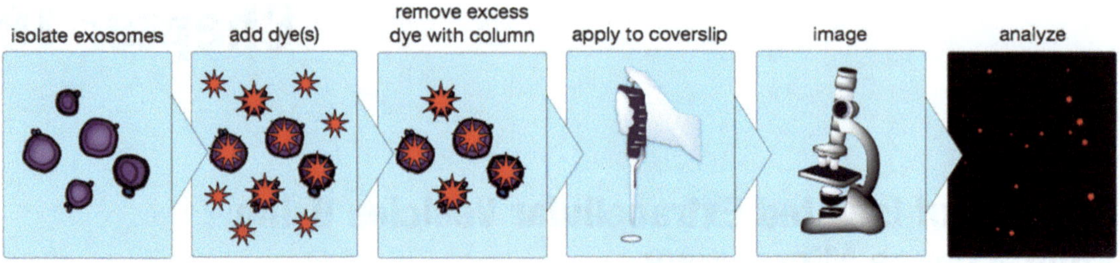

Fig. 1 Conceptual overview of the protocol. Schematic of EV staining and imaging procedure

ques, such as ultracentrifugation [3], result in a diverse mixture of vesicles from different cellular compartments [4] and are often contaminated by nonvesicular components such as extracellular protein complexes [5].

In recent years, high-resolution imaging has contributed greatly to the understanding of cell biology, permitting the detailed visualization of complex processes such as endocytosis [6], or the composition of single objects such as viruses [7]. In particular, total internal reflection fluorescence microscopy (TIRF-M) has been widely applied since it permits a high signal to noise ratio, the detection of single molecules, and a facile calibration of measurements to the single molecule level. These approaches can be efficiently applied to the study of EVs and identify different subpopulation of EVs present in a given sample. Although flow cytometry-based methods are able to characterize EV subpopulations, using flow cytometry for EV detection presents many technical challenges [8, 9] and is much less sensitive than TIRF-M.

Here, we provide a protocol for imaging isolated EVs by fluorescence microscopy. Although we have found TIRF-M to be best suited for these studies, other microscopes such as confocal [10, 11] and STED microscopes [12] can be used as well. We provide detailed instructions for staining EVs with a fluorescent membrane dye and attaching them to a glass slide (Fig. 1). This basic approach can be easily combined with other more specific staining, such as antibodies or nucleic acid dyes to study the heterogeneity of EV composition. Though colocalization does not strictly indicate that a given molecule is encapsulated within the lipid membrane, since it could also be adhered to the outside of the vesicle, use of this technique after various enzymatic treatments could provide this information. Additionally, EVs isolated from cells expressing fluorescent fusion proteins can be visualized using high-resolution microscopy [10, 11]. In conclusion, the high sensitivity of TIRF-M and its ability to make quantitative measurements provides a highly promising approach to study EV composition, as well as other diffraction-limited objects such as viruses or nanoparticles.

2 Materials

Store all materials at room temperature unless otherwise stated.

1. Glass coverslips. # 1.5 round 25 mm glass coverslips (Warner Instruments).
2. Coverslip rack. Teflon rack to vertically hold glass coverslips during cleaning procedures (C-14784, Thermo Fisher Scientific).
3. Coverslip holder for imaging. Attofluor Cell Chamber (A-7816, Thermo Fisher Scientific).
4. Total Internal Reflection Microscopy System. The imaging setup described here is composed of a Zeiss Axio Imager Z2 microscope equipped with a 63× Oil immersion objective (1.46 NA, Carl Zeiss) and a TIRF slider with manual angle and focus controls (Carl Zeiss). The illumination is supplied by solid state and argon ion lasers at 405 nm, 458/488/514 nm, 561 nm and 639 nm (power ~ 2–5 mW at the objective depending on laser). The excitation light is coupled through an acousto-optical tunable filter into a single mode optical fiber, carried to the TIRF slider and reflected into the objective using a single- or multi-band dichroic mirror (various; Semrock). The emission light is collected by the objective, passes the dichroic mirror and a single- or multi-band emission filter (various; Semrock), and is projected onto an electron multiplying charge-coupled device (EMCCD) camera (ImagEM-1K BackThinned EMCCD, Hamamatsu) for a final pixel size of 206 nm. Microscope operation and image acquisition are controlled by the Zen Blue software (Carl Zeiss). Any equivalent setup is suitable for the experiments described in this chapter.
5. EVs (*see* **Note 1**) isolated fresh, stored at 4 °C, or thawed from storage at −80 °C (*see* **Note 2**).
6. Vybrant DiD Cell-Labeling Solution (V-22887, Thermo Fisher Scientific).
7. Other fluorescent dyes (*see* **Note 3**) stored according to respective manufacturer recommendations.
8. Diluted dye stocks (*see* **Note 4**) stored according to respective manufacturer recommendations. For DiD, make a 50 nM stock in DMSO and store at room temperature in the dark.
9. EV Spin Columns, MW 3000 (4484449, Thermo Fisher Scientific) (*see* **Note 5**).
10. Fluorescent beads (*see* **Note 6**). 0.1 μm carboxylated beads (Ex/Em: 505/515 nm; F-8803, Thermo Fisher Scientific) stored at 4 °C.

11. Fluorescent bead solution. Dilute the fluorescent beads (2.10) to obtain ~$4x10^7$ beads in 100 μL of PBS, which according to the manufacturer (*see* 2.10) corresponds to a 10^5-fold dilution. To do this, prepare two consecutive dilutions, first 1:1000 to obtain a 1 mL stock solution, then 1:100 to obtain the working solution. Store all bead solutions at 4 °C or on ice during the experiment.
12. Bottle-top filter unit. Reusable bottle top filter unit (DS0320–5033, Nalgene) with removable membrane of 0.2 μm pore size for vacuum filtration.
13. 100% Ethanol. Filter 1 L of denatured ethanol (9401–03, JT Baker) into a bottle using a vacuum-based bottle-top filter unit (2.10). Fill a glass jar with the filtered ethanol such that it will entirely cover a rack loaded with glass coverslips (2.1 and 2.2).
14. Dulbecco's Phosphate Buffer Saline, no calcium, no magnesium (PBS, Thermo Fisher Scientific).

3 Methods

3.1 Clean Glass Coverslip Preparation

1. Place coverslips in rack (one per sample to be imaged, plus controls; *see* **Note** 7) and submerge rack in filtered 100% ethanol.
2. Sonicate for 20–30 min on low power.
3. Remove rack from ethanol and, without touching coverslips, adsorb excess ethanol by placing rack onto a Kimwipes or other tissue.
4. Place the rack with coverslips in 120 °C oven until dry; a lower temperature is acceptable but will take longer.
5. Immediately before applying sample to be imaged, expose the rack of coverslips to air plasma inside a glow discharge unit operated at 50 mA for 2 min.

3.2 Sample Preparation

1. Prepare one clean 1.5 mL tube of 50 μL PBS (*see* **Note 8**).
2. Prepare one tube of 50 μL EVs. If you are testing other dyes in addition to DiD, prepare one tube of EVs per dye and dye combination to be tested (*see* **Note** 7 on experimental design).
3. Add 1 μL of 50 nM DiD stock to PBS and to EVs. If using other dyes, add to appropriate tubes, making sure to add all dyes in use to control PBS tube.
4. Tap or gently vortex tubes to mix.
5. Incubate all samples for 20 min at 37 °C in the dark.
6. While incubating, prepare spin columns (*see* **Note 5**) according to manufacturer instructions [13], reproduced here:

7. Tap the column to settle the dry gel into the bottom of the spin column.
8. Hydrate the column with 650 μL PBS.
9. Cap, vortex, tap out air bubbles, and leave at RT 5–15 min.
10. Place columns in 2 mL collection tubes and spin at 750 RCF for 2 min at RT. Keep track of the orientation of the column in the rotor.
11. Discard collection tube and immediately apply the EV sample directly to the center of the gel bed at the top of the column IMPORTANT! Do not disturb the gel surface or contact the sides of the column with the pipette tip or reaction mixture.
12. Place column in 1.5 mL elution tube and place in rotor, maintaining same orientation as previous spin.
13. Spin at 750 RCF for 2 min at RT.
14. Discard column, retrieve EV sample from tube.
15. Samples can be imaged immediately or stored at 4 °C in the dark for up to 24 h.

3.3 Sample Addition to Coverslips and Imaging

1. Tape a clean strip of Parafilm onto bench and carefully lay down coverslip(s).
2. If using beads to visualize the focal plane (*see* **Note 6**), add 1 μL working fluorescent bead solution (*see* 2.1.11) to sample and invert several times to mix.
3. Add 50–100 μL sample to each coverslip and wait for 5 min for EVs to adsorb. Monitor the coverslips to ensure that they do not dry out completely, as this will create salt crystals which will interfere with imaging. If necessary, add several drops of PBS. (*see also* **Note 8**).
4. Using tweezers, turn the coverslip on its side and dab against Kimwipes or other tissue to wick off excess liquid, again without drying completely.
5. Place the coverslip carefully into bottom of holder with sample oriented up and make sure it is centered (tap around edges with tweezers, it should not lift on opposite side; *see* **Note 9**) before screwing holder together.
6. Gently add 1 mL PBS directly to coverslip.

3.4 Imaging

1. Place the holder with coverslip and PBS on the microscope stage.
2. Find the focal plane (*see* **Note 6**). In the DiD-stained EV sample, you should see a distribution of diffraction-limited punctae across the field of view (as in Fig. 2, right panel).

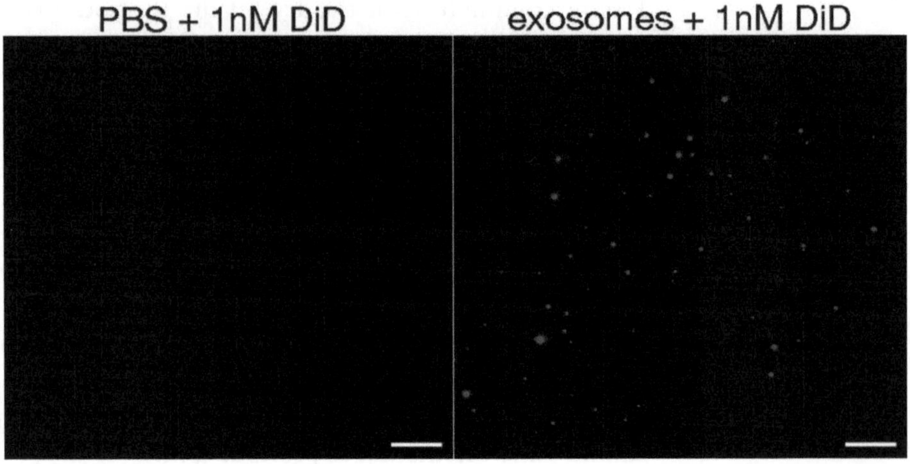

Fig. 2 Imaging of DiD-labeled extracellular vesicles. K562 EVs and PBS control incubated with 1 nM DiD, then imaged in far-red (Em filter 688/48). No signal is visible in the PBS slide while the EVs are clearly visible above background. Images shown are representative of three coverslips, twenty fields of view, in each case. (*Scale bars* = 10 μm)

3. Acquire several images in each channel of interest. Ensure that the PBS + dyes control sample is devoid of signal in each channel (as in Fig. 2, left panel).
4. For colocalization studies, ensure that vesicles labeled with one dye alone do not give signal in other channels (as in Fig. 3, bottom panels) and use caution when interpreting results (*see* **Note 10**).

4 Notes

1. EVs can be derived from any source of interest, i.e., cell culture or biological fluids. We find that EVs collected by standard ultracentrifugation protocols [3] from 250 mL of 50% confluent K562 cells over 24 h are sufficient to cover 50 coverslips with a reasonable concentration of particles/field.
2. EVs stored at 4 °C for under a week look similar to freshly isolated vesicles. For storage at −80 °C we recommend the addition of 20 mM HEPES buffer to stabilize the pH at ~7.4 over freeze-thaw cycles.
3. This protocol specifically outlines use of a lipid dye (DiD) to label EV membranes, but it can be used in combination with other dyes to label EV protein and RNA, or with transgenic cell lines to label specific molecules of interest [10, 11]. DiD in particular is conveniently combined with other probes as it emits in far-red, minimizing chance of bleed-through into other commonly used channels. Based on our tests, the spin

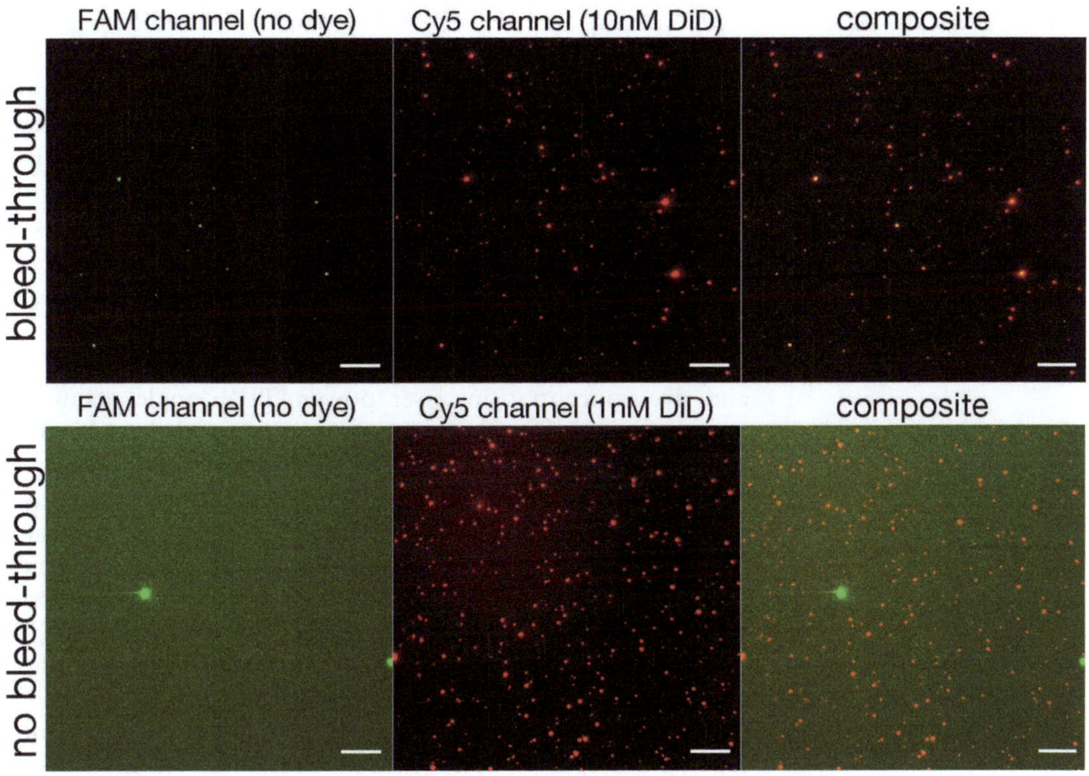

Fig. 3 Imaging of DiD-labeled EVs together with fluorescent beads. Distinguishing EV signal from bleed-through. *Top*: K562 EV sample demonstrating bleed-through between channels. *Left*: green (Em filter 525/50), *Center*: far-red (Em filter 688/48), *Right*: merge. EVs stained with 10 nM DiD appear to colocalize with objects in the green channel (emission filter) despite absence of green fluorophores in the sample. *Bottom*: K562 EV sample demonstrating no bleed-through between channels. *Left*: green (Em filter 525/50), *Center*: far-red (Em filter 688/48), *Right*: merge. Decreasing the concentration of DiD to 1 nM in the staining relieves bleed-through, while GFP-coated beads are still visible. Green channel is deliberately overexposed to demonstrate absence of any signal from DiD-stained objects. Images shown are representative of three coverslips, twenty fields of view, in each case. (*Scale bars* = 10 μm)

columns (2.1.9) did not effectively clear unbound fluorescent antibodies. When using multiple dyes or labeling strategies it is important to perform controls outlined in **Note 7**.

4. We recommend making dye stocks at a concentration equal to the desired final concentration times the volume of the sample, such that each dye can be provided to each sample in 1 μL of stock solution—for example, use a 150 nM stock of DiD with 30 μL EV samples for a final concentration of 5 nM. This is both for convenience and to avoid disrupting vesicle membranes with high concentrations of stock solvent.

5. The purpose of these spin columns is to remove contaminating fluorescent components from the dyed sample (i.e., free or aggregated dye, which should be in complexes smaller than

3 kDa). We found that the spin columns were more effective than dialysis and more convenient than ultracentrifugation.

6. It can be difficult to find the correct focal plane while imaging using signal from the vesicles alone, especially for initial experiments, and so we recommend the use of beads (with Ex/Em spectra distinct from any dyes in use) to aid in finding the correct plane. This is particularly crucial for control experiments with no labeled vesicles (*see* **Note 8**).
7. As EVs are diffraction-limited objects and this technique is very sensitive, there are many potential sources of false positive signal. It is important that for a given set of parameters (EV and dye concentration, laser power, TIRF angle, gain and exposure time) a number of controls is performed, and several images are taken across the coverslip in each channel of interest and at the correct focal plane (*see* **Note 9**). A sample of PBS and dyes only (no EVs) should be prepared and imaged to ensure that the dye itself is not forming micelles or aggregates which escape the spin column, as these would appear as fluorescent punctae indistinguishable from EVs (*see* Fig. 2, left vs. right). If objects are seen in this control, potential fixes are: decreasing the concentration of dye, cleaning and handling coverslips meticulously, filtering PBS to ensure it is free of contaminants, loading the spin column without disturbing the gel bed, or centrifuging dye stocks at high speed to remove precipitates. For colocalization experiments, a separate aliquot of EVs should be labeled with each dye alone and imaged to ensure that there is no bleed-through of fluorescence from the expected channel into another (*see* Fig. 3, top vs. bottom). If bleed-through is seen, exposure time may be lowered in the channel where bleed-through is observed or the concentration of the dye is decreased. Another potential solution is finding a narrower emission band filter for that channel.
8. The manufacturer's protocol [13] for the EV Spin Columns suggests a sample volume of 20–100 μL, but in our hands, samples lower than 40 μL applied to the column do not consistently elute at input volume and often come out 50% more dilute (i.e., 20 μL input yields 30 μL or more stained sample). This should be taken into account when considering desired final sample concentration for imaging.
9. If the coverslip is not flat it will be difficult to keep the focal plane while panning across the sample. If repeated adjustments of the coverslip in the holder do not fix a tilted sample, check that the microscope stage is flat.
10. The diffraction of light limits the resolution of imaging instruments. Therefore objects smaller than the diffraction limit, which approximately corresponds to half of the emitted

wavelength in a fluorescent sample, corresponds to the point spread function of the imaging system. Since a large fraction of EVs is smaller than the diffraction limit (~250 nm) more than one vesicle might be present in a single detected point spread function. Thus, this technique cannot strictly distinguish between one vesicle and two vesicles stuck together. Furthermore, as small EVs are diffraction-limited objects, colocalization does not strictly indicate that a given molecule is encapsulated within the lipid membrane, since it could also be adhered to the outside of the vesicle. Use of this technique in concert with enzymatic protection assays would be necessary to elucidate the contents of EVs.

Acknowledgment

This work was supported by US National Institutes of Health National Human Genome Research Institute Grant P50 HG005550, The Ohio State Cancer Center (P30-CA016058), and an intramural fund (IRP46050-502339) granted to EC.

References

1. Tkach M, Thery C (2016) Communication by extracellular vesicles: where we are and where we need to go. Cell 164(6):1226–1232
2. Revenfeld AL, Baek R, Nielsen MH, Stensballe A, Varming K, Jorgensen M (2014) Diagnostic and prognostic potential of extracellular vesicles in peripheral blood. Clin Ther 36 (6):830–846
3. Thery C, Amigorena S, Raposo G, Clayton A (2006) Isolation and characterization of exosomes from cell culture supernatants and biological fluids. Curr Protoc Cell Biol Chapter 3:Unit 3.22
4. Cocucci E, Meldolesi J (2015) Ectosomes and exosomes: shedding the confusion between extracellular vesicles. Trends Cell Biol 25 (6):364–372
5. Cvjetkovic A, Lotvall J, Lasser C (2014) The influence of rotor type and centrifugation time on the yield and purity of extracellular vesicles. J Extracell Vesicles 3
6. Cocucci E, Aguet F, Boulant S, Kirchhausen T (2012) The first five seconds in the life of a clathrin-coated pit. Cell 150(3):495–507
7. Chou YY, Vafabakhsh R, Doganay S, Gao Q, Ha T, Palese P (2012) One influenza virus particle packages eight unique viral RNAs as shown by FISH analysis. Proc Natl Acad Sci U S A 109(23):9101–9106
8. van der Pol E, Coumans F, Varga Z, Krumrey M, Nieuwland R (2013) Innovation in detection of microparticles and exosomes. J Thromb Haemost 11(Suppl 1):36–45
9. Nolan JP, Moore J (2016) Extracellular vesicles: great potential, many challenges. Cytometry B Clin Cytom 90:324–325
10. Fang Y, Wu N, Gan X, Yan W, Morrell JC, Gould SJ (2007) Higher-order oligomerization targets plasma membrane proteins and HIV gag to exosomes. PLoS Biol 5(6):e158
11. Lai CP, Kim EY, Badr CE, Weissleder R, Mempel TR, Tannous BA, Breakefield XO (2015) Visualization and tracking of tumour extracellular vesicle delivery and RNA translation using multiplexed reporters. Nat Commun 6:7029
12. Koliha N, Wiencek Y, Heider U, Jungst C, Kladt N, Krauthauser S, Johnston IC, Bosio A, Schauss A, Wild S (2016) A novel multiplex bead-based platform highlights the diversity of extracellular vesicles. J Extracell Vesicles 5:29975
13. Exosome Spin Columns. https://tools.thermofisher.com/content/sfs/manuals/MAN0008464_Rev01_PI_08Aug2013.pdf. Accessed 4 Apr 2016

Part V

Labeling and Tracking of Extracellular Vesicles

Chapter 20

In Vivo Tracking of Extracellular Vesicles in Mice Using Fusion Protein Comprising Lactadherin and *Gaussia* Luciferase

Yuki Takahashi, Makiya Nishikawa, and Yoshinobu Takakura

Abstract

Extracellular vesicles (EVs) are cell-derived vesicles comprising a lipid bilayer and are found in body fluids, such as blood, sweat, and urine. As EVs, especially exosomes, function as endogenous intercellular delivery tools, their roles in various biological events have been extensively investigated. In addition, they are expected to become safe and effective drug delivery systems (DDS) because of their intrinsic nature. In the development of EV-based DDS, as well as in the investigation of the biological functions of EVs, it is important to analyze the in vivo behavior of EVs by tracking them. Therefore, we have developed a sensitive EV-labeling method to track EVs in vivo by designing a fusion protein comprising lactadherin (LA) (alias milk fat globule-EGF factor 8), a protein that binds to EV membranes through interaction with phosphatidylserine, and *Gaussia* luciferase (gLuc), a chemiluminescent protein. gLuc-LA-labeled EVs are easily obtained by transfecting EV-producing cells with a gLuc-LA-encoding plasmid vector. Here, we describe methods to label EVs with the fusion protein and to track the labeled EVs in vivo.

Key words *Gaussia* luciferase, Lactadherin, In vivo imaging, Extracellular vesicles, Exosomes

1 Introduction

Extracellular vesicles (EVs) contain nucleic acids and proteins derived from EV-producing cells [1]. They are detected in most, if not all, types of body fluids, such as blood, sweat, urine and milk [2]. There are three main types of EVs: apoptotic bodies, microvesicles (or ectosomes), and exosomes [3]. As microvesicles and exosomes are produced by living cells, their role in cell–cell communication by transporting their cargoes, including nucleic acids, proteins and lipids, has been extensively investigated since the discovery that microRNA (miRNA) contained in exosomes are delivered to and taken up by cells and can modulate the expression of miRNA target genes in them [4]. It has been demonstrated that exosomes and microvesicles play important roles in various events, including immune responses [5, 6], tumor growth and metastases

Winston Patrick Kuo and Shidong Jia (eds.), *Extracellular Vesicles: Methods and Protocols*, Methods in Molecular Biology, vol. 1660, DOI 10.1007/978-1-4939-7253-1_20, © Springer Science+Business Media LLC 2017

[7], cardiovascular diseases [8], viral infections [9], metabolic disorders [10] and neurological disorders [11]. Hereafter, we refer to secreted extracellular vesicles, exosomes, and microvesicles as EVs.

As EVs possess biologically active molecules and target cell specificity, the therapeutic applications of EVs have also been investigated. EVs collected from mesenchymal stem cells display anti-inflammatory effects in various inflammatory diseases, such as graft-versus-host diseases and endotoxin-induced lung injury [12, 13]. The intracerebral administration of exosomes collected from neuroblastoma cells retards Alzheimer's disease progression by promoting amyloid-β clearance in the brain [14, 15]. In addition, the application of EVs as transport vehicles to deliver vaccines has been investigated. As EVs from tumor cells contain tumor-specific proteins, namely antigens, vaccination by administering tumor cell-derived EVs, mainly exosomes, has been investigated [16]. In addition to tumor-derived EVs, those derived from dendritic cells pretreated with antigen have also been used as vaccines because such EVs display major histocompatibility complexes with epitopes [17].

In addition to the direct usage of EVs as therapeutic treatments, EVs are expected to become drug delivery vehicles because of their characteristic function as endogenous delivery carriers [18]. In the development of EV-based therapeutics and drug delivery systems, as well as in the investigation of the biological and physiological roles of EVs, information about the in vivo behavior of EVs is important. To investigate this, tracking EVs by labeling them with a sensitive tracer can provide useful information. EVs labeled with lipophilic fluorescent dyes have been widely used to investigate their in vivo behavior because of the ease of preparation and low cost. However, fluorescent probes are not suitable for quantitative analysis and performing in vivo imaging with fluorescent probes is difficult because of their relatively low sensitivity. In a different approach, probes that emit luminescence are suitable for in vivo imaging because of their high sensitivity and thus are applicable to quantitative analysis. In a previous study, we succeeded in labeling exosomes with *Gaussia* luciferase (gLuc), a reporter protein that emits luminescence in response to its substrate (coelenterazine) by fusing with lactadherin (LA; an EV-tropic protein that binds to EV membrane by interacting with phosphatidylserine) and forming a fusion protein named gLuc-LA (Fig. 1). By simply transfecting EV-producing cells with gLuc-LA-expressing plasmid vectors, gLuc-LA-labeled EVs can be obtained. Using gLuc-LA-labeled exosomes collected from murine melanoma B16BL6 cells, we evaluated the in vivo fate of intravenously administered exosomes and found that they were rapidly cleared from blood circulation and mainly distributed in the liver, spleen and lung [19]. In addition, by administering the labeled exosomes into macrophage-depleted mice, we also demonstrated that the rapid clearance of exosomes was mainly caused by exosome uptake by macrophages [20].

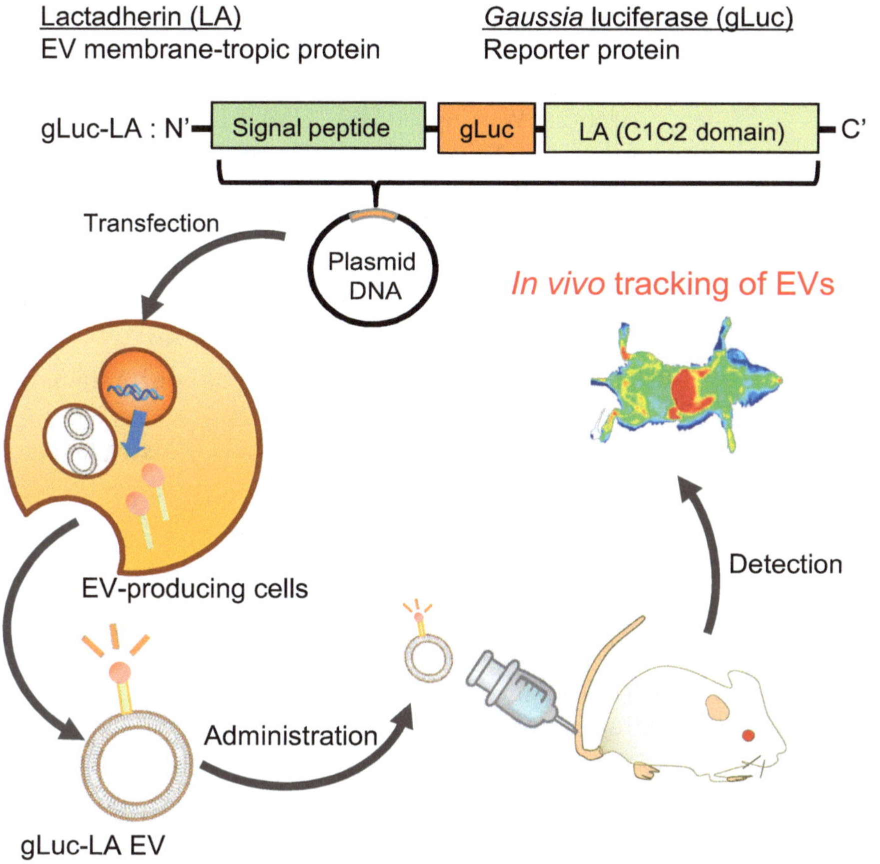

Fig. 1 Schematic representation of the in vivo tracking of EVs using gLuc-LA-labeled EVs (Reproduced in part from [19] with permission from Elsevier)

In this chapter, we describe three protocols for the in vivo tracking of EVs (mainly exosomes): (1) the preparation of labeled EVs, (2) the evaluation of the serum concentration profile of EVs after intravenous injection and (3) the in vivo imaging of EVs. The protocols can be useful in analyzing the in vivo fate of EVs.

2 Materials

2.1 Transfection and EV Collection

1. Plasmid vector: gLuc-LA-expressing vector (*see* **Notes 1** and **2**).
2. EV-producing cells: Any type of cell, e.g., B16-BL6 cells, can be used to produce gLuc-LA-labeled EVs (*see* **Note 3**).
3. Culture medium: Prepare culture medium suitable for the cells used for EV production.
4. EV collection medium (Serum-derived EV-depleted medium): Ultracentrifuge the culture medium at 100,000 × *g* for 2 h and collect supernatant to deplete serum-derived EVs. Sterilize the

collected supernatant by filtering through 0.45-μm filter. Store at 4 °C.

5. Transfection reagent: Polyethylenimine (PEI) "Max" (Polysciences, Warrington, PA, USA). Dissolve 3.23-mg PEI in 8-mL water. Adjust pH to 8.0 by adding NaOH solution. Adjust the volume to 10 mL with water for a final concentration of 0.323-mg PEI/mL. Filter through 0.2-μm membrane for sterilization. Store at −20 °C.
6. NaCl solution (1.5 M, sterile).
7. Centrifuge machine that can hold 50-mL tubes.
8. Ultracentrifugation tube: 33 polycarbonate thick-walled tubes (Hitachi Koki, Tokyo, Japan).
9. Ultracentrifuge machine, e.g., himac CP80WX (Hitachi Koki).
10. Angle-type rotor, e.g., P50AT2 (Hitachi Koki).
11. Phosphate buffered saline (PBS): NaCl (137 mM), KCl (2.7 mM), Na_2HPO_4 (10 mM), KH_2PO_4 (1.76 mM), pH 7.4.
12. Luminometer, e.g., Lumat LB 9507 (EG&G Berthold, Bad Wildbad, Germany).
13. *Gaussia* luciferase assay buffer, e.g., PicaGene™ Dual (Toyo Ink, Tokyo, Japan).
14. Polystyrene tubes for the luminometer.
15. Bradford assay buffer, e.g., Quick Start™ Bradford protein assay (BioRad, Hercules, CA, USA).
16. 96-well assay plate.
17. Absorbance microplate reader, e.g., Multiskan™ FC Microplate Photometer (Thermo Fisher, Waltham, MA, USA).

2.2 Evaluation of Serum EV Concentration Profile After Intravenous Injection

1. Mice.
2. Syringe, e.g., 1-mL syringe.
3. Needle, e.g., 26-gauge, ½-in. needle.
4. Centrifuge machine.
5. Luminometer.
6. *Gaussia* luciferase assay buffer.
7. Polystyrene tubes for luminometer.
8. PBS.

2.3 In Vivo Imaging of Intravenously Administered EVs

1. Mice (*see* **Note 4**).
2. Syringe, e.g., 1-mL syringe.
3. Needle, e.g., 26-gauge, ½-in. needle.
4. Coelenterazine stock solution: Dissolve 10-mg native coelenterazine in 1-mL ethanol containing 20-μL 3 N HCl. Seal and store at −80 °C.

5. Anesthesia: Inhalation anesthesia, e.g., isoflurane, or injection anesthesia, e.g., pentobarbital.
6. Imaging machine, e.g., IVIS Spectrum in vivo imaging system (Caliper Life Science, Hopkinton, MA, USA).
7. Transparent scotch tape.

3 Methods

3.1 Transfection and EV Collection

There are several methods for transfection and EV isolation. In this section, a transfection method using PEI and the isolation of EVs, mainly exosomes, using centrifugation-based sedimentation are described. In previous studies, we showed that cationic liposome-based transfection reagent and exosome collection methods other than ultracentrifugation-based pelleting can also be used to obtain gLuc-LA-labeled exosomes [19, 21]. In addition, EVs can be labeled with fusion proteins formed by fusing LA and a protein of interest other than gLuc, e.g., streptavidin [22].

1. One day before transfection, seed cells at a cell number appropriate for achieving approximately 80% confluency at the time of transfection (*see* **Note 5**).
2. Mix plasmid DNA, PEI solution, and 1.5 M NaCl solution at a final concentration of 40-μg DNA, 320-μL PEI solution, 100-μL 1.5 M NaCl/mL and adjust the volume to 11.1% of the culture medium. For example, prepare 1.11-mL transfection complex to transfect cells cultured with 10-mL culture medium. After mixing plasmid DNA and PEI, vortex it for 30 s and incubate for 15 min at room temperature.
3. Add DNA/PEI complex to the cells.
4. Incubate for 4 h.
5. Wash the transfected cells three times with PBS.
6. Change the medium to EV collection medium and incubate for 1 day (*see* **Note 6**).
7. Collect culture medium and transfer to 50-mL tubes.
8. Centrifuge at 300 × *g* for 10 min.
9. Transfer the supernatant to fresh 50-mL tubes and centrifuge at 2000 × *g* for 20 min.
10. Transfer the supernatant to fresh 50-mL tubes and centrifuge at 10,000 × *g* for 30 min (*see* **Note 7**).
11. Filter the supernatant through 0.2-μm syringe filter (optional step).
12. Transfer to ultracentrifuge tubes and centrifuge at 100,000 × *g* for 1 h.

13. Remove the supernatant, add PBS, and centrifuge at 100,000 × *g* for 1 h. Repeat this step twice (*see* **Note 8**).
14. Resuspend the EV pellet in PBS (*see* **Note 9**).
15. Add an appropriate volume of luciferase assay buffer into polystyrene tubes. Subsequently, add an appropriate volume of EV samples or EV samples diluted with PBS to the assay buffer and mix (*see* **Note 10**).
16. Measure luminescence intensity using the luminometer to estimate labeling efficiency.
17. Measure protein concentration using the Bradford assay kit (*see* **Note 11**).

3.2 Evaluation of Serum EV Concentration Profile of Intravenously Administered EVs

1. Prepare the solution to be administered by diluting the EV samples with PBS (*see* **Note 12**).
2. Attach the needle to the syringe and fill the syringe with EV suspension.
3. Inject into the tail vein of the mice.
4. Collect blood in 1.5-mL or 0.5-mL tubes.
5. Keep the tubes on ice for over 2 h to allow blood clotting.
6. Centrifuge at 3000 × *g* for 20 min.
7. Collect supernatant to obtain serum samples.
8. Dilute serum with PBS (*see* **Note 13**).
9. Add 100-μL luciferase assay buffer to polystyrene tubes. Subsequently, add 10-μL EV sample to the assay buffer and mix.
10. Insert the tube into the luminometer to evaluate luciferase activity by measuring luminescence intensity (Fig. 2).

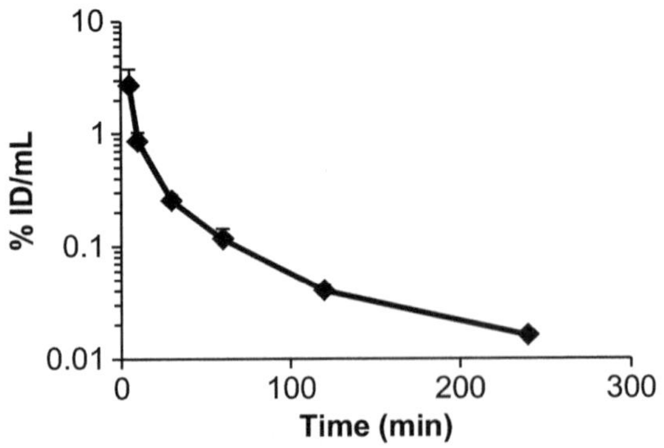

Fig. 2 Serum concentration profile of gLuc-LA-labeled exosomes after intravenous injection. The amount of exosomes in the serum was normalized to the injected dose based on gLuc activity and expressed as a percentage of the injected dose/mL (% ID/mL). The results are expressed as mean ± standard error of mean (S.E.M) of five mice (Reproduced from [19] with permission from Elsevier)

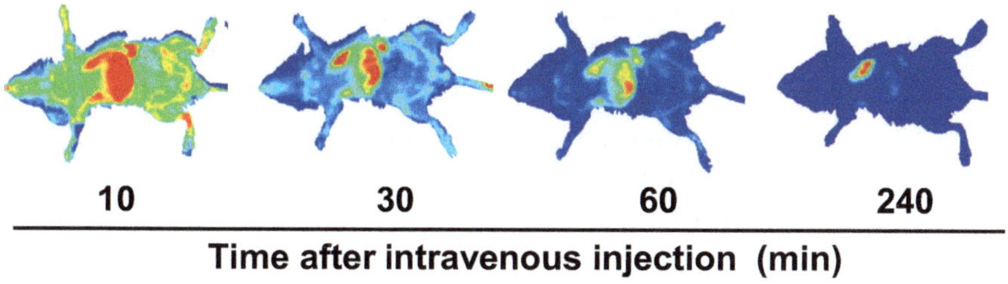

Fig. 3 In vivo imaging of gLuc-LA-labeled exosomes after intravenous injection. Mice received the intravenous injections of gLuc-LA-labeled exosomes collected from B16BL6 cells. gLuc-LA-labeled exosomes were imaged 10, 30, 60 and 240 min after the intravenous injection of exosomes through a bolus intravenous injection of coelenterazine (Reproduced from [19] with permission from Elsevier)

3.3 In Vivo Imaging of Intravenously Administered EVs

1. Prepare the solution to be administered by diluting the exosome suspension with PBS (*see* **Note 14**).
2. Attach the needle to the syringe and fill the syringe with EV suspension.
3. Inject into the tail vein of the mice.
4. Anesthetize the mice (*see* **Note 15**).
5. Immediately before imaging, prepare the imaging injection solution by mixing 10-μL coelenterazine stock solution with 140-μL PBS (*see* **Note 16**).
6. Inject the prepared imaging injection solution into the vein of the mice (*see* **Note 17**).
7. Fix mice onto a black plate by taping their limbs using scotch tape (optional step, *see* **Note 18**).
8. Acquire image using the imaging machine (Fig. 3, *see* **Note 19**).

4 Notes

1. We used pcDNA3.1-based plasmid vector encoding gLuc-LA under the control of a cytomegalovirus promoter. Any type of plasmid vector can be used for the transfection of the gLuc-LA gene, as long as it can produce sufficient gLuc-LA in the transfected EV-producing cells.
2. For efficient transfection, the purity of the plasmid vector is important. Therefore, we recommend the use of plasmid vector purified from *Escherichia coli* using a commercially available plasmid DNA purification kit. We used PureLink HiPure Plasmid Maxiprep Kit (Thermo Fisher).
3. As the amount of EVs secreted varies considerably between different cell types [23], we recommend checking the amount of EVs secreted by the cells of interest.

4. Mice with colored hair, such as C57BL6 mice, are not suitable for in vivo imaging as luminescence emitted from gLuc is attenuated by the hair.
5. To obtain high levels of transgene expression, cells should not be confluent at the time of transfection. We seeded 1.5×10^6 of B16BL6 cells in a 10-cm dish 1 day before transfection.
6. Incubation time can be shortened or extended depending on the properties of the cells. We recommend measuring the effect of incubation time on EV yield and labeling efficiency.
7. At this step, gLuc-LA-labeled vesicles, mainly microvesicles, with relatively high density are collected as pellets. If microvesicles are the EVs of interest, resuspend the pellet with PBS and centrifuge again at $10{,}000 \times g$ for 30 min to wash the microvesicles. Resuspend the pellet and skip to **step 15** in Subheading 3.1.
8. As EV pellets are hardly visible, it may be useful to detect the pellet before starting the removal of supernatant. While removing the supernatant, care must be taken as the EV pellets can easily detach from the wall of centrifugation tubes.
9. At this step, gLuc-LA-labeled vesicles, mainly exosomes are collected. To disrupt the aggregated exosomes in the pellet, repeated pipetting is required. However, as exosomes can get easily adsorbed onto plastics, such as those used in tubes and pipette tips, avoid excessive pipetting as this may reduce the amount of exosomes collected.
10. The required volume of assay buffer and samples is dependent on the *Gaussia* luciferase assay reagent. If the luminescence intensity of the samples is over the detection limit of the luminometer, dilute the EVs with PBS.
11. We generally use 5 μL of exosome sample mixed with 250 μL of Bradford assay buffer.
12. We generally prepare an injection solution containing 0.3–15-μ g exosomes per 200-μL PBS for one mouse.
13. As serum components affect the enzyme reaction of gLuc with its substrate, serum samples should be diluted with PBS before measuring gLuc activity. In our experience, the effect of serum components on gLuc activity is negligible at over threefold dilution with PBS.
14. Based on our experience, approximately five to tenfold higher gLuc activity is desirable for performing in vivo imaging compared with that for evaluating serum EV concentration profile.
15. Both inhalation anesthesia, such as isoflurane and injectable anesthesia, such as pentobarbital, are applicable for anesthetizing the mice.

16. As coelenterazine is unstable in aqueous solution, imaging injection solution should be prepared immediately before injection.
17. As coelenterazine is unstable in injection solution, it is desirable to administer the injection solution as quickly as possible. The use of an intravenous catheter may help to speed up the process.
18. If the anesthetized mice do not move during image acquisition, this step can be skipped.
19. As intravenously administered coelenterazine very quickly disappears from the blood circulation, image acquisition should be initiated as soon as possible after the intravenous injection.

References

1. Yáñez-Mó M, Siljander PR, Andreu Z et al (2015) Biological properties of extracellular vesicles and their physiological functions. J Extracell Vesicles 4:27066. doi:10.3402/jev.v4.27066
2. Boukouris S, Mathivanan S (2015) Exosomes in bodily fluids are a highly stable resource of disease biomarkers. Proteomics Clin Appl 9:358–367. doi:10.1002/prca.201400114
3. Akers JC, Gonda D, Kim R et al (2013) Biogenesis of extracellular vesicles (EV): exosomes, microvesicles, retrovirus-like vesicles, and apoptotic bodies. J Neuro-Oncol 113:1–11. doi:10.1007/s11060-013-1084-8
4. Valadi H, Ekström K, Bossios A et al (2007) Exosome-mediated transfer of mRNAs and microRNAs is a novel mechanism of genetic exchange between cells. Nat Cell Biol 9:654–659. doi:10.1038/ncb1596
5. Turpin D, Truchetet ME, Faustin B et al (2016) Role of extracellular vesicles in autoimmune diseases. Autoimmun Rev 15:174–183. doi:10.1016/j.autrev.2015.11.004
6. Fernández-Messina L, Gutiérrez-Vázquez C, Rivas-García E et al (2015) Immunomodulatory role of microRNAs transferred by extracellular vesicles. Biol Cell 107:61–77. doi:10.1111/boc.201400081
7. Miller IV, Grunewald TG (2015) Tumour-derived exosomes: tiny envelopes for big stories. Biol Cell 107:287–305. doi:10.1111/boc.201400095
8. Das S, Halushka MK (2015) Extracellular vesicle microRNA transfer in cardiovascular disease. Cardiovasc Pathol 24:199–206. doi:10.1016/j.carpath.2015.04.007
9. Alenquer M, Amorim MJ (2015) Exosome biogenesis, regulation, and function in viral infection. Virus 7:5066–5083. doi:10.3390/v7092862
10. Milbank E, Martinez MC, Andriantsitohaina R (2016) Extracellular vesicles: pharmacological modulators of the peripheral and central signals governing obesity. Pharmacol Ther 157:65–83. doi:10.1016/j.pharmthera.2015.11.002
11. Porro C, Trotta T, Panaro MA (2015) Microvesicles in the brain: biomarker, messenger or mediator? J Neuroimmunol 288:70–78. doi:10.1016/j.jneuroim.2015.09.006
12. Zhu YG, Feng XM, Abbott J et al (2014) Human mesenchymal stem cell microvesicles for treatment of Escherichia coli endotoxin-induced acute lung injury in mice. Stem Cells 32:116–125. doi:10.1002/stem.1504
13. Heldring N, Mäger I, Wood MJ et al (2015) Therapeutic potential of multipotent mesenchymal stromal cells and their extracellular vesicles. Hum Gene Ther 26:506–517. doi:10.1089/hum.2015.072
14. Yuyama K, Sun H, Mitsutake S et al (2012) Sphingolipid-modulated exosome secretion promotes clearance of amyloid-β by microglia. J Biol Chem 287:10977–10989. doi:10.1074/jbc.M111.324616
15. Yuyama K, Sun H, Sakai S et al (2014) Decreased amyloid-β pathologies by intracerebral loading of glycosphingolipid-enriched exosomes in Alzheimer model mice. J Biol Chem 289:24488–24498. doi:10.1074/jbc.M114.577213
16. Hao S, Moyana T, Xiang J (2007) Review: cancer immunotherapy by exosome-based vaccines. Cancer Biother Radiopharm 22:692–703. doi:10.1089/cbr.2007.368-R
17. Pitt JM, Charrier M, Viaud S et al (2014) Dendritic cell-derived exosomes as

immunotherapies in the fight against cancer. J Immunol 193:1006–1011. doi:10.4049/jimmunol.1400703
18. Vader P, Mol EA, Pasterkamp G et al (2016) Extracellular vesicles for drug delivery. Adv Drug Deliv Rev 106:148–156. doi:10.1016/j.addr.2016.02.006
19. Takahashi Y, Nishikawa M, Shinotsuka H et al (2013) Visualization and in vivo tracking of the exosomes of murine melanoma B16-BL6 cells in mice after intravenous injection. J Biotechnol 165:77–84. doi:10.1016/j.jbiotec.2013.03.013
20. Imai T, Takahashi Y, Nishikawa M et al (2015) Macrophage-dependent clearance of systemically administered B16BL6-derived exosomes from the blood circulation in mice. J Extracell Vesicles 4:26238. doi:10.3402/jev.v4.26238
21. Yamashita T, Takahashi Y, Nishikawa M et al (2016) Effect of exosome isolation methods on physicochemical properties of exosomes and clearance of exosomes from the blood circulation. Eur J Pharm Biopharm 98:1–8. doi:10.1016/j.ejpb.2015.10.017
22. Morishita M, Takahashi Y, Nishikawa M et al (2015) Quantitative analysis of tissue distribution of the B16BL6-derived exosomes using a streptavidin-lactadherin fusion protein and iodine-125-labeled biotin derivative after intravenous injection in mice. J Pharm Sci 104:705–713. doi:10.1002/jps.24251
23. Yeo RW, Lai RC, Zhang B et al (2013) Mesenchymal stem cell: an efficient mass producer of exosomes for drug delivery. Adv Drug Deliv Rev 65:336–341. doi:10.1016/j.addr.2012.07.001

Chapter 21

Tracking Extracellular Vesicles Delivery and RNA Translation Using Multiplexed Reporters

Anthony Yan-Tang Wu and Charles Pin-Kuang Lai

Abstract

Elucidating extracellular vesicle (EV; e.g., exosomes, microvesicles) delivery and translation of its RNA cargo with an accurate spatiotemporal resolution is critical in helping understand EV's role under normal and pathological conditions. We here describe a multiplexed fluorescent and bioluminescent reporter strategy to simultaneously monitor and quantify EV delivery, as well as EV-RNA translation in EV-recipient cells.

Key words Extracellular vesicles, Exosomes, Microvesicles, Fluorescent proteins, Fluorescence-activated cell sorting, Bioluminescence, *Gaussia* luciferase, RNA translation

1 Introduction

Extracellular vesicles (EVs) are nanosized and cell-released vesicles with a lipid bilayer. These vesicles include exosomes (40–100 nm), microvesicles (100–1000 nm) and apoptotic blebs (50–4000 nm) [1]. Research in the past few years have shown that cells not only communicate with neighboring cells via cellular junctions such as the adherens and gap junctions [2] but also exchange information with local and distant cells through EVs without a direct cell-to-cell contact [1, 2]. Specifically, EVs have been reported to ferry information from EV-donor cells, including fragmented DNA, mRNA, miRNA, siRNA, proteins and even special tropism ligands [1, 3–7].

The ability to visualize and monitor EV provides important spatiotemporal information, and therefore a closer understanding on EV's role under pathological and physiological conditions. Different labeling methods have been applied to track EVs but each with its own limitations. For fluorescent protein conjugated/fused to proteins enriched in EVs such as CD63-GFP and other EV markers

Winston Patrick Kuo and Shidong Jia (eds.), *Extracellular Vesicles: Methods and Protocols*, Methods in Molecular Biology, vol. 1660, DOI 10.1007/978-1-4939-7253-1_21, © Springer Science+Business Media LLC 2017

like CD9, CD81, TSG101 and Alix [8, 9], the signals are restricted to the proposed EV subtypes including exosomes and microvesicles, thereby preventing researchers from studying other EVs devoid of the labeled protein. Lipid and aliphatic dyes such as PKH dyes, on the other hand, can label different subtypes of EVs but with a long in vivo half-life. PKH 2 and PKH 67 (Ex. 490 nm/Em. 502 nm) have a relatively shorter in vivo half-life of 5–8 days and 10–12 days, respectively. Meanwhile, PKH 26 (Ex. 551 nm/Em. 567 nm) has a half-life than 100 days [10, 11]. While the dye's long half-life could be advantageous on labeling cells for long-term tracking as it was intended for, its use on staining EVs may render false-positive detection of EVs due to its nanometer size and half-life. For example, dye residues from the EV staining process without stringent washing steps may be detected as actual EVs when applied to samples. Moreover, the dyes' long-half life may outlast that of EVs, and gets recycled as a non-EV, lipid-attached signals. In addition, the dyes have been reported to self-assemble into micelle structure which may further lead to inaccurate measurement on EV's spatiotemporal properties [12].

Protein S-palmitoylation through a thioester linkage of cysteines side chain to 16-carbon fatty acid palmitic acid allows association of conjugated proteins to the cellular membrane of eukaryotic cells [13, 14]. By fusing fluorescent proteins with the palmitoylation signal, these reporters can be associated with the cellular membrane to enable plasma membrane labeling [15]. Since EVs can be released via direct budding from the plasma membrane to form microvesicles or through inward invagination of endosome to produce multivesicular bodies to make exosomes [16], a palmitoylated fluorescent reporter may serve as an ideal strategy to label multiple EV subtypes [17].

Bioluminescence reporters have also been widely used to study biological process with their good signal intensity and low background noise [18]. Among the luciferases, *Gaussian luciferase* (Gluc) exhibits one of the highest signal intensity (*i.e.*, ~1000-fold higher than *Firefly luciferase*) and a short signal half-life (*i.e.*, flash-type bioluminescence) [19]. With these unique properties, Gluc serves as a suitable reporter to study minute biological events such as EV-delivered RNA translation [17].

Combining these reporters' advantages, we here describe a multiplexed assay with the use of palmitoylated green fluorescent protein (PalmGFP) and membrane-bound Gluc (GlucB) to simultaneously monitor EV uptake and translation of EV-delivered RNA by the recipient cells.

2 Materials

2.1 Reporter Constructs, Extracellular Vesicle Donor, and Extracellular Vesicle Recipient Cells

1. EV reporter constructs: pCSC00572-PalmGFP [17] and pCSCW2-GlucB [12]. To generate the fluorescent EV reporter, a palmitoylation sequence (MLCCMRRTKQ) of growth cone associated protein (GAP43) [20] was genetically fused to the NH_2 terminus of enhanced green fluorescent protein (PalmGFP) and cloned downstream of a cytomegalovirus promoter (CMV) in a second generation lentivirus plasmid [17, 21]. pCSCW2-GlucB was generated as previously described and employed as an EV-RNA translation reporter [22]. Use these plasmids to package PalmGFP or GlucB into lentivectors for subsequent infection and labeling of EV donor cells (*see* Subheading 2.1, **item 2**).
2. Reporter for EV recipient cells: pCSCW2-mCherry was a generous gift from Dr. Miguel Sena-Esteves [21] and packaged into lentivectors (*see* **Note 1**).
3. EV donor cells: Human embryonic kidney 293T (HEK-293T) cells were obtained from American Type Culture Collection (ATCC; Manassas, VA, USA). Culture the cells in Dulbecco's modified Eagle's medium (DMEM) supplemented with 10% fetal bovine serum (FBS), 100 U/mL penicillin and 0.1 mg/mL streptomycin at 37 °C and 5% CO_2. Other cell types may also be used.
4. EV recipient cells: Human Gli36 glioma cell line was a generous gift from Dr. Anthony Capanogni (UCLA, Los Angeles, CA), and transduced with mCherry-encoding lentivectors (Subheading 2.1, **item 2**) to stably express mCherry. Culture the cells as described above (Subheading 2.1, **item 3**). Other cell types may also be used.
5. Filter Unit: 0.2 μm, Nalgene® Rapid-Flow™ Complete Unit, polyethersulfone (PES) membrane, Sterile (Thermo Fisher Scientific, Waltham, MA, USA).
6. EV-depleted FBS: Centrifuge FBS at 100,000 × *g* using an ultracentrifuge set up as described (*see* Subheading 2.2, **items 1–6**) for at least 16 h at 4 °C. Collect the supernatant and filter it through a 0.22 μm PES filter unit (*see* Subheading 2.1, **item 5**) to further reduce FBS-EVs. Aliquot the filtrate (i.e., EV-depleted FBS) into 50 mL canonical tubes and store at −80 °C (*see* **Note 2**).
7. EV-depleted medium: Add 10% EV-depleted FBS to DMEM supplemented with 100 U/mL penicillin, and 0.1 mg/mL streptomycin (*see* **Note 3**).
8. Polybrene: Dissolve polybrene (Sigma-Aldrich, St. Louis, MO) at 10 mg/mL in tissue culture grade water.

9. An epifluorescence or confocal microscope with a 10× or higher power objective lens.
10. BD FACS Aria II SORP (4-Laser) Cell Sorter (BD Biosciences, San Jose, CA, USA) or similar equipment.

2.2 Extracellular Vesicle Isolation Equipment

1. Ultracentrifuge: Optima XE-90 (Beckman Coulter Taiwan Inc., Taipei, Taiwan).
2. Rotor: Type 70 Ti Fixed Angle Rotor (Beckman Coulter) with tube caps (Cat# 331151, Beckman Coulter).
3. Cordless tube topper (Cat #358312, Beckman Coulter), seal former (Cat #348120, Beckman Coulter).
4. Ultracentrifuge tubes: Quick-Seal® tubes (Cat#: 342413, Beckman Coulter).
5. 60 mL syringes.
6. 16 gauge needles.
7. Disc Filter: 0.2 μm polyethersulfone (PES) membrane filter (Pall Singapore Taiwan, Taipei, Taiwan).
8. Double filtered phosphate buffered saline (PBS): 8 g/L NaCl, 0.2 g/L KCl, 1.78 g/L $Na_2HPO_4 \cdot 2\ H_2O$ and 0.27 g/L KH_2PO_4, adjust to pH 7.4 and filter twice through a 0.22 μm disc filter.

2.3 Extracellular Vesicle Fluorescence Assay

1. An epifluorescence or confocal microscope with a 40× or higher power objective lens.
2. Plain microscope glass slides.
3. Coverslips (*see* **Note 4**).
4. A fine tweezers.

2.4 EV Treatment

1. A refrigerated, benchtop centrifuge with a swing-bucket rotor and plate holders to hold multiwell plates (*e.g.* 6-, 12-, 24-well plates).
2. PBS (*see* Subheading 2.2, **item 8**).

2.5 Extracellular Vesicle Uptake Assay

1. BD LSRII Multi-Laser Cytometer (BD Biosciences).
2. 0.05% trypsin–0.53 mM EDTA.
3. PBS (*see* Subheading 2.2, **item 8**).

2.6 Extracellular Vesicle Bioluminescence and EV-RNA Translation Assays

1. A luminometer similar to the Dynex MLX Microtiter Plate Luminometer with an automated injector (Vienna, VA, USA).
2. 96-well white or black microtiter plates.
3. Coelentrazine (CTZ), the GlucB substrate: Coelentrazine native (Cat#: 303; Nanolight, Pinetop, AZ, USA). Add a few drops of HCl to methanol, and use this solution to reconstitute

CTZ to a 5 mg/mL to minimize CTZ oxidation. Aliquot the reconstituted CTZ into microfuge tubes, place in a light-tight container, and store at −20 °C. Prewarm CTZ to room temperature before use.

4. Cycloheximide: Dissolve cycloheximide in water to prepare a 20 mg/mL stock solution.

3 Methods

3.1 Generation of Extracellular Vesicle Donor Cells

1. Seed 0.5 × 10^6 HEK 293T cells per well in a 6-well tissue culture plate with fresh culturing medium. Other cell types may also be used and seeded based on their respective growth rate.
2. Transduce the cells at 70% confluence with GlucB-encoding lentivectors (*see* **Note 5**) and 1 μL polybrene (10 mg/mL) in 3 mL fresh culturing medium per well.
3. Check for GFP expression (from the GlucB reporter) and transduction efficiency using a fluorescence microscope at 48 h post-infection. ≥70–80% of infected cells should be achieved before proceeding to the next step (*see* **Note 6**).
4. Perform fluorescence-activated cell sorting on the infected cells, and select the top 10–35% GlucB-expressing cells (with GFP signal) to propagate as a stable cell line for subsequent steps (*see* **Notes** 7 and **8**).
5. Similar to the procedures described above (Subheading 3.1, **steps 1–4**), infect stable GlucB-expressing cells with lentivectors encoding PalmGFP. Using a fluorescence microscope, ensure that ≥70% of the infected cells exhibits plasma membrane-anchored GFP signal (*i.e.* PalmGFP) in addition to the cytoplasmic GFP signal from the GlucB reporter (*see* **Notes 9** and **10**).

3.2 Extracellular Vesicle Isolation

1. Seed 2 × 10^6 HEK-293T cells stably expressing PalmGFP and GlucB per 150 mm cell culture dish in 20 mL fresh culturing medium.
2. Once the cells reached 50% confluence, replace the culturing medium with 20 mL EV-depleted medium.
3. Incubate the cells for an additional 48 h at 37 °C and 5% CO_2.
4. Collect the conditioned medium into 50 mL canonical tubes, centrifuge at 300 × *g* for 10 min at 4 °C to pellet cell debris, and transfer the supernatant with a serological pipette to new 50 mL canonical tubes (*see* **Note 11**).
5. Centrifuge the medium again at 2000 × *g* for 10 min at 4 °C to pellet additional cell debris.

6. Assemble a 60 mL syringe connected to a 0.8 μm disc filter followed by a 16-gauge needle (*see* **Note 12**).
7. Insert the needle of the assembled filtering apparatus into the opening a Quick-Seal® tube and filter the supernatant.
8. Weigh the tubes using an electronic scale to ensure an equal weigh between each pair of tubes. Remove excess medium or use double filtered PBS to balance the tubes to within 0.1 g (*see* **Note 13**).
9. Seal the tubes with a tube topper and seal formers. Mark one side of the tube with a marker pen for orientation purposes.
10. Place each pair of tubes diagonally opposite to one another into a Type 70 Ti rotor. Note the orientation of the tubes with the tube marking, place a tube spacer on top of each tube, and screw the lid onto the rotor.
11. Centrifuge the samples at 100,000 × *g* for 90 min at 4 °C.
12. Gently remove the tubes from the rotor (*see* **Note 14**).
13. Based on the earlier tube marking, locate EV pellets on the lower corner of the tube facing outward from the rotor's center.
14. While monitoring on the EV pellet, gently remove the tube stem with a razor blade and aspirate most of the supernatant with a Pasteur pipette.
15. Remove the remaining supernatant with a pipettor.
16. Resuspend the EV pellets with 100 μL double filtered PBS per tube using a pipettor.
17. Place the tubes with resuspended EVs covering the initial pellet location on ice for 30 min to increase EV yield.
18. Transfer the EV samples into a 1.5 mL microfuge tube.
19. Determine the amount of EVs per sample using desired quantification methods, including protein concentration assay, nanoparticle tracking analysis and tunable resistive pulse sensing [3].
20. Store the samples at −80 °C.

3.3 Fluorescent Assay on Isolated Extracellular Vesicles

1. To confirm successful fluorescent EV labeling, add 3–5 μL of the isolated EVs onto a microscope slide (*see* **Notes 15** and **16**).
2. Using a tweezers, pick up a coverslip. Place one edge of the coverslip on the microscope slide. Once the coverslip touches the EV solution, gently lower the other edge toward the slide to mount the sample (*see* **Note 17**).
3. Add a drop of immersion medium (e.g., oil or water, depending on the type of immersion objective that will be used) on top of the coverslip.

4. Using a fluorescence microscope with immersion 40× or higher magnification objectives, locate the correct Z-axis for viewing and image PalmGFP-GlucB EVs under the GFP channel (*see* **Note 18**).

3.4 Bioluminescent Assay on Isolated EVs or EV-RNA Translation

1. To detect GlucB signal, prepare working CTZ solution by diluting the stock solution (5 mg/mL) at 1:1000 with PBS. Protect the working CTZ solution from light.
2. Depending on the concentration, EVs or cells can be diluted in PBS at 1:100 to 1:1000, dilutions and added to a 96-well white microtiter plate at 10 μL per well in triplicates (*see* **Note 19**).
3. Adjust the luminometer to autodispense 50 μL CTZ with a 2 s integration time per well (*see* **Note 20**).

3.5 Simultaneous Tracking of EV Delivery and EV-RNA Translation

1. Seed 0.1 × 10^6 mCherry-Gli36 cells per well in a 12-well plate. Prepare a minimum of duplicate wells per time point to track EV delivery and EV-RNA translation. 0, 1, 3, 6, 12 and 24 h post-EV treatment may serve as a good starting time points for optimization.
2. Once the cells reached 80–90% confluence, aspirate the medium, and treat the cells with desired amount of PalmGFP-GlucB-EVs in EV-depleted medium. Centrifuge the plates at 800 × *g* at 4 °C for 90 min, replace the medium with or without cycloheximide (20 μg/mL; a negative control for EV-RNA translation), and return them to 37 °C and 5% CO_2. The time when the centrifugation ended is denoted as "0 h".
3. At the planned time points, collect the conditioned media into separate tubes, and store them at −20 °C.
4. EV uptake analysis: Following the collection of conditioned media, gently wash the cells with PBS and apply 0.05% trypsin–0.53 mM EDTA for 3 min. Resuspend the cells in 3 mL PBS per well, pipette each sample in a separate tube, and store the samples on ice. Immediately perform flow cytometry analysis on the samples using a BD LSRII Multi-Laser Analyzer to detect EV-recipient cells (mCherry) with (EV uptake) or without (no significant EV uptake) GFP signal from the PalmGFP-GlucB-labeled EVs. mCherry signal serves as a positive signal for the recipient mCherry-Gli36 cells. Repeat this step for every time point of the experiment.
5. EV-RNA translation analysis: After the conditioned media have been collected from all of the time points and stored at −20 °C (*see* **Note 21**), thaw all of the samples at once. Resuspend and pipette 200 μL of each sample per well in triplicates to a 96-well white microtiter plate. Perform GlucB bioluminescent assay as

described to detect EV-RNA translation by the recipient cells (Subheading 3.4, **step 3**).

6. Chart the data (i.e., signal vs. time) from the EV delivery (Subheading 3.5, **step 4**) and EV-RNA translation analyses (Subheading 3.5, **step 5**) to investigate the temporal dynamics of EV uptake and RNA translation by the recipient cells.

4 Notes

1. Although it remains to be tried, other fluorescent reporters that do not significantly overlap with the excitation and emission wavelength of GFP (*e.g.* tdTomato) may also serve as a reporter for EV recipient cells
2. Add only 40–45 mL EV-depleted FBS per 50 mL canonical tube to prevent leakage at −80 °C.
3. Depending on the cell type being cultured, 5% EV-depleted FBS may also be tried to reduce FBS usage.
4. No. 1.5 coverslip should work with common 40× or higher power immersion objective lenses. Nevertheless, users should check for optimal coverslip thickness for each objective lenses being used to ensure optimal imaging of fluorescently labeled EVs.
5. A multiplicity of infection (MOI) ≥ 5 is recommended for lentiviral infection on HEK 293T cell. Optimization is recommended when other cell types are used to ensure a high rate of infection.
6. Additional rounds of infection are permitted but cell toxicity should be monitored over cell passages. Overtly high expression of GlucB has been observed to cause cell death over time.
7. Avoid selecting the top 5–10% of GlucB-expressing cells to prevent cell death over cell passages.
8. Freeze the stable GlucB-expressing cells into cryogenic vials at the earliest possible passage as a frozen seed stock.
9. Two rounds of PalmGFP transduction using an optimized MOI is typically required to achieve ≥70% dual-labeled (i.e., PalmGFP and GlucB) EV donor cells.
10. Freeze the stable dual-labeled PalmGFP-GlucB-expressing cells into cryogenic vials at the earliest possible passage as a frozen seed stock.
11. Do not pour the supernatant since cell debris may also be carried over to the new tubes.

12. Membrane disc filters with other pore sizes (0.2, 0.8, 1 μm, etc.) may also be used depending on the desired EV size to be isolated.
13. Do not assume equal tube weight by input volume or by sight. Also note that a minimum volume may need to be reached for each type of ultracentrifuge tube.
14. Sudden physical agitation and/or delayed experimental procedures may cause EV pellets to fall off from the wall of the tubes, thereby reducing EV yield.
15. A 9 mm diameter No. 1.5 round coverslip is sufficient to cover 3–5 μL of EV solution.
16. Pipette the EV-containing solution up and down several times to minimize aggregates.
17. Avoid creating bubbles beneath the coverslip during the sample mounting process.
18. Finding the correct Z-axis where the EV samples are located may be difficult at start. Through the eyepieces of the microscope, first try focusing on the edge of the coverslip to appreciate the depth between the coverslip and the sample, and then locate the EV-containing Z-axis with fine adjustments.
19. Loading too much sample per well may result in signal saturation. It is recommended to try several dilution factors to ensure optimal GlucB reading.
20. Since Gluc is a flash-type bioluminescence with a short half-life, the use of an autodispenser is essential to ensure consistent reading between samples. Pipetting the CTZ substrate manually will result in inconsistent signal output between samples.
21. All of the collected samples should have the same number of freeze/thaw cycles to ensure sampling consistency. GlucB signal has been observed to decrease with every freeze/thaw cycle.

Acknowledgment

This work was supported by the Ministry of Science and Technology of Taiwan (MOST), Taiwan (MOST 104-2320-B-007-005-MY2, C.P.L), and United States NIH grants CA069246 (Xandra O. Breakefield, Bakhos A. Tannous, and Ralph Weissleder), U19CA179563 (X.O.B and Thorsten R. Mempel), Voices Against Brain Cancer (X.O.B. and C.P.L.), and the Canadian Institutes of Health Research (CIHR; C.P.L.).

References

1. Lai CP, Breakefield XO (2012) Role of exosomes/microvesicles in the nervous system and use in emerging therapies. Front Physiol 3:228. doi:10.3389/fphys.2012.00228
2. Kumar NM, Gilula NB (1996) The gap junction communication channel. Cell 84 (3):381–388. doi:10.1016/S0092-8674(00) 81282-9
3. Zaborowski MP, Balaj L, Breakefield XO, Lai CP (2015) Extracellular vesicles: composition, biological relevance, and methods of study. Bioscience 65(8):783–797. doi:10.1093/bio sci/biv084
4. Hoshino A, Costa-Silva B, Shen TL, Rodrigues G, Hashimoto A, Tesic Mark M, Molina H, Kohsaka S, Di Giannatale A, Ceder S, Singh S, Williams C, Soplop N, Uryu K, Pharmer L, King T, Bojmar L, Davies AE, Ararso Y, Zhang T, Zhang H, Hernandez J, Weiss JM, Dumont-Cole VD, Kramer K, Wexler LH, Narendran A, Schwartz GK, Healey JH, Sandstrom P, Labori KJ, Kure EH, Grandgenett PM, Hollingsworth MA, de Sousa M, Kaur S, Jain M, Mallya K, Batra SK, Jarnagin WR, Brady MS, Fodstad O, Muller V, Pantel K, Minn AJ, Bissell MJ, Garcia BA, Kang Y, Rajasekhar VK, Ghajar CM, Matei I, Peinado H, Bromberg J, Lyden D (2015) Tumour exosome integrins determine organotropic metastasis. Nature 527(7578):329–335. doi:10.1038/nature15756
5. EL Andaloussi S, Maeger I, Breakefield XO, Wood MJA (2013) Extracellular vesicles: biology and emerging therapeutic opportunities. Nat Rev Drug Discov 12(5):348–358. doi:10.1038/nrd3978
6. Skog J, Wurdinger T, van Rijn S, Meijer DH, Gainche L, Sena-Esteves M, Curry WT, Carter BS, Krichevsky AM, Breakefield XO (2008) Glioblastoma microvesicles transport RNA and proteins that promote tumour growth and provide diagnostic biomarkers. Nat Cell Biol 10(12):1470–U1209. doi:10.1038/ncb1800
7. Ohno S, Takanashi M, Sudo K, Ueda S, Ishikawa A, Matsuyama N, Fujita K, Mizutani T, Ohgi T, Ochiya T, Gotoh N, Kuroda M (2013) Systemically injected exosomes targeted to EGFR deliver antitumor microRNA to breast cancer cells. Mol Ther 21(1):185–191. doi:10.1038/mt.2012.180
8. Mittelbrunn M, Gutierrez-Vazquez C, Villarroya-Beltri C, Gonzalez S, Sanchez-Cabo F, Gonzalez MA, Bernad A, Sanchez-Madrid F (2011) Unidirectional transfer of microRNA-loaded exosomes from T cells to antigen-presenting cells. Nat Commun 2:282. doi:10.1038/ncomms1285
9. Suetsugu A, Honma K, Saji S, Moriwaki H, Ochiya T, Hoffman RM (2013) Imaging exosome transfer from breast cancer cells to stroma at metastatic sites in orthotopic nude-mouse models. Adv Drug Deliv Rev 65(3):383–390. doi:10.1016/j.addr.2012.08.007
10. Spötl L, Sarti A, Dierich MP, Möst J (1995) Cell membrane labeling with fluorescent dyes for the demonstration of cytokineinduced fusion between monocytes and tumor cells. Cytometry 21:160–169. doi:10.1002/cyto.990210208
11. Lässer C, Alikhani VS, Ekström K, Eldh M, Paredes PT, Bossios A, Sjöstrand M, Gabrielsson S, Lötvall J, Valadi H (2011) Human saliva, plasma and breast milk exosomes contain RNA: uptake by macrophages. J Transl Med 9(1):9
12. Wallace PK, Tario J, Joseph D, Fisher JL, Wallace SS, Ernstoff MS, Muirhead KA (2008) Tracking antigen-driven responses by flow cytometry: monitoring proliferation by dye dilution. Cytometry A 73A(11):1019–1034. doi:10.1002/cyto.a.20619
13. Blaskovic S, Blanc M, van der Goot FG (2013) What does S-palmitoylation do to membrane proteins? FEBS J 280(12):2766–2774. doi:10.1111/febs.12263
14. Aicart-Ramos C, Valero RA, Rodriguez-Crespo I (2011) Protein palmitoylation and subcellular trafficking. Biochim Biophys Acta 1808(12):2981–2994. doi:10.1016/j.bbamem.2011.07.009
15. Zacharias DA, Violin JD, Newton AC, Tsien RY (2002) Partitioning of lipid-modified monomeric GFPs into membrane Microdomains of live cells. Science 296 (5569):913–916. doi:10.1126/science.1068539
16. Fuhrmann G, Herrmann IK, Stevens MM (2015) Cell-derived vesicles for drug therapy and diagnostics: opportunities and challenges. Nano Today 10(3):397–409. doi:10.1016/j.nantod.2015.04.004
17. Lai CP, Kim EY, Badr CE, Weissleder R, Mempel TR, Tannous BA, Breakefield XO (2015) Visualization and tracking of tumour extracellular vesicle delivery and RNA translation using multiplexed reporters. Nat Commun 6:7029. doi:10.1038/ncomms8029
18. Thorne N, Inglese J, Auld DS (2010) Illuminating insights into firefly luciferase and other bioluminescent reporters used in chemical

biology. Chem Biol 17(6):646–657. doi:10.1016/j.chembiol.2010.05.012

19. Tannous BA, Kim DE, Fernandez JL, Weissleder R, Breakefield XO (2005) Codon-optimized Gaussia luciferase cDNA for mammalian gene expression in culture and in vivo. Mol Ther 11(3):435–443. doi:10.1016/j.ymthe.2004.10.016
20. McCabe JB, Berthiaume LG (1999) Functional roles for fatty acylated amino-terminal domains in subcellular localization. Mol Biol Cell 10(11):3771–3786
21. Sena-Esteves M, Tebbets JC, Steffens S, Crombleholme T, Flake AW (2004) Optimized large-scale production of high titer lentivirus vector pseudotypes. J Virol Methods 122(2):131–139. doi:10.1016/j.jviromet.2004.08.017
22. Niers JM, Chen JW, Lewandrowski G, Kerami M, Garanger E, Wojtkiewicz G, Waterman P, Keliher E, Weissleder R, Tannous BA (2012) Single reporter for targeted multimodal in vivo imaging. J Am Chem Soc 134(11):5149–5156. doi:10.1021/ja209868g

Part VI

Downstream Extracellular Vesicle Applications: Genomics and Proteomics

Chapter 22

Extraction and Analysis of Extracellular Vesicle-Associated miRNAs Following Antibody-Based Extracellular Vesicle Capture from Plasma Samples

Davide Zocco and Natasa Zarovni

Abstract

Extracellular vesicle (EV)-associated RNAs (EV-RNA) are under intense investigation due to their potential role in health and disease. Several approaches are currently employed to isolate blood-derived EVs for RNA analysis, most of which are either time-consuming and expensive, such as methods based on EVs physical properties (ultracentrifugation and Optiprep density gradient), or also copurify blood contaminants, mostly protein aggregates and immune complexes, (such as chemical precipitation). In addition, there is a lack of standardized protocols for the extraction of EV-RNA and very little consensus on the technological platforms and normalization tools for assessing the expression levels of different RNA species. These methodological issues complicate the comparison between independent data sets, potentially biasing results and conclusions.

In this book chapter we propose a protocol that might overcome some of the abovementioned issues through antibody-based isolation of blood-derived EVs followed by extraction and expression analysis of small-RNA species (miRNA) by reverse transcriptase quantitative PCR (RT-qPCR). The advantages of immunoaffinity approaches over other isolation methods are multiple and include: (1) the selective enrichment of specific EV subpopulations with restricted tissue/cell origin, (2) reduction of matrix effects and blood contaminants that may confound miRNA profiling from complex biological fluids and (3) easy coupling to conventional quantitative assays (e.g., RT-qPCR). In conclusion, we describe a protocol for standard enrichment and quantitative analysis of EV-miRNAs from blood and we warrant for technological improvements, such as the use of novel biomaterials, surface chemistries, binding agents and assay/sensor design that may further improve it.

Key words Extracellular vesicles, miRNA, Antibody-coated beads, Plasma, RT-qPCR

1 Introduction

In recent years extracellular vesicles (EVs) have emerged as important mediators of biological functions in health and disease [1, 2]. EVs carry a plethora of biomolecules (proteins, lipids, RNA and DNA) that modulate the behavior of the recipient cells and promote tissue-specific effects. For instance, tumor-derived EVs have been reported to hijack the tissue microenvironment to foster

Winston Patrick Kuo and Shidong Jia (eds.), *Extracellular Vesicles: Methods and Protocols*, Methods in Molecular Biology, vol. 1660, DOI 10.1007/978-1-4939-7253-1_22, © Springer Science+Business Media LLC 2017

tumor growth and organotropic metastases [3]. Because of these properties, EVs have been dubbed as promising carriers of biomarkers and nano-sized systems for targeted drug delivery [4]. In this dual role, EVs have entered clinical trials while their implementation in routine clinical and diagnostic practice is hampered by methodological and standardization issues. To date, most of the research efforts in the EV field have been using protocols initially developed on model systems, samples and analytical techniques borrowed from other research fields (e.g., virus research or organelle purification). These typically include using expensive and time-consuming techniques, such as ultracentrifugation (UC) and Optiprep density gradient (ODG), which are not suitable for clinical applications. In addition to the lack of robust and cost-effective isolation tools, there are other significant biological and technical challenges to be considered when developing EV-based diagnostic assays:

1. The choice of the most appropriate body fluid as a source for EV isolation. Blood-derived matrices, such as serum and plasma, are complex biofluids that mostly differ for the relative levels of clotting factors and residual platelets but also for the overall miRNA content and EV composition. Though serum is reported to have higher RNA content with respect to plasma, it is likely that a substantial amount of serum RNA originates from platelet-derived EVs release following coagulation rather than cell lysis [5]. To avoid the confounding effect of platelet-derived EVs when selecting specific EV subpopulations, the research community has selected plasma as the go-to-blood derived matrix for analysis of EV-derived-RNAs [6, 7].
2. The majority (%) of EVs in plasma derives from red and white blood cells and might carry RNA species (e.g., miRNA) shared with tissue-derived EVs, confounding downstream identification of relevant biomarkers. Furthermore the levels of blood cell-derived EVs can be affected by preanalytical parameters (centrifuge type, temperature, sample processing time, etc.) used for blood collection and processing and the patient's state (e.g., inflammation, fatigue or infection), increasing thus intersample and intrasample variability.
3. The blood levels of tissue-derived EVs are often exceedingly low, requiring technologies for EV enrichment and sensitive biomarker detection. Of note, quantitative and stoichiometric analyses of miRNA content of EVs have revealed the average presence of just one copy of abundant miRNA every 100 EVs [8] while the EV occupancy with rare miRNA species of true diagnostic relevance for any pathological condition of interest is likely to be even lower.

Affinity isolation methods might overcome these challenges by enriching for highly pure EV subpopulations with defined tissue/cell origin carrying true liquid biopsy markers, in a cost- and time-effective manner. In this chapter we focus on immunoisolated EVs as reservoirs of miRNAs. The RNA content of EVs, which comprises mRNA and noncoding RNAs such as miRNA, lnc RNA, tRNA and rRNA [2] is of particular diagnostic interest due to a membrane-enclosed protection of RNA cargo that would otherwise be poorly stable and exposed to biofluid RNAses. Although other forms of proteinaceous protection due to association to Ago2 complexes or HDL and LDL particles can be conferred to circulating small RNAs, EV packaging is likely to be a dominant phenomenon [9–11] and is thought to provide the mechanism for enrichment of particular tissue- or disease-specific RNA isoforms [12, 13]. miRNAs isolation from EVs as well as their subsequent quantification with a variety of techniques such as qRT-PCR, miRNA arrays or NGS is well documented [13, 14], and a growing body of evidence associates EV miRNAs with different functional roles in cancer such as metastatic progression or response to therapeutic agents [4]. Therefore, miRNAs fulfill major criteria for appealing cancer biomarkers in a liquid biopsy setting [4]. In this chapter, we report an optimized protocol for the isolation of EVs with antibody-coated beads from plasma and subsequent extraction and analysis of EV-miRNAs. We show that EV-miRNAs can be efficiently extracted from immunocaptured CD9-positive EVs and assessed quantitatively by RT-qPCR (Fig. 1). The expression levels of some EV-miRNAs (miR-16, miR 451) were lower in immunocaptured EVs in comparison to EVs obtained with generic EV capture techniques (ultracentrifugation and chemical precipitation), but others were unchanged (miR-375 and miR-210) and one was enriched (miR-141) (Fig. 1). Electropherogram analysis of EV-RNA showed a measurable peak corresponding to small RNA species from as little as 500 μl of plasma (Fig. 2). The profile was comparable to that obtained from 2 ml of plasma using a column-based kit for generic EV capture (Qiagen ExoRNeasy kit) (Fig. 2). Several EV-associated miRNAs were amplified by RT-qPCR from as little as 100 μl of plasma with no preamplification step (Fig. 3). These data indicate that antibody-based isolation and extraction of EV-derived RNAs is a valuable and, in some circumstances, preferred option when working with low volume, highly complex body fluids.

2 Materials

Prepare all the reagents at room temperature, unless indicated otherwise. Use MilliQ ultrapure water for preparing the solution unless indicated otherwise. Several generic lab disposables are required:

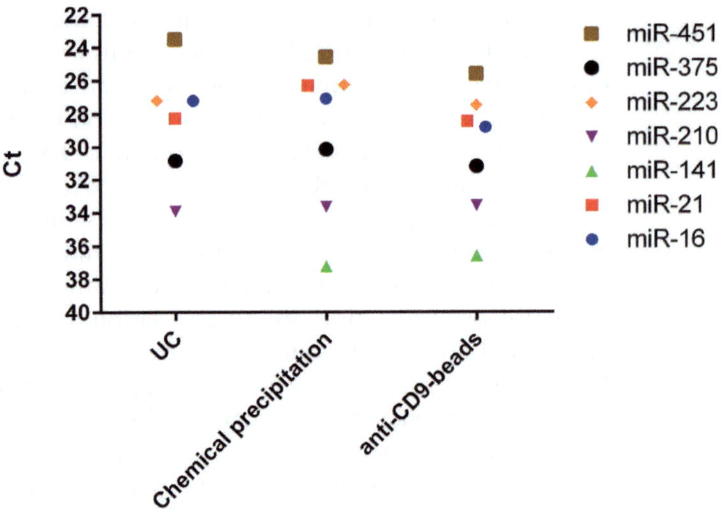

Fig. 1 Detection of EV-miRNAs by RT-qPCR following ultracentrifugation (UC), chemical precipitation and anti-CD9-antibody-based EV capture. EVs were isolated from 500 ml of healthy donor plasma by UC, PEG-based chemical precipitation (ExoQuick, SBI) and anti-CD9-antibody-coated beads. The resulting pellets were lysed with Qiazol (Qiagen) and the RNA was extracted as described in the protocol. MiRNA expression levels were measured using the miRNeasy primers as described (Qiagen)

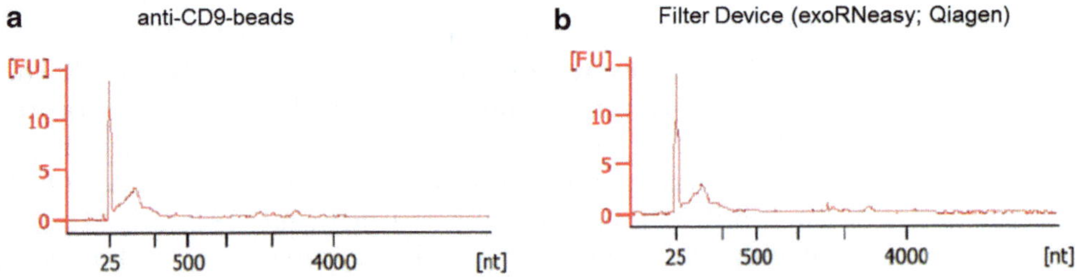

Fig. 2 Electropherogram analysis of EV-RNA extracted after anti-CD9-antibody-based EV capture and a column-based device. (**a**) EV-RNA was extracted from 500 μl of healthy donor plasma after anti-CD9-antibody-based EV capture according to the described protocol and 2 μl were loaded onto a Small-RNA chip (2100 Bioanalyzer; Agilent Genomics) for electropherogram analysis. (**b**) EV-RNA was extracted from 2 ml of the same healthy donor plasma after column-based EV isolation (exoRNeasy; Qiagen) according to the manufacturer's instructions and 2 μl were loaded onto a Small-RNA chip (2100 Bioanalyzer; Agilent Genomics) for electropherogram analysis

1. Gloves.
2. Single-use and/or pipettes with disposable 1 ml tips, with filter when handling PCR reagents.
3. Pipettes 1 ml and 5 ml for reagent preparation.
4. Disposable pipetting reservoirs.
5. 1.5 and 2 ml Eppendorf tubes.

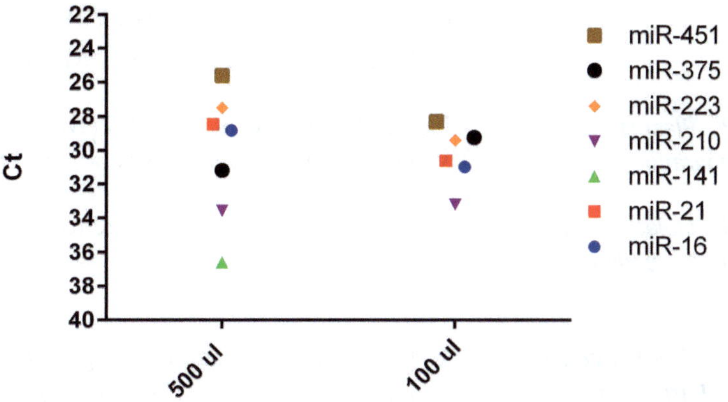

Fig. 3 Detection of EV-miRNAs by RT-qPCR from low plasma volumes. EVs were isolated from 500 μl and 100 μl of healthy donor plasma using anti-CD9-antibody-coated beads. The resulting pellets were lysed with Qiazol (Qiagen) and the RNA was extracted as described in the protocol. MiRNA expression levels were measured using the miRNeasy primers as described (Qiagen)

6. 0.2 ml PCR tubes for RT reaction.
7. 96—384-well plates for RT-qPCR.

2.1 Preparation of Antibody-Coated Beads

1. 0.4 μM aminated latex beads (Magsphere; Cat. No. PS400NM) (*see* **Notes 1–3**).
2. Monoclonal anti-CD9 antibody (HansaBioMed, Cat. No. HBM-CD9B-050).
3. Phosphate-buffered saline (PBS; Sigma).
4. 2-(N-morpholino)ethanesulfonic acid (MES) buffer 25 mM pH = 6: for a 100 ml solution, dissolve 0.53 g of MES powder (Sigma; Cat No. M8250) in 90 ml ultrapure MilliQ water. Use a magnetic stirrer to dissolve the powder at room temperature. Titrate the solution to pH = 6 with sodium hydroxide and bring the volume up to 100 ml with water. The solution can be stored at 4 °C until use.
5. Monosodium phosphate (NaH_2PO_4) 0.1 M pH = 6.2: for a 100 ml solution, dissolve 1.2 g of NaH_2PO_4 powder (Sigma, Cat No. S8282) in 90 ml of 90 ml ultrapure MilliQ water. Check the pH, titrate accordingly to pH = 6.2 and bring the volume up to 100 ml with water. The solution can be stored at 4 °C until use.
6. 1-ethyl-3-(3-dimethylaminopropyl)carbodiimide hydrochloride (EDC) solution: dissolve 50 mg of EDC (Sigma, Cat No. E6383) in 1 ml of NaH2PO4 solution. Avoid light exposure.
7. Blocking solution: Dissolve 5 g of casein (Sigma, Cat No. 3400) in 100 ml of ultrapure MilliQ water using a magnetic stirrer with heating. Once the casein is dissolved, filter the solution with a 0.4 μm filter. Keep the solution at 4 °C until use.

8. Tween (Sigma, Cat. No. P9416).
9. Sodium azide solution 0.1 M (NaN3, Sigma, Cat No. 08591).

2.2 Plasma Preparation

Use the Vacutainer K2-EDTA (BD, Cat. No. 367525) for prospective collection of plasma as reported by Witwer et al. [6].

2.3 Isolation of EVs with Antibody-Coated Beads

1. PBS + Tween 0.05% for gentle washing of EV-bead complexes.

2.4 RNA Extraction and Purification

1. Qiazol Lysis Reagent (Qiagen, Cat No. 79306).
2. Chloroform (Sigma, Cat No. 288306).
3. Ethanol 96% (Sigma, Cat No. 02870).
4. RNA Spin Columns (HansaBioMed, Cat No. HBM-RNA-B20).

2.5 RNA Quality and Quantity Control

1. Agilent Small RNA kit (Agilent Genomics, Cat No. 5067-1548).
2. Qubit RNA HS assay kit (Thermo Fisher Cat. No. Q32852).

2.6 RT-qPCR of EV-miRNAs

RT reactions:

1. miScript RT II kit (Qiagen, Cat. No. 218161).
2. TaqMan microRNA Reverse Transcription kit (Thermo Fisher, Cat. No. 4366596).

   Real-time quantitative PCR: miSCript primer assays (Qiagen) on SYBR Green technology (miScript SYBR Green PCR Kit, Cat. No. 218073, Qiagen) or TaqMan® Array Human MicroRNA A + B Cards Set plus megaplex PreAmp primers (Cat. No. 4444913, Thermo Fisher) for identification of miRNA signatures from blood.
3. Positive control: purified exosomes from healthy donor plasma (HansaBioMed, Cat. No. HBM-PEP-30/2).

3 Methods

3.1 Preparation of Antibody-Coated Beads

All amounts of reagents are indicated for 1 reaction per ml of sample:

1. Resuspend 2.5 μl of aminated latex beads (10% w/v) in 1 ml of MES buffer 25 mM pH 6, centrifuge at 5000 × *g* 10′ at RT.
2. Discard supernatant, wash beads again with 1 ml of MES buffer 25 mM pH 6, centrifuge as indicated above.
3. Discard supernatant, resuspend beads in 1 ml of NaH_2PO_4 0.1 M, pH 6.2 (or PBS 1×) + 1:50 of 50 mg/ml of EDC.

4. Add 1 μg of antibody. Select antibodies directed against EV-expressed epitopes dependent on the scope of the experiment. In this protocol we used a commercially available monoclonal anti-CD9 antibody (HansaBioMed) (*see* **Notes 4** and **5**).
5. Incubate 2 h at RT in the dark under rotation.
6. Centrifuge at 5000 × *g* 10′ at RT and discard the supernatant.
7. Wash bead pellet with 1 ml of MES buffer, centrifuge as above and discard the supernatant.
8. Wash bead pellet with 1 ml of PBS, centrifuge as above and discard the supernatant.
9. Block beads with 1 ml of BSA 0.5% in PBS 1× or Casein 0.5% at RT for 40′ under rotation (*see* **Note 6**).
10. Centrifuge as indicated, discard the supernatant.
11. Wash beads with 1 ml of PBS 1× + 0.05% Tween, centrifuge and discard supernatant.
12. Wash beads with 1 ml of PBS 1×, centrifuge and discard supernatant.
13. Resuspend beads with PBS + 0.01% NaN_3 to reach a final volume of 10 μl. The antibody-coated beads can be stored at 4 °C for up to 6 months. Avoid freezing.

3.2 Plasma Preparation

1. In research and clinical laboratories, plasma is typically used after storage at −20 °C to 80 °C. Plasma samples are thawed prior to analysis at room temperature to avoid the formation of insoluble aggregates. Avoid repeated freeze–thaw cycles. Visually inspect the plasma to identify hemolytic, lipemic, or icteric samples (*see* **Note 7**).
2. Prepare *plasma* by three centrifugation steps to eliminate red blood cells and cellular debris. (1) 10′ at 300 × *g* (save supernatant; discard pellet which contain circulating cells). (2) 20′ at 1200 × *g* (save supernatant; discard pellet which contains cell debris). *Facultative*: 30′ at 10,000 × *g*. The supernatant will contain exosomes (EVs < 200 nm in diameter). The pellet will be enriched by large EVs such as oncosomes, apoptotic bodies, and microvesicles (EVs > 200 nm in diameter) (*see* **Notes 8–11**)

3.3 Isolation of EVs with Antibody-Coated Beads

1. Place 0.1 ml up to 1 ml of sample into low-binding 1.5 ml tubes. Suggested plasma volumes: 0.1 ml up to 0.5 ml for small RNAs analysis; 0.5 ml up to 1 ml for mRNAs analysis.
2. Add PBS 1× to the sample to get a final volume of 1 ml (*see* **Note 12**).
3. Add 10 μl of antibody-coated beads. We recommend adding isotype matched-antibody-coated beads for monitoring non-specific binding (*see* **Note 13**).

4. Incubate 2 h at room temperature (RT) under rotation (*see* **Notes 14** and **15**).
5. Centrifuge at room temperature (RT) for 10′ at 5000 × *g* (*see* **Note 16**).
6. Carefully discard supernatant using a pipette without dislodging the pellet.
7. Add 1 ml of PBS 1× + 0.05% Tween (*see* **Note 17**).
8. Resuspend the bead pellet pipetting up and down 10–15 times.
9. Centrifuge at RT for 10′ at 5000 × *g*.
10. Remove the supernatant carefully. Do not to disturb the pellet.
11. Wash beads once again as indicated above (*see* **Note 18**).

3.4 RNA Extraction and Purification

1. Add 700 μl of Lysis buffer directly onto the bead pellet. The lysis buffer can be any phenol-based solution such as TRIzol, (Invitrogen), Qiazol (Qiagen), or RNA-Bee (Amsbio) (*see* **Note 19**).
2. Dissolve the bead pellet by pipetting up and down. The pellet must be completely dissolved.
3. Incubate 5′ at room temperature.
4. Add 70 μl of 1-bromo-3-chloropropane (BCP) or 140 μl of pure Chloroform.
5. Shake 30 s.
6. Incubate 10 min at room temperature.
7. Incubate 1 min in ice and centrifuge at 12,000 × *g* at 4 °C for 10′ (*see* **Note 20**).
8. Transfer the top (aqueous) phase to a fresh tube.
9. Add 2× of ethanol 96%. Mix by gently inverting 4–5 times. For instance, if the top phase volume is 400 μl, add 800 μl of ethanol 96%.
10. Transfer half volume of the mixture into the spin column (*see* **Note 21**).
11. Spin at 14,000 × *g* for 30″.
12. Discard the flow-throw.
13. Add the remaining volume to the same spin column.
14. Spin at 14,000 × *g* for 30″.
15. Discard the flow-throw.
16. Add 400 μl of RNA Wash buffer.
17. Gently invert the column 3–4 times (*see* **Note 22**).
18. Spin at 14,000 × *g* for 30″.
19. Discard the flow-through.

20. Perform the washing step twice more.
21. Spin for 5 additional minutes at 14,000 × *g* to eliminate ethanol residues from the column.
22. Remove the collection tube and transfer the spin column into a new 1.5 ml Eppendorf tube.
23. Add 15 μl of water or Tris–EDTA buffer pH = 9 for final elution.
24. Incubate 5′ at RT.
25. Spin 2′ at 200 × *g* and 1′ at 14,000 × *g*. Keep the flow-through. The eluted RNA is now ready for downstream analyses or for storage at −80 °C.

3.5 RNA Quality and Quantity Control

1. RNA quality can be assessed by electropherogram analysis as shown in Fig. 2 (Small-RNA chip; Bioanalyzer). The typical EV-RNA profile has a distinctive peak in the range of 20–50 nt (Fig. 2) corresponding to small-RNA species.
2. Solvent contamination (phenol, guanidine hydrochloride, and ethanol) of the sample can be checked by measuring the A260/A230 ratio with NanoDrop (Thermo Fisher). Figure 4 shows

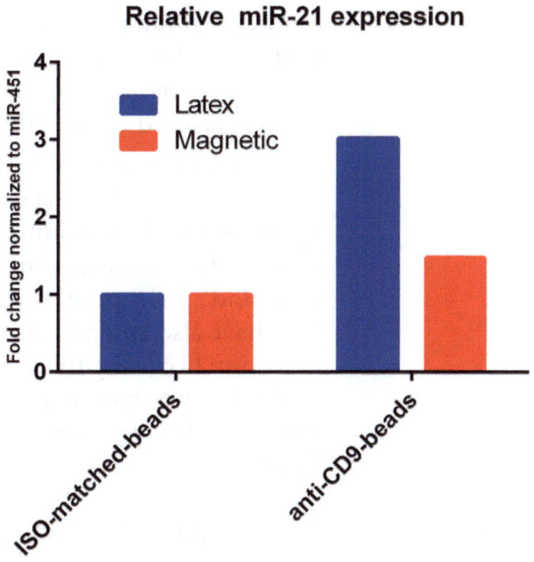

Fig. 4 Relative expression levels of EV-miR-21 after EV capture with two types of antibody-coated beads. EVs were isolated from 500 μl healthy donor plasma by using anti-CD9 and isotype-matched antibody-coated latex and magnetic beads. The resulting pellets were lysed with Qiazol (Qiagen) and the RNA was extracted as described in the protocol. Relative expression levels of miR-21 were measured using the miRNeasy primers (Qiagen) and calculated according to the Pflaff method [17] using miR-451 as housekeeping gene

the effects of solvent contamination on the absorbance profile of RNA samples (Nanodrop).

3. RNA quantity can be measured by fluorescence analysis with Qubit fluorimeter (Thermo Fisher) (*see* **Note 23**).

3.6 RT-qPCR of EV-miRNAs

1. Reverse transcriptase reaction: EV-RNA can be retro-transcribed using most of the commercially available kits according to the manufacturer's instructions (e.g., miScript RT II kit from Qiagen, TaqMan microRNA Reverse Transcription kit from Thermo Fisher). Two or three microliters (or 1–2 ng) of RNA are usually sufficient for the retrotranscription and amplification of most EV-derived miRNAs. When possible, use two orthogonal methods for RNA retro-transcription to reduce technical biases as recommended by the MIQE guidelines [15]. For instance, miScript RT II kit retrotranscribes miRNAs using a proprietary universal primer while the TaqMan microRNA Reverse Transcription kit technology uses a combination of primer extension and target-specific primers.
2. For assessing the levels of low-abundant miRNAs or discovery of new miRNAs in a given body fluid, we recommended to preamplify the cDNA to improve the sensitivity of the assay. We routinely use the Preamp Master Mix kit according to the manufacturer's instruction (Thermo Fisher). We do not recommend adding a preamplification step when measuring the expression levels of abundant miRNAs or for validation of miRNA as biomarkers to avoid technical biases and reduce the turnaround time of the workflow.
3. Real time PCR analysis of EV-miRNAs can be performed with a variety of technologies according to the manufacturer's instructions. We recommend using wet-lab validated PCR primers as the miRNeasy primers (Qiagen) or PrimeTech primers (Bio-Rad). PCR primers with locked nucleic acid (LNA) chemistry have also emerged as the most specific primers for miRNA detection [16]. For the simultaneous analysis of more than 700 miRNAs, we recommend the Megaplex (low-density RT-qPCR) technology on microfluidic cards (Thermo Fisher) (*see* **Note 24**).

3.7 Data Analysis and Limitations of the Protocol

1. The RT-qPCR data can be expressed as absolute raw fluorescence threshold values (Ct/Cq) and as relative expression to one or more normalizer small RNAs according to the Pflaff method [17]. In this protocol, there are three "normalizing" steps: (a) the sample (plasma) volume is kept constant across all the samples; (b) the spike-in exogenous control normalize for the extraction efficiency; (c) equal volumes or amounts, when measurable, of RNA go into the RT reaction. These normalization steps allow for a "first glance" absolute comparison of

miRNA's Cq values within a single RT-qPCR experiment (Fig. 1). We recommend the use of a positive control for normalizing the PCR fluorescence threshold for cross-comparison of multiple RT-qPCR experiments. To this purpose, we extract RNA from commercially available purified exosomes (HansaBioMed).

Relative expression analysis of EV-miRNAs is currently hampered lack of endogenous normalization controls (eg. housekeeping genes). Small nucleolar RNAs like RNU6B and RNU48 are frequently used as normalizing genes in in vitro experiments, but they are inadequate for their low or varying abundance in body fluids. Different small-RNA species are currently under evaluation as universal normalizers for EV-associated RNAs, including piRNA and Y-RNAs [18]. Until new normalizers are found, we recommend to empirically select miRNAs, likely to follow similar biogenesis, metabolism and sorting pathways with target molecules, and not other small-RNA species, as fit-to-purpose housekeeping genes in a fluid/protocol-specific manner [19].

2. Immunocapture-based approaches have some limitations to be considered. Some miRNAs will be enriched (e.g., miR-141 in Fig. 1) but others inevitably lost depending on the EV-antigen selected for the EV capture, the size and/or charge of the EVs and the source of origin. The overall number and amount of miRNAs that can be detected is generally lower than generic approaches such as UC or chemical precipitation as shown in Fig. 1. For this reason, immunocapture approaches are often preferred for diagnostic applications where ease-of-use and robustness of the protocol are more important than comprehensiveness of the analysis. Furthermore, there are no published normalizer genes for immunocapture-based approaches and the few published normalizers for EV-miRNAs found with generic isolation methods do not usually co-segregate with the immune-purified EVs making the search for normalizers even more challenging.

4 Notes

1. Both latex and magnetic beads are suitable for this protocol. We use 0.4 μm latex beads with free amino groups for covalent binding of antibodies (Magsphere Inc). In our hands, these beads have consistently recovered more plasma-derived EVs than magnetic beads, as shown in Fig. 5, justifying the longer turnaround time of the protocol. The covalent coating minimizes the detachment of the antibody from the beads. The surface chemistry can be further improved by reducing steric hindrance with PEG- or multipeptide-based branches [20].

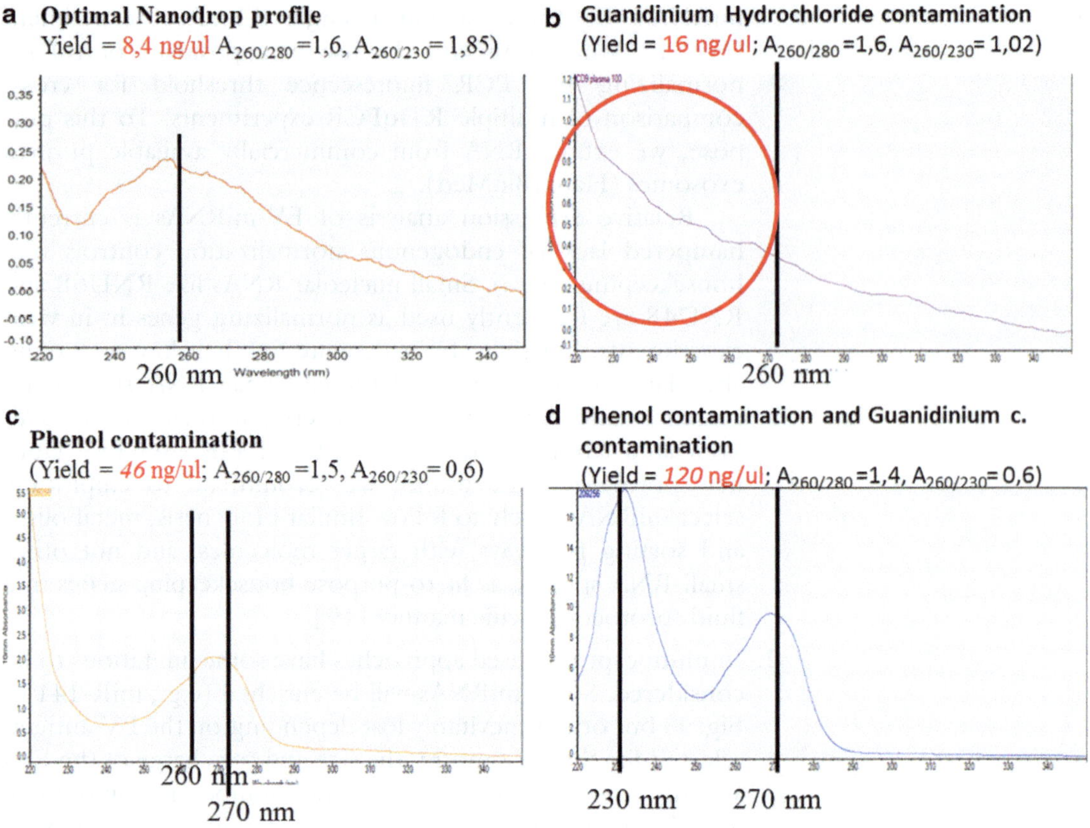

Fig. 5 Absorbance profiles reveal solvent contamination in EV-RNA preparations. (**a**) EV-RNA was prepared according to the described protocol. The Absorbance profile was measured by NanoDrop 2000 and showed the expected peak at 260 nm (Thermo Fisher). (**b**) Absorbance profile of a guanidinium chloride-contaminated RNA sample, showing a large "shoulder" in the 230 nm range. (**c**) Absorbance profile of a phenol-contaminated RNA sample with a distinctive absorbance peak at 270 nm; (**d**) Absorbance profile of a phenol- and guanidinium chloride-contaminated RNA sample with absorbance peaks at 230 and 270 nm

2. EV-bound beads will tend to aggregate more than unbound beads. Always, visually check the degree of (self)-aggregation of the beads pre- and post-incubating with the sample.
3. 2.5 μl of 10% w/v of aminated latex beads (0.4 μm in diameter) contains an approx 3.5×10^9 beads. This bead number exceeds the theoretical number of EVs in plasma given volume of plasma sample (0.1–0.5 ml), as nanovesicle number per ml of plasma is estimated to vary in a range 10^8–10^{10}/ml di plasma as per nanotracking analysis (NTA, Malvern and in-house measurements).
4. The optimal ratio number of antibodies per bead may vary depending on the antibodies selected for beads

functionalization. Avidity of binding that is important feature for pull down applications measures the overall strength of an immobilized antibody-target antigen complex and is dependent on the affinity of the antibody for the epitope (selection of target epitope) and valency of both the antibody (polyclonal vs monoclonal antibody, IgG vs IgM) and antigen (number of copies of antigen per exosome). Empirically determined optimal Ab to particle ratio used in hereby shown examples was 1 μg to 3.5×10^9 latex beads.

5. Binding agents such as aptamers and peptides might also be considered for affinity-isolation of EVs. When available, aptamers against known EV-derived markers are ideal for reducing background affects (IgG recognition from plasma immunocomplexes) and improving scalability and robustness of the protocol (no batch-to-batch or clonal variability, cheap production).
6. In our hands, casein 0.5% was the most effective blocking agent for latex beads. Use a 0.44 μm filters to filter it before incubation. Other commercial blocking agents might also be tested on different bead types (Cantor blocking agent, ethanolamine, etc.).
7. Blood should be collected with a 21-gauge needle or larger, avoiding butterfly systems, and processed within 2 h from collection at room temperature to avoid platelet activation. Apheresis, when possible, is a good way to collect blood as it reduces needle-induced local inflammation and platelet activation [6].
8. Non adequate anticoagulant mixing or hemolysis can alter the overall EV-miRNA content while lipemia may affect the efficiency of EV recovery [6]. Record or discard hemolytic, jaundiced or lipemic samples. We selected EDTA as anticoagulant of reference after performing a comparative analysis of several anticoagulants including heparin and CTAD (a mixture of citrate, theophylline, adenosine, and dipyridamole). We found that platelet activation was more impacted by the centrifugation protocol (speed, time, and temperature), than any of the anticoagulants tested in the study. We always avoided heparin since it may cause EV aggregation, impacting their recovery during isolation and inhibiting downstream PCR analyses. For these reasons, some medications should be recorded in anamnestic dossiers as potentially impacting the amplification of RNA [6].
9. Plasma is usually obtained by centrifugation of blood in the presence of anticoagulants (e.g., EDTA). Speed, length and temperature of the centrifugation step/s are key parameters influencing the levels of (blood cell-derived) EVs and should

be reported and kept constant along the study. Usually a centrifuge step at 1500 × *g* for 20′ at room temperature is sufficient to obtain platelet-poor plasma used in most studies. A second centrifuge step can be added to get platelet-free plasma, if platelet-derived EVs are a confounding factor for downstream analyses. The type of the centrifuge rotor should also be reported and kept constant since it influences the relative pelleting efficiency or K-(clearing)-factor, therefore affecting the EV levels within the sample [6]. Plasma processing with Ficoll-based gradient is also a valuable option to effectively separate red and white blood cells, all potential sources of confounding EV subpopulations. We do not recommend plasma preparation protocols based on sedimentation (aka no centrifuge step) since they are normally associated with a higher risk of platelet activation and contamination from platelet- and red blood cell-derived EVs. Other centrifuge-free protocols for plasma processing are currently under development but they have not released yet.

10. We and others have observed exosome-specific biomarkers, such as Alix and Tsg101 in the 10,000 × *g* pellet, suggesting the presence of exosomes. We speculate that exosomes might cluster into larger macrocomplexes, especially after multiple freeze–thaw cycles [21]. Always record if the sample has been frozen and thawed multiple times and, work, if possible with fresh or newly frozen samples.

11. Plasma samples can be diluted to with phosphate buffer solution to reduce the plasma matrix effect and promote antigen-antibody recognition. We recommend to start with a 1:1 dilution for general purposes and to lower it down to 1:10 or 1:20 if targeting abundant EV populations.

12. PBS may be supplemented with nonionic detergents (Tween and pluronic acid) or dextrans [22] to further reduce the matrix effect and aspecific aggregation of the beads.

13. Uncoated beads and Iso-control antibody-coated beads may be used for testing the antibody-independent binding of EVs or contaminants (protein complexes, LDL molecules, etc.) to the bead surface. Beads will bind to circulating Ago2-protein–RNA complexes, LDL molecules and spurious EVs depending on the surface chemistry and coating (BSA, casein, etc.). Always measure the levels of Ago2 and generic EVs markers (e.g., Alix) on uncoated beads, isotype-matched and marker-specific antibody-coated beads by Western blot analysis [7].

14. In most cases, 2-h incubation at room temperature (RT) is sufficient for effectively capturing EVs from complex body fluids. Longer incubation times might improve the enrichment

of rare EV subpopulations but often cause more nonspecific binding of contaminants or unwanted EVs.

15. Plasma samples are rich of RBC-, platelet and granulocyte-derived EVs. These EVs carry a plethora of biomarkers that often interfere with antibody-based methods for the enrichment of rarer EV subpopulations, such as tissue or tumor-derived EVs. Negative selection of these subpopulation using antibody targeting blood cell biomarkers (e.g., glycophorin A) might reduce interference and improve the positive selection of desired EV populations. This is a facultative step that can be combined to positive selection steps or help fractionating different EV subpopulations from the same sample.
16. Latex beads do not pellet properly at temperatures lower than RT and they can be damaged by centrifugal speeds higher than $5000 \times g$.
17. Small amounts of detergent such as Tween 0.01–0.05% are sufficient to wash away loosely bound plasma contaminants without interfering with the antibody-mediated capture of EVs.
18. Perform the last wash step in a new 1.5 ml low-bind tube (Eppendorf) to minimize background signal coming from aspecific binding to the tube.
19. Spike a synthetic miRNA such as ce-miR-39 or UniSp(1-6) to the lysis buffer to normalize the samples for extraction efficiency
20. Incubation on ice prior to centrifuge helps reducing DNA contamination, which tends to remain in the interphase.
21. Spin columns from various vendors can be used. We recommend spin columns (e.g., HansaBioMed spin columns) that require low elution volumes (10–15 μl max) to concentrate the EV-RNA.
22. This step improves sample purity by washing away solvent droplets (phenol, guanidine hydrochloride) trapped underneath the column lid.
23. Given the extremely low amount of RNA (0.1–10 ng/ml of plasma), absorbance-based technologies, such as Nanodrop (Thermo Fisher), are not sensitive enough for quantitation of EV-RNA. In the absence of fluorescence-based methods (e.g., Qubit, Thermo Fisher), it is common practice to load equal volumes of RNA samples and normalize miRNA's expression levels using appropriate controls in the downstream assays (e.g., RT-qPCR).
24. This isolation/extraction protocol has not been tested for the preparation of RNA samples for RNA sequencing.

References

1. Rak J (2013) Extracellular vesicles – biomarkers and effectors of the cellular interactome in cancer. Front Pharmacol 4:21
2. Yuana Y, Sturk A, Nieuwland R (2013) Extracellular vesicles in physiological and pathological conditions. Blood Rev 27(1):31–39
3. Hoshino A, Costa-Silva B, Shen TL, Rodrigues G, Hashimoto A, Tesic Mark M, Molina H, Kohsaka S, Di Giannatale A, Ceder S, Singh S, Williams C, Soplop N, Uryu K, Pharmer L, King T, Bojmar L, Davies AE, Ararso Y, Zhang T, Zhang H, Hernandez J, Weiss JM, Dumont-Cole VD, Kramer K, Wexler LH, Narendran A, Schwartz GK, Healey JH, Sandstrom P, Labori KJ, Kure EH, Grandgenett PM, Hollingsworth MA, de Sousa M, Kaur S, Jain M, Mallya K, Batra SK, Jarnagin WR, Brady MS, Fodstad O, Muller V, Pantel K, Minn AJ, Bissell MJ, Garcia BA, Kang Y, Rajasekhar VK, Ghajar CM, Matei I, Peinado H, Bromberg J, Lyden D (2015) Tumour exosome integrins determine organotropic metastasis. Nature 527(7578):329–335
4. Zocco D, Ferruzzi P, Cappello F, Kuo WP, Fais S (2014) Extracellular vesicles as shuttles of tumor biomarkers and anti-tumor drugs. Front Oncol 4:267
5. Wang K, Yuan Y, Cho JH, McClarty S, Baxter D, Galas DJ (2012) Comparing the MicroRNA spectrum between serum and plasma. PLoS One 7(7):e41561
6. Witwer KW, Buzás EI, Bemis LT, Bora A, Lässer C, Lötvall J, Nolte-'t Hoen EN, Piper MG, Sivaraman S, Skog J, Théry C, Wauben MH, Hochberg F (2013) Standardization of sample collection, isolation and analysis methods in EV research. J Extracell Vesicles 2
7. Van Deun J, Mestdagh P, Sormunen R, Cocquyt V, Vermaelen K, Vandesompele J, Bracke M, De Wever O, Hendrix A (2014) The impact of disparate isolation methods for EV on downstream RNA profiling. J Extracell Vesicles 3
8. Chevillet JR, Kang Q, Ruf IK, Briggs HA, Vojtech LN, Hughes SM, Cheng HH, Arroyo JD, Meredith EK, Gallichotte EN, Pogosova-Agadjanyan EL, Morrissey C, Stirewalt DL, Hladik F, Yu EY, Higano CS, Tewari M (2014) Quantitative and stoichiometric analysis of the microRNA content of exosomes. PNAS 41:4888–14893
9. Turchinovich A, Weiz L, Langheinz A, Burwinkel B (2011) Characterization of extracellular circulating microRNA. Nucleic Acids Res 39(16):7223–7233
10. Zhang J, Li S, Li L, Li M, Guo C, Yao J, Mi S (2015) Exosomes and exosoman microRNAs: trafficking, sorting and function. Genomics Proteomics Bioinformatics 13(1):17–24
11. Grasedieck S, Sorrentino A, Langer C, Buske C, Döhner H, Mertens D, Kuchenbauer F (2013) Circulating microRNAs in hematological diseases: principles, challenges and perspectives. Blood 121(25):4977–4984
12. Nolte-'t Hoen ENM, Buermans HPJ, Waasdorp M, Stoorvogel W, Wauben MHM, 't Hoen PAC (2012) Deep sequencing of RNA from immune cell-derived vesicles uncovers the selective incorporation of small non-coding RNA biotypes with potential regulatory functions. Nucleic Acids Res 40(18):9272–9285
13. Skog J, Würdinger T, van Rijn S, Meijer DH, Gainche L, Sena-Esteves M et al (2008) Glioblastoma microvesicles transport RNA and proteins that promote tumour growth and provide diagnostic biomarkers. Nat Cell Biol 10 (12):1470–1476
14. Huang X, Yuan T, Tschannen M, Sun Z, Jacob H, Du M et al (2013) Characterization of human plasma-derived exosomal RNAs by deep sequencing. BMC Genomics 14:319
15. Bustin SA, Benes V, Garson JA, Hellemans J, Huggett J, Kubista M, Mueller R, Nolan T, Pfaffl MW, Shipley GL, Vandesompele J, Wittwer CT (2009) The MIQE guidelines: minimum information for publication of quantitative real-time PCR experiments. Clin Chem 55(4):611–622
16. Jensen SG, Lamy P, Rasmussen MH, Ostenfeld MS, Dyrskjøt L, Orntoft TF, Andersen CL (2011) Evaluation of two commercial global miRNA expression profiling platforms for detection of less abundant miRNAs. BMC Genomics 12:435
17. Pfaffl MW (2001) A new mathematical model for relative quantification in real-time RT-PCR. Nucleic Acids Res 29:e45
18. Lunavat TR, Cheng L, Kim DK, Bhadury J, Jang SC, Lässer C, Sharples RA, López MD, Nilsson J, Gho YS, Hill AF, Lötvall J (2015) Small RNA deep sequencing discriminates subsets of extracellular vesicles released by melanoma cells–evidence of unique microRNA cargos. RNA Biol 12(8):810–823
19. Pritchard CC, Kroh E, Wood B, Arroyo JD, Dougherty KJ, Miyaji MM, Tait JF, Tewari M (2012) Blood cell origin of circulating microRNAs: a cautionary note for cancer biomarker studies. Cancer Prev Res (Phila) 5(3):492–497
20. Fields C, Li P, O'Mahony JJ, Lee GU (2016) Advances in affinity ligand-functionalized

nanomaterials for biomagnetic separation. Biotechnol Bioeng 113(1):11–25

21. Akers JC, Ramakrishnan V, Yang I, Hua W, Mao Y, Carter BS, Chen CC (2016) Optimizing preservation of extracellular vesicular miRNAs derived from clinical cerebrospinal fluid. Cancer Biomark 17(2):125–132

22. Yoshioka Y, Kosaka N, Konishi Y, Ohta H, Okamoto H, Sonoda H, Nonaka R, Yamamoto H, Ishii H, Mori M, Furuta K, Nakajima T, Hayashi H, Sugisaki H, Higashimoto H, Kato T, Takeshita F, Ochiya T (2014) Ultra-sensitive liquid biopsy of circulating extracellular vesicles using ExoScreen. Nat Commun 5:3591

Chapter 23

Extracellular Vesicle miRNA Detection Using Molecular Beacons

Won Jong Rhee and Seunga Jeong

Abstract

Many studies have recently focused on extracellular vesicles. Extracellular vesicles including exosomes are nanosized vesicles that are produced and secreted by cells. Extracellular vesicles contain various types of nucleic acid biomarkers, including microRNAs (miRNAs) from their parent cells. As high concentrations of extracellular vesicles are found in most body fluids, miRNA detection of exosomes in body fluids, including blood and urine, enables early and easy disease diagnosis. In this context, the in situ detection method of extracellular vesicles miRNA using molecular beacons is described here.

Key words Extracellular vesicle, Exosome, miRNA, Diagnosis, Molecular beacon

1 Introduction

Cells secrete various kinds of vesicles known as extracellular vesicles (EVs) into the extracellular space. EVs cover a broad range of secreted vesicles, including exosomes, microvesicles, shedding vesicles and apoptotic bodies. Among these different cell vesicles, exosomes have recently been of particular interest. Exosomes are nanosized extracellular vesicles of 30–100 nm in diameter [1–4]. Extracellular vesicles are constantly produced and secreted by most cells and enter the circulatory system. High concentrations of extracellular vesicles can be found in most body fluids, including blood, urine, and breast milk [5–7]. Extracellular vesicles are thought to be extracellular organelles that contain precious biomolecules, including proteins, lipids and nucleic acids. They also mediate intercellular communications among tissues and reflect the state of our bodies [5, 6, 8–10]. Therefore, the detection of extracellular vesicles biomolecules in body fluids offers great opportunities for the diagnosis of diseases. Among extracellular vesicle nucleic acids, there have been extensive studies on utilizing microRNA (miRNA) as biomarkers for disease diagnosis. miRNAs are small (19–25 nucleotides in length) noncoding RNAs that regulate gene

Winston Patrick Kuo and Shidong Jia (eds.), *Extracellular Vesicles: Methods and Protocols*, Methods in Molecular Biology, vol. 1660, DOI 10.1007/978-1-4939-7253-1_23, © Springer Science+Business Media LLC 2017

expression posttranscriptionally. Therefore, miRNAs play important roles in regulating cellular processes and are critically associated with the pathogenesis of cancer and other diseases [11–18]. Recent studies demonstrated that miRNAs are present in circulating extracellular vesicles and transferred to other cells, altering the function of those target cells [1, 19]. In addition, miRNAs are stable in extracellular vesicles and protected by the lipid bilayer from environmental stresses that cause miRNA degradation. In this context, extracellular vesicles' miRNAs can be used as biomarkers for diagnosing and predicting various diseases, including cancers [18]. The standard and common method for the detection of miRNAs in extracellular vesicles is based on PCR (polymerase chain reaction). However, the method requires combinations of multiple experimental processes, including extracellular vesicle lysis, miRNA isolation, cDNA synthesis and real-time PCR analysis. Therefore, it is important to have a simple, accurate, fast, and inexpensive technique for the detection of specific miRNAs in extracellular vesicles. A recent study has shown that miRNA in extracellular vesicles, especially exosomes produced from breast cancer cells can be successfully detected using a molecular beacon (MB) [20]. A MB is a nanosized oligonucleotide probe that has a fluorophore and a quencher at each end. Herein, we introduce a method that enables the in situ detection of miRNAs in extracellular vesicles (mainly focused on exosomes) without collapse (*see* **Note 1**).

2 Materials

2.1 Extracellular Vesicle-Free FBS

1. TLA-100.3 fixed angle ultracentrifuge rotor (Optima TL-100, Beckman Coulter, USA).
2. 0.22 μm syringe filter (GVS, USA).

2.2 Extracellular Vesicle Preparation in the Cell

1. DMEM for MCF-7 breast cancer cell line culture medium (Biowest, France).
2. Fetal bovine serum (FBS, Biowest, France).
3. Penicillin and streptomycin (Life Technologies, USA).
4. TLA-100.3 fixed angle ultracentrifuge rotor (Optima TL-100, Beckman Coulter, USA).
5. 0.22 μm syringe filter (GVS, USA).

2.3 Extracellular Vesicle Isolation

1. Total Exosome Isolation™ solution (Invitrogen, USA).
2. ExoQuick-TC™ (EXOTC50A-1, System Biosciences, USA).
3. Differential ultracentrifugation (Optima TL-100, Beckman Coulter, USA).
4. TLA-100.3 fixed ultracentrifuge angle rotor (Optima TL-100, Beckman Coulter, USA).

2.4 Extracellular Vesicle Quantification by ELISA (Enzyme-Linked Immunosorbent Assay)

1. Exosome ELISA kit (EXOEL-CD63A-1, System Biosciences, USA).
2. Exosome-specific anti-CD63 (EXOAB-CD63A-1, System Biosciences, USA).
3. HRP-conjugated anti-rabbit antibody (EXOAB-CD63A-1, System Biosciences, USA).
4. Varioskan™ Flash Multimode Reader (Thermo Scientific, USA).

2.5 Detection of miRNA-21 (miR-21) in Extracellular Vesicles Using MB

1. miR-21 MB:
   5′-Cy3-GCGCGTCAACATCAGTCTGATAAGC-TACGCGC-BHQ2–3′
   (Integrated DNA Technologies, USA).
2. Varioskan™ Flash Multimode Reader (Thermo Scientific, USA).
3. Black 384-well microplate (Corning, USA).

2.6 MB Delivery into Extracellular Vesicle Using Streptolysin O (SLO)

1. Streptolysin O (Sigma, USA).
2. Tris(2-carboxyethyl)phosphine (Thermo Scientific, USA).

3 Methods

3.1 Extracellular Vesicle-Free FBS Preparation

1. Add FBS into polycarbonate ultracentrifugation tubes (*see* **Note 2**).
2. Ultracentrifuge at 120,000 × *g* for 10 h at 4 °C (*see* **Note 3**).
3. Filtrate supernatant with 0.22 μm syringe filter for sterilization.
4. Store at 4 °C before use.

3.2 Extracellular Vesicle Production

1. Seed cells into 100 mm dish or T75 culture flask in DMEM supplemented with 10% (v/v) FBS.
2. Culture cells before they are 70% confluent.
3. Remove culture medium and wash cells twice with PBS.
4. Add 10 mL of DMEM supplemented with 10% exosome-free FBS.
5. Culture cells for an additional 2 days and collect culture medium for exosome preparation (*see* **Note 4**).

3.3 Extracellular Vesicle Isolation Using Total Exosome Isolation

Extracellular vesicle (especially exosomes) in culture medium as well as body fluids can be isolated by precipitation methods using commercially available kits or ultracentrifugation. Here, one ultracentrifugation method and two precipitation methods are introduced for extracellular vesicle isolation. Precipitation protocols are performed according to manufacturer's protocols.

1. Collect the culture media containing exosomes and other extracellular vesicles.
2. Centrifuge at 2,000 × *g* for 30 min to remove cells and cell debris (*see* **Note 5**).
3. Transfer supernatant to a sterile tube and add the appropriate volume of Total Exosome Isolation reagent as follows:

Culture medium volume	Reagent volume
1 mL	0.5 mL
10 mL	5 mL

4. Mix by vortexing or pipetting, and incubate overnight at 4 °C.
5. Centrifuge at 10,000 × *g* for 1 h at 4 °C.
6. Remove supernatant and resuspend pellet by adding 20–100 μL of sterile PBS (volume depends on the amount of extracellular vesicle).
7. Store isolated extracellular vesicle at 4 °C before use.
8. For preservation longer than a week, store at −20 °C.

3.4 Extracellular Vesicle Isolation Using ExoQuick-TC

1. Collect the culture media containing extracellular vesicles.
2. Centrifuge at 3,000 × *g* for 15 min to remove cells and cell debris (*see* **Note 5**).
3. Transfer supernatant to a sterile tube and add the appropriate volume of ExoQuick-TC as follows:

Culture medium volume	Reagent volume
10 mL	5 mL

4. Mix sample by inverting 3 times and store overnight at 4 °C. The mixture does not require rotation during this incubation period.
5. Centrifuge the mixture at 1,500 × *g* for 30 min at 4 °C.
6. Remove supernatant. In this step, make sure not to disrupt the extracellular vesicle pellet.
7. Centrifuge at 1,500 × *g* for an additional 5 min to remove all traces of fluids.
8. Resuspend pellet by adding 100 μL of sterile PBS (volume depends on the amount of extracellular vesicles).
9. Store isolated extracellular vesicles at 4 °C before use.
10. For preservation longer than a week, store at −20 °C.

3.5 Extracellular Vesicle Isolation by Differential Centrifugation

1. Collect the culture media containing extracellular vesicles.
2. Centrifuge at 300 × *g* for 10 min in at 4 °C min to remove cells.
3. Transfer supernatant to a sterile tube and centrifuge at 2,000 × *g* for 10 min at 4 °C to remove cell debris.
4. Transfer supernatant to a sterile vessel and centrifuge at 10,000 × *g* for 30 min at 4 °C.
5. Add an equal volume of supernatant into polycarbonate tubes (*see* **Note 6**).
6. Centrifuge at 100,000 × *g* for 2 h at 4 °C using a TLA-100.3 fixed angle rotor.
7. Remove the supernatant and resuspend pellets in 1× filtered PBS for washing. If you have the same samples divided in different tubes, resuspend pellets sequentially and collect into one tube.
8. Centrifuge at 100,000 × *g* for 2 h at 4 °C using a TLA-100.3 fixed angle rotor.
9. Remove the supernatant and resuspend pellet (mainly exosomes) in filtered PBS. The volume of PBS required for resuspension should be empirically determined and depends on the amount of extracellular vesicles in the pellet.
10. Extracellular vesicles can be stored at 4 °C before use.
11. For preservations longer than a week, store at −20 °C or −80 °C.

3.6 Extracellular Vesicle Quantification by ELISA

1. Before starting, make sure to warm the Super-sensitive TMB ELISA substrate to room temperature.
2. Make several diluted extracellular vesicle samples from isolated extracellular vesicles with included binding buffer (*see* **Note** 7).
3. Add 50 μL of each prepared diluted extracellular vesicle sample to a microtiter plate well provided by the manufacturer and cover plate with sealing film/cover.
4. Incubate the plate at 37 °C overnight.
5. After incubation, empty the well and wash it 3 times for 5 min each with 100 μL 1× washing buffer.
6. Dilute anti-CD63 antibody to 1:100 with 1× blocking buffer.
7. Add 50 μL of diluted anti-CD63 antibody solution into each well in the plate, and incubate for 1 h at 37 °C.
8. Wash the plate 3 times for 5 min each with 100 μL 1× washing buffer.
9. Dilute HRP-conjugated anti-rabbit antibody to 1:5,000 with 1× blocking buffer.

10. Add 50 μL of diluted anti-rabbit antibody solution and incubate for 1 h at 37 °C.
11. Wash the plate 3 times for 5 min each with 100 μL 1× washing buffer.
12. Add 50 μL of Super-sensitive TMB substrate and incubate for 45 min at room temperature.
13. Add 50 μL of stop buffer.
14. Measure ODs (optical densities) of samples using spectrophotometer at 450 nm.
15. Compare ODs with standards for calculating extracellular vesicle concentrations.

3.7 In Situ Detection of Extracellular Vesicle miRNA Using MB

1. For *in situ* detection of extracellular vesicle miRNA, the concentrations of extracellular vesicles and MBs used for detection should be empirically determined (*see* **Note 8**).
2. Add a certain volume of isolated extracellular vesicles into 1× RNase-free PBS. For a negative control (to account for background signal without extracellular vesicles), add PBS (instead of extracellular vesicle samples) at the same volume.
3. Add MBs into the mixture. The MB final concentration is 100 nM and volume of reactant is 50 μL. Higher MB concentrations may increase the hybridization signal, but it also increases background fluorescence if the MB structure is not appropriate.
4. Incubate extracellular vesicles and MBs for 1–2 h at 37 °C. In most cases, 1 h is enough to observe detection signals. For the extracellular vesicles with low concentration of target miRNAs or the concentration of the extracellular vesicle itself is low, further incubation of up to 2 h may help increase hybridization signals. This incubation time should be determined on a case-by-case basis.
5. After incubation, measure the fluorescence of samples with a Varioskan™ Flash Reader. For Cy3 fluorescent dye-labeled MBs, measure the absorbance at 550 nm wavelength for excitation and 570 nm wavelengths for emission.
6. Use the negative control (MBs without extracellular vesicles) to account for background signal. Signal-to-background ratio can be calculated by dividing the detection fluorescence intensity by the background intensity.

3.8 Enhanced Delivery of MBs into Extracellular Vesicles Using Streptolysin O (SLO)

1. To activate SLO, add 5 mM of tris(2-carboxyethyl)phosphine to 2 U/mL of SLO in PBS for 1 h at 37 °C. The concentration of SLO can be optimized before use.
2. Add a certain volume of isolated extracellular vesicles (*see* Subheading 3.7) into 1× RNase-free PBS. As mentioned above,

for a negative control (to account for background signal without extracellular vesicles), add PBS (instead of extracellular vesicle samples) at the same volume.

3. Add MBs and activated SLO into the mixture. The final concentrations of MB and SLO are 100 nM and 0.2 U/mL, respectively. The final volume of reactant is 50 μL.
4. Incubate extracellular vesicles and MBs for 1–2 h at 37 °C. The incubation time depends on the concentration of extracellular vesicles and its miRNAs as mentioned earlier.
5. After incubation, measure the fluorescence of samples by a Varioskan™ Flash Reader. For Cy3 fluorescent dye-labeled MBs, measure the absorbance at 550 nm wavelength for excitation and 570 nm wavelengths for emission.
6. Use the negative control (MBs without extracellular vesicles but with SLO) to account for background signal. Signal-to-background ratio can be calculated by dividing the detection fluorescence intensity by the background intensity.

4 Notes

1. The methods described in this chapter are mainly applicable to exosomes among extracellular vesicles. Some of the methods may not be applicable to other types of extracellular vesicles.
2. Extracellular vesicles are abundant in serum. Likewise, FBS contains a high amount of bovine extracellular vesicles and these bovine extracellular vesicles should be removed before addition to culture medium for extracellular vesicle production. To do this, remove extracellular vesicles in FBS by ultracentrifugation.
3. Use a higher centrifugal force for the extracellular vesicle removal in FBS than that for the extracellular vesicle isolation in culture medium.
4. The extracellular vesicle production time may depend on cell type, concentration, and growth rate. Make sure that apoptotic bodies are not generated by cell death during cell cultivation.
5. In some cases, you may filter the supernatant with a 0.22 μm syringe filter before adding the reagent.
6. Weigh entire tube including supernatant for a balanced loading of ultracentrifuge rotor.
7. For ExoQuick-TC isolated extracellular vesicles, you may make $1/10^3$–$1/10^4$ dilutions. For differential centrifugation isolated extracellular vesicles, you may make 1/5 and 1/10 dilutions.

8. In reference [20], 0.2–6.0 × 10^{10} particles/μL of extracellular vesicle and 100 nM of MB are used for detection. Total volume is adjusted to 50 μL with 1× RNase-free PBS. All experimental steps should be protected from light.

Acknowledgment

This research was supported by Basic Science Research Program through the National Research Foundation of Korea (NRF) funded by the Ministry of Education (NRF-2014R1A1A2054839).

References

1. Valadi H, Ekstrom K, Bossios A, Sjostrand M, Lee JJ, Lotvall JO (2007) Exosome mediated transfer of mRNAs and microRNAs is a novel mechanism of genetic exchange between cells. Nat Cell Biol 9:654–659
2. Laulagnier K, Motta C, Hamdi S, Roy S, Fauvelle F, Pageaux JF et al (2004) Mast cell and dendritic cell-derived exosomes display a specific lipid composition and an unusual membrane organization. Biochem J 380:161–171
3. Hogan MC, Manganelli L, Woollard JR, Masyuk AI, Masyuk TV, Tammachote R et al (2009) Characterization of PKD protein-positive exosome-like vesicles. J Am Soc Nephrol 20:278–288
4. Zhou R, O'Hara SP, Chen XM (2011) MicroRNA regulation of innate immune responses in epithelial cells. Cell Mol Immunol 8:371–379
5. Hu G, Drescher KM, Chen XM (2012) Exosomal miRNAs: biological properties and therapeutic potential. Front Genet 3:article 56
6. Raposo G, Stoorvogel W (2013) Extracellular vesicles: exosomes, microvesicles, and friends. J Cell Biol 200:373–383
7. Vlassov AV, Magdaleno S, Setterquist R, Conrad R (2012) Exosomes: current knowledge of their composition, biological functions, and diagnostic and therapeutic potentials. Biochim Biophys Acta 1820:940–948
8. Silverman JM, Reiner NE (2011) Exosomes and other microvesicles in infection biology: organelles with unanticipated phenotypes. Cell Microbiol 13:1–9
9. Abusamra AJ, Zhong Z, Zheng X, Li M, Ichim TE, Chin JL et al (2005) Tumor exosomes expressing Fas ligand mediate CD8+ T-cell apoptosis. Blood Cells Mol Dis 35:169–173
10. Yu S, Liu C, Su K, Wang J, Liu Y, Zhang L et al (2007) Tumor exosomes inhibit differentiation of bone marrow dendritic cells. J Immunol 178:6867–6875
11. Jeffrey SS (2008) Cancer biomarker profiling with microRNAs. Nat Biotechnol 26:400–401
12. Alevizos I, Illei GG (2010) MicroRNAs as biomarkers in rheumatic diseases. Nat Rev Rheumatol 6:391–398
13. Wahlquist C, Jeong D, Rojas-Muñoz A, Kho C, Lee A, Mitsuyama S et al (2014) Inhibition of miR-25 improves cardiac contractility in the failing heart. Nature 24:531–535
14. Spizzo R, Nicoloso MS, Croce CM, Calin GA (2009) SnapShot: microRNAs in cancer. Cell 137:586–586.e1
15. Esquela-Kerscher A, Slack FJ (2006) Oncomirs-microRNAs with a role in cancer. Nat Rev Cancer 6:259–269
16. Niwa R, Zhou F, Li C, Slack FJ (2008) The expression of the Alzheimer's amyloid precursor protein-like gene is regulated by developmental timing microRNAs and their targets in Caenorhabditis elegans. Dev Biol 315:418–425
17. Sarasin-Filipowicz M, Krol J, Markiewicz I, Heim MH, Filipowicz W (2009) Decreased levels of microRNA miR-122 in individuals with hepatitis C responding poorly to interferon therapy. Nat Med 15:31–33
18. Urbich C, Kuehbacher A, Dimmeler S (2008) Role of microRNAs in vascular diseases, inflammation, and angiogenesis. Cardiovasc Res 79:581–588
19. Rabinowits G, Gerçel-Taylor C, Day JM, Taylor DD, Kloecker GH (2009) Exosomal microRNA: a diagnostic marker for lung cancer. Clin Lung Cancer 10:42e6
20. Lee JH, Kim JA, Kwon MH, Kang JY, Rhee WJ (2015) In situ single step detection of exosome microRNA using molecular beacon. Biomaterials 54:116–125

Chapter 24

Rapid Isolation of Extracellular Vesicles from Blood Plasma with Size-Exclusion Chromatography Followed by Mass Spectrometry-Based Proteomic Profiling

Simion Kreimer and Alexander R. Ivanov

Abstract

The presented procedure allows rapid isolation of extracellular vesicles (EVs) from plasma using size-exclusion chromatography (SEC). Additionally, an approach for reducing the lipid and salt content of the EV isolate in preparation for mass spectrometry (MS)-based proteomic analysis is presented. An example setup for proteomic profiling of the processed samples by nanoflow liquid chromatography coupled to tandem mass spectrometry (nLC-MS/MS) is also presented. Approximately 1000 protein groups in blood plasma-derived EVs can be identified and quantitated following this procedure and using the described instrumentation.

Key words Extracellular vesicles, EVs, Microparticles, Exosomes, Proteomics, Size-exclusion chromatography, Mass spectrometry, Nano-LC-MS, Sample preparation, Proteomic profiling

1 Introduction

Extracellular vesicles (EVs) are now the accepted nomenclature for phospholipid bilayer enclosed particles (exosomes, microparticles, etc.) secreted by most, likely all, cells [1, 2]. EVs are found in all physiological fluids and are implicated in disease progression, immune response, inflammation, intercellular communication and other biological phenomena. EVs have diagnostic potential as "liquid biopsies," because disease-associated EVs can be detected in physiological fluids without invasive surgery [3–5]. Conventional EV isolation techniques by ultracentrifugation and density gradient centrifugation [6] are low-throughput and impractical in a clinical setting. Alternatively, size-exclusion chromatography (SEC) can be used to isolate EVs in minutes without specialized equipment [7]. SEC uses porous low binding beads to allow large particles (i.e., EVs), which cannot enter the pores, to readily pass through the bed of the stationary phase, while smaller sample constituents (e.g.,

Winston Patrick Kuo and Shidong Jia (eds.), *Extracellular Vesicles: Methods and Protocols*, Methods in Molecular Biology, vol. 1660, DOI 10.1007/978-1-4939-7253-1_24,

proteins) are slowed by the tortuous path through the pores. SEC is a low-resolution chromatographic technique, and SEC-based EV isolations may also contain other particles of similar size (30–800 nm) and low levels of plasma protein complexes and aggregates [3]. While higher purity isolation can be obtained using more time- and resource-consuming procedures, the presented approach is capable of rapid enrichment of EVs from blood plasma and may be applied to other highly complex physiological fluids, and samples of other types (e.g., cell culture media).

The described protocol greatly accelerates EV isolation and allows a reasonable depth of proteomic profiling. The SEC columns for gravity-driven EV isolation are easy to pack, and demonstrate reproducible performance. Figure 1 compares the gene ontology (GO) terms between the proteins identified in the early eluting 4–6 mL and 6–6.5 mL fractions obtained using a 10-mL SEC column. The increase in the representation of GO terms associated with plasma proteins in the later eluting fraction indicates that the content of plasma proteins is increased, and EVs and plasma proteins are indeed separated by size. The described Bligh–Dyer liquid–liquid extraction [8] separates the lipid and protein components of the collected fraction while reducing the salt content. This improves

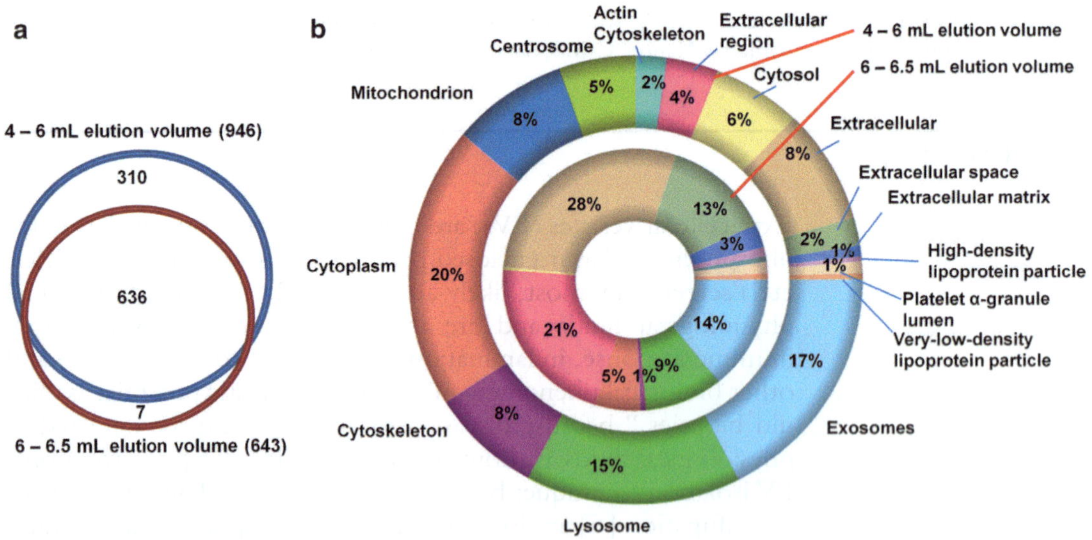

Fig. 1 Baseline proteome and Gene Ontology (GO) term enrichment comparison between the 4–6 mL elution volumes and 6–6.5 mL elution volumes. Frozen, cell-free, plasma was processed using the described workflow. Nearly 1000 protein groups were identified (<1% FDR) and quantified by MaxQuant. The overlap in protein identifications is presented in the Venn diagram (panel **a**). The GO cellular localization terms correspond to enriched proteins (2× intensity compared to the other fraction; panel **b**). The additional proteins identified in the 4–6 mL elution volume contribute to increased intracellular, and Exosome GO localization terms (panel **b**). The results are consistent with reduced concentration of plasma proteins in the earlier eluting fraction which improves detection of EV proteins and deeper proteomic profiling

the reproducibility of nLC-MS/MS-based proteomic profiling and extends the life span of the LC columns. The resulting lipid-rich fractions can be used for lipidomic profiling of EV isolates.

2 Materials

Use HPLC-grade solvents (water, isopropanol) and sterile Dulbecco's phosphate buffered saline (DPBS), pH 7.4 (Sigma-Aldrich, St. Louis MO) for SEC column preparation and separation of plasma. Use LC-MS grade water, acetonitrile, and formic acid for LC-MS analysis (Sigma-Aldrich). Filter the solvents and buffers by passing them through a 0.22 μm filter before the SEC preparation to mitigate possible contamination with the microparticulate material. Prepare the single use SEC column within hours of planned isolation by packing an empty 10 mL centrifuge column (e.g., Pierce pt. # 89898 10 mL columns with a 13 mL reservoir) with a slurry of cross-linked Sepharose CL-2B particles (Sigma-Aldrich pt.# CL2B300) shipped in 50% isopropanol.

Rinse the 5–10 mL disposable syringe with a Luer-lock adapter (e.g., BD pt.#.309604) with filtered DPBS. Condition the 30 mm diameter 0.8 μm pore cellulose acetate membrane filters with the Luer adapter (e.g., Cameo syringe filters Sigma-Aldrich pt. #741,675 or #741864) with sterile DPBS before use.

Blood drawing system: Use BD Vacutainer® UltraTouch™ Push Button Blood Collection Set consisting of a 21 gauge 0.75″ needle, 7″ tubing (pt#367393) and 10 mL EDTA-coated Vacutainer® tube (pt# 367856) (BD Life Sciences, Franklin Lakes, NJ) for drawing blood.

Sample preparation reagents: Use HPLC grade chloroform, methanol, water, 2,2,2-triflouroethanol, and highest purity iodoacetamide (IAA), Tris(2-carboxyethyl)phosphine hydrochloride (TCEP), formic acid and other reagents from Sigma-Aldrich, unless specified otherwise. Use MS-grade Trypsin/Lys-C mix from Promega (Madison, WI pt.#V5073) or similar. Use low retention Eppendorf tubes from VWR (Radnor, PA) or other sources.

3 Methods

All steps should be performed at room temperature unless otherwise indicated. It is advised to rinse the Eppendorf tubes with LC-MS grade methanol and dry in a clean environment before use. NOTE: Use freshly prepared 50 mM TCEP and 100 mM IAA stock solutions. Store the IAA solution in the dark as light degrades the reagent.

3.1 Preparation of Single-Use SEC Columns

1. Open the top cap and the bottom stopper and rinse the empty 10 mL column with 10 mL of 50% aqueous isopropanol.
2. Pour 12 mL of the CL-2B slurry and allow the slurry to settle down and condense as the beads are retained by the filter (30 μm pore size) and the solvent flows through.
3. If the bed volume is below the 10 mL mark on the side of the column, add more slurry until the bed volume is at 10 mL. If the bed volume exceeds 10 mL, the excess slurry can be scooped out using a clean spatula.
4. Fill the 13 mL reservoir with 50% isopropanol and allow all of the liquid to flow through. Repeat once more with 50% isopropanol and twice more with water.
5. Condition the column with DPBS allowing a total of 50 mL to flow through. Seal the top and bottom of the column so that it does not dry out; it is advised to prepare the column within a few hours of isolation. NOTE: If the stationary phase becomes dry, then cracks may form and lead to channeling effects. Drying can also change the porosity of the stationary phase, which can reduce the separation performance of the columns resulting in increased plasma protein contamination of the EV isolate.

3.2 SEC Isolation of EVs

1. Collect blood by percutaneous cubital venipuncture drawn into an EDTA-containing Vacutainer (BD Biosciences) and discard the first bolus to reduce artifacts from piercing the skin.
2. Collect 8 mL of plasma in the 10 mL EDTA-coated Vacutainer.
3. Centrifuge at 2000 × *g* for 5 min. During centrifugation or before starting the procedure, rinse with DPBS the 5 mL syringe and disk membrane filter to be used.
4. Discard the top 100–200 μL of plasma.
5. Aspirate 2 mL of plasma in a Luer-lock 5–10 mL syringe (e.g., BD pt.# 309,604). NOTE: Avoid any contact of the sample with the syringe plunger.
6. Attach the DPBS rinsed 0.8 μm cellulose acetate disk filter (Sterlitech, Kent, WA) to the syringe.
7. Open the column stopper and slowly inject 1 mL of plasma through the filter onto a SEC column so that the bed of the stationary phase is not disturbed.
8. Allow the injected volume to enter the stationary phase; the yellow hue of the plasma can easily be observed through the container wall. Close the stopper and immediately proceed to **step 9** to minimize longitudinal sample diffusion.

9. Add 10 mL of DPBS into the reservoir, open the stopper, and start collecting the eluate. Use 2 mL Eppendorf tubes with a labeled volume scale (or similar) to measure the eluate volume.
10. The EV population generally elutes by gravity flow in the 2–7 mL elution volume, and the peak is between 4 and 6 mL while the protein fraction starts to elute at around 6 mL. Collect the fraction required for analysis. Use the fraction of 4–6 mL that contains a high level of EVs with a reduced level of plasma proteins for LC-MS proteomic profiling. Close the stopper after collecting the EV-rich fraction(s). The collected EV isolate can be now used for qualitative and quantitative imaging techniques (microscopy, nano-flow cytometry, etc.).
11. The column, syringe, filter, and blood collection system can now be discarded following appropriate biohazard protocols. For quantitative comparison it is advised to estimate vesicle concentration in each sample using nanopore, brownian motion (e.g. NanoSight, Malvern), or dynamic light scattering, and normalize loading based on these measurements. Plasma composition is heterogeneous and normalization based on net protein concentration may not represent EV levels due to plasma protein contamination.
12. The samples can now be either frozen and stored at −80 °C, or processed immediately.

3.3 Lipid Depletion Using Bligh–Dyer Liquid–Liquid Extraction Followed by Tryptic Digestion of the EV Isolate

1. If frozen, thaw the samples at room temperature.
2. Concentrate the samples in the SpeedVac to a volume of approximately 200 μL.
3. Add 400 μL of methanol and 200 μL of chloroform. Vortex for 1 min and then sonicate in the ultrasonication water bath for 5 min at room temperature.
4. Add 200 μL of chloroform and vortex for 1 min.
5. Add 200 μL of water and vortex for 1 min.
6. Centrifuge at 14,000 × *g* for 10 min at room temperature.
7. Transfer the sample tube into a −20 °C freezer and incubate for 4 h (the cold temperature and centrifugation are used to focus the precipitated proteins in the interface between the two phases).
8. Centrifuge at 14,000 × *g* for 10 min at room temperature.
9. If a complementary metabolomics or lipidomics analysis is desired, the chloroform (lower) phase should be transferred into a clean Eppendorf container; otherwise, it can be discarded.
10. Carefully discard the methanol/water (upper) phase.

11. Lyophilize the remaining protein pellet to dryness in a SpeedVac.
12. Add 50 μL of 30 mM ammonium bicarbonate pH 8.0 in 50% aqueous 2,2,2-triflouroethanol to solubilize the precipitated proteins, sonicate for 5 min in the water ultrasonication bath at room temperature. NOTE: Protein concentration of the sample can be assayed after this step using a sample aliquot.
13. Add 6 μL of the 50 mM TCEP in 125 mM ammonium bicarbonate, pH 8 (to the TCEP concentration of approximately 5 mM in the sample) and incubate at 37 °C for 30 min. Check the pH using a submicroliter aliquot and a pH strip and make sure that the pH of the sample is approximately 7–8 after adding TCEP.
14. Add 7 μL of freshly prepared 100 mM IAA in 125 mM ammonium bicarbonate, pH 8 (to the IAA concentration of approximately 10 mM in the sample) and incubate in the shaker for 45 min in the dark.
15. Add 457 μL of 25 mM ammonium bicarbonate pH 7.5 and check the pH.
16. Confirm that the pH is approximately 7–8 and add 2 μg of Trypsin/Lys-C Mix, MS-grade (Promega), corresponding to the 1:20–1:50 enzyme to protein ratio.
17. Incubate in the shaker at 40 °C for 16 h or overnight.
18. Add formic acid to the concentration of 0.1% to quench digestion. NOTE: Peptide concentration of the sample can be assayed after this step using a sample aliquot.
19. Concentrate to the volume of approximately 50 μL using a SpeedVac.
20. Inject 1 μL of the resulting digest on to the nano-LC-MS column. Such injection volume will correspond to the amount of EV isolate derived from 20 μL of plasma. The sample amount to subject to molecular profiling can be adjusted appropriately according to the goals of the study.

3.4 nLC-MS/MS Analysis

Due to the presence of residual salts and lipids, it is advised to desalt online with a trapping column before LC-MS analysis. An example LC-MS protocol is presented below:

Column: a 25 cm long bed packed with 3 μm Magic C18 AQ, pore size 200 Å stationary phase in a fused silica capillary of the 75 μm internal diameter and 365 μm outer diameter with a porous Kasil frit polymerized in-house [9] or acquired from commercial sources (e.g., New Objective pt. # IF360-75-50-N-5). To generate stable nanoelectrospray, use electrospray voltage of 1.9–2.4 kV applied through a distal coated fused capillary tip 20 μm ID pulled to 10 μm opening (New

Objective pt.# FS360-20-10-D) connected to the end of the column with a 250 μm Teflon sleeve (Thermo Fisher Scientific, pt.# 160486).

Mass spectrometry: Use Top 15 data-dependent acquisition (DDA) on a QExactive with the following settings. MS1: automated gain control (AGC) 1e6, maximum ion accumulation time (IT): 30 ms, resolving power at *m/z* 200 (R): 70 k. MS2: AGC 1e5, maximum IT: 50 ms, R: 17.5 k. Dynamic Exclusion 60 s. Isolation window: 2 *m/z* and optimized HCD energy for the instrument used (e.g., 28). Use the signal for dodecamethylcyclohexasiloxane ions (445.12 *m/z*) as an internal calibrant in lock-mass mode.

Liquid chromatography: Use the mobile phase consisting of Solvent A (0.1% formic acid in water) and Solvent B (0.1% formic acid in acetonitrile). Inject 1 μL of the sample directly onto the column at a 250 nL/min flow rate and 2% B.

Gradient (250 nL/min): desalting for 15 min at 2%B; linear ramp to 42% B over 90 min; linear ramp to 95% B over 15 min, 10 min hold at 95%B; drop to 2%B in 1 min, hold at 2%B for 14 min. NOTE: Longer linear gradients can be used to improve the depth of proteomic profiling.

Data analysis: Use MaxQuant (v.1.5.3.28 or later) or similar database search software. Use the most current SwissProt/UniProt human protein sequence database appended with sequences of common contaminants (e.g., "the common Repository of Adventitious Proteins, cRAP", http://www.thegpm.org/crap/). Apply the following settings for the first search: 20 ppm allowed MS1 and MS2 mass tolerance; enzyme specificity: tryptic; carbamidomethylation of cysteine as a static modification. Main search settings: 4.5 ppm precursor mass and 8 ppm fragment mass errors; carbamidomethylation of cysteine set as a static modification, oxidation of methionine set as a variable modification; digestion specificity: semi-tryptic with two allowed missed cleavages. Filter the results to the peptide-level FDR of 1% or less.

Acknowledgments

This work was supported by the ASMS 2015 Research Award (ARI), NIH 1R01GM120272-01 (ARI), and the Dana-Farber Cancer Institute/Northeastern University Joint Seed Funding Program in Cancer Drug Development (ARI). The authors thank Dr. Ionita Ghiran (Beth Israel Deaconess Medical Center) for fruitful discussions and his assistance with blood collection.

References

1. Lotvall J, Hill AF, Hochberg F, Buzas EI, Di Vizio D, Gardiner C, Gho YS, Kurochkin IV, Mathivanan S, Quesenberry P, Sahoo S, Tahara H, Wauben MH, Witwer KW, Thery C (2014) Minimal experimental requirements for definition of extracellular vesicles and their functions: a position statement from the International Society for Extracellular Vesicles. J Extracell Vesicles 3:26913. doi:10.3402/jev.v3.26913
2. Gould SJ, Raposo G (2013) As we wait: coping with an imperfect nomenclature for extracellular vesicles. J Extracell Vesicles 2. doi:10.3402/jev.v2i0.20389
3. Kreimer S, Belov AM, Ghiran I, Murthy SK, Frank DA, Ivanov AR (2015) Mass spectrometry-based molecular characterization of extracellular vesicles: lipidomics and proteomics. J Proteome Res 14(6):2367–2384
4. Colombo M, Raposo G, Thery C (2014) Biogenesis, secretion, and intercellular interactions of exosomes and other extracellular vesicles. Annu Rev Cell Dev Biol 30:255–289. doi:10.1146/annurev-cellbio-101512-122326
5. Thery C (2015) Cancer: diagnosis by extracellular vesicles. Nature 523(7559):161–162. doi:10.1038/nature14626
6. Tauro BJ, Greening DW, Mathias RA, Ji H, Mathivanan S, Scott AM, Simpson RJ (2012) Comparison of ultracentrifugation, density gradient separation, and immunoaffinity capture methods for isolating human colon cancer cell line LIM1863-derived exosomes. Methods 56(2):293–304. doi:10.1016/j.ymeth.2012.01.002
7. Boing AN, van der Pol E, Grootemaat AE, Coumans FA, Sturk A, Nieuwland R (2014) Single-step isolation of extracellular vesicles by size-exclusion chromatography. J Extracell Vesicles 3. doi:10.3402/jev.v3.23430
8. Bligh EG, Dyer WJ (1959) A rapid method of total lipid extraction and purification. Can J Biochem Physiol 37(8):911–917
9. Cortes HJ, Pfeiffer CD, Richter BE, Stevens TS (1987) Porous ceramic bed supports for fused silica packed capillary columns used in liquid chromatography. J High Resol Chromatogr 10(8):446–448

Chapter 25

An Adaptable Polyethylene Glycol-Based Workflow for Proteomic Analysis of Extracellular Vesicles

Stephanie N. Hurwitz and David G. Meckes Jr.

Abstract

Extracellular vesicles (EVs), including exosomes are endocytically derived nanovesicles expelled from cells that contain molecular information in the form of lipids, proteins, and nucleic acids. Transfer of this information to other cells in local or distant microenvironments facilitates cell-to-cell communication. Importantly, diseased cells release exosomes containing specific cargo that may contribute to pathology and can be harnessed for diagnostic or prognostic use. The broad potential medical utility of exosomes has fueled rapidly expanding research on understanding the composition and functions of exosomes in normal and pathological conditions. Here, we provide a complete workflow for purifying exosome-sized vesicles from biological fluids for in-depth proteomic analyses. Moreover, this polyethylene glycol-based method is efficient, highly adaptable, and compatible with a variety of downstream applications.

Key words Exosomes, Microvesicles, Extracellular vesicles, Mass spectrometry, Proteomics, Polyethylene glycol, Ultracentrifuge

1 Introduction

Exosomes are small virus-sized vesicles released from cells that have recently attracted a striking amount of interest from diverse research fields for their role in a number of biological processes [1]. The endocytic origin of exosomes distinguishes these extracellular vesicles (EVs) from vesicles released from the cell surface (shedding microvesicles) or from dying cells (apoptotic bodies). Recent discoveries have alluded to the likelihood of further exosomal subpopulations that contain distinct protein cargo and may contribute to diverse functions in a biological system [2].

To better understand the heterogeneous mixture of exosomal populations secreted from a variety of cells and the content they carry, many studies have employed mass spectrometry approaches to

The original version of this chapter was revised. An erratum to this chapter can be found at DOI 10.1007/978-1-4939-7253-1_36

Winston Patrick Kuo and Shidong Jia (eds.), *Extracellular Vesicles: Methods and Protocols*, Methods in Molecular Biology, vol. 1660, DOI 10.1007/978-1-4939-7253-1_25, © Springer Science+Business Media LLC 2017

characterize the proteome of vesicles. These methods can be particularly informative following in vitro manipulation of endocytic pathway-associated proteins, or following immunoisolation of targeted vesicle subpopulations. Recent studies have further examined modulation of the exosomal proteome in disease [3–5]. However, there are several challenges faced in the preparation of exosomal protein for downstream mass spectrometry. Most notably, amassing sufficient protein quantities from these vesicles often requires harvesting large volumes of medium, a process which is cumbersome and time-consuming. Additionally, it is increasingly well accepted that most exosome enrichment methods frequently result in contaminating EV populations and serum proteins. In the past decade, emerging research in the field of EVs has progressively contributed to public genomic and proteomic databases in efforts to collaboratively advance understanding of vesicular content and mechanisms of biogenesis [6–9]. Unfortunately, diverse methods are currently used to purify exosomes for these collective databases, posing perplexities in comparing and interpreting data from different studies.

Historically, the differential centrifugation method has been the most-widely utilized technique for enriching exosomes [10]. However, further purification steps, such as sucrose or iodixanol density gradients are required to rid vesicle preparations of non-exosomal proteins [11]. Moreover, in the context of large volumes of cell culture medium, differential centrifugation can be a daunting procedure, requiring numerous ultracentrifugation steps to simply pellet vesicles before further purification steps. Other simple methods such as commercial polymer-based kits are considerably more expensive and do not achieve sufficient levels of vesicle purity [12].

Here, we provide an adaptable method of exosome purification that combines preliminary large volume centrifugation spins with a polyethylene glycol concentration step followed by a single ultracentrifugation to pellet vesicles. Resuspension of the crude vesicle pellet in a small volume facilitates further purification on an iodixanol density gradient, resulting in a highly pure extraction of exosomes that can be used for downstream analyses. We suggest this workflow, combined with the sensitive method of in-gel fractionation and digestion proposed here, provides a relatively efficient, inexpensive, and reproducible technique for the purification of exosomes for proteomic analyses.

2 Materials

All solutions used for exosome purification should be diluted in ultrapure HPLC-grade or Milli-Q filtered water. Ultrapure water used to make solutions for protein digestion should be autoclaved before use.

2.1 Polyethylene Glycol-Based Extracellular Vesicle Enrichment

1. 50 mL conical tubes.
2. 0.22 μm filter (optional).
3. 1 L centrifuge bottles.
4. 2× polyethylene glycol (PEG) solution: 16% polyethylene glycol 6000 (w/v), 1 M NaCl, and ultrapure water (*see* **Note 1**).
5. Phosphate buffered saline (PBS): 137 mM NaCl, 2.7 mM KCl, 10 mM Na_2HPO_4 and 2 mM KH_2PO_4, pH 7.4 (*see* **Note 2**).
6. 1.0 mL or 38.5 mL ultracentrifugation tubes (Beckman Coulter, #347287, #326823).
7. Optional: 250 mL centrifugation bottles.

2.2 Optiprep Density Gradient Purification

1. Optiprep (60% iodixanol in sterile water) solution (Sigma, D1556).
2. 0.25 M sucrose, 10 mM Tris–HCl, pH 7.4.
3. 12 mL ultracentrifuge tubes (Beckman Coulter, #331372).
4. 5 mL syringes.
5. 18-guage needles.
6. Fraction recovery system (Beckman Coulter, #343890).
7. 1.5 mL microcentrifuge tubes.
8. Refractometer.
9. Strong lysis buffer (SLB): 5% SDS, 10 mM EDTA, 120 mM Tris–HCl pH 6.8, 2.5% β-mercaptoethanol and 8 M urea (*see* **Note 3**).
10. Aspirating pipet (optional).

2.3 Protein Quantification

1. EZQ™ protein quantification kit (Invitrogen, R33200).
2. Plastic staining tray (or tip box lid).
3. Methanol.
4. Heat sealable plastic.
5. Heat sealer.
6. 10% methanol–7% acetic acid buffer.
7. UV light.

2.4 Colloidal Coomassie Staining

1. 4–20% precast SDS-PAGE gels (*see* **Note 4**).
2. SDS-PAGE electrophoresis running buffer: 25 mM Tris, 192 mM glycine, and 0.1% SDS, pH 8.3.
3. 5× SDS loading sample buffer: 250 mM Tris–HCl pH 6.8, 10% SDS, 50% glycerol, 5% β-mercaptoethanol, 0.5 M DTT and 1 mg/mL bromophenol blue.
4. Fixing solution: 50% ethanol–7% acetic acid solution.
5. Super clean gel staining container.

6. Ultrapure water.
7. Staining Solution: 0.82 mM Coomassie Brilliant Blue G-250, 34% methanol, 3% phosphoric acid, and 1.28 M ammonium sulfate in ultrapure water (*see* **Note 5**).
8. Kimwipes™.

2.5 In-Gel Trypsin Digestion

1. Grid cutter (Gel Company, MEE1.5-5-48).
2. Light box.
3. Scalpel or razor blade.
4. Pick or forceps.
5. ThermoMixer.
6. SpeedVac unit with refrigerated vapor trap.
7. Sonicator.
8. Solution 1: ultrapure water.
9. Solution 2: 50 mM ammonium bicarbonate in 50% methanol (*see* **Note 6**).
10. Solution 3: 50 mM ammonium bicarbonate in 50% acetonitrile.
11. Solution 4: 100% acetonitrile.
12. Solution 5: 25 mM dithiothreitol in 50 mM ammonium bicarbonate (*see* **Note 7**).
13. Solution 6: 55 mM iodoacetamide in 50 mM ammonium bicarbonate (in a dark bottle).
14. Solution 7: 0.01% ProteaseMax [Promega] in 50 mM ammonium bicarbonate (*see* **Note 8**).
15. Solution 8: 12 ng/μL Trypsin Gold [Promega, V5280] in solution 7.
16. 10% trifluoroacetic acid (TFA).

3 Methods

3.1 PEG-Based Exosome Enrichment

1. Collect conditioned cell culture medium from which exosomes will be harvested in 50 mL conical tubes. Larger centrifugation bottles can be used for greater volumes of medium.
2. Centrifuge 500 × *g* for 5 min at 4 °C to pellet and discard cells. Transfer supernatant to new conical tube and centrifuge 2000 × *g* for 10 min at 4 °C to pellet and discard cellular debris and large EVs.
3. Transfer and centrifuge supernatant at 10,000 × *g* for 30 min at 4 °C to pellet microvesicles (*see* **Note 9**).

4. Optional: Filter supernatant through 0.22 μm filter and pool all processed medium into a clean container. Store at 4 °C for up to 1 week (*see* **Note 10**).
5. Add an equal volume of 2× PEG solution [16% PEG average M_n 6000, 1 M NaCl, ultrapure water]. For large volumes, medium can be incubated with PEG solution in 250 mL or 1 L bottles. Invert several times to mix and incubate overnight at 4 °C.
6. Pellet vesicles by low-speed centrifugation at 3200 × *g* for 1 h. For large volumes (250–1000 mL), centrifuge 10,000 × *g* for 1 h at 4 °C to ensure complete pelleting.
7. Decant the supernatant and leave bottles inverted for 5–10 min, tapping frequently to rid the sides of residual liquid. Resuspend vesicle pellet in 0.5–35 mL particle-free filtered PBS (*see* **Note 11**). This pellet should resuspend fairly easily.
8. Transfer vesicle/PBS suspension to 1 mL or 38.5 mL ultracentrifuge tube.
9. Ultracentrifuge sample at 100,000 × *g* for 2 h at 4 °C.
10. Resuspend pellet in 500 μL of particle-free PBS. It is important to ensure complete suspension of vesicles, as the previous ultracentrifugation step will yield a tight pellet (*see* **Note 12**).

3.2 Optiprep Density Gradient Purification

1. Dilute iodixanol (Optiprep) stock solution (60% w/v) in 0.25 M sucrose (10 mM Tris–HCl, pH 7.4) buffer to make 5, 10, 20 and 40% iodixanol solutions (*see* **Note 13**).
2. Pipet 3 mL of 40% iodixanol solution onto the bottom of a 12 mL ultracentrifugation tube to begin forming the density gradient (Fig. 1a).
3. Using a syringe and 18-guage needle, carefully layer 3 mL of 20% iodixanol solution over the bottom layer. When adding density layers, touch the bevel of the needle to the side of the tube just above the liquid line and slowly drop the solution, preventing undesired mixture of the layers at the interface.
4. Using the same technique, layer 3 mL of 10% iodixanol solution over the previous layers.
5. Add 2.5 mL of the 5% iodixanol solution on top.
6. Carefully pipet the 500 μL EV/PBS resuspension on top of the 5% iodixanol solution. If necessary, add particle-free PBS dropwise to the top layer in order to balance tubes.
7. Ultracentrifuge gradients at 100,000 × *g* for 18 h at 4 °C. Set the acceleration and the deceleration to their minimum respective rates. The ultracentrifuge should slow down without the use of the break to prevent disruption of density layers at the beginning and end of the centrifugation.

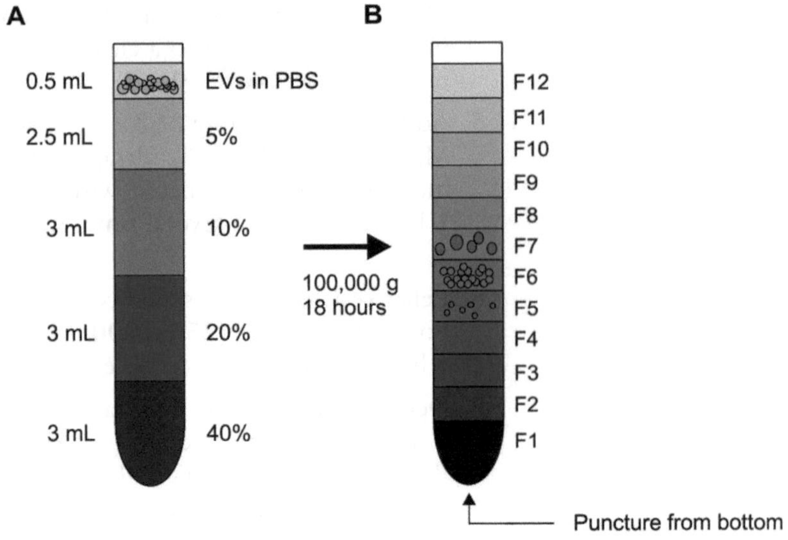

Fig. 1 Iodixanol density gradient purification of exosomes. (**a**) Prepare gradient by layering 3 mL of 40, 20, and 10% iodixanol solutions in a 12 mL ultracentrifuge tube. Add 2.5 mL of 5% iodixanol and pipet 0.5 mL crude vesicle resuspension on *top*. (**b**) Ultracentrifuge at 100,000 × *g* for 18 h to obtain continuous density gradient and then collect 1 mL fractions from the *bottom*

8. Label 1.5 mL microcentrifuge tubes to correspond to fractions 1 through 12 and clearly mark a 1 mL volume line on the side of each tube. Remove tubes from rotor buckets gently. Place the gradient tube on top of a fraction recovery system, piercing the bottom with a needle, and collect 1 mL fractions in successive 1.5 mL tubes (Fig. 1b).
9. Measure refractive indices of each fraction by pipetting 30 μL of each fraction onto a refractometer (*see* **Note 14**). Estimated densities can be calculated by comparing refractive indices to known corresponding iodixanol densities.
10. Transfer 1 mL fractions to new 12 mL ultracentrifugation tubes. Add 6 mL particle-free PBS and mix by pipetting up and down slowly. Add an additional 4–5 mL of PBS to the top and again mix carefully.
11. Ultracentrifuge samples at 100,000 × *g* for 2 h at 4 °C to remove iodixanol and pellet EVs (*see* **Note 15**).
12. Decant supernatant, invert tubes for 5 min, and tap to dry. Remove any residual liquid from the sides of tubes with an aspirating pipet.
13. Lyse EVs by adding 30–50 μL of strong lysis buffer (*see* **Note 16**).
14. Transfer samples to new microcentrifuge tubes and store at −20 to −80 °C until further processing.

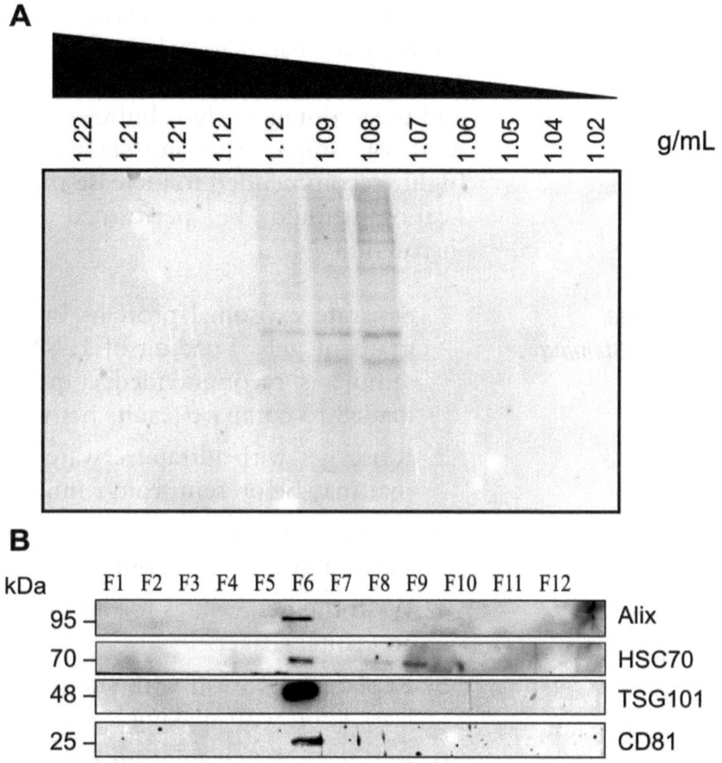

Fig. 2 Immunoblot analysis of Optiprep density gradient-purified fractions. (**a**) Ponceau S stain of Optiprep density gradient fractions run on an SDS-PAGE gel. Visible protein was seen primarily in fractions 6–8, corresponding to densities of 1.08–1.12 g/mL. (**b**) Twelve fractions were probed with antibodies against common exosomal protein markers Alix, HSC70, TSG101, and tetraspanin CD81. Exosomal protein was observed predominately in fraction 6. However, depending on the preparation or exposure, signal for exosomal markers may also detected in fractions 5 and 7

3.3 Western Blot Analysis

Fractions with exosomal protein to be digested for mass spectrometry should be confirmed by immunoblot analysis before protein quantification. Load equal volumes of lysed sample from each fraction 1–12 into a 10% SDS-PAGE gel for electrophoresis and western blot analysis (Fig. 2a). To detect most markers, exosomal protein from the equivalent of 25–40 mL of cell-conditioned medium is recommended (about 1–2 μg). For less sensitive antibodies, a greater volume of sample may be required. Fractions for exosome proteomic analysis should be confirmed with four demonstrated exosomal protein markers, including one tetraspanin protein, as recommended by EV researchers (Fig. 2b) [13] (*see* **Note 17**).

3.4 Exosomal Protein Quantification

Once fractions containing pure exosomal protein have been confirmed by western blot analysis, these specific fractions can be processed further for downstream proteomic analysis. For sensitive quantification of low exosomal protein concentrations, a fluorescence-based assay is recommended. The Invitrogen EZQ™ Protein Quantification Kit is also compatible with the detergent and urea-containing lysis buffer used previously, and only requires 3 μL of sample. Spotting standards and samples in triplicate is highly recommended to increase the accuracy of the assay. Quantification should be performed according to manufacturer's instructions.

3.5 Colloidal Coomassie Staining

1. Separate exosomal proteins by SDS:PAGE in 4–20% precast gradient gels. Loading of 15–25 μg of exosomal protein per sample is recommended. Equal mass of protein should be loaded to compare results between samples.
2. Rinse gel with ultrapure water to remove any contaminants that may be present from running the gel (*see* **Note 18**).
3. Incubate the gel with fixing solution for 1 h to overnight with shaking in a staining container acceptable for proteomics work.
4. Wash the gel 3 times for 30 min each with ultrapure water on an orbital shaker.
5. Replace final wash with staining solution and incubate for at least 1.5 h with shaking.
6. Remove staining solution and destain the gel in ultrapure water for 30 min on orbital shaker.
7. Drain and replace water, then add Kimwipes to one side of the container and continue shaking overnight.

3.6 In-Gel Trypsin Digestion

1. Gently place the gel on the small (cleaned) light box in the hood and excise bands of interest from gel using a grid cutter (Fig. 3) (*see* **Note 19**).
2. Place the gel cubes (25 per section) into 1.5 mL microcentrifuge tubes (*see* **Note 20**). If needed, samples can be frozen at −20 °C before further processing.
3. Add 1 mL per section of Solution 1 (ultrapure water) to each sample.
4. Incubate with gentle shaking for 1–5 min at room temperature.
5. Aspirate and discard the water (*see* **Note 21**).
6. Add 1 mL per section of Solution 2 (50 mM ammonium bicarbonate in 50% methanol) and incubate for 1 h with gentle shaking as before.
7. Remove Solution 2, and repeat several times as necessary (*see* **Note 22**) until the blue stain has been fully removed.
8. Wash with 500 μL per section of Solution 3 (50 mM ammonium bicarbonate in 50% acetonitrile) for 5 min with shaking at room temperature.

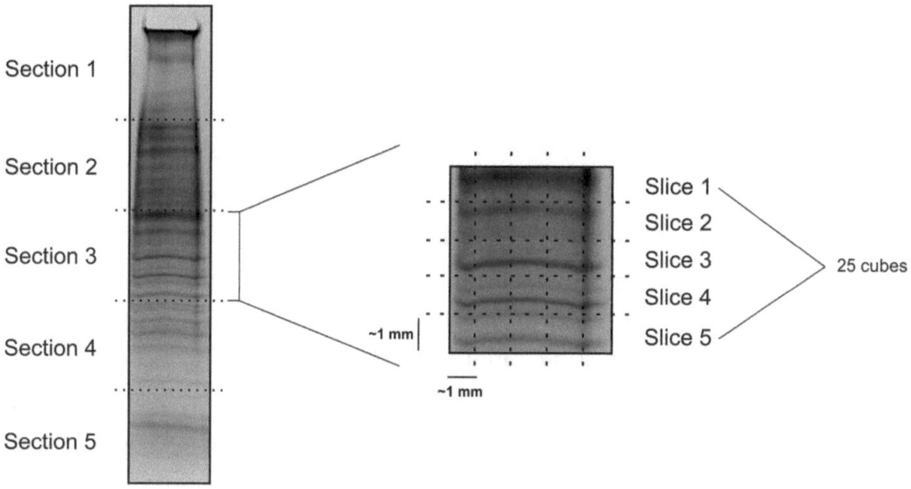

Fig. 3 Fractionation of gel lane following SDS-PAGE protein separation. Destained gel lane containing Optiprep-purified exosomal protein should be partitioned into five sections using a clean scalpel or razor blade. Each section should subsequently be cut into five slices both horizontally and vertically, creating 25 $1 \times 1 \times 1$ mm^3 cubes to be digested together

9. Aspirate Solution 3 and add 250–500 μL per section of Solution 4 (100% acetonitrile) for 1 min (*see* **Note 23**).
10. Remove as much of the acetonitrile as possible. Further dry the cubes in the SpeedVac for 10 min.
11. Add 500 μL per section of Solution 5 (25 mM dithiothreitol in 50 mM ammonium bicarbonate) and incubate at 56 °C for 20 min in a ThermoMixer set to 500 RPM (*see* **Note 24**). Be sure the solution covers all gel cubes while shaking. If needed, add more volume of Solution 5.
12. Remove Solution 5 and add the same volume of Solution 6 (55 mM iodoacetamide in 50 mM ammonium bicarbonate).
13. Incubate in the dark at room temperature for 20 min.
14. Remove Solution 6, and wash with 1 mL per section of ultrapure water by shaking gently for 1 min. Discard supernatant and repeat twice.
15. Remove water and wash with 500 μL per section of Solution 3 for 5 min with shaking.
16. Discard supernatant and dehydrate gel cubes with 250–500 μL per section of Solution 4 for 5 min with shaking.
17. Remove supernatant and then dry the gel cubes again in the SpeedVac for 10 min at room temperature. Once dry, samples can be stored at −20 °C until further processing.
18. Cool gel cubes on ice.

19. Add 300 μL per section of Solution 8 [12 ng/μL Trypsin Gold in solution 7 (0.01% ProteaseMax in 50 mM ammonium bicarbonate)].
20. Incubate for 5 min on ice.
21. Overlay with 150 μL per section of Solution 7 and place in ThermoMixer for 3 h at 37 °C with shaking at 500 RPM.
22. Add TFA to a 0.5% final concentration (*see* **Note 25**).
23. Mix and keep tubes on ice for 10 min to inactivate the trypsin.
24. Add 100 μL of Solution 1, and then 100 μL of Solution 4 to each tube containing the gel cubes.
25. Mix briefly and sonicate for 10 min (*see* **Note 26**).
26. Transfer digestion reaction with extracted peptides into a new 0.6 mL Axygen tube.
27. Dry samples in SpeedVac to completion at room temperature (~2–3 h). Do not overdry, as samples may be lost. If needed, samples can be stored at −80 °C after fully drying until further processing.
28. Resuspend peptides in appropriate solution for LC-MS/MS analyses (*see* **Note 27**).

3.7 Summary

We have provided a complete workflow for efficiently harvesting highly pure exosome-sized extracellular vesicles from large quantities of culture media for downstream mass spectrometry analyses. It is important to note that vesicles harvested using this method are of sufficient quantity and purity for other purposes such as electron microscopy, nanoparticle tracking, RNA analyses, and biological assays [12, 14]. Additionally, researchers will find this method highly adaptable to the desired application. For example, the ExtraPEG protocol can be used as a precursor to size exclusion chromatography, ultrafiltration, immunoaffinity purification, flow cytometry, or floatation gradients. The simplicity and versatility of the ExtraPEG method make it an excellent method to incorporate into EV research.

4 Notes

1. Make PEG solution 1 h before using to ensure that the polyethylene glycol is fully dissolved. To make a 1 L solution, mix 160 g of polyethylene glycol (average M_n 6000) with 200 mL of 5 M NaCl and bring up to 1 L in ultrapure water. Store at 4 °C.
2. PBS used for vesicle resuspensions should be particle free. Sterile cell-culture grade PBS should be filtered through a 0.22 μm filter and stored at room temperature to prevent salt precipitation. If available, a nanoparticle tracking instrument

(Malvern) can be used periodically to confirm the absence of nanoparticles in the solution.

3. A stock solution of SDS, EDTA, and Tris–HCl may be kept at room temperature. To make stock solution, add 5 mL of 10% filtered SDS, 0.2 mL of 500 mM EDTA and 1.2 mL of 1 M Tris–HCl, pH 6.8. Immediately before lysing samples, a final concentration of 2.5% BME and 8 M urea should be added. Addition of a protease inhibitor is recommended as well.
4. To minimize keratin and other contaminants, always use clean equipment. It is highly recommended to run samples in precast gels.
5. To make stock staining solution, combine 340 mL of methanol, 30 mL of phosphoric acid, and 170 g of ammonium sulfate. Mix 0.7 g of Coomassie Brilliant Blue G-250 in 75 mL of methanol and then add to the solution above. Bring the total volume up to 1 L with ultrapure water. Investigators can also substitute in commercially available staining reagents and protocols that are mass spectrometry-compatible.
6. Always make ammonium bicarbonate solutions fresh to ensure correct and consistent pH.
7. Always make fresh DTT-containing solution 5, as this reagent is not stable in liquid.
8. It is best to make fresh Solution 7 and Solution 8. However, if wanted, aliquots of Solution 8 can be stored in the −80 °C freezer until use. Solution 7 contains a surfactant (ProteaseMax, Promega) to facilitate the removal of peptides from the gel slices, and Solution 8 contains solution 7 with Trypsin Gold digest enzyme (Promega, # V5280). The 0.01% ProteaseMax solution (solution 7) should be kept on ice and used within several hours. This solution is not very stable, so any excess should be discarded and not frozen for reuse.
9. Following the 2,000 × g or 10,000 × g centrifugation step, medium can be stored at 4 °C for up to 1 week. If multiple harvests are performed, cell-conditioned medium can be pooled in a sterile container until further batch processed. If centrifuging larger volumes of medium, 250 mL bottles can be used. If retention of microvesicles is desired for downstream analyses, the 10,000 × g centrifugation step can be bypassed here, or the pellet can be saved for parallel processing.
10. Filtering the medium can aid in enrichment of small (less than 220 nm) vesicles and prevent unwanted bacterial growth during storage for several days.
11. Pellets from small volumes of starting material (e.g., 50 mL) can be resuspended in 1 mL of PBS for a small-scale ultracentrifugation wash. Resuspended pellets can also be combined

into one ultracentrifugation tube to concentrate vesicles. For example, 100 mL of medium can be centrifuged following overnight PEG incubation in two 50 mL conical tubes and resulting pellets can be resuspended in 0.5 mL of PBS each, then combined in a 1 mL ultracentrifugation tube. For larger volumes (e.g., 1 L), it is recommended to resuspend pellets in 35 mL of PBS and transfer to a larger ultracentrifugation tube.

12. Following the 100,000 × *g* ultracentrifugation step, decant the supernatant from the tubes and invert, tapping frequently to dry the sides of the tubes. Add 500 μL of PBS, cover the tubes with Parafilm, and vortex for 15 s. Rock or shake tubes for 15–20 min to disrupt the pellet and help resuspension. Vortex again briefly. Finally, centrifuge tubes for 1 min at a low speed (500–1000 × *g*) to collect all material at the bottom of the tube.

13. To make 50 mL of 40% iodixanol solution, combine 33.33 mL of the stock (60%) solution with 16.67 of 0.25 M sucrose in Tris buffer. Mix well, then perform serial dilutions with 1:1 volumes of iodixanol solutions with the sucrose buffer.

14. Exosomes should reside in fractions reflecting densities of 1.08–1.13 g/mL

15. Once consistent density fractions containing exosomes have been identified by western blotting (*see* Subheading 3.3), the final ultracentrifugation wash may be omitted for fractions that are not needed.

16. Again, cover tubes with Parafilm and vortex vigorously for 15 s. Shake or rock for 15–20 min, vortex again, and centrifuge for 1 min at 500–1,000 × *g* to collect all material at the bottom of the tube.

17. In the immunoblot shown in Fig. 2b, exosomal markers are primarily detected in fraction 6. Depending on the origin of EVs and exposure of the blot, exosomal proteins may be detected in neighboring fractions as well.

18. As keratin protein is abundant in dust particles and on the surfaces of researchers' skin, hair, and some articles of clothing, careful avoidance of keratin contamination in samples is essential. As mass spectrometry instruments are able to detect proteins at very low levels, the smallest amount of keratin contamination can corrupt accurate protein detection of less abundant sample peptides. Clean nitrile gloves and lab coats covering all skin exposed in proximity to samples is necessary when handling any equipment. Hair should be pulled back while processing samples. Sample processing should be performed in a HEPA-filtered laminar-airflow hood, and all work surfaces, tools, and bottles should be wiped down with 70% ethanol before use. Glass and plasticware should be reserved

for proteomic use only and rinsed with ultrapure water before usage.

19. Cut into five equal sections, using a scalpel or razor blade. Isolate each section before cutting with the grid cutter or razor blade to form five 5 × 1 × 1 ($l \times w \times h$) mm^3 slices. Then, further cut each 5 × 1 × 1 mm slice into five 1 × 1 × 1 cubes. If cutting isolated gel bands out with clean scalpel, optimal gel slices are 5–6 × 1–2 × 1 mm^3 in size. Slices should be further cut into five 1 × 1 × 1 mm^3cubes. Importantly, this protocol assumes processing of 1 gel section (25 cubes). If a greater degree of fractionation is desired, individual slices can be processed. Volumes should be adjusted accordingly. For example, in **step 3**, 200 μL of Solution 1 should be added per slice instead of 1 mL.
20. High quality microfuge tubes (Axygen or Eppendorf) should be used for digestion and peptide storage. Tubes may be prerinsed with 70% acetonitrile 0.1% formic acid and allowed to dry in a laminar flow hood before use. The goal is to prevent chemical leakage that may interfere with the mass spectrometer and to avoid peptides sticking to the walls of the tubes. Each tube should contain one section of gel, containing 25 cubes of gel each.
21. Use long fine gel-loading tips to prevent aspiration and loss of gel plugs.
22. This step destains and partially dehydrates the gel and may take as many as ten washes depending on the levels of protein and stain in the gel plug. Destaining may take as long as several hours up to 2–3 days. Gel cubes can be destained overnight while shaking to save time.
23. The dehydrated cubes should now be translucent, whitish, and firm. Hold the gel cubes up to the light to ensure no blue stain can be seen. If stain remains, rehydrate in water and repeat washes with Solution 2 (*see* **step 6**).
24. This protocol assumes 15–25 μg of protein was run into a gel. For different amounts of protein, more (or more concentrated) trypsin and other reagents should be considered.
25. From the 10% TFA stock, add 23.6 μL per 450 μL sample.
26. Be sure to clean and replace water in the sonicator bath with ultrapure water prior to sonication to avoid contamination of samples.
27. Prior to loading samples into the mass spectrometer, the solution containing peptides should be centrifuged for 10 min at 16,000 × g and transferred to a new tube. This step removes the degraded surfactant. Some proteomics facilities will complete this centrifugation step following submission. If not,

following **step 23**, aspirate the digestion solution and place in new 1.5 mL microcentrifuge tubes. Add **500** μL per section of Solution 1 to the remaining gel pieces, sonicate for 5 min, and add to the previously aspirated digestion solution. Centrifuge at 16,000 × *g* for 10 min to remove the enzyme. Transfer (avoiding the site of pellet) the digestion reaction with extracted peptides into a new 0.6 mL Axygen tube, and processed to **step 27**.

Acknowledgments

Special thanks to Mark Rider for the development of the ExtraPEG protocol used in the described methods, to Xia Liu for careful editing of the methods provided, and to Marius Kostelic for the acquisition of Coomassie-stained images. This work was supported by grants from the Florida Department of Health (6AZ11 and 4BB05) and the National Institutes of Health (CA204621 and CA188941) awarded to D.G.M.

References

1. Meckes DG, Raab-Traub N (2011) Microvesicles and viral infection. J Virol 85(24):12844–12854. doi:10.1128/JVI.05853-11
2. Kowal J, Arras G, Colombo M, Jouve M, Morath JP, Primdal-Bengtson B, Dingli F, Loew D, Tkach M, Théry C (2016) Proteomic comparison defines novel markers to characterize heterogeneous populations of extracellular vesicle subtypes. Proc Natl Acad Sci U S A 113(8): E968–E977. doi:10.1073/pnas.1521230113
3. Meckes DG, Gunawardena HP, Dekroon RM, Heaton PR, Edwards RH, Ozgur S, Griffith JD, Damania B, Raab-Traub N (2013) Modulation of B-cell exosome proteins by gamma herpesvirus infection. Proc Natl Acad Sci U S A 110(31):E2925–E2933. doi:10.1073/pnas.1303906110
4. Meckes DG (2015) Exosomal communication goes viral. J Virol 89(10):5200–5203. doi:10.1128/JVI.02470-14
5. Meckes DG, Shair KH, Marquitz AR, Kung CP, Edwards RH, Raab-Traub N (2010) Human tumor virus utilizes exosomes for intercellular communication. Proc Natl Acad Sci U S A 107(47):20370–20375. doi:10.1073/pnas.1014194107
6. Keerthikumar S, Chisanga D, Ariyaratne D, Al Saffar H, Anand S, Zhao K, Samuel M, Pathan M, Jois M, Chilamkurti N, Gangoda L, Mathivanan S (2015) ExoCarta: a web-based compendium of exosomal cargo. J Mol Biol. doi:10.1016/j.jmb.2015.09.019
7. Mathivanan S, Simpson RJ (2009) ExoCarta: a compendium of exosomal proteins and RNA. Proteomics 9(21):4997–5000. doi:10.1002/pmic.200900351
8. Simpson RJ, Kalra H, Mathivanan S (2012) ExoCarta as a resource for exosomal research. J Extracell Vesicles 1. doi:10.3402/jev.v1i0.18374
9. Kalra H, Simpson RJ, Ji H, Aikawa E, Altevogt P, Askenase P, Bond VC, Borràs FE, Breakefield X, Budnik V, Buzas E, Camussi G, Clayton A, Cocucci E, Falcon-Perez JM, Gabrielsson S, Gho YS, Gupta D, Harsha HC, Hendrix A, Hill AF, Inal JM, Jenster G, Krämer-Albers EM, Lim SK, Llorente A, Lötvall J, Marcilla A, Mincheva-Nilsson L, Nazarenko I, Nieuwland R, Nolte-'t Hoen EN, Pandey A, Patel T, Piper MG, Pluchino S, Prasad TS, Rajendran L, Raposo G, Record M, Reid GE, Sánchez-Madrid F, Schiffelers RM, Siljander P, Stensballe A, Stoorvogel W, Taylor D, Thery C, Valadi H, van Balkom BW, Vázquez J, Vidal M, Wauben MH, Yáñez-Mó M, Zoeller M, Mathivanan S (2012) Vesiclepedia: a compendium for extracellular vesicles with continuous community annotation. PLoS Biol 10(12): e1001450. doi:10.1371/journal.pbio.1001450

10. Théry C, Amigorena S, Raposo G, Clayton A (2006) Isolation and characterization of exosomes from cell culture supernatants and biological fluids. Curr Protoc Cell Biol Chapter 3:Unit 3.22. doi:10.1002/0471143030.cb0322s30
11. Abramowicz A, Widlak P, Pietrowska M (2016) Proteomic analysis of exosomal cargo: the challenge of high purity vesicle isolation. Mol BioSyst 12(5):1407–1419. doi:10.1039/c6mb00082g
12. Rider MA, Hurwitz SN, Meckes DG (2016) ExtraPEG: a polyethylene glycol-based method for enrichment of extracellular vesicles. Sci Rep 6:23978. doi:10.1038/srep23978. http://www.nature.com/articles/srep23978#supplementary-information
13. Lötvall J, Hill AF, Hochberg F, Buzás EI, Di Vizio D, Gardiner C, Gho YS, Kurochkin IV, Mathivanan S, Quesenberry P, Sahoo S, Tahara H, Wauben MH, Witwer KW, Théry C (2014) Minimal experimental requirements for definition of extracellular vesicles and their functions: a position statement from the International Society for Extracellular Vesicles. J Extracell Vesicles 3:26913
14. Hurwitz SN, Conlon MM, Rider MA, Brownstein NC, Meckes DG (2016) Nanoparticle analysis sheds budding insights into genetic drivers of extracellular vesicle biogenesis. J Extracell Vesicles 5:31295

Part VII

Isolation of Extracellular Vesicles from Biofluids

Chapter 26

Protocol for Exosome Isolation from Small Volume of Ovarian Follicular Fluid: Evaluation of Ultracentrifugation and Commercial Kits

Shlomit Kenigsberg, Brandon A. Wyse, Clifford L. Librach, and Juliano C. da Silveira

Abstract

The ovarian follicular fluid (FF) is a complex fluid that constitutes the microenvironment of developing follicles and contains factors secreted by the surrounding cells and blood plasma compounds that cross the "blood–follicle barrier." Upon oocyte retrieval (in human, bovine, and equine) the follicular fluid is normally discarded and represents a repertoire of cellular messages exchanged during follicle development, thus providing a suitable sample for performing oocyte quality diagnostics. Several studies report on the presence of extracellular vesicles (EVs) in FF from human, bovine and equine. Here, we describe the process of FF collection from human and bovine and the enrichment and isolation of EVs that we termed folliculosomes (FFEs), using available commercial kits as well as the traditional ultracentrifugation methods.

Key words Follicular fluid, Extracellular vesicles, Exosomes, Folliculosomes, Ultracentrifugation

1 Introduction

The follicular fluid (FF) provides the microenvironment for the developing oocyte [1–4]. It is the product of blood plasma constituents that cross the "blood–follicle" barrier and of the secretory activity of the surrounding granulosa (GCs), cumulus (CCs) and theca cells (TC). It contains nutrients necessary for the maturing egg, growth factors, hormones, enzymes, lipids, glycoproteins [5–9], and other compound solutes such as albumin and immunoglobulins [10, 11]. Extracellular vesicles (EVs) are a heterogeneous group of particles ranging in size from 30 to 1000 nm. A subset of these particles, exosomes, is defined as small extracellular vesicles in the 30–150 nm size range. These particles originate from the endosomal pathway and are secreted after binding of the multivesicular bodies (MVB) to the cell membrane [12–15]. Extracellular vesicles (EVs) are present in every body fluid tested so far [12,

Winston Patrick Kuo and Shidong Jia (eds.), *Extracellular Vesicles: Methods and Protocols*, Methods in Molecular Biology, vol. 1660,
DOI 10.1007/978-1-4939-7253-1_26,

16–19], including human, bovine and equine FF [20–22]. It has been suggested that these EVs have a role in folliculogenesis and the maturation of the oocyte [23–25]. We have coined these nanoparticles as folliculosomes (FFEs). Other examples of EVs found in reproductive fluids are epididymosomes and prostasomes in the male genital tract [26–28], and uterosomes and oviductosomes in the female genital tract [29–31].

A collection of clinical samples, such as FF requires the active participation of the physician, nurses and the in vitro fertilization (IVF) laboratory staff. The biological material (fluids and cells (FF, GCs, and CCs) collected during the ovum retrieval after the removal of the oocyte, is considered "biological waste material"; however, appropriate Research Ethical Board (REB) approval and patient consents for the sample collection are always required. Retrieval of the oocytes is performed under ultrasound guidance (*see* Fig. 1a) by sucking the follicular fluid which contains a

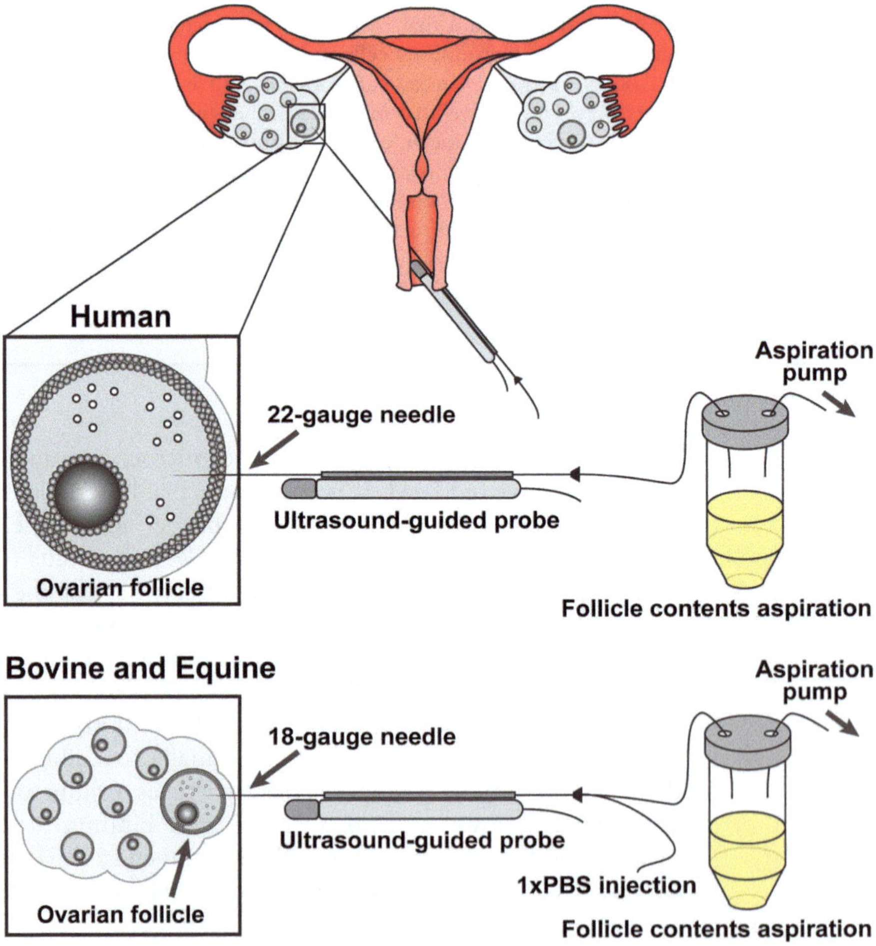

Fig. 1 Procedure of follicle content retrieval. (**a**) Procedure as performed in human subjects. (**b**) Procedure as performed in bovine and mares

cumulus–oocyte complex (COC). After removal of the oocyte from the COC (either by enzymatic or mechanical removal) by the embryologist [32] the remaining fluid (FF) containing the GCs can be collected for research. In common practice, FF from all follicles is pooled together; however, it is more useful if FF from individual follicles can be collected which can be challenging.

There are several challenges in studying exosomes from body fluids. One such challenge is the small volume often available in clinical samples. FF from an individual immature follicle can be as little as 100 μl in humans. Notably, the presence of high abundance proteins, such as immunoglobulins and albumin may entrap the EVs and can mask less abundant proteins and interfere with downstream analyses, such as Western blotting (WB) and mass spectrometry (MS), as observed in other body fluids [14, 33]. Therefore, it is recommended to adjust and optimize the different FFEs isolation steps, and a careful interpretation of the results is warranted.

The traditional method of particle isolation by ultracentrifugation (UC) and density gradient (DG) has been extensively covered in other book chapters and papers [34–36]. Other methods include precipitation by various ratios of polyethylene Glycol (PEG), size exclusion chromatography (SEC), affinity chromatography and magnetic bead separation. Immuno-capture techniques using magnetic beads have also been extensively covered by Oksvold et al. [37]. The readers should be aware that many published protocols describing EV isolation are suitable for isolation from conditioned media which does not reflect the complexity of body fluids that contain protein complexes, glycol and lipoproteins, cells debris, and more. In addition, the exact population of EVs isolated with the different techniques varies depending on isolation method used [18, 33, 38]. Notably, cancer cell lines produce large amounts of EVs, which is not the case for biological fluids [38]. Therefore, special steps need to be taken to ensure high purity and high-quality isolations for research purposes [39].

Our group performed extensive comparison analysis evaluating the ability of the traditional methods and several commercial kits to efficiently isolate exosomes from small volumes (>1 ml) of FF samples. The methods and kits include precipitation with Exoquick™ [21, 40], Immunoaffinity columns from several companies Exoeasy™ (Qiagen, Toronto, ON, Canada), Exosome isolation kit™ (Norgen, Thorold, ON, Canada), EXOcap™ (JSR Micro, Inc., Sunnyvale, CA [41], DiagoExo™ (101BIO, Palo Alto, CA, USA) [42, 43], and ME™ Kit (New England Biopeptides (NEP), Gardner, MA, USA) [44]. Finally, we assessed the size exclusion columns techniques qEV Single (Izon Scientific, Christchurch, NZ) [45, 46]. In addition, after enrichment for exosomes with either UC or Exoquick™, we have also performed immune-isolation using latex and magnetic beads coupled with antibodies for CD9, CD81 and CD63 universal exosome markers to isolate pure exosome population. However, the quantities of exosomes were too low for further analysis, and these results are not discussed in here.

We found that the isolated FFEs samples contain high quantities of nonrelated protein such as albumin and immunoglobulins that masked our proteomic studies. Therefore, we modified and optimized the workflow for FFEs isolation to reduce the amount of the carryover protein. This workflow is likely to be useful for any other type of biological fluid.

In this chapter we will describe the use of commercial kits that are suitable alternatives to UC and the steps needed to clear the FF samples before FFEs isolation in order to reduce protein complexes contaminants. It is important to note that some of the methods for isolating exosomes are enriching for EVs and not necessarily yield a pure exosome population. It is important to include appropriate controls and to ensure that free-floating RNA–protein complexes are excluded. In fact, when running SDS-protein gels followed by staining (GelCode™ or Coomassie Brilliant Blue™), it is clear that the carryover proteins are actually more concentrated in the final preparation, leading to very challenging mass spectrometry assays. To address this issue, the use of serum abundant protein depletion kits results in losing the majority of the EVs because the serum proteins form a complex with the EVs. We also found that many of the antibodies that are used to identify the universal markers cross-react with the heavy and light chain of immunoglobulins and thus not necessarily detecting pure exosome proteins.

The purpose of this chapter is to provide detailed protocols for the collection and processing of clinical samples from human FF, as well as FF from experimental models, such as the bovine and equine models, to enrich and isolate exosomes from these fluids.

1.1 General Notes

Working with biological fluids requires the proper equipment to prevent exposure to biohazardous and blood-borne pathogens. Collection of bovine or equine FF requires the participation of a certified veterinarian to perform the sampling and guarantee animal well-being. Working with animals requires animal-handling experience and appropriate facilities, as well as ethics approval. To prevent contamination of FF with blood or external follicular contaminants, FF is collected in a separate tube before ovarian follicle collapse (*see* Fig. 1b). This procedure allows collection of FF and contents separately during the in vivo procedure.

1. Follow biosafety regulation for working with hazardous biological material (*see* **Note 1**).
2. Follow the waste disposal regulations when disposing of biological waste materials (*see* **Note 2**).
3. Avoid FF samples that appear bloody as it might include exosomes of hematopoietic origin.

4. Follicular fluid, as other biological fluids contain highly abundant proteins that can trap the EVs, special attention should be given to include appropriate controls—therefore the flow-through and the left-over supernatant fluids should be kept.
5. Isolated granulosa cells (GCs, somatic cells) can be used as the tissue of origin and can be isolated in parallel to exosomes isolation.

2 Materials

2.1 Follicular Fluid Collection

1. Stereomicroscope.
2. Table top centrifuge with fixed-angle rotor mini for 1.5 ml tubes.
3. Table top centrifuge with swing-out rotor for 15 and 50 ml tubes.
4. Barrier filter tips to avoid biological contamination.
5. 15 ml polypropylene centrifuge tubes.
6. 50 ml polypropylene centrifuge tubes.
7. 1.5 ml polypropylene tubes, low protein binding, capable of processing at 12,000 × *g* (*see* **Note 3**).
8. Sterile syringe filters, 0.22 μm for hard to filter fluids, low protein binding polyethersulfone (PES) membrane (*see* **Note 4**).
9. Dulbecco's Phosphate Buffered Saline, without calcium chloride and magnesium chloride (DPBS×1), filtered (*see* **Note 5**).
10. Do not add sodium azide to any reagents or solution.

2.2 Ultra-centrifugation

1. Ultracentrifuge and rotors (e.g., Beckman Coulter Optima L-90K Ultracentrifuge and 70Ti rotor) (*see* **Note 6**). Ultracentrifuge tubes should be used according to sample volume.

2.3 Commercial Kits

1. Table 1 specifies the reagents and kits described in this section. All protocols are available online. We added a modified protocols for the Exoquick™ (Subheading 3.5, **step 1**) and qEV-Single (Subheading 3.5, **step 2**).
2. For Polyethylene glycol (PEG) removal use Sephadex spin columns—Exosome Spin Columns MW 3000 (Cat. #4484449 or SpinTrap™ G-25 Desalting Columns Cat. #45-001-527 (Thermo Fisher Scientific).
3. When precipitating exosomes from FF plasma or serum, the resulting pellet may be difficult to resuspend due to precipitated fibrin and other fibrinogens (*see* **Note 7**).
4. Size exclusion columns eliminate the need for the three spins and filter and will provide an exosome sample clean of co-purifying proteins

Table 1
Follicular fluid derived extracellular vesicles (FFEs) commercial kits and isolation methods

Name	Company	Method of isolation	End product	RNA isolation method	Starting volume	Protein yield (a)	Process time (six samples)	Cost$/ sample	References
Ultra-centrifugation	N/A	Ultra-centrifugation	Intact EVs +	(b)	High	6 ml	8–12 h	N/A	[36, 38]
Exoquick™	System Biosciences (SBI)	Precipitation polymer-based	Exosomes precipitation	(b, c)	High	20,000	2-ON	2–5$	[21, 40]
exoEasy Maxi Kit™	Qiagen	Membrane affinity spin columns	Intact EVs	(b, c)	4–32 ml	High	<3 h	30–50 $	Kenigsberg et al. (2016) unpublished
Plasma/Serum Exosome Purification Kit	Norgen Biotech	Membrane affinity spin columns	Intact EVs	(b, f)	Medium	50–1000 μl	<3 h	15–30 $	Kenigsberg et al. (2016) unpublished
DiagExo Human Body Fluid Exosome Isolation kit	101Bio	Membrane affinity spin columns	Intact EVs	Exosomal RNA and Protein Extraction kit	High	10 tests 100–500 μl (d)	<3 h	30–50 $	[42, 43]
ExoCap™ for serum and plasma Exosome Isolation and Enrichment kits	JSR Micro Inc.	Magnetic beads columns (e)	Intact sub-population EVs	ExoCap NA Elution Buffer	High	20 tests 100–500 μl	<3 h	30–50 $	[41]
The ME™ kit (plasma/serum)	New England Peptide (NEP)	The Vn96™ peptide (f)	Intact sub-population EVs	(b)	Medium	20 tests 100–500 μl (g)	<3 h	30–50 $	[44]
qEV™ columns (original or single)	Izon Sciences	Size Exclusion chromatography	Pure Intact EVs	(b)	100–500 μl	None in the EVs fractions	<2 h	20–30 $	[45, 46]

Summary of kits and methods used for FF isolation. (a) Indicate the amount of carryover protein as observed by gel staining. (b) Combination of Qiazol™ to lyse the exosomes and RNAeasy micro kit for RNA extraction for small quantities. (c) Company's exosomal RNA extraction kits or direct RNA isolation also available. (d) Each 1 ml—can produce ~100 μl pure exosome, 50–100 ng RNA or ~200 μg protein. (e) Magnetic beads coated with a polymer and cofactors which specifically bind to exosomes (proprietary). (f) The Vn96™ peptide binds to canonical Heat Shock Proteins (HSPs) that are found on exosomes and EVs, particularly from cells that are under stress such as cancer cells. Kit include positive and negative controls. (g) A dilution of 1:10 is recommended for starting volume

2.4 Controls

1. *Negative control*—FF depleted from exosomes ("FF-depleted") can be prepared as indicated in Subheading 3.6.
2. *Negative control*—Exosome-depleted Fetal Bovine serum ("FBS-depleted") can be purchased or prepared as indicated in Subheading 3.6.
3. *Positive control*—FBS (diluted 1:10 in DPBS×1).
4. *Positive control*—commercial or pre-prepared exosomes isolate (from serum or conditioned media) (*see* **Note 8**).
5. Neat FF sample (diluted 1:10 in DPBS×1).

3 Methods

Collection of FF is performed during in vitro follicle aspiration under ultrasound guidance for human samples, or in vivo follicular aspiration from slaughterhouse ovaries in bovines and mares [47, 48]. Follow the regulation for working with biological material. Take extra cautious when dealing with human fluids and use the appropriate protection gear. Is recommended to process the FF before freezing to prevent particles aggregation. We recommend to perform protein concentration analysis by (Bradford assay or BCA) and assessing the size distribution and concentration of the isolated particles using Nanoparticle Tracking Analysis (NTA) [50], and Tunable Resistive Pulse Sensing (TRPS) while using the appropriate controls.

3.1 Human Follicles Sample Collection

1. Figure 2a describes the process of in vitro follicle aspiration performed in human subjects. After the embryologist removes the egg and discards the "waste biological material" the FF can be collected from the petri plate (*see* **Note 9**).
2. Preferably, ask to collect an individual follicle from each ovary without any added buffer (*see* **Notes 10** and **11**).
3. After the removal of the oocyte by the embryologist, collect the FF (which include GCs, *see* **Note 12**) into a 15 ml tube (individual collection) or 50 ml tube (pooled collection).
4. If the collection of an individual follicle is not possible, collect the pooled FF from all follicles (can be up to 100 ml) into 50 ml tubes and process immediately to Subheading 3.3 to avoid cell clotting.
5. Place on ice and process as soon as possible.
6. *Optional*: Place the FF after collection into an incubator at 38.5 °C for 10–20 min to allow loose follicular cells to precipitate before processing the FF.
7. Proceed to Subheading 3.3 (Sample processing).

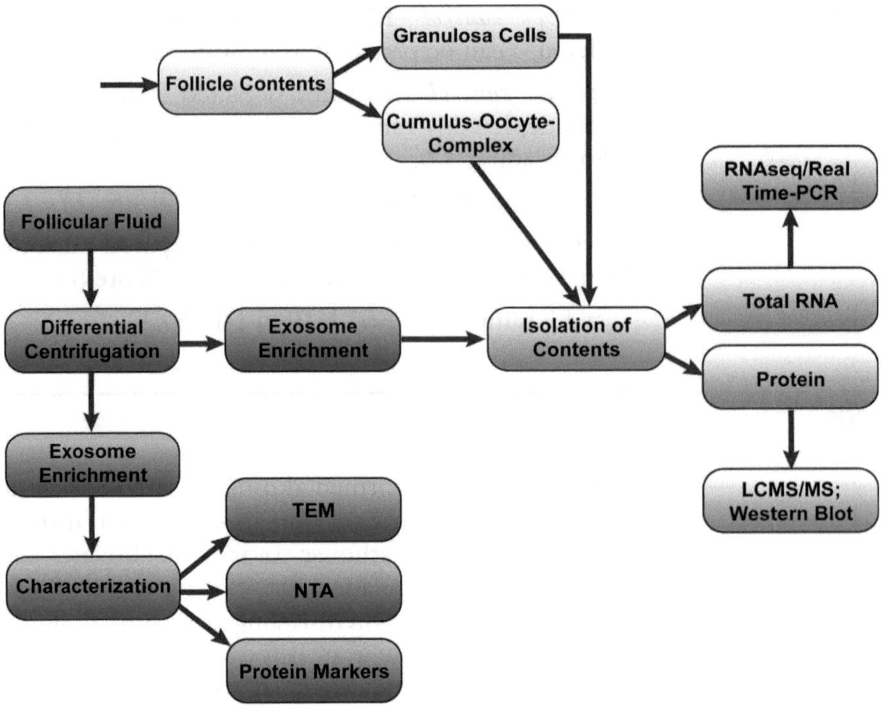

Fig. 2 Flowcharts for FFEs Isolation from ovarian follicular fluid from human, bovine and mares

3.2 Equine and Bovine Follicles Sample Collection

1. Figure 2b describe the process of in vivo follicle aspiration from slaughterhouse or post-mortem ovaries. The collection can be performed using a 12-G needle in equine and 18-G needle in bovine [49].
2. For in vitro collection transfer the ovaries to the laboratory in a saline solution at room temperature (RT).
3. Under the stereomicroscope, dissect the ovary on top of a petri plate using a curette.
4. After dissection, rupture the individual follicles and collect the fluid into a 15 ml tube or 1.5 ml microtube (*see* **Note 13**).
5. *Optional*: place the FF after collection into an incubator at 38.5 °C for 10–20 min to allow loose follicular cells to precipitate before processing the FF.
6. Transfer the FF to the lab on ice and process as soon as possible after COC analysis.
7. If collection of an individual follicle is not possible, collect the pooled FF from all follicles (can be up to 100 ml) into 50 ml tubes and process immediately to Subheading 3.3 to avoid cell clotting.

3.3 Clearing Protocol (Sample Processing)

To eliminate cells, cell debris, and reduce precipitation of protein aggregates, differential centrifugation steps are needed, followed by a filtration step. This clearing step is recommended before freezing to prevent aggregation of exosomes and large protein complexes. Freeze–thaw cycles did not influence the morphology or number of EVs isolated in our hands and is in agreement with the current literature [14]; however, it might affect the membrane composition [39].

1. Centrifuge the fluids upon arrival to the lab at 700 × *g* for 10 min at 4 °C to pellet the GCs.
2. Collect the supernatant without disturbing the pellet, leaving 300–500 μl on top of the pellet and transfer to new 15 or 50 ml tube.
3. The bloody pellet at the bottom of the tube contains GCs; collect the cell pellet immediately (*see* **Note 14** and support protocol 3.6).
4. The following steps are not needed when using size exclusion columns—proceed to isolation.
5. Centrifuge the fluid at 2000 × *g* for 10 min to remove dead cells. Collect the supernatant (leaving ~50 μl in the bottom) and transfer to microtubes capable of reaching 12,000 × *g* speed.
6. Centrifuge at 12,000 × *g* for 30 min at 4 °C to remove protein aggregates. Collect the supernatant without disturbing the pellet (leaving ~50 μl at the bottom).
7. Filter the fluid through 0.22 μM syringe filter to eliminate particles that are larger than 200 nm.
8. The sample can be kept at 4 °C for up to a week or in aliquots (ideally 500 μl) stored at −80 °C for longer storage time.
9. In cases where a collection of individual follicles is not possible, the results of pooling of multiple follicles is a large volume of FF (up to 100 ml), which is then need to be concentrated before using any kit (*see* **Note 15**).

3.4 Ultracentrifugation Procedure

The traditional method of ultracentrifugation method is in use for the isolation of extracellular vesicle, including exosomes.

1. Thaw samples at 37 °C and keep at RT until centrifugation.
2. Add cleared FF to the ultracentrifuge tubes. Weight the tubes carefully to balance them.
3. After assembling the tubes in the rotor wait for the centrifuge to achieve the right temperature and pressure before starting.
4. To isolate FFEs, use a fixed-angle rotor (70 Ti-Beckman), and spin at 120,000 × *g* for 70 min.

5. Remove the supernatant carefully, leave ~100 μl at the bottom of the tube and resuspend the pellet carefully in 1 ml of 1×DPBS (*see* **Note 16**). Add PBS to top it up to 6 ml.
6. Balance the tubes and repeat the centrifugation step to increase purity.
7. Remove the supernatant carefully, resuspend in 1×DPBS, and proceed with further analysis or aliquot and store at −80 °C.

3.5 Commercial Isolation Techniques (Summarized in Table 1)

Table 1 specifies the different kits that were compared for the isolation of FFEs from FF (*see* Fig. 3). In our hands, all the commercial kits provide enough biological material for downstream analysis. There are pros and cons to each exosomes isolation method. A careful approach should be taken when comparing results between different methods. First, as seen in Fig. 4 the amount/concentration of protein does not necessarily reflect the quantity of exosomes isolated but rather the amount of carryover proteins in the sample. It is likely that each method or kit might even isolate different subsets of EVs and co-precipitating proteins, all of which may lead to qualitative and quantitative variability.

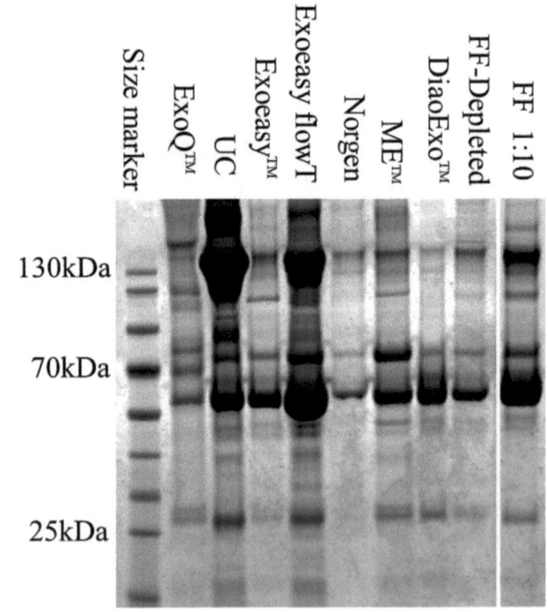

Fig. 3 Follicular fluid exosomes isolated with the different methods and kits as specified in Table 1. ExoQuick (ExQ), Ultracentrifugation (UC), Exoeasy, Norgen, ME kit, DiagExo, Negative control (FF-depleted) and neat FF 1: METHOD: 4–12% SDS-PAGE gel was run at 165 V for 1 h. The gel was staining using GelCode (ThermoFisher) for 1 h, washed 3× double-distilled water (DDW) for 15 min and image was taken using GelDoc system (DNR, Kiryat Anavin, Israel)

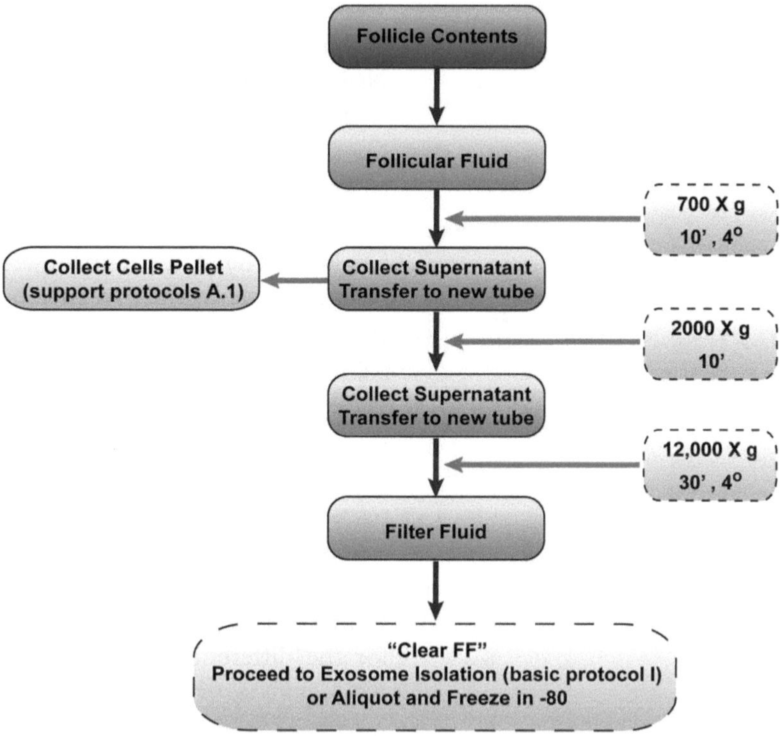

Fig. 4 Flowchart of the clearing protocol prior to FFEs isolation

Therefore, it is important to use appropriate controls both in Western blots and in RNA analysis.

1. FFEs isolation by precipitation

   Polyethylene glycol (PEG) is a water-soluble, volume excluding polymer that forces less-soluble components such as vesicles out of solution by sequestering water molecules. This allows exosome precipitation by short, low-speed centrifugation [33, 36, 50]. The advantages using PEG precipitation are (1) the use of simple low-speed centrifuge, which gentler compared to UC, (2) the absence of other organic materials and (3) the ability to precipitate at neutral pH and high ionic concentrations. However, the presence of the polymer may not be compatible with downstream analyses such as transmission electron microscopy (TEM). Effective isolation of extracellular nanoparticle in the desired size distribution can be achieved by careful manipulation of the PEG concentration [50, 51]. The most popular PEGs solutions used for isolation are the ExoQuick™ (System Biosciences (SBI), Palo Alto, CA, USA) and the Total Exosome Isolation Reagent™ (ThermoFisher Scientific, Burlington, ON, Canada). We describe here our modified protocol for FFEs isolation using the ExoQuick™ (SBI).

(a) Add ExoQuick™ to the FF in a 1:4 ratio (v/v). For example: add 125 μl of ExoQuick™ to 500 μl of FF (*see* **Note 17**). Mix well by inverting ten times or flicking the tube. Do not vortex.

(b) Incubate at 4 °C for 2 h. The tubes do not need to be rotated.

(c) Centrifuge ExoQuick/FF mixture at 1500 × *g* for 30 min (*see* **Note 18**). After centrifugation, the pellet may appear yellowish due to the presence of plasma proteins.

(d) Remove the supernatant and keep it is as negative control. Aliquot in 100 μl and store at −80 °C.

(e) To further reduce carry over protein a second precipitation step is recommended (*see* **Note 19**). Add 100 μl of 1×DPBS to the pellet and mix until the pellet is fully resuspended. If the pellet is not easily suspended, gradually add 20% (v/v) of 1×DPBS until the fluid is clear.

(f) Depending on the starting volume and the downstream analysis, the exosome suspension can be split at this step for the different downstream analysis—size distribution, RNA analysis or proteomics.

(g) Add ExoQuick^{TM} (1:4 v/v) and mix well by inverting ten times or flicking the tube. Do not vortex.

(h) Refrigerate Overnight at 4 °C without rotation.

(i) Centrifuge the ExoQuick/FF mixture at 1500 × *g* for 30 min. The FFEs pellet after the second spin should be smaller than the first pellet, indicating there are less carry-over proteins.

(j) Discard the supernatant. The pellet contains enriched FFEs population. Depending on the downstream applications—the exosome pellet can be resuspended in 50 μl of 1×DPBS for size distribution analysis or functional assays, resuspended in RNA extraction buffer or protein extraction buffer.

(k) The removal of the PEG is required for electron microscopy, size distribution analysis or functional assays. Sephadex Spin columns are recommended (*see* **Note 20**).

(l) Resuspend the pellet in the appropriate buffer for RNA extraction (*see* **Note 21**) or protein extraction [40].

2. FFEs isolation by Size-exclusion chromatography

Size exclusion chromatography separates molecules based on their size [52]. It allows the fractionation of vesicles and contaminating proteins without depletion kits, the addition of polymers (like PEG), and allows the elution of EVs in a buffer appropriate for downstream analysis. This method leads to a

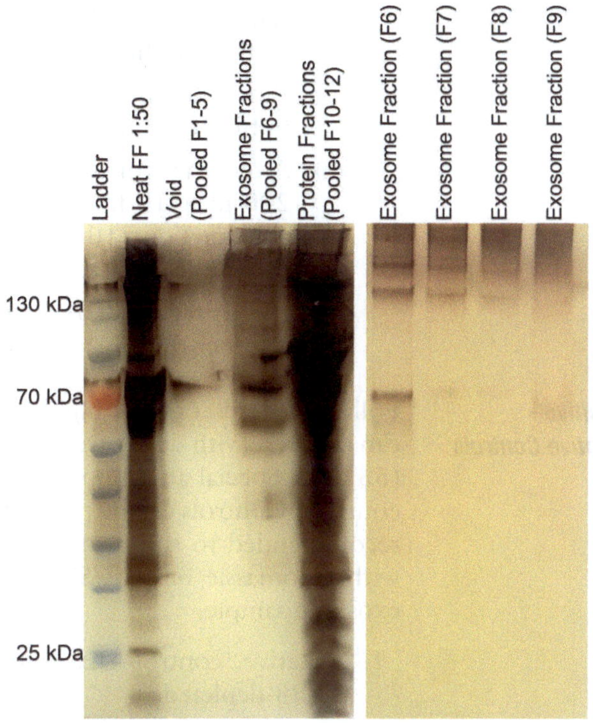

Fig. 5 Follicular fluid exosomes isolated with qEVsingle columns stained with SilverQuest silver stain. (**a**) Neat FF 1:50—Follicular fluid unpurified using qEV columns, Void represents the protein present before eluting fractions of exosomes, pooled exosomes fractions (fractions 6–9), Protein fractions were pooled from fractions 10–12. (**b**) Individual exosome fractions 6–9. METHOD: 4–12% SDS-PAGE gel was run at 165 V for 1 h. The proteins were fixed with Fixative Solution (40% ethanol, 10% acetic acid) for 1 h and stained using SilverQuest (Invitrogen) according to the manufacturer's protocol

more pure EV fraction that is compatible with TEM, Western blot, Mass spectrometry and RNA extraction. We describe here our modified protocol using the qEV single columns (Izon Scientific) (Fig. 5).

(a) Allow the column to warm to room temperature.

(b) Equilibrate the column with at least 10 ml of your buffer of choice (1×DPBS, 50 mM ammonium bicarbonate, etc.). Use only freshly filtered buffer to avoid precipitates that may clog the column.

(c) While the column is equilibrating, spin the FF at 2000 × g for 10 min at 4 °C to remove cell debris. The additional clearing steps outlined in Subheading 3.3 are not necessary.

(d) Load 100–150 μl of the sample onto the column and begin collecting 200 μl fractions. The first five fractions can be collected together (1000 µl) as this is the void volume of the column and will contain no EVs.

(e) When the last of the sample enters the beads, top up with buffer to keep consistent head pressure.

(f) After the void volume (1000 µl), collect a total of 200 µl × 10 fractions. Fractions 6–9 will contain FFEs, and fractions 10 and above will contain contaminating protein.

(g) FFEs Fractions (6–9) can be pooled to achieve maximum number of FFEs (*see* **Notes 22** and **23**).

3.6 Positive and Negative Controls

Follicular fluid contains highly abundant serum-like proteins that can interfere with downstream applications and affect the final yield [53, 54]. Special attention should be given to include appropriate controls. Controls can be prepared or purchased commercially. It is recommended to run protein profile gel (SDS-PAGE gel, stained with Coomassie blue or Silver stain) to ensure the purity of the exosome sample.

1. Negative control—Exosome-depleted Fetal Bovine serum ("FBS-depleted").
   (a) Centrifuge fetal bovine serum at 120,000 × *g* for 12–16 h at 4 °C in a fixed angle rotor. Carefully remove supernatant without disrupting the pellet formed after centrifugation. Mix the exosomes-free supernatant thoroughly, aliquot and store at −80 °C to avoid freeze–thaw.
2. Negative control—Follicular fluid depleted from exosomes ("FF-depleted").
   (a) Centrifuge 50 ml of FF at 120,000 × *g* for 12–16 h at 4 °C in a fixed angle rotor. Carefully remove supernatant without disrupting the pellet formed after centrifugation. Mix thoroughly the supernatant and aliquot the exosomes-free FF in 500 µl aliquots at −80 °C to avoid freeze–thaw.
3. Positive control—commercial or preprepared exosomes isolate (from serum or conditioned media) (*see* **Note 8**).
4. Neat sample (diluted 1:10 in 1×DPBS).

4 Support Protocols

4.1 Granulosa Cells (GCs) Collection

1. Materials
   (a) 1×DPBS
   (b) Dulbecco's Modified Eagle's Medium and Ham's Mixture F-12 (DMEM/F-12, (ThermoFisher).
   (c) Ficoll-Paque PLUS (GE Healthcare Life Sciences).

(d) Lysis-EZ K, Red Blood Cell (RBC) Lysis Buffer, 2× Strength Solution (BioBasic) or other RBC lysis buffer. Alternatively, pure and sterile double-distilled water (DDW) can be used.

(e) Barrier Filter tips to avoid biological contamination.

(f) 15 ml polypropylene centrifuge tubes.

(g) Disposable transfer pipets, 5 ml, individually wrapped, Sterile.

2. Methods

After removing the FF for EVs collection, leaving 3–5 ml on top of the pellet, the white buffy coat on top on the pellet should be collected immediately and transferred to new tubes containing 1×DBPS as described below (*see* **Note 24**).

(a) Prepare 3 × 15 ml tubes with 3 ml 1×DPBS each. Clearly labels the tubes "wash" (wash tubes I).

(b) Prepare 2 × 15 ml tubes with Ficoll-Paque solution. Clearly label the tubes "gradient" for the density gradient (gradient tubes).

(c) Transfer the pellet to one of the "wash tubes." Resuspend well but gently by pipetting. Transfer half of the fluid to each of the other "wash tubes" and top to 5 ml with 1×DPBS (*see* **Note 25**).

(d) Using the disposable transfer pipets carefully and slowly dispense down the side of the tube without breaking the Ficoll gradient, thus creating a 50% Ficoll–PBS gradient solution.

(e) Set the centrifuge ramp (startup) and brake to the lowest possible speed to avoid breaking the pellet and centrifuge at 700 × *g* for 20 min.

(f) Meanwhile, prepare 2 × 10 ml tubes with 1×DPBS each (Tube III).

(g) Aspirate the cellular layer above at the Ficoll–PBS interface (*see* **Note 26**) and transfer to the 1XDPBS (Tube III) to wash the cells after the Ficoll-Paque density gradient.

(h) Centrifuge at 700 × *g* for 10 min at 4 °C to pellet the washed GLCs. Discard the supernatant.

(i) Add 1 ml of 1×DPBS to the pellet and resuspend gently, but thoroughly to break any clumps. Add 1 ml of 2× Lysis-EZx red blood cell lysis buffer (BioBasic) to further eliminate any residual RBCs (*see* **Note 27**).

(j) Incubate for 3 min at RT.

(k) Add 8 ml of plain media (DMEM/F12) and centrifuge at 700 × *g* for 8 min.

(l) Wash the pellet with 5 ml of plain media, centrifuge at 700 × *g* for 8 min and place on ice. The cells are now ready for culture, RNA or protein extraction.

4.2 Cumulus Cells (CCs) Collection

1. Materials

   (a) 1×DPBS

   (b) Dulbecco's Modified Eagle's Medium and Ham's Mixture F-12 (DMEM F-12, ThermoFisher).

   (c) Lysis-EZ K, Red Blood Cell (RBC) Lysis Buffer, 2× Strength Solution (BioBasic) or other RBC lysis buffer. Alternatively, pure and sterile double-distilled water (DDW) can be used.

   (d) Barrier Filter tips to avoid biological contamination.

   (e) 1.5 ml Eppendorf tubes.

2. Methods

   The CCs can be collected by microdissection from oocytes prior to hyaluronidase application. The method of collection should be discussed with the embryologist and depends on laboratory practice.

   (a) Place in 1 ml 1×DPBS and transfer to the lab on ice.

   (b) Add 1 ml of 2× Lysis-EZ red blood cell lysis buffer to further eliminate any residual RBCs (*see* **Note 27**).

   (c) Incubate for 3 min at RT.

   (d) Add 8 ml of plain media (DMEM/F12) and centrifuge at 700 × *g* for 8 min.

   (e) Discard 7 ml of the supernatant, resuspend, and transfer to a 1.5 ml tube. Centrifuge at 700 × *g* for 8 min and place on ice. The cells are now ready for culture, RNA or protein extraction.

4.3 Exosome-Free Fetal Calf Serum (Exo-FBS) Preparation

For negative control is all experiments Exosome-depleted fetal bovine serum can be used: either commercially purchased or prepared by UC.

1. Centrifuge fetal bovine serum at 120,000 × *g* for 12–16 h at 4 °C in a fixed angle rotor.
2. Carefully remove the supernatant (FBS depleted) without disrupting the pellet formed after centrifugation.
3. Mix the supernatant thoroughly and aliquot in 500 μl.
4. Avoid freeze–thaw.

5 Notes

1. *See* Occupational Safety and Health Standards from USA Department of Labor section (1910–1930) www.osha.gov/pls/oshaweb/owadisp.show_document.
2. Biosafety: Laboratory Biosafety includes all containment principles, techniques, equipment, and practices that are implemented to prevent the unintentional exposure to biological agents and toxins, or other accidental release (WHO/CDS/EPR, 2006).
3. The polypropylene microtubes help minimize protein loss and are also DNase/RNase free.
4. Some FF or serum samples are viscous and hard to filter. It is recommended to use low protein binding polyethersulfone membrane (PES), suitable for viscous fluids such as Whatman® GD/X syringe filters (GE Healthcare Life Sciences).
5. Calcium and magnesium facilitate cell adhesion and clumping, there for it is recommended to use DPBS. Prepare 1×DPBS and filtered through 0.22 μm membrane to eliminate nanoparticles.
6. Different ultracentrifuges and rotors can be used to obtain extracellular vesicles. Thus, it is recommended to verify if the rotor type and the speed are adequate to obtain a pellet of the desired extracellular vesicles according to http://www.beckman.com/centrifugation and [35].
7. Thromboplastin-D is used to pretreat plasma samples to make them compatible with ExoQuick™ exosome precipitation. Store unopened vials at 2–8 °C. Reconstitute with preservative-free distilled/deionized water according to the Thromboplastin-D vial label, swirl gently and let the vial stand undisturbed for 15 min at room temperature (RT). Do not invert the vial or mix vigorously. Reconstitution fluid is available if the quality of water is questionable. After reconstitution, the solution is stable for 7 days at 2–8 °C, 8 h at 37 °C. Store at 2–8 °C when not in use.
8. Controls—negative: Exo-free FBS; Positive: Exosome rich positive control from Bioreactor Media or serum/plasma, are all commercially available.
9. Each laboratory has its own regulations; the transfer of discarded material should be discussed with your facility.
10. IVF Follicle Flush Buffer (HEPES-buffered solution) is used for flushing ovarian follicles during oocyte collection. Collecting an individual follicle from each ovary provide a biological duplicate.

11. In general, mature follicle can yield 1.5–4 ml of fluid. Immature follicle will yield <500 μl. If pooling the FF from all retrieved follicles, the volume can reach to 100 ml and will include the IVF Follicle Flush Buffer. After clearing the fluid (Subheading 3.3), the fluid can be concentrated.
12. The FF contains granulosa cells (GCs) that can be collected and used for culture. *See* Subheading 3.3 for instructions.
13. In general, a total of 500–3000 μl of follicular fluid can be obtained from a growing equine follicle. In bovine, ovarian follicles are smaller than in equine and it is possible to obtain a total of 25–250 μl of follicular fluid. The addition of 1×DPBS should not affect the yield or purity of exosomes.
14. The white layer on top of the bloody pellet is the buffy coat that contains the GCs; refer to support protocol 3.6 for GCs collection. The cells need to be transferred to 1×DPBS and washed immediately to avoid clotting.
15. Pierce™ Protein Concentrator PES, 100 K MWCO, 2–6 ml (ThermoFisher). More information on the selection of centrifugal concentrators can be found in "Pall Corporation: Selection Guide: Separation Products for Centrifugal and Tangential Flow Filtration".
16. The pellet might be invisible. Resuspended carefully but thoroughly the fluid.
17. Different species might have different FFEs concentration and/or proteins and lipid content in the FF. It might be needed to optimize the starting volume and adjust the ExoQuick™ accordingly.
18. Centrifugation may be performed at either room temperature or 4 °C with similar results.
19. We have found that a second precipitation step reduces the protein carryover and improve the quality of downstream applications.
20. Based on the 8000 molecular weight range of PEG, the G25 retains the polymer, while the exosomes are excluded in the void volume (https://www.systembio.com/downloads/Remove-EXOQ-polymers.pdf).
21. Exosomes are might be difficult to lyse and require harsher conditions than cells—we recommend using 700 μl of Qiazol and proceed to total RNA isolation kits (Qiagen).
22. We recommend to perform protein concentration analysis for the different fraction to validate the purity of the exosomes by (Bradford assay or BCA).
23. Size distribution analysis can be performed using the Nanoparticle Tracking Analysis (NTA) or Tunable Resistive Pulse Sensing (TRPS) [55].

24. The white buffy coat is more viscous than the red blood cells (RBC) pellet. Collect only the white top portion of the bloody pellet without disturbing the pellet itself
25. In cases where the buffy-coat pellet is very large (>2 ml) and very bloody (when a large number of oocyte/follicles were collected or in pathological cases) consider using higher volume of 1×DPBS to dilute the cells pellet. Double the volume by adding two more wash tubes and two more gradient tubes.
26. The GCs are forming a white ring floating at about ¾ of the fluid volume of the tube.
27. Pure DDW can be used to eliminate the residual RBCs—add 100 μl of DDW to the pellet, resuspend well and leave for 3 min in RT. 8 ml of plain media to stop the reaction.

Acknowledgments

The authors would like to thank Arshia Azizeddin and Bahar Beharouzi for technical assistance, and CReATe IVF laboratory staff for help with sample collection.

Conflict of interest: Shlomit Kenigsberg and Brandon A. Wyse were research associates at Create Fertility Centre while conducting this research and writing the manuscript. Clifford L. Librach is the Medical Director and Owner of Create Fertility Centre. The authors have nothing else to declare.

References

1. Fortune JE, Rivera GM, Yang MY (2004) Follicular development: the role of the follicular microenvironment in selection of the dominant follicle. Anim Reprod Sci 82-83:109–126
2. Rodgers RJ, Irving-Rodgers HF (2010) Formation of the ovarian follicular antrum and follicular fluid. Biol Reprod 82(6):1021–1029
3. Revelli A et al (2009) Follicular fluid content and oocyte quality: from single biochemical markers to metabolomics. Reprod Biol Endocrinol 7:40
4. Baka S, Malamitsi-Puchner A (2006) Novel follicular fluid factors influencing oocyte developmental potential in IVF: a review. Reprod Biomed Online 12(4):500–506
5. Ambekar AS et al (2013) Proteomic analysis of human follicular fluid: a new perspective towards understanding folliculogenesis. J Proteome 87:68–77
6. Dunning KR, Russell DL, Robker RL (2014) Lipids and oocyte developmental competence: the role of fatty acids and beta-oxidation. Reproduction 148(1):R15–R27
7. Hsieh M, Zamah AM, Conti M (2009) Epidermal growth factor-like growth factors in the follicular fluid: role in oocyte development and maturation. Semin Reprod Med 27 (1):52–61
8. O'Gorman A et al (2013) Metabolic profiling of human follicular fluid identifies potential biomarkers of oocyte developmental competence. Reproduction 146(4):389–395
9. Leroy JL et al (2004) Metabolite and ionic composition of follicular fluid from different-sized follicles and their relationship to serum concentrations in dairy cows. Anim Reprod Sci 80(3–4):201–211
10. Siu MK, Cheng CY (2012) The blood-follicle barrier (BFB) in disease and in ovarian function. Adv Exp Med Biol 763:186–192
11. Mantzavinos T et al (1993) Immunoglobulins IgG, IgA, IgM, complement C3, C4 and

ferritin and transferrin levels in serum and follicular fluid in IVF patients. Clin Exp Obstet Gynecol 20(1):32–36
12. Vlassov AV et al (2012) Exosomes: current knowledge of their composition, biological functions, and diagnostic and therapeutic potentials. Biochim Biophys Acta 1820 (7):940–948
13. van der Pol E et al (2016) Recent developments in the nomenclature, presence, isolation, detection and clinical impact of extracellular vesicles. J Thromb Haemost 14(1):48–56
14. Witwer KW et al (2013) Standardization of sample collection, isolation and analysis methods in extracellular vesicle research. J Extracell Vesicles 2. doi:10.3402/jev.v2i0.20360
15. Raposo G, Stoorvogel W (2013) Extracellular vesicles: exosomes, microvesicles, and friends. J Cell Biol 200(4):373–383
16. Lasser C et al (2011) Human saliva, plasma and breast milk exosomes contain RNA: uptake by macrophages. J Transl Med 9:9
17. Keller S et al (2011) Body fluid derived exosomes as a novel template for clinical diagnostics. J Transl Med 9:86
18. Revenfeld AL et al (2014) Diagnostic and prognostic potential of extracellular vesicles in peripheral blood. Clin Ther 36(6):830–846
19. Yanez-Mo M et al (2015) Biological properties of extracellular vesicles and their physiological functions. J Extracell Vesicles 4:27066
20. da Silveira JC et al (2012) Cell-secreted vesicles in equine ovarian follicular fluid contain miRNAs and proteins: a possible new form of cell communication within the ovarian follicle. Biol Reprod 86(3):71
21. Sohel MM et al (2013) Exosomal and non-exosomal transport of extra-cellular microRNAs in follicular fluid: implications for bovine oocyte developmental competence. PLoS One 8(11):e78505
22. Santonocito M et al (2014) Molecular characterization of exosomes and their microRNA cargo in human follicular fluid: bioinformatic analysis reveals that exosomal microRNAs control pathways involved in follicular maturation. Fertil Steril 102(6):1751–1761.e1
23. da Silveira JC et al (2014) Regulation of ACVR1 and ID2 by cell-secreted exosomes during follicle maturation in the mare. Reprod Biol Endocrinol 12:44
24. Machtinger R, Laurent LC, Baccarelli AA (2015) Extracellular vesicles: roles in gamete maturation, fertilization and embryo implantation. Hum Reprod Update 22(2): 182–193
25. Di Pietro C (2016) Exosome-mediated communication in the ovarian follicle. J Assist Reprod Genet 33(3):303–311
26. Saez F, Frenette G, Sullivan R (2003) Epididymosomes and prostasomes: their roles in post-testicular maturation of the sperm cells. J Androl 24(2):149–154
27. Sullivan R et al (2005) Role of exosomes in sperm maturation during the transit along the male reproductive tract. Blood Cells Mol Dis 35(1):1–10
28. Hoog JL, Lotvall J (2015) Diversity of extracellular vesicles in human ejaculates revealed by cryo-electron microscopy. J Extracell Vesicles 4:28680
29. Martin-DeLeon PA (2016) Uterosomes: exosomal cargo during the estrus cycle and interaction with sperm. Front Biosci (Schol Ed) 8:115–122
30. Ng YH et al (2013) Endometrial exosomes/microvesicles in the uterine microenvironment: a new paradigm for embryo-endometrial cross talk at implantation. PLoS One 8(3):e58502
31. Al-Dossary AA et al (2015) Oviductosome-sperm membrane interaction in cargo delivery. J Biol Chem 290(29):17710–17723
32. Rienzi L et al (2012) The oocyte. Hum Reprod 27(Suppl 1):i2–21
33. Lane RE et al (2015) Analysis of exosome purification methods using a model liposome system and tunable-resistive pulse sensing. Sci Rep 5:7639
34. Thery C et al (2006) Isolation and characterization of exosomes from cell culture supernatants and biological fluids. Curr Protoc Cell Biol. Chapter 3:Unit 3.22
35. Livshts MA et al (2015) Isolation of exosomes by differential centrifugation: theoretical analysis of a commonly used protocol. Sci Rep 5:17319
36. Taylor DD, Shah S (2015) Methods of isolating extracellular vesicles impact down-stream analyses of their cargoes. Methods 87:3–10
37. Oksvold MP, Neurauter A, Pedersen KW (2015) Magnetic bead-based isolation of exosomes. Methods Mol Biol 1218:465–481
38. Tauro BJ et al (2012) Comparison of ultracentrifugation, density gradient separation, and immunoaffinity capture methods for isolating human colon cancer cell line LIM1863-derived exosomes. Methods 56(2):293–304
39. Yuana Y et al (2015) Handling and storage of human body fluids for analysis of extracellular vesicles. J Extracell Vesicles 4:29260
40. Van Deun J et al (2014) The impact of disparate isolation methods for extracellular vesicles

on downstream RNA profiling. J Extracell Vesicles 3. doi:10.3402/jev.v3.24858

41. Heinzelman P et al (2016) Nanoscale extracellular vesicle analysis in Alzheimer's disease diagnosis and therapy. Int J Alzheimers Dis 2016:8053139
42. Osteikoetxea X et al (2016) Extracellular vesicles in cardiovascular disease: are they Jedi or Sith? J Physiol 594(11):2881–2894
43. Chen L et al (2015) Suppression of fibrogenic signaling in hepatic stellate cells by Twist1-dependent microRNA-214 expression: role of exosomes in horizontal transfer of Twist1. Am J Physiol Gastrointest Liver Physiol 309(6): G491–G499
44. Ghosh A et al (2014) Rapid isolation of extracellular vesicles from cell culture and biological fluids using a synthetic peptide with specific affinity for heat shock proteins. PLoS One 9 (10):e110443
45. Hong CS et al (2016) Isolation of biologically active and morphologically intact exosomes from plasma of patients with cancer. J Extracell Vesicles 5:29289
46. Vogel R et al (2016) A standardized method to determine the concentration of extracellular vesicles using tunable resistive pulse sensing. J Extracell Vesicles 5:31242
47. Pieterse MC et al (1991) Transvaginal ultrasound guided follicular aspiration of bovine oocytes. Theriogenology 35(4):857–862
48. Ludwig AK et al (2006) Perioperative and post-operative complications of transvaginal ultrasound-guided oocyte retrieval: prospective study of >1000 oocyte retrievals. Hum Reprod 21(12):3235–3240
49. Carnevale EM, Maclellan LJ (2006) Collection, evaluation, and use of oocytes in equine assisted reproduction. Vet Clin North Am Equine Pract 22(3):843–856
50. Rider MA, Hurwitz SN, Meckes DG Jr (2016) ExtraPEG: a polyethylene glycol-based method for enrichment of extracellular vesicles. Sci rep 6:23978
51. Weng Y et al (2016) Effective isolation of exosomes with polyethylene glycol from cell culture supernatant for in-depth proteome profiling. Analyst 141(15):4640–4646
52. Baranyai T et al (2015) Isolation of exosomes from blood plasma: qualitative and quantitative comparison of ultracentrifugation and size exclusion chromatography methods. PLoS One 10(12):e0145686
53. Fahiminiya S et al (2011) Proteomic analysis of mare follicular fluid during late follicle development. Proteome Sci 9:54
54. Muller L et al (2014) Isolation of biologically-active exosomes from human plasma. J Immunol Methods 411:55–65
55. van der Pol E et al (2014) Particle size distribution of exosomes and microvesicles determined by transmission electron microscopy, flow cytometry, nanoparticle tracking analysis, and resistive pulse sensing. J Thromb Haemost 12(7):1182–1192

Chapter 27

Isolation of Extracellular Vesicles in Saliva Using Density Gradient Ultracentrifugation

Kazuya Iwai, Satoshi Yamamoto, Mitsutaka Yoshida, and Kiyotaka Shiba

Abstract

This chapter describes a method for isolating human salivary extracellular vesicles (EVs) using density gradient ultracentrifugation. Standard protocols established for isolation of EVs from blood or a conditioned medium of cultured cells do not work for whole saliva, due to its viscosity. Therefore, procedures including a pretreatment step and utilizing iodixanol as a gradient material enable EVs to be concentrated to a 1.1 g/ml density. This protocol is compatible with both swing and angle rotors. By employing an angle rotor, which enables high g-force, the centrifugation time was reduced to 4 h from the 17 h required when using a swing rotor.

Key words Exosome, Extracellular vesicles, EVs, Iodixanol, Angle rotor, Whole saliva

1 Introduction

Sources of extracellular vesicles (EVs) for diagnostic purposes are not limited to blood and include saliva [1], urine [1], mammary secretions (milk) [2], prostatic fluid [3], seminal fluid [1], cerebrospinal fluid [4], ascites [5], pleural effusion [6], bronchoalveolar fluid [7], aqueous humor [8] and even exhaled breath [9]. Among these sources, saliva is obtained in a noninvasive way, and whole saliva can be obtained in a normal, ambient environment, without the use of a syringe or the help of medical professionals [10, 11]. Information provided by salivary EVs is used to diagnose both oral and systemic diseases, making saliva a commonly researched source for EVs.

During the developmental stage of diagnostic systems, EVs should be collected in the purist form possible. Although many procedures have been proposed to purify EVs through simple manipulations [12–14], isolation by density gradient centrifugation

Winston Patrick Kuo and Shidong Jia (eds.), *Extracellular Vesicles: Methods and Protocols*, Methods in Molecular Biology, vol. 1660, DOI 10.1007/978-1-4939-7253-1_27, © Springer Science+Business Media LLC 2017

is the gold standard method because it fractionates molecules based on density [15]. Because EVs or membranous vesicles have low densities compared to other biomolecules, such as proteins and nucleic acids, EVs are fractioned to approximately a 1.1 g/ml density during density gradient centrifugation [15]. This method is used to obtain highly enriched EVs from blood or conditioned mediums of cultured cells; however, simple application of this protocol to saliva does not work as described in the previous study [16]. During this study, when crude EVs were prepared from saliva using sucrose-based density gradient centrifugation, maker proteins for EVs were not recovered from the fraction with a density of 1.1 g/ml, which could be due to the viscous nature of the crude EV fractions prepared from saliva.

In the study, a modified protocol for gradient-based isolation of salivary EVs was established [16]. This protocol includes a pretreatment step and iodixanol as the gradient material and uses centrifugation with an angle rotor, instead of a swing rotor, to retain the linear gradient after a run (Fig. 1). By employing an angle rotor, the centrifugation time was reduced to 4 h from the 17 h required when using a swing rotor.

2 Materials

2.1 Chemicals

1. Phosphate buffer saline (PBS): 137 mM of NaCl, 2.68 mM of KCl, 8.10 mM of Na_2HPO_4, and 1.47 mM of KH_2PO_4, pH 7.4.
2. A stock solution of 1 M 4-(2-hydroxyethyl)-1-piperazine ethanesulfonic acid (HEPES) with a pH of 7.2, adjusted using NaOH.
3. A stock solution of 47% (W/V) iodixanol (OptiPrep™) (*see* **Note 1**) in 20 mM of HEPS and 37%, 28%, and 18% (W/V) iodixanol solutions in 20 mM of HEPES.

2.2 Plastic Tubes

1. 50 ml plastic tube (*see* **Note 2**).
2. 10 ml plastic tube for sonication (*see* **Note 3**).
3. 4.7 ml polypropylene tube, 13 × 48 mm (OptiSeal™, Beckman).
4. 3.2 ml polycarbonate thick-walled tube, 13 × 56 mm (Beckman).
5. 1.8 ml low-binding plastic tube (*see* **Note 4**).

2.3 Centrifuges and Rotors

1. High-speed centrifuge (*see* **Note 5**).
2. Ultrahigh-speed centrifuge with a TLA-110 rotor (Beckman).

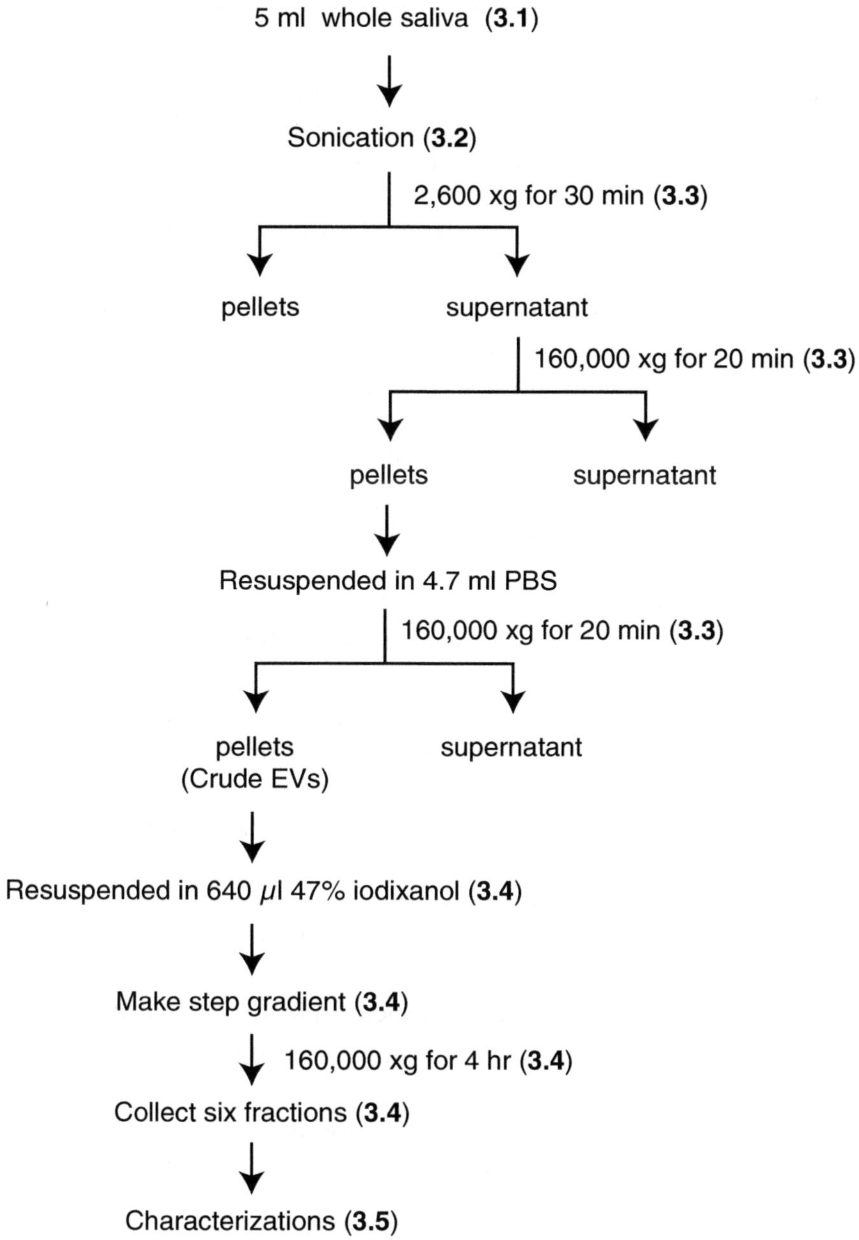

Fig. 1 Flowchart for the EV isolation procedure

2.4 Other Instruments

1. Closed-type sonication system (*see* **Note 6**).
2. Refractometer (*see* **Note 7**).
3. Nanoparticle tracking analyzer (optional) (*see* **Note 8**).
4. Ubbelohde viscometer (optional) (*see* **Note 9**).

3 Methods

3.1 Saliva Collection

1. Five milliliters of whole saliva is collected in a consistent manner and placed in 50 ml plastic tubes by spitting. The samples are kept on ice (*see* **Note 10**).

3.2 Pretreatment (See Note 11)

1. The 5 ml of saliva are transferred into three 15 ml hard plastic tubes (*see* **Note 3**).
2. The tubes are set in a tube holder for use with the Bioruptor®.
3. The tube holder is set in a Bioruptor®, which had been precooled using a cycling cooling device or ice to 4 °C.
4. The power selection is set to M, and a timer is set to the option "30 s on and 60 s off."
5. The tubes are sonicated for 10 min (seven on–off cycles), and a power of three is ensured by adjusting the tune dial.

3.3 Preparation of Crude EVs

1. The sonicated saliva is collected in a 50 ml tube.
2. The tubes are centrifuged at 2600 × *g* for 30 min at 4 °C (*see* **Note 5**).
3. The supernatant is transferred using a micropipette with a gel-loading round tip into a 4.7 ml polypropylene tube until the tube is full.
4. The supernatant is centrifuged at 160,000 × *g* for 20 min at 4 °C (*see* **Note 12**).
5. The supernatant is removed using a micropipette with a gel-loading round tip to ensure the pelleted crude EVs would not be sucked in.
6. The palette is resuspended in 4.7 ml of PBS by pipetting.
7. The palette is centrifuged at 160,000 × *g* for 20 min at 4 °C (*see* **Note 12**).
8. The supernatant is removed using a micropipette with a gel-loading round tip to ensure the pelleted crude EVs would not be sucked in.

3.4 Fractionation of Crude EVs Based on Densities (See Note 13)

1. The upper portion of the 4.7 ml tube is cut to perform a resuspension.
2. The crude EVs are resuspended in 640 μl of a 47% iodixanol solution (*see* **Note 14**).
3. Then, 640 μl of the 47% iodixanol solution is added to the bottom of the 3.2 ml polycarbonate thick-walled tube (Fig. 2).
4. The 640 μl solution of crude EVs prepared in **step 1** is overlaid (Fig. 2).

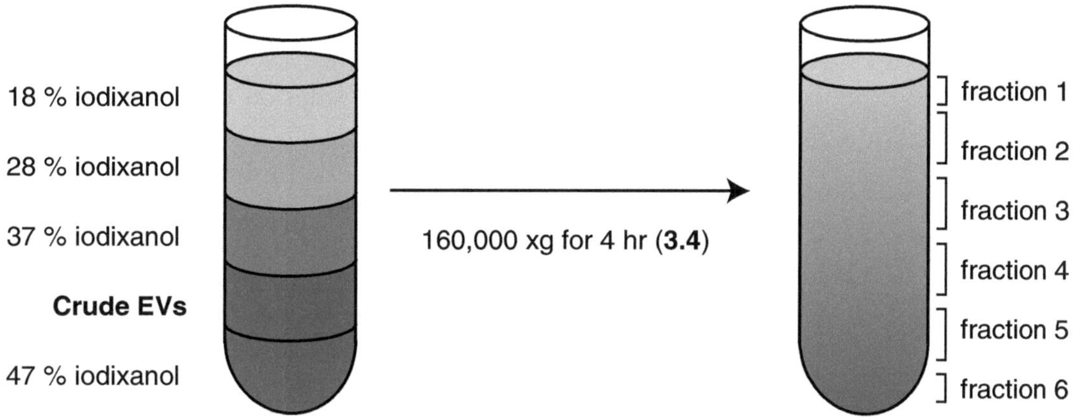

Fig. 2 Step gradients of iodixanol solutions and fractions after centrifugation

5. Six-hundred and forty microliters of the 37% iodixanol solution is overlaid (Fig. 2).
6. Six-hundred and forty microliters of the 28% iodixanol solution is overlaid (Fig. 2).
7. Six-hundred and forty microliters of the 18% iodixanol solution is overlaid (Fig. 2).
8. Centrifuged at 160,000 × *g* for 4 h at 4 °C.
9. From the top fraction, 320 μl is extracted and placed in a 1.8 ml low-binding plastic tube (fraction 1) (Fig. 2).
10. Then, 640 μl × 4 of continuous fractions are transferred into four 1.8 ml low-binding plastic tubes (fractions 2–5) (Fig. 2).
11. Finally, 320 μl of the last fraction is transferred into a 1.8 ml low-binding plastic tube (fraction 6) (Fig. 2).
12. The densities of the six fractions are determined using a refractometer (*see* **Note 7**).
13. The pellets are washed with PBS, if necessary (*see* **Note 15**).

3.5 NTA Analysis [17] (See Note 8)

1. Each fraction is diluted with PBS to ensure the concentrations of particles would fall between 2 and 10 × 10^8 particles/ml.
2. Videos are taken for NTA analyses (*see* **Note 16**).
3. The densities of the samples after dilution are determined.
4. The viscosities of the iodixanol solutions are measured, and the densities corresponded to the calculated values in 4 using an Ubbelohde viscometer (*see* **Note 17**).
5. The number and size of particles are analyzed using the NTA method and the viscosity values determined in 4.

4 Notes

1. A sterile, endotoxin tested 60% solution of OptiPrep™ produced by Axis-Shield was used.
2. Any type of plastic tube should work. In this study, 50 Cellstar® tubes (Greiner Bio-One) were used.
3. Hard plastic tubes should be used to maximize sonication efficiency. This study used a 15 ml TPX® polymethylpentene tube (Diagenode Inc.). The manufacturer recommends treating samples less than 2 ml in the 15 ml tube to maximize sonication efficiency.
4. Once the EVs were purified, they became very sticky and easily adsorbed by the surface of the plastic tubes. Therefore, Eppendorf-type tubes that have reduced protein binding should be used.
5. Any type of centrifuge should work. This study used model 5500 by Kubota.
6. Bioruptor® (Diagenode Inc.) was used.
7. RX-5000α by Atago Co. Ltd. was used.
8. This study routinely used the NanoSight LM10 (Malvern Instruments Ltd). Other types of nanoparticle analyzers should work. For EV characterization, a resistive pulse-sensing analyzer (qNano, iZON Science) was used; however, to measure resistive pulse sensing during the final step of the purification process, EVs should be washed with PBS to remove iodixanol.
9. When the fractionated EVs are quantified using the nanotracking analysis (NTS) method (NanoSight) without washing, precise viscosity values of the samples must be entered.
10. In this published study, test subjects were asked not to eat, drink, or perform strenuous physical exercise for at least 1 h prior to saliva collection. Saliva samples were collected between 9:00 a.m. and 11:00 a.m. These sampling conditions can be changed.
11. Many filtrating treatments also work as pretreatments. This study used a 5 μm pore metal filter (13 mm in diameter, Optnics Precision Co. Ltd.).
12. A combination of a centrifuge, Optima MAX-TL, and a rotor, TLA-110, were used. Other combinations should work; however, g values and time should be determined by considering the maximum speed of rotation and the k-factor of the rotor [18].
13. In an alternative protocol, EVs from 15 ml samples of whole saliva were purified using a swing rotor and a continuous gradient of iodixanol.

14. The usage of iodixanol is strongly recommended. In the conditions described here, sucrose and iodixanol (Nycodenz™) do not give robust results for salivary EVs.
15. Removal of residual iodixanol is preferable for further characterization. Residual iodixanol was found to affect the results of the NTA analyses and protein quantification (MCA method). However, as noted in **Note 4**, purified EVs stick to the surfaces of plastic tubes, and the washing step removes nonnegligible amounts of EVs, especially at the 5 ml scale of purification. For western blotting analyses or RNA analyses, this washing step can be omitted.
16. When using NanoSight LM10 (NTA 2.3 software), a camera level of 14 and a threshold of four are routinely used. Videos are recorded for 30 s at least five times. In each capture, EV samples within measured cells should be renewed by pushing the syringe. This study used 100 nm silica microspheres (Polysciences, Inc.) for calibration [19].
17. Measured using a glass Ubbelohde viscometer (SIBATA) at an ambient temperature, according to the manufacturer's instructions ($n = 3$).

Acknowledgments

This work was supported by the "Project for Private Universities: Matching Fund Subsidy" from the Ministry of Education, Culture, Sports, Science, and Technology and by a grant from the Vehicle Racing Commemorative Foundation.

References

1. Marzesco AM, Janich P, Wilsch-Brauninger M, Dubreuil V, Langenfeld K, Corbeil D, Huttner WB (2005) Release of extracellular membrane particles carrying the stem cell marker prominin-1 (CD133) from neural progenitors and other epithelial cells. J Cell Sci 118:2849–2858
2. Zonneveld MI, Brisson AR, van Herwijnen MJ, Tan S, van de Lest CH, Redegeld FA, Garssen J, Wauben MH, Nolte-'t Hoen EN (2014) Recovery of extracellular vesicles from human breast milk is influenced by sample collection and vesicle isolation procedures. J Extracell Vesicles 3:24215
3. Ronquist G, Hedstrom M (1977) Restoration of detergent-inactivated adenosine triphosphatase activity of human prostatic fluid with concanavalin A. Biochim Biophys Acta 483:483–486
4. Chiasserini D, van Weering JR, Piersma SR, Pham TV, Malekzadeh A, Teunissen CE, de Wit H, Jimenez CR (2014) Proteomic analysis of cerebrospinal fluid extracellular vesicles: a comprehensive dataset. J Proteome 106C:191–204
5. Press JZ, Reyes M, Pitteri SJ, Pennil C, Garcia R, Goff BA, Hanash SM, Swisher EM (2012) Microparticles from ovarian carcinomas are shed into ascites and promote cell migration. Int J Gynecol Cancer 22:546–552
6. Roca E, Lacroix R, Judicone C, Laroumagne S, Robert S, Cointe S, Muller A, Kaspi E, Roll P, Brisson AR, Tantucci C, Astoul P, Dignat-George F (2016) Detection of EpCAM-positive microparticles in pleural fluid: a new approach to mini-invasively identify patients with malignant pleural effusions. Oncotarget 7:3357–3366

7. Torregrosa Paredes P, Esser J, Admyre C, Nord M, Rahman QK, Lukic A, Radmark O, Gronneberg R, Grunewald J, Eklund A, Scheynius A, Gabrielsson S (2012) Bronchoalveolar lavage fluid exosomes contribute to cytokine and leukotriene production in allergic asthma. Allergy 67:911–919
8. Kang GY, Bang JY, Choi AJ, Yoon J, Lee WC, Choi S, Yoon S, Kim HC, Baek JH, Park HS, Lim HJ, Chung H (2014) Exosomal proteins in the aqueous humor as novel biomarkers in patients with neovascular age-related macular degeneration. J Proteome Res 13:581–595
9. Sinha A, Yadav AK, Chakraborty S, Kabra SK, Lodha R, Kumar M, Kulshreshtha A, Sethi T, Pandey R, Malik G, Laddha S, Mukhopadhyay A, Dash D, Ghosh B, Agrawal A (2013) Exosome-enclosed microRNAs in exhaled breath hold potential for biomarker discovery in patients with pulmonary diseases. J Allergy Clin Immunol 132:219–222
10. Wong DT (2006) Salivary diagnostics powered by nanotechnologies, proteomics and genomics. J Am Dent Assoc 137:313–321
11. Pfaffe T, Cooper-White J, Beyerlein P, Kostner K, Punyadeera C (2011) Diagnostic potential of saliva: current state and future applications. Clin Chem 57:675–687
12. Ogawa Y, Taketomi Y, Murakami M, Tsujimoto M, Yanoshita R (2013) Small RNA transcriptomes of two types of exosomes in human whole saliva determined by next generation sequencing. Biol Pharm Bull 36:66–75
13. Zlotogorski-Hurvitz A, Dayan D, Chaushu G, Korvala J, Salo T, Sormunen R, Vered M (2015) Human saliva-derived exosomes: comparing methods of isolation. J Histochem Cytochem 63:181–189
14. Yang J, Wei F, Schafer C, Wong DT (2014) Detection of tumor cell-specific mRNA and protein in exosome-like microvesicles from blood and saliva. PLoS One 9:e110641
15. Thery C, Amigorena S, Raposo G, Clayton A (2006) Isolation and characterization of exosomes from cell culture supernatants and biological fluids. Curr Protoc Cell Biol. Chapter 3:Unit 3.22
16. Iwai K, Minamisawa T, Suga K, Yajima Y, Shiba K (2016) Isolation of human salivary extracellular vesicles by iodixanol density gradient ultracentrifugation and their characterizations. J Extracell Vesicles 5:30829
17. Carr B, Hole P, Malloy A, Nelson P, Wright M, Smith J (2009) Applications of nanoparticle tracking analysis in nanoparticle research – a mini-review. Eur J Parenter Pharm Sci 14:45–50
18. Livshts MA, Khomyakova E, Evtushenko EG, Lazarev VN, Kulemin NA, Semina SE, Generozov EV, Govorun VM (2015) Isolation of exosomes by differential centrifugation: theoretical analysis of a commonly used protocol. Sci Rep 5:17319
19. Gardiner C, Ferreira YJ, Dragovic RA, Redman CW, Sargent IL (2013) Extracellular vesicle sizing and enumeration by nanoparticle tracking analysis. J Extracell Vesicles 2:19671

Chapter 28

Isolation of Extracellular Vesicles from Breast Milk

Xin Wang

Abstract

Isolation of extracellular vesicles (EVs) from breast milk enables the downstream analysis of the contents, including various proteins, lipids, and RNAs (messenger RNAs and microRNAs). Here, we present the method of centrifugation with ExoQuick™ precipitation to isolate EVs from breast milk. It has been proven to be a rapid and effective alternative compared to conventional differential centrifugation method.

Key words Breast milk, Extracellular vesicles, Centrifugation, ExoQuick™ precipitation

1 Introduction

Extracellular vesicles (EVs) have been identified in human breast milk [1, 2] as well as in bovine and porcine milk [3, 4]. EVs in breast milk serve as bioactive cargo to transmit proteins, lipids and genetic information from mother to infant, and are essential for the development of infant's immune system [2]. The gold standard and most commonly used method for EVs isolation from breast milk is differential centrifugation, which consists of multiple steps of spins. The limitations of this method include its lengthy (4–5 h), its requirement of high speed ultracentrifuge and its inconsistent recovery of EVs [5]. Recently, several alternative methods for EVs isolation from breast milk have been emerging, including sucrose gradient centrifugation, microfiltration technologies, antibody-coated magnetic beads, microfluidic devices, and ExoQuick™ [5]. ExoQuick™ is a proprietary polymer that gently precipitates EVs, and it has been demonstrated as a rapid and effective alternative method to gold standard [6].

Winston Patrick Kuo and Shidong Jia (eds.), *Extracellular Vesicles: Methods and Protocols*, Methods in Molecular Biology, vol. 1660, DOI 10.1007/978-1-4939-7253-1_28, © Springer Science+Business Media LLC 2017

2 Materials

1. Sterile 50 ml tubes.
2. Styrofoam box and heating package.
3. Benchtop centrifuge.
4. 1.5 ml tubes.
5. 0.45 μm polyvinylidene difluoride (PVDF) filter.
6. ExoQuick™ Exosome Precipitation Solution (catalog # EXOQ5A-1, SBI, CA, USA).
7. Nuclease-free water or PBS.

3 Methods

3.1 Breast Milk Collection

1. Collect breast milk samples (20–50 ml) into sterile tubes.
2. Transport milk to the laboratory as soon as collection finishes. Use Styrofoam box and heating package during the transportation to prevent milk from cooling down.
3. Store samples at −80 °C.

3.2 Sample Preparation

1. Centrifuge aliquot of raw milk (1.5 ml) at 2000 × *g* for 10 min (*see* **Note 1**).
2. Remove fat globules (upper layer).
3. Centrifuge supernatant at 12,000 × *g* for 30 min and discard the cellular debris at the bottom of the tube (*see* **Note 2**).
4. Further filter through a 0.45 μm PVDF filter to eliminate cells and cellular debris.

3.3 EV Isolation

1. Mix 1 ml of supernatant with 500 μl of ExoQuick™ Exosome Precipitation Solution (SBI, CA, USA).
2. Incubate at 4 °C for 12 h for fully precipitation.
3. Centrifuge the ExoQuick™–supernatant mixture at 5000 × *g* for 30 min to obtain a beige EV pellet at the bottom of the tube.
4. Discard supernatant by careful aspiration with a pipette.
5. Resuspend pellet in 250 μl nuclease-free water or PBS.
6. Isolated EVs are ready for downstream analysis. Isolated EVs can be kept at 4 °C for up to 1 week, or at −20 °C or lower for long-term storage.

4 Notes

1. Storage of breast milk in refrigerator (4 °C) or freezer (−20 or −80 °C) induces cell death and causes contamination to the composition of milk EVs [6]. Therefore, it's recommended to prepare milk sample right after collection. Keep only the milk supernatant (called whey or milk serum), which is free of fat globules and cell debris, before freezing. Data indicates freezing only the milk whey leads to more efficient recovery and more reliable analysis of EVs compared to freezing the raw milk [6].
2. Sucrose gradient centrifugation method is more suitable for purification of EVs with native morphology intact and higher purity [7].

References

1. Kosaka N, Izumi H, Sekine K, Ochiya T (2010) MicroRNA as a new immune-regulatory agent in breast milk. Silence 1(1):7
2. Zhou Q, Li M, Wang X, Li Q, Wang T, Zhu Q et al (2011) Immune-related microRNAs are abundant in breast milk exosomes. Int J Biol Sci 8(1):118–123
3. Izumi H, Tsuda M, Sato Y, Kosaka N, Ochiya T, Iwamoto H et al (2015) Bovine milk exosomes contain microRNA and mRNA and are taken up by human macrophages. J Dairy Sci 98 (5):2920–2933
4. Gu Y, Li M, Wang T, Liang Y, Zhong Z, Wang X et al (2012) Lactation-related microRNA expression profiles of porcine breast milk exosomes. PLoS One 7(8):e43691
5. Momen-Heravi F, Balaj L, Alian S, Mantel PY, Halleck AE, Trachtenberg AJ et al (2013) Current methods for the isolation of extracellular vesicles. Biol Chem 394(10):1253–1262
6. Zonneveld MI, Brisson AR, van Herwijnen MJ, Tan S, van de Lest CH, Redegeld FA et al (2014) Recovery of extracellular vesicles from human breast milk is influenced by sample collection and vesicle isolation procedures. J Extracell Vesicles 3:24215
7. Yamada T, Inoshima Y, Matsuda T, Ishiguro N (2012) Comparison of methods for isolating exosomes from bovine milk. J Vet Med Sci 74 (11):1523–1525

Chapter 29

An Integrated Double-Filtration Microfluidic Device for Detection of Extracellular Vesicles from Urine for Bladder Cancer Diagnosis

Li-Guo Liang, Ye-Feng Sheng, Sherry Zhou, Fatih Inci, Lanjuan Li, Utkan Demirci, and ShuQi Wang

Abstract

Extracellular vesicles (EVs) are present in a variety of bodily fluids and they play an important role in cellular communications and signal transduction mechanisms. Studies have shown that the number of EVs and EV-associated biomarkers (i.e., proteins, nucleic acids and lipids) can be used to aid clinical diagnosis. Although ultracentrifugation is commonly used for EV isolation, it is not practical for clinical settings. Here, we developed an integrated double-filtration device that isolated and enriched EVs from urine, and subsequently detected/quantified EVs from urine via microchip ELISA. Results showed that the concentration of EVs was significantly elevated compared to healthy controls. Receiver operating characteristic analysis demonstrated that this integrated EV quantification device had a sensitivity of 81.3% at a specificity of 90% (16 bladder cancer patients and eight healthy controls). Thus, this integrated device shows great potential to supplement urine cytology for diagnosis of bladder cancer in point-of-care (POC) settings.

Key words Filtration, Extracellular vesicles (EVs), Microchip ELISA, Bladder cancer, Diagnostics

1 Introduction

Bladder cancer is the top two malignancies of the genitourinary system, with an annual incidence of 20.1 per 100,000 men and women in the US [1]. The 5-year survival rate of in situ and localized bladder cancer is approximately 77.5%, whereas it drops to 34.5–5.2% for regional and distant bladder cancer [2]. Since the majority of bladder cancer patients (~80–90%) experience painless hematuria, or additionally with urination discomfort, bladder cancer is often underdiagnosed, potentially leading to development of muscle-invasive tumor. Thus, high-risk populations should be routinely screened for bladder cancer to avoid serious clinical outcomes such as radical cystoprostatectomy. However, cytology, cystoscopy and other diagnostic approaches such as BTA stat and BTA TRAK

Winston Patrick Kuo and Shidong Jia (eds.), *Extracellular Vesicles: Methods and Protocols*, Methods in Molecular Biology, vol. 1660, DOI 10.1007/978-1-4939-7253-1_29,

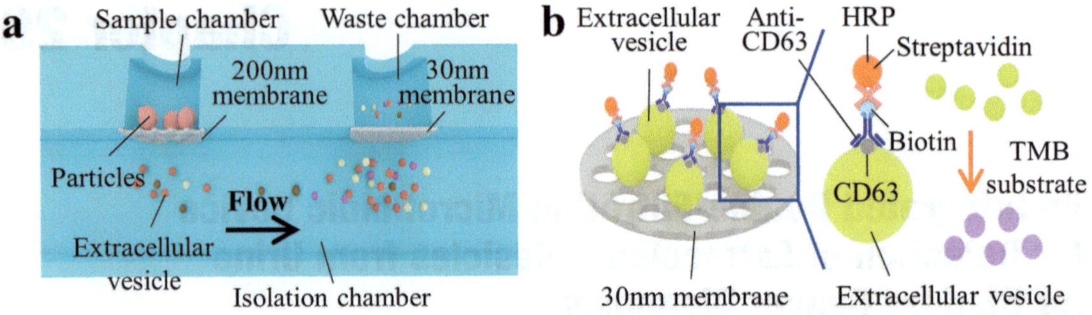

Fig. 1 Working principle of an integrated EV detection device. (**a**) Illustration of the double-filtration process for isolation and enrichment of EVs from urine. (**b**) Illustration of microchip-ELISA for detection/quantification of EVs

tests, ImmunoCyt and UroVysion cannot meet the requirement for early diagnosis or screening of bladder cancer [3, 4].

EVs are microvesicles derived from cells with a diameter ranging approximately 40–100 nm [5]. Studies have shown that EVs play an important role in cellular communication, especially for cancer metastasis [6, 7], and that the number of EVs and EV-associated biomarkers such as proteins, nucleic acids and lipids can be used to aid diagnosis of cancer [8, 9]. Although ultracentrifugation can be used to isolate EVs from biological samples, the requirement for a prohibitive ultracentrifuge and its time-consuming process constitute a major bottle for clinical translation of EV-related diagnostics [10–12]. Here, we present an integrated double-filtration device that can isolate, enrich, and quantify EVs from urine via microchip ELISA for differentiating bladder cancer patients from healthy controls (Fig. 1). This double-filtration microfluidic device eliminates the need for an expensive, bulky ultracentrifuge for EV-related research and potentially clinical diagnosis in a less-equipped laboratory. Since simple, rapid and inexpensive diagnostics are preferably needed for resource-limited settings [13, 14], this integrated EV diagnostic platform holds great potential for POC testing.

2 Materials

All solutions are prepared and stored as recommended by manufacturers.

2.1 Cell Culture

1. Roswell Park Memorial Institute (RPMI) 1640 basal media.
2. 10% heat-inactivated fetal bovine serum (FBS).
3. 100 U/mL penicillin.
4. 100 μg/mL streptomycin.

5. Hanks' Balanced Salt Solution, With Ca^{2+} and Mg^{2+}(137.93 mM NaCl, 5.33 mM KCl, 4.17 mM $NaHCO_3$, 1.26 mM $CaCl_2$, 0.493 mM $MgCl_2$, 0.407 mM $MgSO_4$, 0.441 mM KH_2PO_4, 0.338 mM Na_2HPO_4 and 5.56 mM D-glucose).
6. Trypsin–EDTA Solution (0.25% trypsin, 0.02% EDTA and 0.02% Phenol Red in 0.01 M phosphate buffer saline, PBS).
7. PBS, pH 7.0 (Sangon Biotech, Shanghai, China).

2.2 Urine Samples

1. Urine samples are collected from bladder cancer patients ($n = 16$) and healthy donors ($n = 8$).

2.3 Devices

1. Poly(methyl methacrylate) (PMMA) (3M Company, St. Paul, MN).
2. Double-sided adhesive (DSA).
3. Nuclepore track-etched membrane (GE Healthcare Life Science, Shanghai, China).
4. Laser cutter (VSL 2.30, Universal Laser System Inc., Scottsdale, AZ).
5. Micropump (TS-2A, Longer, Baoding, China).
6. Epoxy adhesive (DP 460NS, 3M, Shanghai, China).

2.4 ELISA

1. Biotinylated anti-CD63 (1 mg/mL) (Abcam, Cambridge, MA).
2. Streptavidin-labeled horseradish peroxidase (HRP) (1 mg/mL) (Abcam, Cambridge, MA).
3. Alexa Fluor 488-labeled anti-CD9 (1 mg/mL) (Abcam, Cambridge, MA).
4. 100 and 500 nm fluorescent particles (Ocean Nanotech, Springdale, CA).
5. Tubing Φ 0.8 mm $\times$ 1.85 mm (Longer, Baoding, China).
6. Tetramethyl benzidine (TMB) Chromogenic Reagent (Sangon Biotech, Shanghai, China).
7. 0.22 μm filtering device (Sangon Biotech, Shanghai, China).
8. PBS (pH 7.0) (Sangon Biotech, Shanghai, China).
9. BCA kit (Sangon Biotech, Shanghai, China).
10. Nuclease-free H_2O (Sangon Biotech, Shanghai, China).
11. Fluorescence microscope (Leica DM4000, German).
12. Scanning electron microscope (SEM) (SU-8010, Hitachi, Tokyo, Japan).
13. 96-well ELISA plate (Coring, Shanghai, China).
14. Spectrophotometer (Molecular Devices, San Francesco, CA).

3 Methods

3.1 Device Fabrication

1. Four layers of PMMA with a thickness of 2, 1, 1 and 2 mm, are excised to dimensions of 20 × 40 × 6 mm (Fig. 2).
2. Two circular openings with a diameter of 1.8 mm are excised on the on the first PMMA (2 mm in thickness) layer to form an inlet and outlet (*see* **Note 1**).
3. Two circular openings with a diameter of 10 mm are excised on the second PMMA layer (1 mm in thickness) to form a sample chamber and waste chamber.
4. The third PMMA layer (1 mm in thickness) had two openings with a diameter of 10 mm, which are connected through a microchannel with a width of 1.5 mm.
5. The fourth PMMA layer (2 mm in thickness), which had no openings, is used to form the base of the device.
6. Two Whatman membranes with a pore size of 200 and 30 nm in diameter are included between the second and third PMMA layer.
7. Four layers of DSA (50 μm in thickness) with corresponding configurations are excised using a laser cutter to assemble the double-filtration device using a non-photolithography method [15] (*see* **Note 2**).
8. The formed double-filtration device had two circular chambers (78.5 μL) above the membranes and two circular chambers (78.5 μL) below (*see* **Note 3**).

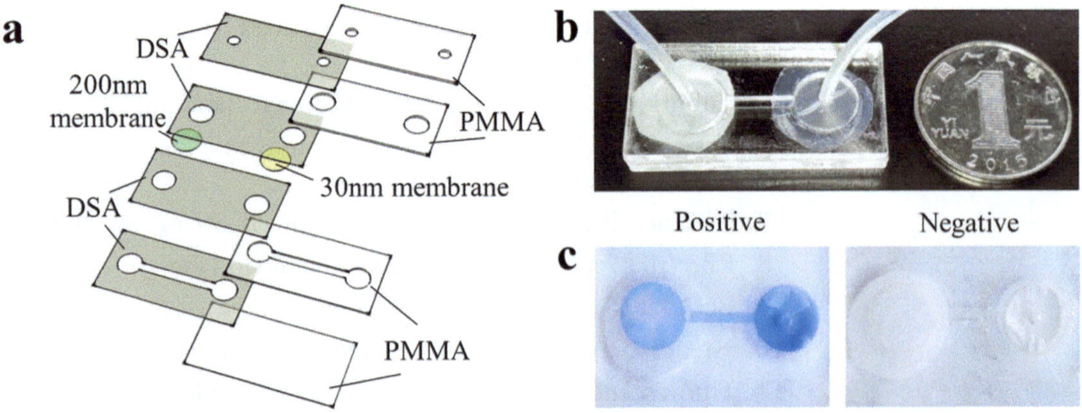

Fig. 2 Assemble of an integrated EV detection device. (**a**) Layer-by-layer structure of the device. (**b**) Actual image of the device. (**c**) Images of the device after EV detection using microchip ELISA

3.2 Cell Culture

1. Human bladder cancer cell line T24 is cultured in RPMI 1640 medium supplemented with 10% heat-inactivated FBS, 100 U/mL penicillin, and 100 μg/mL streptomycin at 37 °C in a humid atmosphere with 5% CO_2 (*see* **Note 4**).
2. T24 cells are harvested on day 3, and the number of cells is counted using a hemocytometer.
3. The supernatant is collected after centrifuging 100 mL of T24 cell culture with a cell density of 10^6 cells/mL at room temperature for 15 min at 2000 × *g* (*see* **Note 5**).

3.3 EV Isolation Using Ultracentrifugation

3.3.1 EVs Isolation from Cell Medium

1. 100 mL of cell culture media with a cell density of 10^6 cells/mL is harvested after T24 cell culture for 4 days.
2. 100 mL of cell culture media is centrifuged at 2000 × *g* for 15 min.
3. To pellet EVs, the supernatant is centrifuged at 100,000 × *g* at 4 °C for 70 min in an SW32 Ti rotor.
4. The supernatant is discarded carefully.
5. 200 μL of PBS is used to rinse the pellets gently and then discarded with caution.
6. Crude EV-containing pellets are suspended in 1 mL of PBS.
7. Isolated EVs are kept at −80 °C for storage until use.

3.3.2 EVs Isolation from Urine

1. 10 mL of urine is centrifuged at 2000 × *g* at room temperature for 15 min to remove cells (*see* **Note 5**).
2. The supernatant is carefully removed.
3. The supernatant is centrifuged at 100,000 × *g* at 4 °C for 70 min in an SW32 Ti rotor.
4. Isolated EVs are kept at −80 °C for storage until use.

3.4 EV Isolation Using Double Filtration

1. Urine samples or T24 culture are centrifuged at 2000 × *g* for 15 min to remove cells and cell debris (*see* **Note 5**).
2. 8 mL of urine sample or cell medium is continuously injected into a double-filtration device at a flow rate of 40 μL/min (Fig. 3a).
3. 400 μL of PBS is injected into the device at a flow rate of 40 μL/min for three times.

3.5 Microplate ELISA

1. EVs isolated from T24 cell cultures by ultracentrifugation are serially diluted to 10^6, 3.3×10^5, 1.1×10^5, 3.7×10^4 and 1.23×10^4 AU.
2. 100 μL of EV suspension is added to each well of a 96-well ELISA plate and incubated 4 °C for 2 h.
3. 100 of μL BSA (1%) is added to the 96-well ELISA plate and incubated for 1 h at 37 °C (*see* **Note 6**).

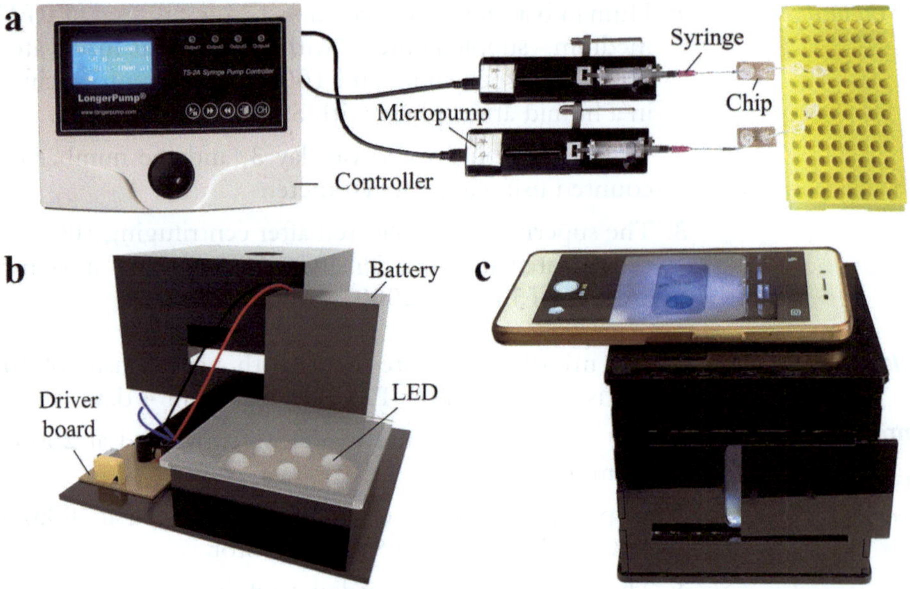

Fig. 3 Experimental setup for isolation, enrichment, and quantification of EVs. (**a**) Image of device operation with the aid of a micropump. Samples and reagents were flowed into the double-filtration device at a flow rate of 40 μL/min. (**b**) Schematic of the detection setup for cell phone imaging. The battery-operated imaging setup was illuminated using an LED. (**c**) Actual image of the cell phone-based detection setup

4. 200 μL of PBS (pH 7.0) is used to wash the plate for three times.
5. 100 μL of biotinylated anti-CD63 antibody (1:200) is added to the plate and incubated at 37 °C for 1 h (*see* **Note 6**).
6. The plate is washed with 200 μL PBS per well (pH 7.0) for three times.
7. 100 μL of streptavidin-labeled HRP (1:2000, 100 μL per well) is added and incubated for 1 h at 37 °C (*see* **Note 6**).
8. The plate is washed using 200 μL of PBS for three times.
9. 100 μL of TMB is added into each well and incubated at 37 °C in darkness (*see* **Note 6**).
10. 50 μL of stopping solution is then added to each well to terminate color development.
11. The color intensity of each well is measured using a spectrophotometer at a wavelength of 450 nm.

3.6 Microchip ELISA

1. EVs are isolated from urine using the double-filtration device as described in Subheading 3.4.
2. 300 of μL of biotinylated anti-CD63 antibody (1:200) is injected into the device at a flow rate of 40 μL/min and then incubated at 25 °C for 1 h (Fig. 3a).

3. 300 μL of PBS is injected three times to the device at a flow rate of 40 μL/min.
4. 500 μL of air is injected once to completely remove PBS (*see* **Note 7**).
5. 300 μL of streptavidin-labeled HRP (1:2000) is injected to the device at a flow rate of 40 μL/min and then incubated at 37 °C for 1 h (*see* **Note 8**).
6. The device is washed using PBS and air as **steps 3** and **4** as described above.
7. 300 μL of TMB is injected into the device and then incubated at 37 °C for 10 min in darkness (*see* **Notes 8–10**).
8. Color development is imaged using a smartphone (OPPO R7, Guangdong, China) in an in-house imaging box (Fig. 3b and c) (*see* **Notes 11** and **12**).
9. Cellphone images are then transferred to a laptop for image analysis using ImageJ.
10. R channel values of blue color is extracted and used to plot the standard curve at the function of EV concentrations, as previously reported [16, 17].

4 Notes

1. The power of the laser cutter used for cutting PMMA is 75% at a speed of 10%.
2. The power of the laser cutter used for cutting DSA is 10% at a speed of 10%.
3. The surroundings of the double-filtration device are sealed using Epoxy adhesive to avoid leakage or evaporation.
4. A commercial 0.22 μm filter device is used to remove potential bacterial contamination in culture media.
5. Urine samples and cell culture are centrifuged at a speed of 2000 × *g* for 15 minutes to remove cells and cell debris.
6. The 96-well plate is covered using Parafilm.
7. Air wash is performed following PBS washing.
8. Devices are placed in a wet box during incubation steps at 37 °C wherever applied to avoid evaporation.
9. Foil is used to cover the double-filtration device during the injection and incubation of streptavidin-labeled HRP and TMB.

10. TMB solution is warmed up to room temperature before use.
11. The double-filtration device with developed blue color is imaged immediately after 10-min incubation of TMB to avoid signal saturation.
12. Blue color developed in double-filtration devices is imaged in a homemade black box with predefined lightening conditions to avoid the interference with ambient lights. EVs isolated from T24 cell culture by ultracentrifugation.

In this study, we present an integrated device that can isolate, enrich, and quantify EV from urine samples without referring to ultracentrifugation. Based on size-exclusion, this device isolated EVs with a size ranging from 30–200 nm, and meanwhile enriched EVs in a defined small volume (165 μL) from a starting volume of 8 mL. In addition, we developed a 96-well plate ELISA for detection of EV using anti-CD63 labeling (Fig. 4a). This ELISA detection strategy was further miniaturized into the double-filtration device for on-chip detection using a cellphone (Fig. 4b) [16, 17]. This integrated EV quantification platform was validated using 16 urine samples from bladder cancer patients and eight urine samples from healthy controls. Box-whisker analysis clearly shows that the concentration of EVs was significantly elevated compared to healthy controls (Fig. 4c), indicating that the concentration of urine EVs can be potentially used as a diagnostic biomarker to differentiate bladder cancer patients from healthy controls. Receiver operating characteristic analysis demonstrates that this integrated EV quantification device had a sensitivity of 81.3%, at a specificity of 90% (Fig. 4d). This integrated EV microchip offers an inexpensive option for translating EV-related research or clinical diagnosis for less-equipped laboratories. Nevertheless, clinical validation of this integrated device in a large population would be helpful to assess the diagnostic value of urine EV as a biomarker for detection of bladder cancer or other urogenital cancers.

Acknowledgments

Dr. Wang acknowledges the support from the Ministry of Science and Technology of the People's Republic of China (2016YFC1101302) from China. Dr. Demirci would like to acknowledge R01 AI093282, R01 GM108584, R01 DE02497101, R01 AI081534, R21 AI113117, R21 AI110277, U54 EB015408, DOD LC150650 11976867, and Canary Center seed grant.

Competing financial interests: Dr. U. Demirci is a founder of, and has an equity interest in: (1) DxNow Inc., a company that is developing microfluidic and imaging technologies for point-of-

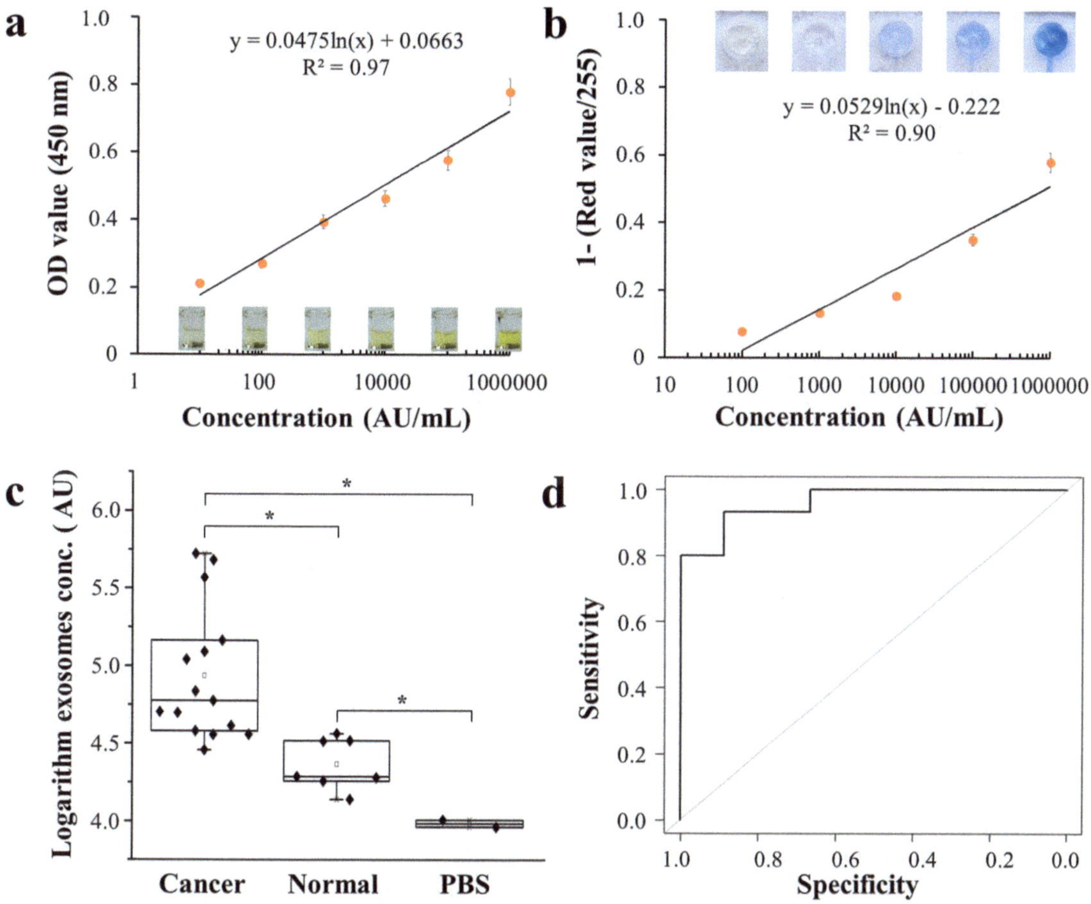

Fig. 4 Quantification of EVs using 96-well plate and microchip ELISA. (**a**) Standard curve of 96-well plate ELISA for detection of EVs. EVs were isolated from T24 cell culture using ultracentrifugation, and the lowest detection point was defined 10 artificial unit (AU). (**b**) Standard curve of microchip ELISA for detection of EVs. EVs were isolated from T24 cell culture using an integrated double-filtration device. Images were taken using a cell phone and R channel values of *blue color* were extracted using ImageJ and then plotted at the function of EV concentrations. (**c**) Box-whisker analysis of log-transformed EV concentrations of 24 urine samples (bladder cancer = 16, healthy controls = 8). The concentration of urinary EVs from bladder cancer was significantly higher that from healthy controls ($p < 0.05$). (**d**) Receiver operating characteristic (ROC) analysis of microchip ELISA for differentiating bladder cancer patients from healthy controls, and the microchip ELISA had a sensitivity of 81.3% at the specificity set to 90.0%

care diagnostic solutions, and (2) Koek Biotech, a company that is developing microfluidic IVF technologies for clinical solutions. U.D.'s interests were viewed and managed in accordance with the conflict of interest policies.

References

1. Siegel RL, Miller KD, Jemal A (2015) Cancer statistics, 2015. CA Cancer J Clin 65(1):5–29
2. National Cancer Institute (2016) Cancer of the urinary bladder. http://seer.cancer.gov/statfacts/html/urinb.html. Accessed 1 May 2016
3. Kaufman DS, Shipley WU, Feldman AS (2009) Bladder cancer. Lancet 374(9685):239–249

4. Tetu B, Tiguert R, Harel F, Fradet Y (2005) ImmunoCyt/uCyt+ improves the sensitivity of urine cytology in patients followed for urothelial carcinoma. Mod Pathol 18(1):83–89
5. Jia S, Zocco D, Samuels ML, Chou MF, Chammas R, Skog J, Zarovni N, Momen-Heravi F, Kuo WP (2014) Emerging technologies in extracellular vesicle-based molecular diagnostics. Expert Rev Mol Diagn 14(3):307–321
6. Costa-Silva B, Aiello NM, Ocean AJ, Singh S, Zhang HY, Thakur BK, Becker A, Hoshino A, Mark MT, Molina H, Xiang J, Zhang T, Theilen TM, Garcia-Santos G, Williams C, Ararso Y, Huang YJ, Rodrigues G, Shen TL, Labori KJ, Lothe IMB, Kure EH, Hernandez J, Doussot A, Ebbesen SH, Grandgenett PM, Hollingsworth MA, Jain M, Mallya K, Batra SK, Jarnagin WR, Schwartz RE, Matei I, Peinado H, Stanger BZ, Bromberg J, Lyden D (2015) Pancreatic cancer exosomes initiate premetastatic niche formation in the liver. Nat Cell Biol 17(6):816
7. Franzen CA, Blackwell RH, Todorovic V, Greco KA, Foreman KE, Flanigan RC, Kuo PC, Gupta GN (2015) Urothelial cells undergo epithelial-to-mesenchymal transition after exposure to muscle invasive bladder cancer exosomes. Oncogene 4:e163
8. Melo SA, Luecke LB, Kahlert C, Fernandez AF, Gammon ST, Kaye J, LeBleu VS, Mittendorf EA, Weitz J, Rahbari N, Reissfelder C, Pilarsky C, Fraga MF, Piwnica-Worms D, Kalluri R (2015) Glypican-1 identifies cancer exosomes and detects early pancreatic cancer. Nature 523(7559):177–182
9. Nawaz M, Camussi G, Valadi H, Nazarenko I, Ekstrom K, Wang XQ, Principe S, Shah N, Ashraf NM, Fatima F, Neder L, Kislinger T (2014) The emerging role of extracellular vesicles as biomarkers for urogenital cancers. Nat Rev Urol 11(12):688–701
10. Liga A, Vliegenthart ADB, Oosthuyzen W, Dear JW, Kersaudy-Kerhoas M (2015) Exosome isolation: a microfluidic road-map. Lab Chip 15(11):2388–2394
11. Sunkara V, Woo HK, Cho YK (2016) Emerging techniques in the isolation and characterization of extracellular vesicles and their roles in cancer diagnostics and prognostics. Analyst 141(2):371–381
12. Yoshioka Y, Kosaka N, Konishi Y, Ohta H, Okamoto H, Sonoda H, Nonaka R, Yamamoto H, Ishii H, Mori M, Furuta K, Nakajima T, Hayashi H, Sugisaki H, Higashimoto H, Kato T, Takeshita F, Ochiya T (2014) Ultra-sensitive liquid biopsy of circulating extracellular vesicles using ExoScreen. Nat Commun 5:3591
13. Wang S, Xu F, Demirci U (2010) Advances in developing HIV-1 viral load assays for resource-limited settings. Biotechnol Adv 28 (6):770–781
14. Wang SQ, Inci F, De Libero G, Singhal A, Demirci U (2013) Point-of-care assays for tuberculosis: role of nanotechnology/microfluidics. Biotechnol Adv 31(4):438–449
15. Wang S, Sarenac D, Chen MH, Huang SH, Giguel FF, Kuritzkes DR, Demirci U (2012) Simple filter microchip for rapid separation of plasma and viruses from whole blood. Int J Nanomedicine 7:5019–5028
16. Wang S, Tasoglu S, Chen PZ, Chen M, Akbas R, Wach S, Ozdemir CI, Gurkan UA, Giguel FF, Kuritzkes DR, Demirci U (2014) Micro-a-fluidics ELISA for rapid CD4 cell count at the point-of-care. Sci Rep 4:3796
17. Wang S, Zhao X, Khimji I, Akbas R, Qiu W, Edwards D, Cramer DW, Ye B, Demirci U (2011) Integration of cell phone imaging with microchip ELISA to detect ovarian cancer HE4 biomarker in urine at the point-of-care. Lab Chip 11(20):3411–3418

Part VIII

Extracellular Vesicles Isolated from Cell Culture, Parasites and Stem Cells

Chapter 30

Electric Field-Induced Disruption and Releasing Viable Content from Extracellular Vesicles

Chris Wang, Austin Wang, Fang Wei, David T.W. Wong, and Michael Tu

Abstract

In order to more fully elucidate the biogenesis of exosomes, a type of extracellular vesicles (EVs) and understand the role of EVs in disease processes, it is necessary to develop methods to capture EVs and induce the unloading of viable cargo. Traditionally, ultracentrifugation followed by chemical based EV lysis techniques is used to isolate these extracellular vesicles and release their internal content. Here, we describe a novel technique for capturing and releasing exosomal content through magnetic bead-based EV extraction coupled with electric field induced release and measurement (EFIRM). The usage of low-voltage electric fields allows the EV to be lysed without chemical treatment, and surface immobilized probes can allow for the rapid capture of the content of lysed EVs. EFIRM as an integrated EV lysing and analysis system offers great potential for the investigation of EVs in clinical and basic science contexts.

Key words Exosomes, Extracellular vesicles, Electric field induced release and measurement (EFIRM), Biomarkers, Diagnostics

1 Introduction

Exosomes are 30–100 nm microvesicular structures, a type of extracellular vesicles (EV) that have been found to be associated with the function of intercellular communication between different organ systems of the body [1]. As Zhang et al. [2] have noted, the transfer of oncogenic proteins and of nucleic acids via EVs modulates the activity of recipient cells contribute to roles in tumor growth, progression, metastasis, and drug resistance [3]. Consequently, it is of particular interest to develop methods to precisely unload the internal cargo of EVs and precisely analyze their internal molecular content [4]. The techniques for unloading the exosomal cargo involve usage of chemical agents for lysing EV samples or

Chris Wang and Austin Wang are joint first authors.

Winston Patrick Kuo and Shidong Jia (eds.), *Extracellular Vesicles: Methods and Protocols*, Methods in Molecular Biology, vol. 1660, DOI 10.1007/978-1-4939-7253-1_30, © Springer Science+Business Media LLC 2017

usage of the electric field induced release and measurement (EFIRM) technique. EFIRM is a lysis buffer-free technique for rapidly unloading EV cargo and analyzing their internal content through the application of a low-voltage cyclic square wave (*see* Fig. 1) [5]. There have been previous demonstrations of the use of the EFIRM technique for the analysis of mRNA targets from murine models of pancreatic and lung cancer [6–8]. During lysis buffer method incubation times, unloaded content released from EV samples degrade over time. Wei et al. [5] used transmission electron microscopy to visualize the effect of both the chemical lysis buffer and the electric field based lysis method on CD63 coated EVs, demonstrating that both methods worked identically (*see* Fig. 1). The degradation profiles of exosomal mRNA was also demonstrated to be the same in both the chemical lysis and EFIRM lysis methods (*see* Fig. 1). One advantage of the EFIRM technique to note is that EFIRM integrates the EV lysis procedure together with the high-sensitivity detection (and capture) of molecular content from the EV (*see* Fig. 1); this feature can allow for reducing the potential of content degradation when EV cargo is exposed to the extravesicular milieu. This chapter details the methods utilized in our work for the performance of EV cargo unloading using both chemical-based and EFIRM-based techniques. Since EFIRM is able to immediately capture the lysed content from EVs, this feature is also included in the protocol.

Successful exosomal content analysis includes and requires both an initial extraction of EVs from a biofluid preceding the lysis process [9]. There are two primary EV extraction techniques that will be described: the first to be discussed is the ultracentrifugation method, which allows for sized-based purification of EVs with minimally contaminated pellets of EVs. The second approach to be discussed is a magnetic bead-based EV precipitation technique, which utilizes surface specific marker antibodies (such as anti-CD63 antibody) to capture EVs. This bead-based approach allows for EV extraction with a lower sample volume and can be performed quickly, but does not allow for fractionation of microvesicles by size. EFIRM and chemical lysis techniques have both been used with both EV extraction methods. The various extraction and unloading techniques that can be used in conjunction with EFIRM must be weighed by the individual researcher based on the study goals.

2 Materials

2.1 Ultracentrifugation-Based Extracellular Vesicle Extraction

1. Culture medium specific to cells of interest.
2. Atmosphere of 95% air and 5% CO_2.
3. Phosphate-buffered saline (PBS).
4. 50 mL polypropylene centrifuge tubes.

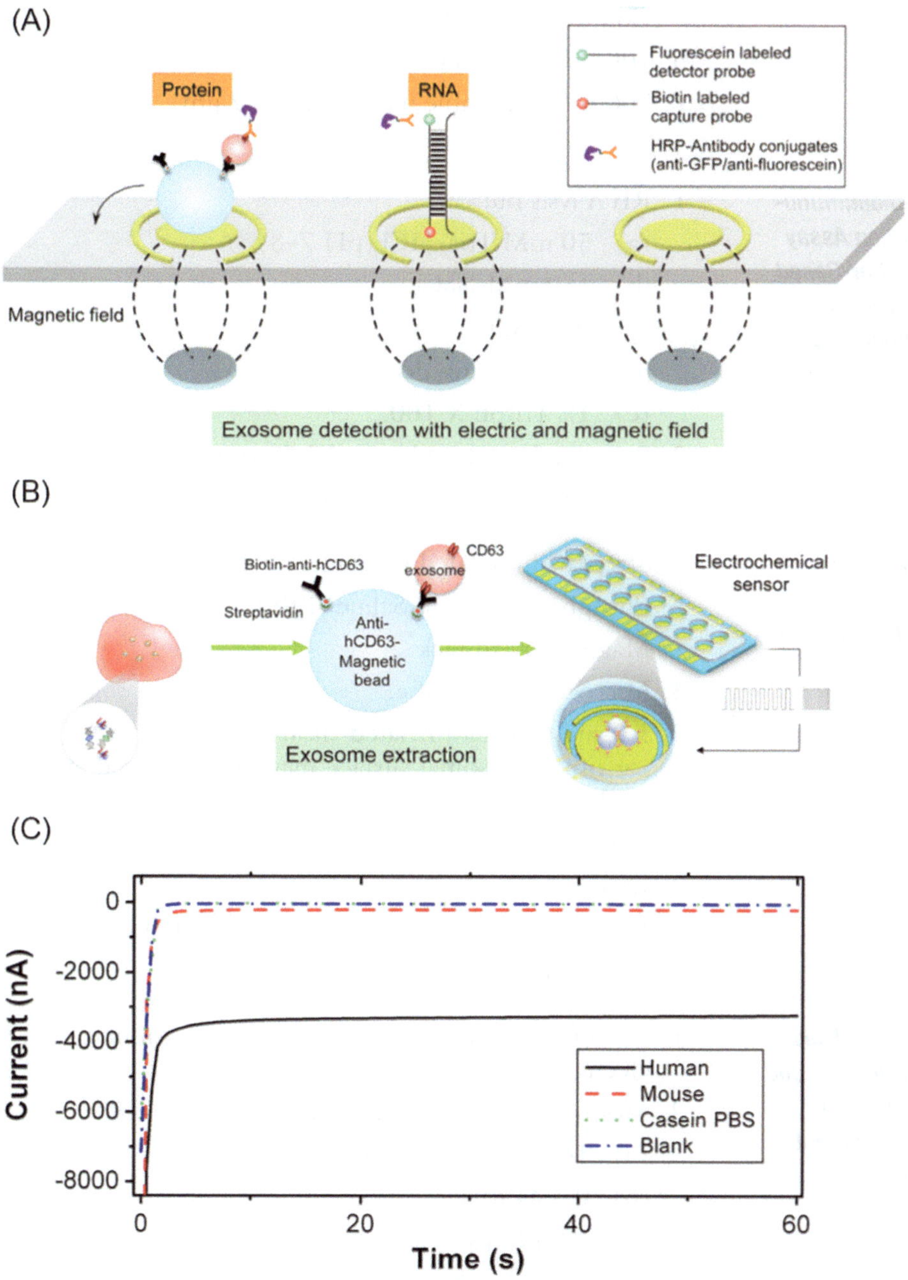

Fig. 1 Components of EFIRM method. (**a**) Method of extracting exosomes from biofluid using anti-human CD63-coated magnetic microparticles and then unloading exosome cargo using cyclic square waves applied to the particle–exosome complex. (**b**) Scheme of electrode biosensor used for detecting RNA/DNA/protein targets from the released exosomes. (**c**) Representative example of amperometric readout from the EFIRM methodology, where larger current magnitude corresponds to higher levels of a biomolecule. Figure is from Wei et al. (reproduced with permission from [5])

5. Counter-top centrifuge.
6. 100 mL polycarbonate tubes.
7. 500 mL disposable filter unit (0.22 μm).

2.2 Radioimmunoprecipitation Assay (RIPA) Buffer-Based Extracellular Vesicle Cargo Unloading

1. RIPA lysis buffer.
   (a) 50 mM Tris–HCl (pH 7–8).
   (b) 150 mM NaCl.
   (c) 0.1% SDS.
   (d) 0.5% sodium deoxycholate.
   (e) 1% Triton X-100.
2. Protease and phosphatase inhibitor cocktail tablet.
3. 100 mL polycarbonate tube.

2.3 Magnetic Bead-Based Extracellular Vesicle Extraction

1. Streptavidin-coated magnetic microparticles.
2. Phosphate Buffered Saline (PBS).
3. Magnetic bead rack.
4. Microcentrifuge tube.
5. 1.0 mg/mL biotinylated mouse anti-human CD63 antibody (or other exosomal surface marker designated appropriate for study).
6. Microcentrifuge tube rotator rack.
7. Casein-PBS.
8. Biofluid of interest (e.g., serum, saliva and cell culture media).
9. 1 M Tris–HCl.

2.4 Electric Field Induced Release and Measurement of Exosomal Content (EFIRM)

1. Gold three-electrode system (working, counter, and reference electrodes).
2. Monomer–DNA probe mixture.
   (a) 100 nM DNA probe.
   (b) 0.3 M KCl.
   (c) 10 mM pyrrole.
   (d) Ultrapure water.
3. Microcentrifuge tube.
4. Electrochemical potentiostat.
5. Distilled water.
6. Nitrogen gas.
7. Detector probe (hybridizes to the DNA probe).
8. Magnet array of neodymium magnets.

9. 150 unit/mL anti-fluorescein horseradish peroxidase 1:1000 dilution (HRP).
10. Casein-PBS.
11. 3,3′,5,5′-tetramethylbenzidine (TMB) substrate solution.

3 Methods

Carry out all procedures at room temperature unless otherwise specified.

3.1 Ultra-centrifugation-Based Extracellular Vesicle Extraction from Cell Culture

1. Grow cells of interest in flasks with culture medium.
2. Place in an incubator at 37 °C and at an atmosphere of 95% air and 5% CO_2 until the cells reach 55–75% confluency. The time period of this step depends on cell type.
3. Remove the culture medium,
4. Wash with phosphate-buffered saline (PBS), and finally.
5. Replace medium with a similar volume of EV-free medium.
6. Return the cells to the incubator (*see* **Note 1**).
7. Collect the conditioned media with a pipette. Transfer the filtered media to 50 mL polypropylene centrifuge tubes. Centrifuge cells at 300 × *g* for 10 min and collect the conditioned medium supernatant with a pipette into new 50 mL polypropylene centrifuge tubes (*see* **Note 2**).
8. Centrifuge the collected media at 2000 × *g* for 20 min. Pipette the supernatant and transfer to 100 mL polycarbonate tubes (maximum volume 68 mL) for use with a countertop centrifuge (*see* **Note 2**).
9. Mark one side of each tube with a marker (do not use anything permanent), orient the tube in the rotor with the mark facing up, and centrifuge the medium at 16,500 × *g* for 20 min. Transfer the supernatant to a 500 mL disposable filter unit (0.22 μm). Transfer filtered media to new 100 mL polycarbonate tubes (*see* **Note 3**).
10. Before centrifugation, mark the approximate location on the tubes of where the pellet would be. Then, centrifuge at 120,000 × *g* for 70 min and then remove the supernatant completely by aspiration. The product of this step will be EVs (*see* **Note 4**).
11. Allow the pellet to dry in the tube, then immediately resuspend the pellet in each tube in 1 mL of ice-cold PBS, using a micropipette. Pool the resuspended pellets from all the tubes containing materials from the same cells in a single ultracentrifuge tube, and then add PBS to fill the tube completely (*see* **Note 4**).

12. Centrifuge at 100,000 × *g* again for 70 min, and then remove the supernatant as completely as possible by aspiration. Proceed with caution as to not lose pellet.
13. *Extracellular vesicles*

    To resuspend the pellet (i.e., EVs), gently add a small volume (~160 μL/68 mL of starting volume) of PBS and allow the pellet to sit in the PBS for ~15–30 min at room temperature. Time will depend on size of pellet. Afterward, gently swirl the PBS over the pellet until it fully dissolves. Aliquot your resuspended EVs into smaller volumes ~50 μL and store at −80 °C for long-term storage (up to 1 year) and for downstream applications. Avoid repeated freezing and thawing.

3.2 Magnetic Bead-Based Extracellular Vesicle Extraction

1. Pipette a well-mixed solution of 5 μL of Streptavidin-coated magnetic microparticles into 495 μL of PBS buffer in a microcentrifuge tube to resuspend the beads using a magnetic rack to hold multiple tubes.
2. Wash and resuspend the beads by letting the tubes sit on the rack for 1 min, then using a pipette tip to carefully remove the supernatant buffer without disturbing the beads.
3. Place the tubes on a regular rack without magnets at the side.
4. Add 500 μL of PBS into the tubes and mix the solution with the beads using the pipette, then put the tubes back on the magnetic rack to again separate the beads from the solution.
5. Perform this removal of buffer for a total of three times.
6. Move the tubes to the nonmagnetized portion of the magnetic rack and resuspend the beads into 490 μL of PBS buffer. Pipette 5 μL of biotinylated mouse anti-human CD63 antibody at 1.0 mg/mL stock concentration into the mixture of beads. Use the pipette to mix the beads and antibody in solution.
7. After placing the microcentrifuge tubes with the mixture on a sample rotator, set the parameters for the sample rotator for reciprocal rotation at 90° tilting for 5 s and vibrating at 5° for 1 s. Rotate the sample–bead mixture tubes for 30 min at RT.
8. After 30 min of rotation at RT, place the tubes back in the magnetic rack for 5 min. Perform three washes of beads by removing the liquid phase using a micropipette and wash with 500 μL of PBS. After the triple wash, resuspend the beads in 490 μL of casein–PBS and place on the unmagnetized portion of the rack.
9. Label each tube with targeted sample ID. After pipetting a 10 μL sample of serum or saliva into the microcentrifuge tube, use the pipette to mix the sample and beads.

10. Place the tubes with sample and anti-human CD63 antibody beads on rotator and rotate for 2 h at RT, using the same rotator parameters as described in **step 7**.
11. Perform a triple wash by magnetizing to separate the beads from solution, removing liquid phase with micropipette, and resuspending beads in 500 μL of Tris–HCl buffer. The resultant beads are now bound to the EVs and are ready for the electrical field release and measurement.

3.3 Radioimmunoprecipitation Assay (RIPA) Buffer-Based Extracellular Vesicle Cargo Unloading

1. Dissolve EV pellet in RIPA lysis buffer (~160 μL/68 mL of starting volume for cell culture based EV extract) containing protease and phosphatase inhibitors (1 inhibitor tablet each/10 mL of RIPA). Pipette thoroughly, followed by vortex. To further lyse EVs, sonicate the sample in a water bath 3× for 5 min at room temperature with vortex in between each sonication step. Centrifuge sample at 13,000 × *g* for 5 min at room temperature and transfer supernatant to a 100 mL polycarbonate tube. Store at −80 °C for long-term storage.

3.4 Electric Field Induced Release and Measurement of Extracellular Vesicle Content (EFIRM)

3.4.1 Precoating of Electrode with a DNA Probe

1. For standard electric field-induced release and measurement (EFIRM) experiments, a gold three-electrode system consisting of working, counter, and reference electrodes is used as the base for the cargo unloading procedure.
2. In a microcentrifuge tube, prepare a stock mixture that will consist of 100 nM DNA probe, 0.3 M KCl, and 10 mM pyrrole in ultrapure water in a microcentrifuge tube. Vortex the mixture gently to ensure an adequate mixture of the different components of the solution (*see* **Note 5**).
3. Pipette 60 μL of the prepared monomer DNA probe mixture and ensure the monomer mixture provides adequate coverage of the individual working, counter, and reference electrodes. Inspect the seals and dividers of the electrodes used to ensure that there is no cross contamination between different electrodes.
4. Apply a cyclic square wave (CSW) electric field profile to each electrode surface to initiate the formation of a conducting polymer surface coated with capture probes. The polymerization process is initiated by first applying a +350 mV potential for 9 s and then immediately changing the potential to +950 mV for 1 s. This CSW profile is applied to each electrode for a total of ten cycles, which means in total the application of the electric field will be 100 s of applied electric field.
5. After applying the electric field profile for 100 s, rinse each electrode surface three times with distilled water and dry with nitrogen gas. This procedure is performed to ensure that unbound monomeric units from the mixture are not remaining on the electrode and will not interfere with the cargo unloading phase.

3.4.2 Extracellular Vesicle Cargo Unloading

1. Pipette 5 μL of 1 μM of detector probe into 495 μL of the EV extract mixture. Mix the detector probe solution with the EV extract by repeatedly pipetting. Pipette 60 μL of the new mixture into each well. Under each electrode, there should be a magnet array that consists of neodymium magnets aligned to each corresponding electrode sensor (*see* **Note 6** and Fig. 1).
2. After adding the mixture to each well, inspect to ensure that mixture completely covers each electrode. Tap the side of the plate to ensure an even distribution of liquid. Apply 20 cycles of electric field across the electrode, which each cycle consisting of 9 s at −300 mV potential and 1 s at +200 mV potential (for a total of 200 s). During this process, the exosomal content will be released and hybridized onto the surface probe.
3. Wash off the unbound contents in the mixture by rinsing each electrode surface three times with distilled water. Dry the electrodes with nitrogen gas.

3.4.3 Reporter Antibody and Readout

1. Pipette 60 μL of 150 unit/mL anti-fluorescein antibody conjugated to horseradish peroxidase (HRP 1:1000 dilution) into a tube. Add Casein/(PBS) to dilute the mixture.
2. Apply an electric field to conjugate the complex anti-fluorescein HRP to the probe surface. Apply a-200 mV potential for 1 s and +500 mV potential for 1 s for five cycles (total 10 s).
3. Wash the electrode surface by rinsing three times with distilled water. Following wash, dry with nitrogen gas.
4. Pipette 60 μL of 3,3′,5,5′-tetramethylbenzidine (TMB) substrate solution onto the surface of each well. Measure the electrode current at −200 mV for 60 s to perform amperometric readout.

4 Notes

1. Cells that grow fast may be overconfluent and start dying after 48 h. In this case, take the conditioned medium after 24 h. Cells that are very sensitive may not like growing in EV-free medium or in FBS-free medium. In this case, take the supernatant after 24 h.
2. Ensure that none of the pellet is collected or contaminates the supernatant. Use a pipette and leave behind 2 mL of media.
3. Ensure that none of the pellet is collected or contaminates the supernatant. There will probably not be a visible pellet at this step. When removing the supernatant with the pipette, hold the tube at an angle, so that the pellet is always covered with

supernatant, and stop removing supernatant when half a centimeter of liquid is still covering the pellet.

4. For this high-speed centrifugation, all tubes should be at least three-quarters full. Use PBS to fill tubes. Centrifugation time is calculated to allow a full hour at 100,000 × *g*, i.e., ~10 min to reach 100,000 × *g* plus 1 h at the final speed. A longer time (up to 3 h) will not damage the EVs.
5. The probe utilized should be designed to be complementary with the sequence that you desire to capture from the EV. The GAPDH probe, for example, consists of the sequence 5′-Biotin-AGGTCCACCACTGACACGTTG. If one seeks to only lyse the EV without capturing the sequence, one can skip to the cargo unloading step of this section.
6. In the case of detecting GAPDH as a reference, the detector probe 5′-GCAGTGGGGACACGGAAGGCC-Fluorescein-3′ was used. The inclusion of a detector probe can be neglected if one seeks to only lyse the EV without performing detection using the electrode.

Acknowledgments

Research reported in this publication was supported by the National Institute of Dental & Craniofacial Research of the National Institutes of Health under Award Number T90DE022734. The content is solely the responsibility of the authors and does not necessarily represent the official views of the National Institutes of Health.

Disclosures: David Wong is cofounder of RNAmeTRIX Inc., a molecular diagnostic company. He holds equity in RNAmeTRIX, and serves as a company Director and Scientific Advisor. The University of California also holds equity in RNAmeTRIX. Intellectual property that David Wong invented and which was patented by the University of California and has been licensed to RNAmeTRIX. Additionally, he is a consultant to PeriRx.

References

1. Wei F, Patel P, Liao W et al (2009) Electrochemical sensor for multiplex biomarkers detection. Clin Cancer Res 15:4446–4452
2. Zhang X, Yuan X, Shi H et al (2015) Exosomes in cancer: small particle, big player. J Hematol Oncol 8:83
3. Camussi G, Deregibus MC, Bruno S et al (2010) Exosomes/microvesicles as a mechanism of cell-to-cell communication. Kidney Int 78:838–848
4. Zlotogorski-Hurvitz A, Dayan D, Chaushu G et al (2015) Human saliva-derived exosomes: comparing methods of isolation. J Histochem Cytochem 63:181–189
5. Wei F, Yang J, Wong DTW (2013) Detection of exosomal biomarker by electric field-induced release and measurement (EFIRM). Biosens Bioelectron 44:115–121
6. Lau C, Kim Y, Chia D et al (2013) Role of pancreatic cancer-derived exosomes in salivary

biomarker development. J Biol Chem 288:26888–26897

7. Tu M, Chia D, Wei F et al (2016) Liquid biopsy for detection of actionable oncogenic mutations in human cancers and electric field induced release and measurement liquid biopsy (eLB). Analyst 141(2):393–402
8. Wei F, Lin C-C, Joon A et al (2014) Noninvasive saliva-based *EGFR* gene mutation detection in patients with lung cancer. Am J Respir Crit Care Med 190:1117–1126
9. Théry C, Amigorena S, Raposo G et al (2001) Isolation and characterization of exosomes from cell culture supernatants and biological fluids. In: Current protocols in cell biology. John Wiley & Sons, Inc., New York

Chapter 31

Production and Characterization of Extracellular Vesicles in Malaria

Smart Mbagwu, Michael Walch, Luis Filgueira, and Pierre-Yves Mantel

Abstract

Growing attention is drawn toward the role of extracellular vesicles (EVs) in infectious diseases. EVs, which are small vesicles released by cells, are involved in cellular communication, immune regulation, and pathogenesis. EVs act as messenger carrying functional cargoes, including RNA, DNA, lipids and proteins from a donor cell to regulate the function of a recipient cell. In malaria, EVs play a key role in regulating the progression from the blood to the transmission stage by promoting the switch between asexual and sexual stages that are taken up by mosquitoes. In addition to their role in parasite communication, EVs modulate the immune system and regulate endothelial cell function.

In this chapter, we describe protocols to isolate, purify and characterize EVs derived from *Plasmodium falciparum* infected red blood cell culture.

Key words Malaria, Extracellular vesicles, *Plasmodium falciparum*, Cellular communication

1 Introduction

Malaria is an infectious disease that is caused by a protozoan parasite belonging to the *Plasmodium* species. There are five different plasmodium species out of which the most virulent is *Plasmodium falciparum* (*P.f.*) [1]. This parasite is transmitted by an infected female Anopheles mosquito during a blood meal. People living in tropical areas are more susceptible to the infection. The symptoms usually associated with the disease include fever, fatigue and vomiting. Its advanced state could result in coma, seizures and even death. Every year approximately, 400,000 deaths are recorded following malaria infection which often times may be uncomplicated [2].

Malaria infection consists of different stages in the human host with first a liver stage and subsequently an erythrocytic stage (red blood cells), respectively. After a bite by an infected mosquito, the sporozoites in the saliva of the mosquito travel through the blood stream to reach the liver cells (hepatocytes) where they undergo

Winston Patrick Kuo and Shidong Jia (eds.), *Extracellular Vesicles: Methods and Protocols*, Methods in Molecular Biology, vol. 1660, DOI 10.1007/978-1-4939-7253-1_31, © Springer Science+Business Media LLC 2017

asexual reproduction to produce thousands of merozoites. This stage is usually asymptomatic and spans for about 8–30 days. The merozoites released from the rupturing of the hepatocytes infect the red blood cells (RBCs) and marks the onset of the erythrocytic stage. Within the RBCs, they replicate further asexually during a 48-h cycle producing new infective merozoites. This leads to multiple cycles of continuous infection of new RBCs, as the infected RBCs (iRBCs) burst to release new merozoites infecting other uninfected RBCs. The blood stage is characterized by symptoms such as fever, cold and shivering. During the blood stage a small proportion of parasites differentiate into gametocytes, which are transmitted to the mosquito [1].

Malaria infection can lead to severe and even fatal disease. Severe malaria is accompanied by the following complications; metabolic acidosis, severe anemia, and cerebral malaria [3]. The factors determining the outcome and extent of the disease are yet to be explored. Although the role of extracellular vesicles (EVs) in malaria infection has not yet been well defined, studies have demonstrated an increase in the amount of EVs derived from RBCs in the plasma of patients with malaria culminating in severe malaria [4–6]. From these studies, the capacity of the malaria infected RBC derived EVs to activate immune cells is suggestive of its role in the pathogenesis of the disease [7, 8]. EVs are small vesicles released by iRBCs; their exact nature is still unknown. However, proteomics analysis revealed that the isolated EVs contain plasmodium parasite proteins including known vaccine candidates [9]. Most of the identified plasmodial proteins are exported proteins transferred by the parasite into the RBC cytosol and membrane. Remarkably when added to in vitro *P.f.* cultures, EVs promote the differentiation toward gametocytes, which constitute the transmission stage [4, 10, 11]. Therefore EVs might be essential for the completion of the malaria life cycle. EVs can transfer nucleic acids across the parasite population to transfer drug resistance [12]. This constitutes a novel way of cellular communication between parasites. The exact cargo present in the EVs that is involved in the differentiation toward gametocytes is not known. We have identified that EVs carry small RNAs including functional human miRNAs. When cocultured with EVs, endothelial cells are able to take up the miRNAs. Remarkably, the miRNAs were still functional and regulate gene expression of target genes in the endothelial cells. Ultimately the miRNAs modulate the endothelial barrier function [13].

The discoveries of EVs in malaria create new avenues of research for cell to cell communication, host–pathogen interactions and to provide new therapeutic approaches. However further work is needful in order to fully decipher the role of iRBC-EVs during malaria in cellular communication and host–pathogen interaction.

In this chapter, we describe protocols to produce and isolate EVs from cell culture of malaria iRBCs.

2 Materials

Prepare all solutions using ultrapure water (prepared by purifying deionized water, to attain a sensitivity of 18 MΩ cm at 25 °C) and analytical grade reagents. Prepare and store all reagents at room temperature (unless indicated otherwise). Carefully follow all waste disposal regulations when disposing of waste materials.

2.1 Preparation of EV-Depleted Serum

1. Human serum (blood bank or commercial) or Albumax II (ThermoFisher Scientific, *see* **Notes 1** and **2**).
2. Ultracentrifugation tubes: Thinwall Ultra-Clear^{TM}, 38.5 ml, 25 × 89 mm (Beckman Coulter).
3. Nalgene Sterile Filter Units with PES Membrane 0.1 μm.

2.2 Cell Culture and Production of EVs

1. *Plasmodium falciparum* such as 3D7 (www.beiresources.org) (*see* **Note 3**).
2. 37 °C incubator.
3. Malaria gas (5% O_2, 5% CO_2 and 90% N_2) or candle jar (*see* **Note 4**).
4. 225 cm^2 culture flasks with plug caps (Corning).
5. Centrifuge.
6. Phosphate-buffered saline (PBS).
7. Giemsa staining solutions (Sigma-Aldrich, *see* **Note 5**).
8. Incomplete culture medium: 10.4 g RPMI 1640 (with L-Gln, no $NaHCO_3$) (company), 24 ml 7.5% $NaHCO_3$ (Sigma) 25 ml 1 M HEPES (Sigma), 10 ml 20% glucose (Sigma), 0.5 ml 25 mg/ml hypoxanthine (Sigma), 50 μg/ml gentamicin (Sigma), 900 ml ddH_2O. Adjust pH to 7.4, adjust water to 1 L, filter to sterilize, aliquot and store at 4 °C (*see* **Note 6**).
9. Complete culture medium: take above incomplete medium and add heat-inactivated human serum (55 °C for at least 45 min, then freeze and store at −20 °C to a final concentration of 10%) (*see* **Note 7**).
10. Fresh (less than 1-week old) human RBCs, wash the blood twice with incomplete medium and resuspend at 50% hematocrit in complete culture medium.

2.3 Purification of EVs

1. Ultracentrifuges such as Optima XPN or another ultracentrifuge.
2. JA10 Rotor, Fixed Angle, Aluminum, 6 × 500 ml, 10,000 rpm, 17,700 × *g*.
3. SW32Ti Rotor Pkg, Swinging Bucket, 6 × 38.5 ml, 32,000 rpm, 175,000 × *g*.

4. Ultracentrifugation tube, Thinwall Ultra-Clear™, 38.5 ml, 25 × 89 mm (344058 Beckman Coulter).
5. Vivacell 250 unit with vivacell 250 PES inserts 100,000 Da (Sartorius).
6. Nalgene Sterile Filter Units with PES Membrane 0.8 μm.
7. 50 ml tubes.
8. 60% (w/w) sucrose solution in H_2O:

   60 g protease-free sucrose.

   2.4 g Tris base.

   60 ml of ultrapure H2O.

   Adjust the pH with 1 N HCl.

   Adjust the volume to 100 ml with H_2O.

   Sterilize by passing through a 0.22 μm filter. Store at 4 °C.

2.4 Size Fractionation of Cell Extracts Using a Fast Protein Liquid Chromatography System

1. Fast protein liquid chromatography (FPLC) system, such as Äkta (GE Healthcare) or BioLogic LP (Bio-Rad) systems.
2. Sephacryl S-500 resin (GE Healthcare).
3. PBS.
4. Ultrapure water.
5. For the standard curve:

   BSA (67 kDa; GE Healthcare).

   Tyrosine (0.181 kDa; Sigma-Aldrich).

   Gel Filtration Chromatography Standard (Bio-Rad).

3 Methods

3.1 Preparation of EV-Depleted Serum

Human serum is needed for iRBC culture. Human serum contains EVs. It is therefore essential to process the serum before using it with cell culture (*see* **Note 7**).

Use sterile technique throughout the following procedures.

1. To prepare EV-depleted human serum, mix five batches of human serum taken from healthy donors.
2. Transfer the serum into ultracentrifugation tubes and centrifuge at 100,000 × *g* for 2 h at 4 °C. Collect the supernatant, which contains the EV-depleted serum and discard the pellet.
3. Collect the supernatant and pass it through a 0.1 μM filter to remove remaining EVs.
4. The serum is now ready to be used or can be aliquoted in 50 ml tubes and stored at −20 °C for over a year for later usage.

3.2 Cell Culture of iRBCs and Production of EVs

It is essential to keep a healthy growing culture in order to avoid cell lysis. Maintain culture at a parasitemia between 5 and 10%, and a 4% hematocrit (*see* **Note 8**). Importantly, keep the parasitemia under 10% to limit the accumulation of toxic debris. Perform a Giemsa smear of the culture on a daily basis to monitor the parasitemia and look for signs of lysis. Ideally the conditioned media is collected every day and replaced by fresh culture medium to maintain optimal growth conditions. Although parasites grow well under static conditions, growing the cultures in suspension under gentle agitation (55 rpm) might improve the yield and production of EVs by infected RBCs, particularly when using culture volumes larger than 10 ml.

1. Prepare the culture medium in advance allowing to warm at 37 °C.
2. Start a parasite culture as described by [14].
3. Prepare 100 ml RPMI + 10% EV-depleted human serum. To which you add 8 ml of 50% hematocrit fresh RBCs to iRBC and transfer to a 225 cm^2 tissue culture flask.
4. Gently resuspend iRBC pellets by pipetting up and down.
5. Gas the flask with 3% O_2, 4% CO_2 and 93% N_2 for 30 s using a sterile, blunt-ended needle connected via a filter to the gas cylinder by plastic tubing. Tightly seal the tissue culture flask and place in an incubator set to 37 °C.
6. Maintain a healthy culture (<10% parasitemia) at 4% hematocrit.
7. To reach optimal growth of the parasites, the culture medium should be replaced every 24 h. Resuspend the cells by pipetting up and down with a 50 ml pipette, then transfer to 50 ml tubes.
8. Spin down the 50 ml tubes for 15 min at 600 × g in a centrifuge with low brake.
9. Carefully collect the supernatant without disturbing the cell pellet and proceed to the EV purification protocol with the supernatant.
10. Prepare a Giemsa-stained smear of the pellet and count the proportion of parasite-infected red blood cells with a microscope at a 63× magnification under oil immersion. Dilute the iRBCs with fresh uninfected blood and fresh medium accordingly to reach a parasitemia of 0.4–1%, assuming the parasites will reach 5–10% after 48 h of culture (*see* **Note 5**).

3.3 Purification of EVs

EVs are isolated and enriched by a combination of sequential rounds of centrifugation to remove contaminating cells and debris. The EVs are further purified by filtration. The EVs are then concentrated overnight by pressure ultrafiltration at 4 °C. Finally the EVs are collected by ultracentrifugation and sucrose gradient

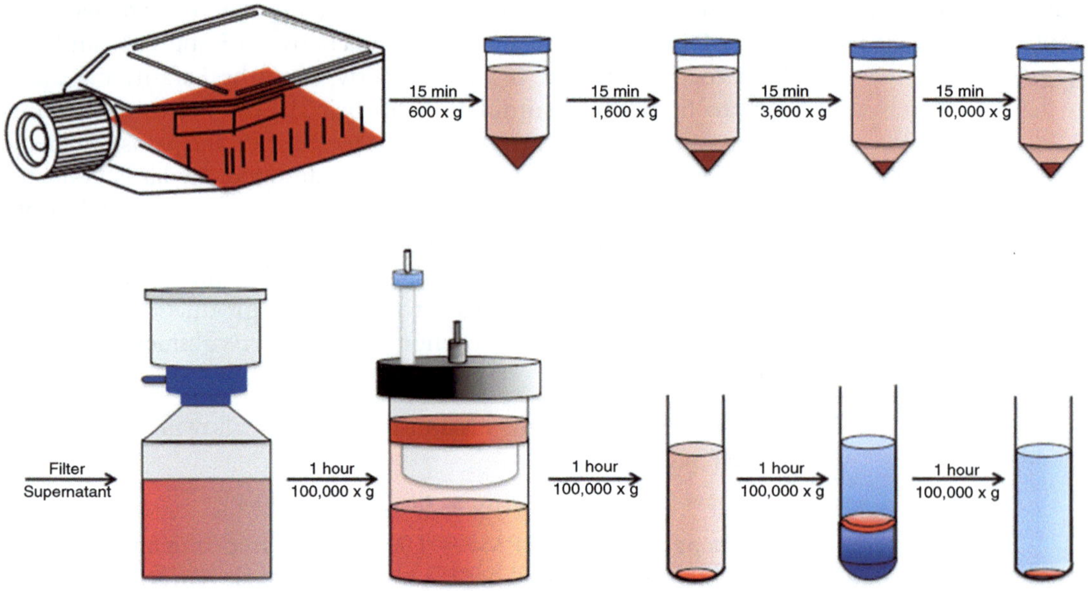

Fig. 1 Schematic of the protocol. During the first part of the protocol, the cells and debris are eliminated by centrifugation at increasing speed. The supernatant is then filtered through a 0.8 μm filter unit, before being concentrated in a pressure concentrator. Finally the EVs are pelleted by ultracentrifugation

(Fig. 1). The purity and the concentration of the resulting samples can be controlled by NanoSight, western blot analysis, imaging flow cytometry and by transmission electron microscopy (*see* **Note 9**). For some experiments you will need to prepare sterile EVs (*see* **Note 10**).

1. To remove cells and debris, pellet cell suspensions at 600 × *g* in 50-ml tubes for 15 min at room temperature. Always spin at low brake to avoid disruption of the pellet.
2. Carefully transfer the supernatant without disturbing the pellet to new 50 ml tubes and spin down at 1600 × *g* with low brake. Discard the pellet.
3. Carefully transfer the supernatant to new 50 ml Falcon tubes and spin down at 3600 × *g* for 15 min with low brake. From now on, it is recommended to work at 4 °C.
4. Carefully transfer the supernatant to a new 50 ml Falcon tube and spin down at 10,000 × *g* for 15 min.
5. Carefully collect the supernatant and pass it through a 0.8 μm filter to further purify the EVs and collect the filtrate.
6. Concentrate the EVs in order to decrease the volume of supernatant to process through ultracentrifugation. Proceed directly to **step 8** if you work with smaller volume of supernatant.

7. To concentrate the EVs, transfer up to 250 ml of the supernatant into a Vivacell 250 (100 kDa molecular weight cutoff). Apply approximately 3–4 bar of pressure and incubate at 4 °C to concentrate the supernatant. Usually the supernatant volume can be decreased by a factor 3–5.
8. Transfer approximately 35 ml of supernatants to ultracentrifugation tubes and spin 1 h at 100,000 × *g* in an ultracentrifuge.
9. Remove carefully the supernatant and discard it (use several times the same tube to increase the size of the pellet). The pellet should be visible at the bottom of the tube as dark brown.
10. Wash the pellet once with 35 ml of PBS and spin at 100,000 × *g*.
11. Discard the supernatant and carefully resuspend the pellet in 27 ml of PBS. The diluted EVs are then gently layered on top of 5 ml of 60% sucrose cushion (*see* **Note 11**). Be careful not to disturb the interphase, centrifuge for 1 h at 100,000 × *g* with low acceleration and low brake at 4 °C.
12. Collect the interphase on top of the sucrose (the interphase should be red-brownish).
13. Carefully collect the interphase and transfer to a new ultracentrifugation tube. Fill up the tube to approximately 35 ml and spin at 100,000 × *g* for 1 h at 4 °C.
14. Discard the supernatant and resuspend pellet in 35 ml of PBS, spin down at 100,000 × *g* for 1 h at 4 °C.
15. Repeat **step 14**.
16. Resuspend the pellet well by pipetting up and down in 1 ml of PBS.
17. The amount of vesicles can be determined by NanoSight or protein content (Bradford assay).
18. The EVs can be stored at −80 °C for over a year, avoid freeze–thaw. Discard the batch as soon as you notice signs of lysis.

3.4 Size Fractionation of Purified EVs Using a Fast Protein Liquid Chromatography System

Size-exclusion chromatography is a very efficient approach to fractionate complex biological fluids such as cell culture supernatants, serum or plasma based on the size of their molecular components. The higher molecular weight molecules or the larger molecular complex will elute faster, since smaller molecules are trapped in the pores constituting the matrix and take therefore more time to migrate through the column. The larger molecules simply pass by the pores because those molecules are too large to enter the pores. This approach allows to distinguish extracellular RNA found in serum inside EVs or in association with protein lipid complexes

[15]. In fact EVs are larger than protein–lipid complex and can easily be separated from other protein lipid complexes.

Size-Exclusion Chromatography of Plasma and Serum.

1. Pack sephacryl S-500 resin (GE Healthcare) in a chromatography column (0.9 × 30 cm, 19.1 ml bed volume).
2. Attach all columns to the FPLC system in a cold room.
3. Before injection, equilibrate the column with 25 ml of PBS solution at 0.5 ml/ min at 4 °C.
4. Use a syringe to load the sample (0.5 ml of 2 mg/ml of EVs resuspended in PBS, supernatant of *P.f.* infected RBCs or undiluted fresh serum) into a superloop attached to a FPLC workstation.
5. Inject the column with the sample and elute at 4 °C for approximately 1 h with PBS solution (pH 7.4) at a flow rate of 0.5 ml/ min.
6. Collect 25 1-ml-fractions. Flush the column with 75 ml of PBS solution at 0.5 ml/min (3.75 column volumes) between serum and plasma fractionation to eliminate carryover.
7. Store fractions at −20 °C before use.
8. Determine protein concentration by BCA assay using BSA for calibration or absorbance at 280 OD for BSA and tyrosine (Fig. 2a) and samples (EVs, serum and Parasite SN) (Fig. 2b).
9. The fractions can be used for protein or nucleic acid analysis as well as in functional assays.
10. Protein molecular weight standards included thyroglobulin (669 kDa; GE Healthcare), BSA (67 kDa; GE Healthcare), ribonuclease A (13.7 kDa; GE Healthcare), and tyrosine (0.181 kDa; Sigma-Aldrich).

4 Notes

1. Be aware that human serum may contain pathogens, therefore for safety reason it is important to use serum that has been tested for blood-borne viruses (e.g., hepatitis B and C, HIV). We use serum from O^{Rh+} blood group individuals because it will not react with the O^{Rh+} blood used for culturing the parasites. Alternatively you can use serum from the AB blood group individuals since they will not have antibodies against the A and B groups. Human serum can be purchased from commercial companies or it can be obtained from a registered and licensed blood service center.
2. Parasites can also be grown in Albumax added to the culture medium instead of human serum. Albumax allows robust

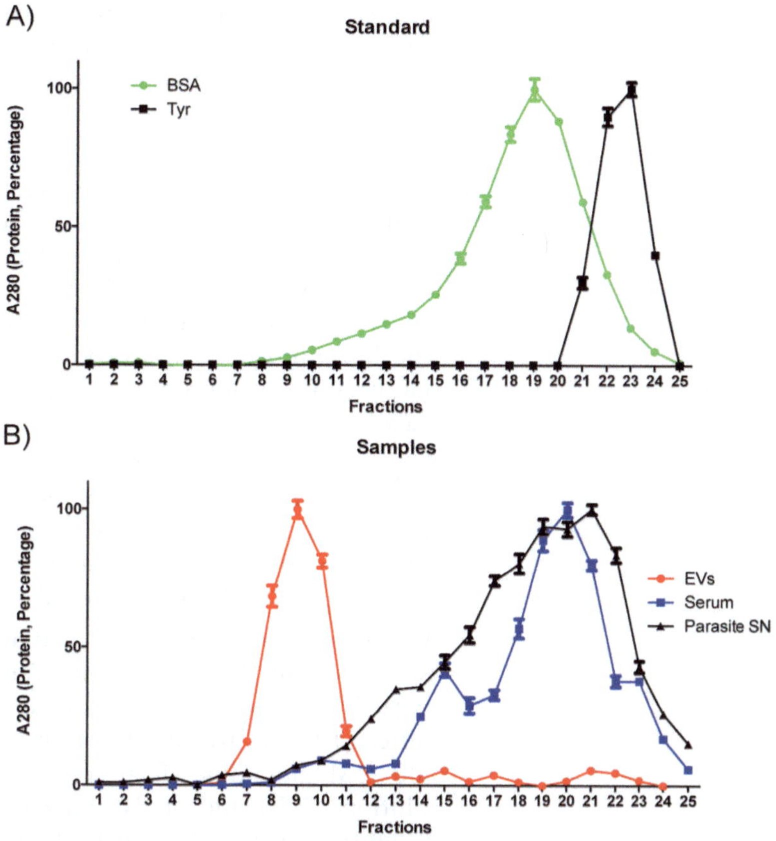

Fig. 2 Purification of EVs from parasite culture supernatants by fast protein liquid chromatography. (**a**) The profile as measured by absorption at 280 nm of eluted protein from the FPLC is shown for the standards: BSA (*green line*) and tyrosine (*black line*). (**b**) The profile of eluted purified EVs (*red line*), Unfractionated serum (*blue line*) and parasite culture supernatant (*black line*)

growth of the cells similar to human serum and does not seem to affect EV release. One advantage of working with Albumax is to avoid donor variations that can appear while working with different human serum donors. However, Albumax interferes with parasite biology; for example, the export of virulence factors to the red blood cell membrane does not occur. Therefore, we prefer to work with human serum. For our stocks, we mix serums from five different donors to reduce donor variability (*see* Subheading 3.1).

3. *P. falciparum* is a human pathogen. When working with malaria iRBCs make sure that the appropriate biosafety measures and authorizations are in place. Depending on local regulations, *P. falciparum* is either a biohazard safety level II or III organism. Therefore personnel should receive appropriate training before handling the parasites. An infection with

infected red blood cells can be fatal if not treated. In the laboratory, the most common route of infection is through contamination of an open wound or a needle-stick injury. Therefore appropriate equipment and safety measures should be observed, following home institution policies.

4. An alternative method to gassing the malaria culture is the candle jar. The candle jar has to be an airtight container in glass or plastic (Tupperware). Place the flask containing the parasite cultures in the candle jar and loosen the screw cap. Add a burning candle inside the container and close the container. The burning candle will produce low oxygen while burning out, creating an appropriate O_2-poor environment optimal for malaria parasite growth.
5. Prepare a thick film from parasite culture and let it dry properly. Prepare a fresh 2% Giemsa solution in PBS. Place the slides in the Giemsa solution. Rinse very carefully and gently with water. Then leave the slide in an upright position to dry. Observe with immersion oil and objective at 63×.
6. Alternatively you can prepare medium from commercial RPMI 1640. Remove 62.25 ml from a 500 ml bottle of RPMI1640 (with 20 mM sodium bicarbonate, and with 25 mM HEPES buffer [12.5 ml to 500 if not already added]). Add: 5 ml of 200 mM glutamine (final conc. 2 mM), 5 ml 20% glucose solution, 1.25 ml of 10 mg/ml gentamicin (final conc. 25 μg/ml), 1 ml 1 M NaOH (to achieve pH 7.2–7.4), 50 ml Human serum.
7. Human serum contains a high amount of EVs derived from many different cell types. It is therefore mandatory to use EV-depleted serum for all the experiments particularly when interested in RNA or protein composition of EVs. An alternative to serum could the usage of Albumax (*see* **Note 2**).
8. We have observed that the ratio of red blood cell derived EVs to uninfected RBC derived EVs is of about 15:1 in in vitro culture. Therefore, the amount of EVs dramatically increases in infected cultures. In order to increase the proportion of iRBC-derived EVs it is possible to culture the parasites at lower hematocrit (HCT), and therefore at higher parasitemia. This can be achieved by increasing the volume of medium or decreasing the volume of RBCs. For example at an HCT of 0.5%, the parasites should be able to grow at parasitemia of up to 40%. Be aware that some strains are more capable of growing at higher parasitemia than others.
9. During the optimization of this protocol, samples can be collected from each purification step, including each centrifugation and the sucrose cushion. The composition of the samples can be analyzed by a combination of ImageStream, microscopy

(from blood smears), transmission electron microscopy and western blot to determine the purity of EVs.

10. Do as much as possible in the TC hood (balancing the tubes for Ultracentrifugation) to keep it semisterile. For experiments that require sterile EVs use sterilized ultracentrifugation tubes. To obtain sterile tubes you may autoclave polyallomer, polyethylene, and glass tubes, or you may use OptiSeal; Quick-Seal; and Ultra-Clear,thin-wall. Polyallomer which are disposable-format tubes should be discarded after use.

11. We use normal H_2O for the preparation of the sucrose solutions, instead of D_2O (heavy water). The sucrose concentration giving the same density is lower in D_2O than in H_2O, since D_2O is a lot more expensive than normal water. EVs will float up to the interphase between the 60% and water layers.

    In order to keep the sucrose cushion intact we use a freeze–thaw approach in which the 5 ml of 60% sucrose are added to the ultracentrifugation tubes. The tubes are then transferred to −80 °C for 15 min for allowing freezing of the solution. The diluted EVs can finally easily be layered on top of the frozen sucrose cushion.

Acknowledgments

This work was supported by grants from the Novartis Foundation for Medical-Biological Research, from the Research Pool and Fund of the University of Fribourg, as well as from the Gottfried and Julia Bangerter-Rhyner foundation (to M.W. and P.Y.M.).

References

1. Miller LH, Baruch DI, Marsh K, Doumbo OK (2002) The pathogenic basis of malaria. Nature 415(6872):673–679. doi:10.1038/415673a
2. WHO (2012) WHO Malaria Report 2012
3. Schofield L, Grau GE (2005) Immunological processes in malaria pathogenesis. Nat Rev Immunol 5(9):722–735. doi:10.1038/nri1686
4. Mantel PY, Marti M (2014) The role of extracellular vesicles in Plasmodium and other protozoan parasites. Cell Microbiol 16(3):344–354. doi:10.1111/cmi.12259
5. Nantakomol D, Dondorp AM, Krudsood S, Udomsangpetch R, Pattanapanyasat K, Combes V, Grau GE, White NJ, Viriyavejakul P, Day NP, Chotivanich K (2011) Circulating red cell-derived microparticles in human malaria. J Infect Dis 203(5):700–706. doi:10.1093/infdis/jiq104
6. Pankoui Mfonkeu JB, Gouado I, Fotso Kuate H, Zambou O, Amvam Zollo PH, Grau GE, Combes V (2010) Elevated cell-specific microparticles are a biological marker for cerebral dysfunctions in human severe malaria. PLoS One 5(10):e13415. doi:10.1371/journal.pone.0013415
7. Combes V, Coltel N, Alibert M, van Eck M, Raymond C, Juhan-Vague I, Grau GE, Chimini G (2005) ABCA1 gene deletion protects against cerebral malaria: potential pathogenic role of microparticles in neuropathology. Am J Pathol 166(1):295–302. doi:10.1016/S0002-9440(10)62253-5
8. Couper KN, Barnes T, Hafalla JC, Combes V, Ryffel B, Secher T, Grau GE, Riley EM, de Souza JB (2010) Parasite-derived plasma microparticles contribute significantly to malaria infection-induced inflammation

through potent macrophage stimulation. PLoS Pathog 6(1):e1000744. doi:10.1371/journal.ppat.1000744

9. Mantel PY, Hoang AN, Goldowitz I, Potashnikova D, Hamza B, Vorobjev I, Ghiran I, Toner M, Irimia D, Ivanov AR, Barteneva N, Marti M (2013) Malaria-infected erythrocyte-derived microvesicles mediate cellular communication within the parasite population and with the host immune system. Cell Host Microbe 13(5):521–534. doi:10.1016/j.chom.2013.04.009
10. Pelle KG, Ahouidi AD, Mantel PY (2013) Role of microvesicles in malaria infections. Med Sci 29(11):960–962. doi:10.1051/medsci/20132911010
11. Ankarklev J, Brancucci NM, Goldowitz I, Mantel PY, Marti M (2014) Sex: how malaria parasites get turned on. Curr Biol 24(9):R368–R370. doi:10.1016/j.cub.2014.03.046
12. Regev-Rudzki N, Wilson DW, Carvalho TG, Sisquella X, Coleman BM, Rug M, Bursac D, Angrisano F, Gee M, Hill AF, Baum J, Cowman AF (2013) Cell-cell communication between malaria-infected red blood cells via exosome-like vesicles. Cell 153(5):1120–1133. doi:10.1016/j.cell.2013.04.029
13. Mantel PY, Hjelmqvist D, Walch M, Kharoubi-Hess S, Nilsson S, Ravel D, Ribeiro M, Grüring C, Ma S, Padmanabhan P, Trachtenberg AJ, Ankarklev J, Brancucci NM, Huttenhower C, Duraisingh MT, Ghiran I, Kuo WP, Filgueira L, Martinelli R, Marti M (2016) Infected erythrocyte-derived extracellular vesicles alter vascular function via regulatory Ago2-miRNA complexes in malaria. Nat Commun 7:12727
14. Horowitz A, Riley EM (2013) Activation of human NK cells by Plasmodium-infected red blood cells. Methods Mol Biol 923:447–464. doi:10.1007/978-1-62703-026-7_31
15. Arroyo JD, Chevillet JR, Kroh EM, Ruf IK, Pritchard CC, Gibson DF, Mitchell PS, Bennett CF, Pogosova-Agadjanyan EL, Stirewalt DL, Tait JF, Tewari M (2011) Argonaute2 complexes carry a population of circulating microRNAs independent of vesicles in human plasma. Proc Natl Acad Sci U S A 108(12):5003–5008. doi:10.1073/pnas.1019055108

Chapter 32

Isolation of Extracellular Vesicles from Stem Cells

Zixin Chen, Yongjun Li, Hong Yu, Yan Shen, Chengwei Ju, Genshan Ma, Yutao Liu, Il-man Kim, Neal L. Weintraub, and Yaoliang Tang

Abstract

Transplanted induced pluripotent stem cells (IPSC) and embryonic stem cells (ESC) exhibit enhanced survival in ischemic tissues and promote survival of neighboring cells via paracrine effects. Recent studies indicate that stem cells can secrete extracellular vesicles (EV), which can shuttle noncoding RNA between cells and facilitate intercellular signaling and communication between donor stem cells and recipient tissues. Direct transplantation of IPSC-derived EV (IPSC-EV) is highly effective at promoting survival and preventing apoptosis of cardiomyocytes in a mouse model of acute myocardial ischemia–reperfusion (MI/R). Here, we describe a feasible protocol to purify EV from cultured IPSC.

Key words Induced pluripotent stem cells, Extracellular vesicles

1 Introduction

Acute myocardial ischemia–reperfusion (MI/R) injury leads to cardiomyocyte necrosis and apoptosis [1]. Because the adult mammalian heart has very limited regenerative capacity, loss of cardiomyocytes leads to scar formation, left ventricular remodeling and heart failure [2]. Recent evidence indicates that the transplantation of stem cells into the heart can promote cardiac repair via paracrine effects [3–7]. Stem cells can secret extracellular vesicles (EV) [8, 9], which carry noncoding RNAs, including microRNA (miRNA) and long noncoding RNA (lncRNA), which control gene expression via posttranscriptional repression [10–12] or epigenetic regulation [13]. In stem cells, noncoding RNAs have been show to play critical roles in modulating cell differentiation, proliferation, and survival [14, 15]. EV play an important role in cell–cell communication via delivering noncoding RNAs. We have reported that EV-mediated paracrine effects contribute for beneficial effect of cardiac progenitor cell transplantation [8]. Recently, we reported that administration of purified IPSC-EV can efficiently protects cardiomyocytes

Winston Patrick Kuo and Shidong Jia (eds.), *Extracellular Vesicles: Methods and Protocols*, Methods in Molecular Biology, vol. 1660,
DOI 10.1007/978-1-4939-7253-1_32,

from oxidative stress induced apoptosis in vitro and acute myocardial apoptosis in mouse model of acute MI/R [9].

Numerous methods have been developed to purify EV from the culture media of stem cells, such as ultracentrifugation, Polyethylene Glycol(PEG) based purification, antibody-based purification, et al. Ultracentrifugation method is the traditional method for isolating EV, although PEG purification is a more feasible approach. In this protocol, we describe a PEG4000-based protocol in which a large amount of EV can be purified from small number of cultured IPSC.

2 Materials

2.1 Equipment

1. SW 28 Ti Rotor, Swinging Bucket, Aluminum, 6 × 38.5 ml, 141,000 × *g* (Beckman Coulter, 342,204).
2. Centrifuge Tubes, Ultra-Clear™, 38.5 ml, 25 × 89 mm (Beckman Coulter, 344,058).
3. Optima™ L-90K Ultracentrifuge (Beckman Coulter).
4. Transmission electron microscope (TEM) (JEOL JEM-1230 TEM, Peabody, MA).
5. ZetaView PMX 110 (Particle Metrix, Meerbusch, Germany) and corresponding software ZetaView 8.02.28.

2.2 Reagent Setup

1. Normal mouse ES cell culture medium: DMEM high glucose, 15% ES cell qualified FBS, 1× nonessential amino acids, 1× β-mercaptoethanol, 1× antibiotic–antimycotic, 1× L-glutamine and 1000 Units/ml LIF.
2. EV-depleted FBS: Add ES cell qualified FBS equally to six Beckman Coulter Ultra-Clear centrifuge tubes (38.5 ml) with an adapter, then place centrifuge tubes with adapters into an Optima™ L-90K Ultracentrifuge with SW28 Ti Rotor and spin at 120,000 × *g* for 18 h at 4 °C, finally, collect the supernatant of each centrifuge tube, aliquot into 50 ml tubes and store at −20 °C.
3. EV depleted mouse ES cell culture medium: DMEM high glucose, 15% EV depleted ES cell qualified FBS, 1× nonessential amino acids, 1× β-mercaptoethanol, 1× antibiotic–antimycotic, 1× L-glutamine, and 1000 Units/ml LIF.

3 Methods

3.1 Cell Culture Using EV Depleted Culture Media

1. Thaw mouse IPSC cells into gelatin coated 100 mm dish, and culture with normal ES cell culture medium (Fig. 1, *see* **Note 1**).

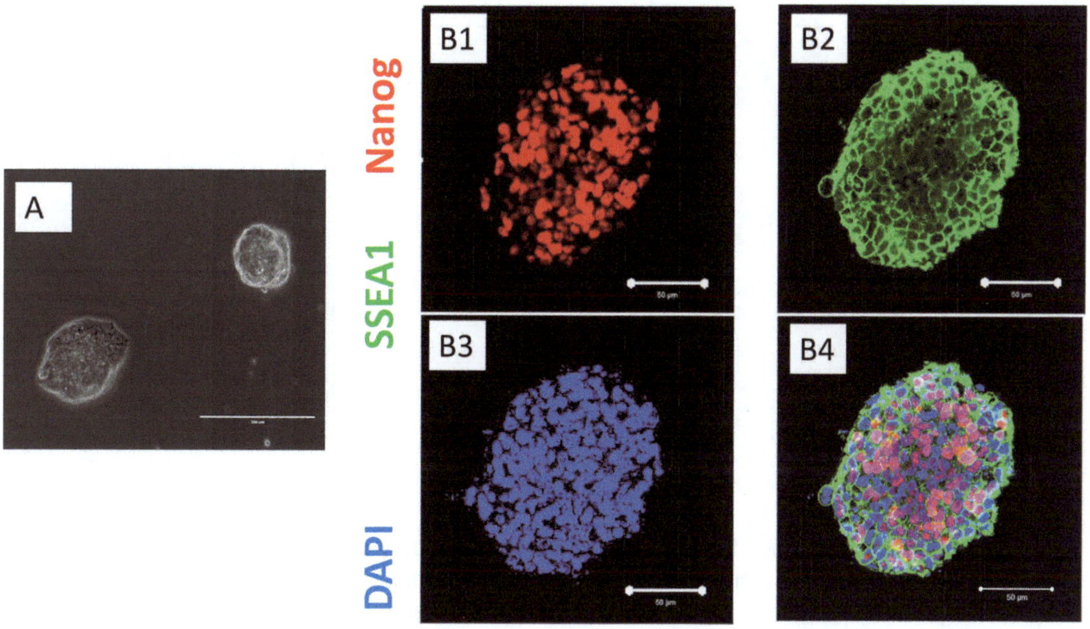

Fig. 1 Characterization of mouse induced pluripotent stem cells (IPSC). (**a**) Mouse IPSC cultured on dishes. (**b1**–**b4**) Immunofluorescent staining for the classic stem cell markers SSEA-1 (*green*) and Nanog (*red*); nuclei were counterstained with Draq5 (*blue*)

2. Dissociate mouse IPSC with HyQTase™ and passage in situ every 72 h until 70% confluent.
3. Replace mouse ES cell culture medium with EV depleted mouse ES cell culture medium.
4. Collect supernatant after 48 h, centrifuge at 1000 rpm for 10 min at 4 °C to remove cells.
5. Pass the supernatant through 0.22 μm filters to remove cell debris.

3.2 Purification of EV by Precipitation Using PEG (See Note 2)

1. Transfer 0.22 μm filtered supernatants to a 15 ml centrifuge container and add polyethylene glycol (PEG) 4000 (to final concentration of 8.5%) and NaCl (to final concentration of 0.4 M) to the supernatant and mix well by inverting.
2. Store at 4 °C overnight.
3. Centrifuge the centrifuge container at 1358 × *g* for 15 min at 4 °C to precipitate EV pellet (white).
4. Remove the supernatant, wash the pellet with 1 ml PBS one time gently.
5. Resuspend the white pellet with PBS (50 μl PBS/5 × 10^6 cells) by vigorous pipetting (*see* **Note 3**).
6. Aliquot the resuspended EV into 0.2 ml PCR tubes, and store at −80 °C.

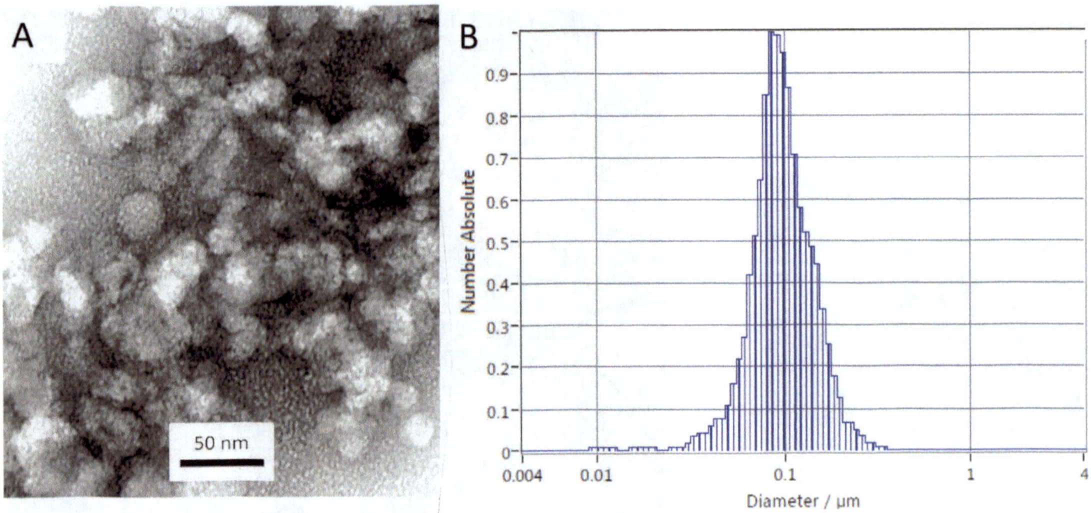

Fig. 2 Characterization of mouse IPSC-EV. (**a**) Electron micrograph image of IPSC-derived EV. The image shows a mass of round-shaped vesicles that resemble a cluster of grapes. Scale bar = 50 nm; (**b**) particle size distribution in purified pellets consistent with size range of EV (average size 100 nm), measured by ZetaView® Particle Tracking Analyzer

3.3 Imaging IPSC-EV Using Transmission Electron Microscopy (TEM)

1. Place 3 µl of PBS containing EVs on formvar carbon-coated 200-mesh copper electron microscopy grids.
2. Incubate for 5 min at room temperature (RT).
3. Perform standard uranyl acetate staining, wash the grid with three exchanges of PBS, and semidry the grid at room temperature.
4. Image the grid using TEM (Fig. 2a, *see* **Note 4**).

3.4 Measuring the Size Distribution Using Nanoparticle Tracking Analysis (NTA)

1. Dilute isolated EV samples with 1× PBS buffer to measure the particle size and concentration (Fig. 2b).
2. Record NTA measurement and analyze at 11 positions.
3. Calibrate the ZetaView system using 100 nm polystyrene particles.
4. Measure Zeta potential using 0.05× PBS instead of 1× PBS to adjust conductivity to approximately 500 µS/cm (*see* **Note 5**).

4 Notes

1. It is important to prevent IPSC differentiation under feeder-free culture conditions. We typically add PD0325901 (1 µM final concentration, MEK inhibitor, StemGent, Cat. 04-0006) and CHIR99021 (3 µM final concentration, GSK3 inhibitor, StemGent, 04-0004) to mouse ES cell culture medium to prevent spontaneous differentiation of IPSC.

2. EV can also be purified by ultracentrifugation as described in studies of embryonic stem cell- [16] and human CD34+ stem cell-derived EV [17] . Briefly, centrifuge the culture media at 217 × *g* for 5 min to remove the cells; pass the media through 0.22 μm syringe filters to remove cell debris, and use ultracentrifugation to pellet EV (100,000 × *g* for 1 h) on a 30% sucrose-D_2O solution (density ~ 1.127 g/cm^3).
3. Quantify the cell number to determine the amount of PBS needed to dissolve EV by using HyQTase cell detachment reagent to digest the cultured IPSC into single cells post media harvest.
4. Purified IPSC-EV are sticky and typically resemble a mass of round-shaped cluster of grapes under electron microscopy. Sonication is commonly used to dissociate aggregated EV [18].
5. The IPSC-EV are not stable and have a tendency to aggregate, which can be measured by zeta potential, a physical property of EV stability. We found that the zeta potential for IPS-EV is −15.44 mV at 23 °C. Nanoparticles are stable at absolute zeta potential values ranging from −31 to −40 mV [19]. The measured zeta potential indicates that these EV are not very stable in solution at 23 °C. Temperature also affects zeta potential positively; the zeta potential for IPSC-derived exosomes increases to −23.93 mV at 37 °C, indicating that EV might become more stable in higher temperature environments [9].

Acknowledgment

This work was supported NIH grants AR070029 and HL086555 (to Y.T.), and HL076684 and HL62984 (to N.L.W.).

References

1. Phillips MI et al (2002) Vigilant vector: heart-specific promoter in an adeno-associated virus vector for cardioprotection. Hypertension 39 (2 Pt 2):651–655
2. Phillips MI, Tang YL, Pinkernell K (2008) Stem cell therapy for heart failure: the science and current progress. Futur Cardiol 4 (3):285–298
3. Tang YL et al (2005) Paracrine action enhances the effects of autologous mesenchymal stem cell transplantation on vascular regeneration in rat model of myocardial infarction. Ann Thorac Surg 80(1):229–236. discussion 236–7
4. Tang YL et al (2004) Autologous mesenchymal stem cell transplantation induce VEGF and neovascularization in ischemic myocardium. Regul Pept 117(1):3–10
5. Pan A, Weintraub NL, Tang Y (2014) Enhancing stem cell survival in an ischemic heart by CRISPR-dCas9-based gene regulation. Med Hypotheses 83(6):702–705
6. Chen L et al (2013) Two-step protocol for isolation and culture of cardiospheres. Methods Mol Biol 1036:75–80
7. Tang YL et al (2013) Cardiac-derived stem cell-based therapy for heart failure: progress and clinical applications. Exp Biol Med (Maywood) 238(3):294–300
8. Chen L et al (2013) Cardiac progenitor-derived Exosomes protect ischemic myocardium from acute ischemia/reperfusion injury. Biochem Biophys Res Commun 431 (3):566–571

9. Wang Y et al (2015) Exosomes/microvesicles from induced pluripotent stem cells deliver cardioprotective miRNAs and prevent cardiomyocyte apoptosis in the ischemic myocardium. Int J Cardiol 192:61–69
10. Zhang L et al (2014) MiR-92a regulates viability and angiogenesis of endothelial cells under oxidative stress. Biochem Biophys Res Commun 446(4):952–958
11. Zhang L et al (2014) miR-92a inhibits vascular smooth muscle cell apoptosis: role of the MKK4-JNK pathway. Apoptosis 19(6):975–983
12. Tang Y et al (2015) MicroRNA-150 protects the mouse heart from ischaemic injury by regulating cell death. Cardiovasc Res 106(3):387–397
13. Tang Y et al (2012) Cross talk between the Notch signaling and noncoding RNA on the fate of stem cells. Prog Mol Biol Transl Sci 111:175–193
14. Archer K et al (2015) Long non-coding RNAs as master regulators in cardiovascular diseases. Int J Mol Sci 16(10):23651–23667
15. Bayoumi AS et al (2016) Crosstalk between long noncoding RNAs and microRNAs in health and disease. Int J Mol Sci 17(3):356
16. Khan M et al (2015) Embryonic stem cell-derived exosomes promote endogenous repair mechanisms and enhance cardiac function following myocardial infarction. Circ Res 117 (1):52–64
17. Sahoo S et al (2011) Exosomes from human CD34(+) stem cells mediate their proangiogenic paracrine activity. Circ Res 109 (7):724–728
18. Ansa-Addo EA et al (2010) Human plasma membrane-derived vesicles halt proliferation and induce differentiation of THP-1 acute monocytic leukemia cells. J Immunol 185 (9):5236–5246
19. Ostolska I, Wisniewska M (2014) Application of the zeta potential measurements to explanation of colloidal CrO stability mechanism in the presence of the ionic polyamino acids. Colloid Polym Sci 292(10):2453–2464

Part IX

Extracellular Vesicles in Mouse Models

Chapter 33

The Use of Peripheral Extracellular Vesicles for Identification of Molecular Biomarkers in a Solid Tumor Mouse Model

Noemí García-Romero, Gorjana Rackov, Cristobal Belda-Iniesta, and Ángel Ayuso-Sacido

Abstract

Extracellular vesicles (EVs) have been increasingly recognized as a potential source of disease biomarkers, since they contain a multitude of biologically active protein, DNA and RNA species, and they can be retrieved from circulating blood of patients. Here, we describe a protocol for DNA extraction from exosomes, shedding microvesicles and apoptotic bodies isolated from peripheral blood in a mouse xenograft model of solid tumor. In this model, human DNA isolated from tumor-derived EVs can be readily distinguished from the one of the hosts, which is of particular interest for studies aimed at molecular characterization of tumor biomarkers.

Key words Extracellular vesicles, Glioblastoma, Xenograft, Biomarkers, Peripheral blood

1 Introduction

Historically, the release of cellular material encapsulated in membrane vesicles was considered as a way of discarding harmful molecules or useless signaling mediators, as exemplified by the transferrin receptor removal during erythropoiesis [1, 2]. Recent data, nonetheless, suggest that these extracellular vesicles (EVs) are more than mere "molecular trash". Rather, they are now considered as units of information, which play important roles in medium and long-range intercellular communication. EV cargo contains a multitude of bioactive molecules such as lipids, nucleic acids (DNA, mRNA and miRNA) and cellular proteins, including oncogenes and active oncoproteins originating from malignant cells [3]. Moreover, EVs can be isolated from diverse body fluids: blood, urine, saliva, bronchoalveolar fluid, amniotic fluid, breast milk,

Noemí García-Romero and Gorjana Rackov contributed equally to this work.

Winston Patrick Kuo and Shidong Jia (eds.), *Extracellular Vesicles: Methods and Protocols*, Methods in Molecular Biology, vol. 1660, DOI 10.1007/978-1-4939-7253-1_33, © Springer Science+Business Media LLC 2017

semen, cerebrospinal fluid and bile [4, 5], thereby offering a unique source of disease-related biomarkers.

Molecular characterization of malignant cells is of great importance for disease diagnosis and prediction of its progress, as well as for evaluation of therapy effectiveness. However, cancer cells can only be accessed through invasive procedures such as biopsy or surgical removal of tumor tissue, which complicates frequent or long-term monitoring of molecular markers of tumor progression. In addition, tumors are often composed of heterogeneous cell populations such that different regions of tumor mass are bearing different oncogenic mutations, hence the biopsy may fail to provide representative molecular profile. To bypass these obstacles, EVs isolated from peripheral blood can be analyzed for the presence of tumor-associated biomarkers, such as mutated DNA, mRNA or proteins. Here, we describe a method to isolate EVs from peripheral blood and detect tumor-associated biomarkers in a mouse model xenotransplanted with human glioma cancer stem cells (CSCs) (Fig. 1). Of interest, cargo present in EVs derived from human tumor cells can be readily distinguished from those of the host, which makes this model suitable for studying tumor-associated biomarkers present within EVs in other types of xenograft models as well.

2 Materials

2.1 Cancer Stem Cell (CSCs) Isolation from Solid Tumor and Culture

All materials and solutions coming in contact with the cells must be sterile, and proper sterile procedures should be employed.

1. Scalpels (immerse in 70% ethanol).
2. 100 × 15-mm petri dishes (BD Falcon).
3. Glass Pasteur pipettes.
4. Hank's Balanced Salt Solution (HBSS), with or without calcium and magnesium.
5. PIPES Buffer.
6. L-cysteine–EDTA solution. Preparation: In 500 ml water (MilliQ), dissolve 2.04 g EDTA (55 mM final concentration) and 4.829 g L-cysteine hydrochloride (14 mM final concentration); sterilize using 0.22-μm filter.
7. Papain solution (10 mg/ml).
8. Solution for enzymatic digestion: mix 3.1 ml Pipes Buffer, 1.4 ml papain (10 mg/ml), and 0.5 ml L-cysteine–EDTA solution. Heat to 37 °C.
9. DNase I.
10. Defined Proliferative Medium: Dulbecco's modified Eagle's medium/F12 (DMEM/F-12) supplemented with nonessential

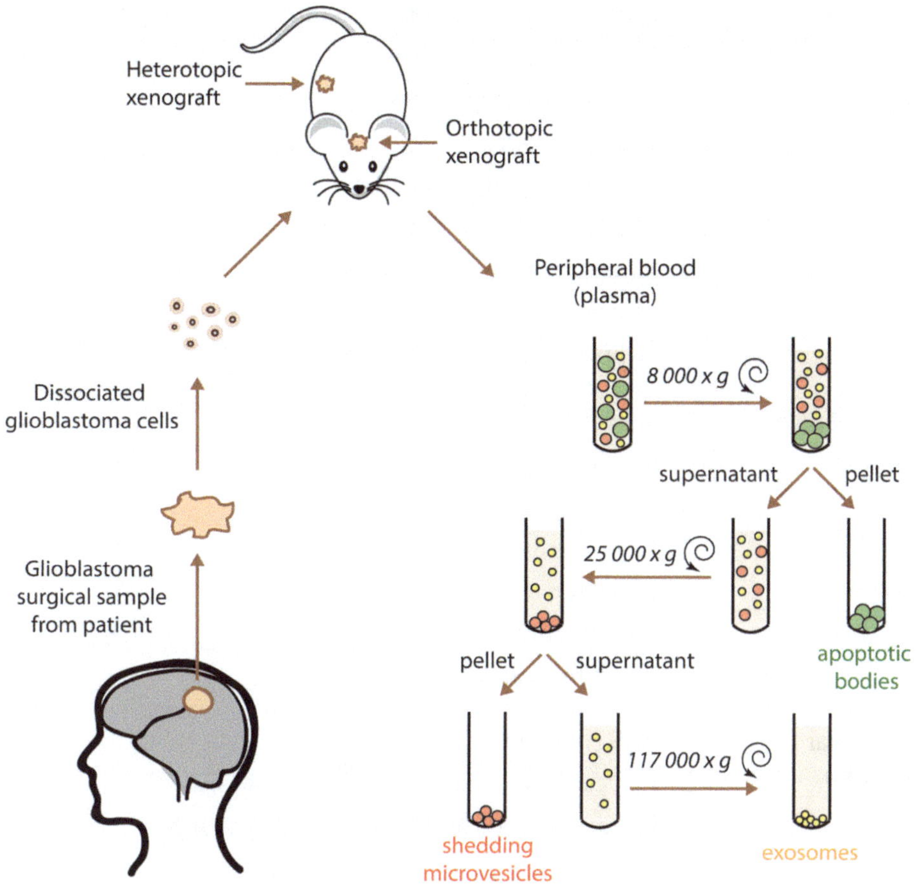

Fig. 1 Schematic view of the glioblastoma xenograft model and isolation of peripheral blood EVs by differential centrifugation

amino acids (10 mM), HEPES (1 M), D-glucose (45%), bovine serum albumin (BSA)-fraction V (7.5%), sodium pyruvate (100 mM), L-glutamine (200 mM), antibiotic–antimycotic (100×), N_2 supplement (100×), hydrocortisone (1 μg/μl), triiodothyronine (100 μg/ml), epidermal growth factor (EGF) (25 ng/μl), basic fibroblast growth factor (bFGF) (25 ng/μl), and heparin (1 μg/μl).

11. Fetal bovine serum (FBS).

2.2 Xenograft Model

1. Nine-week-old athymic nude mice of approximately 20 g weight (Charles River).
2. Hemocytometer.
3. Isofluorane anesthesia (3.5% in 70% N_2O: 30% O_2).
4. Stereotaxic frame (Stoelting).
5. Ophthalmic ointment.
6. 70% ethanol, H_2O_2, Betadine, and surgical sterile instruments.

7. PBS (Phosphate Buffered Saline).
8. High-speed micro drill.
9. Hamilton syringe.
10. 1 cm^3 syringe.
11. 21 and 22 gauge needles.
12. Manual or programmable Syringe Pump.
13. Surgical glue (Histoacryl®).
14. Matrigel™ Basement Membrane Matrix.
15. Caliper to measure tumor dimension.

2.3 Peripheral Blood Isolation

1. Surgical platform.
2. Surgical scissors.
3. 1.5 ml microcentrifuge tubes.
4. 1 cm^3 syringe.
5. 23 gauge needles.
6. 70% ethanol.
7. Benchtop centrifuge.

2.4 Extracellular Vesicle Isolation

1. PBS (phosphate buffered saline).
2. Polycarbonate or polyallomer tube.
3. Ultracentrifuge and swinging-bucket rotor.

3 Methods

3.1 CSCs Isolation from Solid Tumor and Cell Culture

Cancer stem cell (CSC) hypothesis postulates that a subpopulation of tumor cells is responsible for tumor initiation, invasive growth and metastasis formation. These cells possess the capacity of self-renewal and they generate heterogeneous cancer cell lineages within tumor. In most cases, existing therapies do not eradicate CSCs, which remain quiescent and retain the ability to regenerate tumor. There are several methods for CSC identification and isolation: (a) assessment of their ability to grow as floating spheres in serum-free medium, (b) fluorescence-activated cell sorting (FACS) according to CSC-specific cell surface markers, (c) detection of side population phenotype by Hoechst 33342 exclusion and (d) aldehyde dehydrogenase (ALDH) activity assay (Fig. 2).

1. Wash biopsy specimens in HBSS with Ca^{2+} and Mg^{2+}.
2. Remove necrotic areas and mechanically dissociate the tissue using two scalpels in a Petri Dish (*see* **Note 1**).
3. Add four volumes of HBSS without Ca^{2+} and Mg^{2+} and transfer to a 15-ml tube.

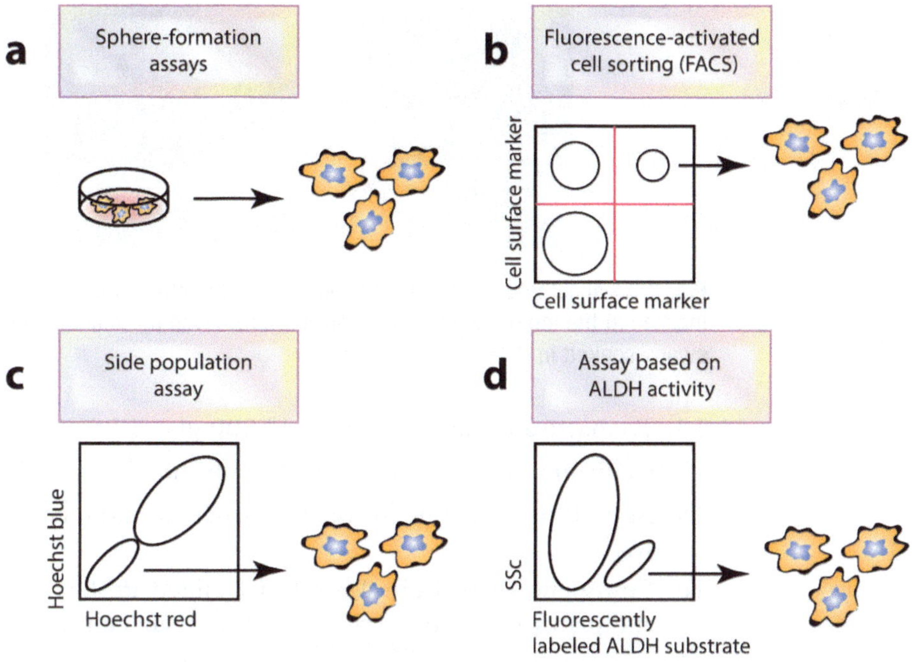

Fig. 2 Approaches for CSC identification and isolation. (**a**) CSCs form clonal spheres when plated in soft agar. (**b**) CSCs express cell-specific markers that distinguish them from non-CSC populations. (**c**) Side population analysis by FACS. The Hoechst 33342 dye is measured at two wavelengths for detection of Hoechst blue and red fluorescence. (**d**) To determine ALDH activity, cells are stained with fluorescent ALDH substrate and analyzed by flow cytometry. CSCs are characterized by low side scatter and high ALDH activity

4. Wait 2 min or until the precipitate has settled, and remove the supernatant.
5. Repeat **steps 3** and **4** one more time.
6. Add one volume of Pipes Buffer, resuspend gently and spin 15 s at 200 × *g*.
7. Remove the supernatant and resuspend the pellet in 2.5 ml of the solution for enzymatic digestion.
8. Add 10 U/ml of DNAseI and incubate at 37 °C for 30 min with rotation.
9. Centrifuge 5 min at 1500 rpm at RT and aspirate the supernatant.
10. Prepare three Pasteur pipettes with decreasing diameters of fire-polished tips.
11. To dissociate into single cells, gently pipette the sample starting with the pipette of the largest diameter, and then switch to the other two pipettes of decreasing diameter. Pipette slowly up and down ten times with each pipette.
12. Resuspend the cells in 9 ml of Defined Proliferative Media and filter through 70-μm cell strainer.

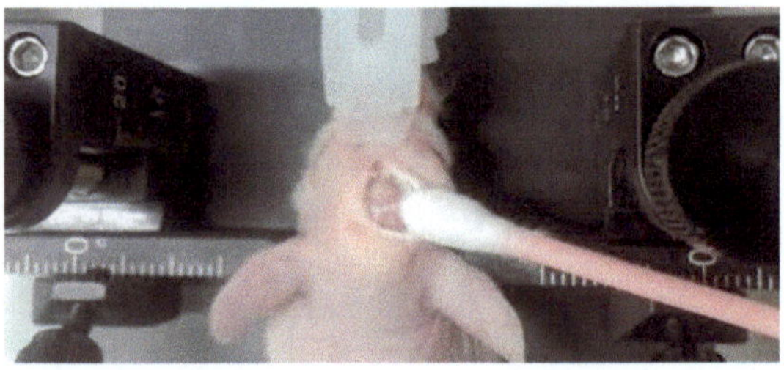

Fig. 3 An anesthetized mouse positioned in a stereotaxic frame with a midline incision at the top of the skull. Meninges are removed by cleaning the area with swabs soaked in H_2O_2

13. To inactivate papain activity, add 500 μl of FBS.
14. Centrifuge at RT 10 min at 1500 rpm and discard supernatant.
15. Add 10 ml of Defined Proliferative Media and repeat previous step.
16. Resuspend the cells in 4.5 ml of Defined Proliferative Media and plates 1.5 ml per well in a 6-well tissue culture plate.
17. Incubate the cells in a humidified 37 °C, 5% O_2 and 5% CO_2 incubator. Change the medium every 3 days.

3.2 Xenograft Model

3.2.1 Orthotopic Model (Brain Location)

1. Immediately prior to implantation, disaggregate the cells and resuspend 10^5 cells in Defined Proliferative Media (*see* **Note 2**).
2. Position the isofluorane-anesthetized mice in a stereotaxic frame with a mouth holder (*see* **Note 3**).
3. Apply ophthalmic ointment to the mice eyes to keep the moist.
4. Make a midline incision overlying the top of the skull (Fig. 3).
5. Clean the area with H_2O_2 swabs to remove the meninges.
6. Mark the position of the injection site: the striatum of the right brain hemisphere (0 mm anterior and 2.5 mm lateral to the bregma; 3.5 mm intraparenchymal).
7. Drill a burr-hole at this point through the skull without breaking the dura.
8. Clean the area with Betadine solution.
9. Inject 5 μl of CSC suspension into the hole with a Hamilton syringe (22 G needle) over 5 min (*see* **Note 4**).
10. Withdraw the needle after 5 min to avoid misplacing the cells.
11. Close the wound with surgical glue (*see* **Note 5**).
12. Return the mouse to its cage after it recovers from anesthesia (*see* **Note 6**).

3.2.2 Heterotopic Model (Flanks Location)

1. Disaggregate the cells by gentle mixing, resuspend 2–3 $\times$ 10^6 cells in 1:1 Defined Proliferative media and Matrigel (*see* **Note 7**) and place on ice.
2. Inject 250 μl of the mixture into the subcutaneous space in the mice flank, using a 1 cm^3 syringe with 21 G needle (*see* **Note 8**).
3. Tumor volumes need to be measured with a caliper at least twice a week.

3.3 Peripheral Blood Collection

1. Observe mice once or twice per week for tumor growth and symptoms. Blood should be collected after approximately 12 weeks in orthotopic and 4 weeks in heterotopic model (*see* **Note 9**).
2. This procedure requires general anesthesia and posterior animal sacrifice.
3. Place the mouse ventral side up at the working platform, disinfect the abdomen with 70% ethanol and make a ventral midline incision (2 cm) with surgical scissors.
4. Remove the skin aside to expose the abdominal cavity.
5. Find the inferior vena cava, which is located between the kidneys, lying laterally on the right side of abdominal aorta.
6. Use a 1 cm^3 syringe with a 23 G needle and introduce carefully into the vein. Apply suction and wait until blood appears in the syringe (*see* **Note 10**). Repeat until blood flow ceases (up to 1 ml can be obtained) (*see* **Note 11**).
7. Centrifuge 10 min at 1500 $\times$ *g* to obtain plasma (*see* **Note 12**). Store plasma at −80 °C or continue with the protocol.

3.4 Isolation of Extracellular Vesicles

Extracellular vesicles (EVs) are categorized into three main classes: (1) shedding microvesicles (derived by outward budding of plasma membrane), (2) exosomes (derived by inward budding of endosomal membrane) and (3) apoptotic bodies (originating from apoptotic blebs during apoptosis) [6]. All three types of EVs contain DNA, which can be used for biomarker detection. Recently, several methods for isolation and purification of EVs have been developed, including antibody-coated beads (Exosome-Streptavidin for Isolation/Detection from Invitrogen, ExoPureimmunobeads™ from BioVision), microfluidic devices, precipitation technologies (total exosome isolation from Invitrogen, ExoQuick™ from SystemBiosciences, miRCURYTM from EXIQON, ExoPure™ Reagent from BioVision), or filtration technologies. Most of these platforms are optimized for exosome isolation, they are not suitable for isolation of large amounts of EVs and captured EVs may not retain functionality after elution from the beads [7]. Here, we describe differential centrifugation protocol, which is the gold standard for EV isolation and purification. This protocol involves several centrifugation and ultracentrifugation steps, although it may lead to inconsistencies in recovery of EVs because of different biofluid viscosity.

1. Centrifuge plasma samples 5 min at 1200 × *g* to remove cellular debris (*see* **Note 13**).
2. Transfer the supernatant to a specific centrifuge tube (*see* **Note 14**) and centrifuge for 20 min at 8000 × *g*.
3. Transfer the supernatant to a new tube and resuspend the pellet in PBS to obtain apoptotic bodies (*see* **Note 15**).
4. Centrifuge this supernatant 20 min at 25,000 × *g*. Transfer the supernatant to a new tube and resuspend the pellet in PBS to obtain shedding microvesicles.
5. Ultracentrifuge the remaining supernatant 90 min at 117,000 × *g* to obtain exosomes.
6. Repeat the ultracentrifugation step and store the EV samples at −20 °C until use or continue with the protocol (*see* **Note 16**).

3.5 gDNA Biomarkers

1. Treat EV samples with DNase I (54 Kunitz/ml) for 30 min at 37 °C to remove DNA contamination.
2. Perform Polymerase Chain Reaction (PCR) of a housekeeping gene to ensure DNase I activity and analyze PCR products on 1.8% agarose gel (*see* **Note 17**).
3. Extract DNA from each EVs fraction using a DNeasy Blood and Tissue Kit (Qiagen) using manufacturer's recommendations.
4. Detect gene mutations, minority alleles, point mutations, small deletions or amplifications, and fusion genes with clinical or biological value.

4 Notes

1. Fresh human tissue must be processed within 12 h after extraction.
2. Volume and cell concentration can be adjusted depending on the assay.
3. All animal experiments must be approved by and performed according to the guidelines of the Institutional Animal Care Committee.
4. Clean the syringe with 70% ethanol, MilliQ H_20 and PBS before and after its use, by drawing up multiple times.
5. Instead of surgical glue, sterile silk sutures could be used.
6. Mice should be placed in an electric blanket to recover from anesthesia in order to maintain corporal temperature.

7. All experiments with Matrigel need to be carried out at 4 °C, since it solidifies at room temperature and at −20 °C. Matrigel should be thawed out in ice or at 4 °C overnight. Matrigel that has solidified cannot be reused. Syringes and material that contacts the Matrigel must be kept cold. When pipetting Matrigel, do not make any bubbles.
8. Anesthesia is recommended but not required for this procedure; its application facilitates the experiment and minimizes stress in mice.
9. Tumor growth depends on multiple factors such as origin of CSCs, animal model of choice, growth factors, etc.
10. If blood stops flowing, rotate the needle and pull it out lightly.
11. In case that vein collapses, an alternative assay can be performed, such as cardiac puncture [8].
12. There is no consensus regarding to the choice between serum and plasma [9], although some authors have postulated that more EVs can be recovered from serum than plasma [10].
13. All steps must be performed in a laminar flow cell culture hood and all centrifugations must be performed at 4 °C. *g*-force data are estimated for an Optima-LE 90K ultracentrifuge, 50.2 Ti rotor, Beckman Coulter. To convert *g* force to RPM, the following formula should be used: Relative Centrifuge Force (RCF) or *g*-force $= 1.12 \times R \times (\text{RPM}/1000)^2$. *R*—rotor radius (mm), RPM—revolutions per minute.
14. Polycarbonate or pollyallomer sterile tubes are needed depending of the ultracentrifuge rotor used.
15. Volume of PBS to resuspend the pellet depends on the starting volume and EVs concentration.
16. It is desirable to store in small aliquots, since one freeze and thaw cycle reduces the number of EVs up to 15% [11].
17. No amplification products should be detected in agarose gel. If so, repeat DNase I treatment, to remove DNA contamination.

Acknowledgment

This work was supported in part by grants from Health Research Fund (Fondo de Investigaciones Sanitarias, FIS), Carlos III Health Institute (PI10/01069, CP11/00147, and PI14/00077), Plan Estatal I + D + I 2013–2016, and Ministerio de Economía y Competitividad (RTC-2015-3846-1) (A.A.S.). The funders had no role in study design, data collection and analysis, decision to publish, or preparation of the manuscript. Noemí García-Romero and Gorjana Rackov contributed equally to this work.

References

1. Harding C, Heuser J, Stahl P (1983) Receptor-mediated endocytosis of transferrin and recycling of the transferrin receptor in rat reticulocytes. J Cell Biol 97:329–339
2. Pan BT, Teng K, Wu C et al (1985) Electron microscopic evidence for externalization of the transferrin receptor in vesicular form in sheep reticulocytes. J Cell Biol 101:942–948
3. Al-Nedawi K, Meehan B, Rak J (2009) Microvesicles: messengers and mediators of tumor progression. Cell Cycle 8:2014–2018
4. Chaput N, Théry C (2011) Exosomes: immune properties and potential clinical implementations. Semin Immunopathol 33 (5):419–440
5. Raposo G, Stoorvogel W (2013) Extracellular vesicles: exosomes, microvesicles, and friends. J Cell Biol 200:373–383
6. Borges FT, LA R, Schor N (2013) Extracellular vesicles: structure, function, and potential clinical uses in renal diseases. Braz J Med Biol Res 46:824–830
7. Momen-Heravi F, Balaj L, Alian S et al (2013) Current methods for the isolation of extracellular vesicles. Biol Chem 394:1253–1262
8. Doeing DC, Borowicz JL, Crockett ET (2003) Gender dimorphism in differential peripheral blood leukocyte counts in mice using cardiac, tail, foot, and saphenous vein puncture methods. BMC Clin Pathol 3:3
9. Witwer KW, Buzás EI, Bemis LT et al (2013) Standardization of sample collection, isolation and analysis methods in extracellular vesicle research. J Extracell Vesicles 2:1–25
10. James GN, Thoi LL, McManus LM, Reimann TA (1982) Isolation of human platelet membrane microparticles from plasma and serum. Blood 60:834–841
11. Jayachandran M, Miller VM, Heit JA, Owen WG (2012) Methodology for isolation, identification and characterization of microvesicles in peripheral blood. J Immunol Methods 375:207–214

Part X

Therapeutic Applications of Extracellular Vesicles

Chapter 34

Therapeutic Applications of Extracellular Vesicles: Perspectives from Newborn Medicine

Gareth R. Willis, Stella Kourembanas, and S. Alex Mitsialis

Abstract

With the advancements in antenatal steroid therapies and surfactant replacement, current clinical practices in neonatal intensive care units allow the survival of infants at very low gestational age. Despite these advances, there continues to be significant morbidity associated with extreme preterm birth that includes both short-term and long-term cardiorespiratory impairment. With no effective single therapy in preventing or treating developmental lung injuries, the need for new tools to treat and reduce risk of complications associated with extreme preterm birth is urgent. Stem cell-based therapies, in particular therapies utilizing mesenchymal stem (stromal) cells (MSCs), have shown promise in a number of animal models of lung pathologies relevant to neonatology. Recent studies in this field have consolidated the concept that the therapeutic mechanism of MSC action is paracrine, and this led to wide acceptance of the concept that the delivery of the MSC secretome rather than live cells may provide an alternative therapeutic approach for many complex diseases. Here, we summarize the significance and application of cell-free based therapies in preclinical models of neonatal lung injury. We emphasize the development of extracellular vesicle (EV)-based therapeutics and focus on the challenges that remain to be addressed before their application to clinical practice.

Key words Extracellular vesicles (EVs), Exosomes, Therapeutics, Newborn medicine, Bronchopulmonary dysplasia (BPD), Mesenchymal stem cells (MSCs), Exosomes, Preclinical

1 Introduction

Infants born at very low gestational age often suffer from several diseases associated with premature birth, including periventricular leukomalacia (PVL), retinopathy of prematurity (ROP), hypoxic–ischemic encephalopathy (HIE) and bronchopulmonary dysplasia (BPD) [1]. Despite medical care advances in perinatal–neonatal medicine, cardiorespiratory impairments prevail in constituting a major adverse outcome in neonatal intensive care unit survivors. A recent study reporting on the 20-year survival and outcome trend of >34,000 preterm infants, reported that the prevalence of several morbidities has decreased, however the frequency of BPD has increased [2]. Specifically, between 2009 and 2011, BPD rates

Winston Patrick Kuo and Shidong Jia (eds.), *Extracellular Vesicles: Methods and Protocols*, Methods in Molecular Biology, vol. 1660, DOI 10.1007/978-1-4939-7253-1_34,

increased across infants of 22–27 weeks of gestational age. BPD remains one of the most common complications of preterm birth occurring in ~35% of infants born at ≤28 weeks gestation, affecting in excess of 10,000 infants every year in the USA [3].

BPD occurs predominantly in preterm infants receiving oxygen therapy and mechanical ventilation. It is characterized by lung growth arrest, diminished alveolar and blood vessel development and impaired pulmonary function [4]. Several BPD outcomes remain firmly associated with long-term pulmonary complications, such as abnormal pulmonary function test results and airway hyper-responsiveness. An increasing number of young adults, who were former preterm infants often present with emphysema-like features of lung disease [5]. Notably, secondary pulmonary hypertension has been associated with moderate to severe cases of BPD and is linked to increased mortality [6, 7]. Thus, BPD is no longer considered a disease of the neonatal period, but a complex condition that can afford lifelong consequences. To date, effective treatments remain absent. Inhaled nitric oxide, antioxidant therapies and vitamin A are just some of the alternative drug approaches that have been investigated in BPD but have either failed or demonstrated a minimal effect on clinical outcomes. Ultimately, there is a persistent need to deliver an effective treatment strategy for the prevention and management of BPD.

Herein, we summarize current knowledge on the therapeutic applications of stem/progenitor cells, in particular mesenchymal stem cells (MSCs), and focus on their protective role in experimental models of neonatal lung injury. We review MSC therapeutic mechanisms of action with emphasis on the development of extracellular vesicle (EV)-based therapeutics and focus on the challenges that remain to be addressed before their application to clinical practice.

2 Stem Cell-Based Therapies in Newborn Medicine

Although antenatal steroid therapies and surfactant replacement have increased the survival rates of infants born at low gestational age, these infants remain vulnerable to the detrimental effects of oxidative stress and mechanical ventilation, and as a result are prone to develop lung and brain diseases. BPD is characterized by significant pulmonary architecture simplification and developmental arrest, and several investigators have noted a loss of stem/progenitor cell populations that may underlie the observed tissue simplification [1, 8, 9]. Such findings provided the basis for the application of stem cell-based therapies to ameliorate the injury associated with preterm birth. Transplantation of different stem cell types, such as MSCs, endothelial progenitor cells (EPCs) and human amnion epithelial cells (AECs) have shown promise in preclinical models for the prevention and/or treatment of BPD and other major sequelae of prematurity. Such therapies have been effective in providing neuroprotection, neuroregeneration, and improvement of functional outcomes in preclinical models of neonatal hypoxia–ischemia, cerebral palsy and stroke, reviewed in Refs. [1, 3, 10–14].

EPCs are a subpopulation of mononuclear cells found in the bloodstream that possess the ability to differentiate into endothelial cells. With notable lung vascular growth arrest and a depleted level of EPCs in infants with BPD [9], researchers hypothesized that EPC therapy would be of benefit to BPD infants. Baker and colleagues showed that administration of conditioned media (CM) derived from late-outgrowth endothelial colony-forming cells (ECFCs), a type of circulating EPC, promoted angiogenesis in vitro and prevented pulmonary hypertension [15]. However, although this therapeutic approach decreased right ventricular hypertrophy, no improvement in alveolarization was found, which in turn may limit its therapeutic potential. Cells derived from other organs such as the human amniotic epithelium (AECs) have also been explored as therapeutic in preclinical models of BPD. These cells are considered "stem-cell like" as they possess differentiating capabilities (differentiating into mesodermal, ectodermal, and endodermal lineages in vitro) [16]. In an in utero ovine model of ventilation-induced BPD, Hodges and coworkers found that intravenous (IV) plus intratracheal (IT) delivery of AECs successfully ameliorated lung structure and reduced fibrosis with limited evidence of cell engraftment and transdifferentiation events, concordant with the concept that paracrine mechanisms of protection are also involved in these processes [17, 18].

3 MSC Based Therapies for BPD

There is a burgeoning awareness in the potential therapeutic applications of MSCs, since their immunomodulatory capabilities have shown promise in preclinical models of BPD and neonatal hyperoxia. Although MSC-based therapies have shown promise in numerous disease states, herein we focus on the use of MSCs in BPD. To date, preclinical BPD studies have predominantly focused on stem cells harvested from bone marrow, umbilical cord blood and adipose tissue. More often, these cells are tested in the hyperoxia-induced neonatal BPD murine and rat models, with results that show an improvement in lung development and a notable suppression of inflammatory mediators such as tumor necrosis factor-α (Tnfα) and interleukiun-6 (Il6) [19]. In preclinical models of hyperoxia-induced BPD, our research group [20] and others [21–23] have shown that administration of bone marrow derived-MSCs (BMSCs), can protect against the

BPD changes in lung architecture and vascular remodeling as well as improve exercise capacity and increase survival rates compared to control animals in rodent models.

Intrigued by the observation that MSC injections result in robust physiologic responses despite the absence of significant engraftment of donor cells in the recipient lung, our group demonstrated that the cell-free media conditioned by MSCs afforded superior protection over BMSCs in preserving the lung architecture and preventing alveolar loss [20, 24]. Several other groups using rodent BMSCs or MSCs derived from umbilical cord blood and their cell-free CM have shown similar results [19, 25]. Notably, a bolus IV dose of cell-free BMSC-CM delivered after 14 days of hyperoxia exposure (75% O_2), when the disease had already developed, inhibited lung inflammation, reversed pulmonary hypertension and the pruning of the distal vascular tree and markedly ameliorated lung fibrosis, collagen deposition, and alveolar destruction, although the lungs manifested some degree of emphysema-like features compared with normoxic controls [24].

The active moieties that confer the therapeutic efficacy of the MSC secretome remain unclear but likely include secreted proteins, nucleic acids, and membrane components. Such bioactive material maybe selectively packaged in EVs which comprise a major phenomenon of the MSC secretome, in turn acting as potent bioactive signaling vectors (as discussed below). The MSC secretome represents an exciting and innovative approach to therapeutic interventions for lung diseases because it bypasses safety concerns associated with live cell treatments. We must keep firmly in mind that generalizations in this relatively early stage of mechanistic understanding can be misleading. For example, Islam et al, argue for mechanisms inherent in live MSCs rather than merely their secreted products as possibly active in certain models [26]. This is based on the demonstration that mitochondrial transfer from MSCs to resident cells confers protection against lung injury; albeit the mechanism(s)-of-action are likely to be multifactorial. Taken together, the clinical application of stem cell therapies for neonatal lung disease represents a great promise in the near-term horizon, although much preclinical work remains to be done.

Considering broader lung-associated applications, administration of BMSCs or their secretome has been shown to be effective in ameliorating injury and/or reestablishing homeostasis in several preclinical models of lung disease. Among others, these include pulmonary hypertension [27], acute lung injury [28], pulmonary fibrosis [24], chronic obstructive pulmonary disorder (COPD) [29], and asthma [30, 31]. Combined, the above studies support a cytoprotective role for MSCs and/or the MSC secretome in preventing and reversing injury in several neonatal and adult rodent models of lung disease. MSCs exhibit numerous properties , making them ideal candidates for cell therapy. Firstly, they represent an ethically noncontroversial and easily accessible source of multipotent stem cells contrasted to embryonic stem cells [3]. Furthermore, their safety, immunomodulatory potential and regenerative capacity has been previously demonstrated in approved clinical protocols, where MSC therapy is used during allogenic hematopoietic stem cell transplantation, to reduce the incidence or severity of acute graft-versus-host disease (GvHD) [32]. Moreover, MSC-based therapies have already been demonstrated to be effective in other preclinical models of neonatal diseases, including hypoxic–ischemic encephalopathy. They are the most commonly used adult stem cell in regenerative medicine, where in excess of 500 MSC-based clinical trials appear in the database of the US National Institutes of Health to date. To date, a Phase I dose escalation clinical trial has been conducted looking at the administration of MSCs for the treatment of BPD [33]. Although the phase I study design does not allow conclusions to be drawn in regards to the efficacy of MSCs, this study is important for its demonstration of feasibility and short-term safety. The precise mechanism(s) of how MSCs exert beneficial effects remain elusive. A ubiquitous theme with stem cell based-therapies to date is that administered MSCs in lung disease appears to be immunomodulatory and paracrine in nature, where CM has been shown to be as efficacious as, or even more potent than cell transplantation.

4 Cell-Free Based Therapies: Extracellular Vesicles

It has been shown that the therapeutic capacity of MSCs is comprised in their secretome, and our recent studies demonstrated that the therapeutic vector is represented by the EVs they release. EVs are small, submicron plasma membrane enclosed particles (Fig. 1). Their production is widespread, as vesicles are released by bacteria, eukaryotes, and archaea. EVs, and particularly the class generated through the endocytic pathway (exosomes), were originally assumed to represent just a mechanism for the cell to jettison unwanted moieties. They are currently not only postulated to play a pivotal role in health and disease but also to represent reagents for therapeutic applications, leading thus to an active research theme for potential diagnostic and prognostic applications across multiple disciplines. As the focus of a relatively new and multidisciplinary research community, the nomenclature, clarification, and detection of these vesicles remains abstract. Although the International Society for Extracellular Vesicles (ISEV) is pivotal in the process to clarify these uncertainties, typically EVs have been categorized into several classes (including exosomes, microvesicles (MVs) and apoptotic bodies). These vesicle subsets were named based on their biogenesis and biophysical properties, such as size, density, lipid composition, subcellular origin (i.e., membrane derived or

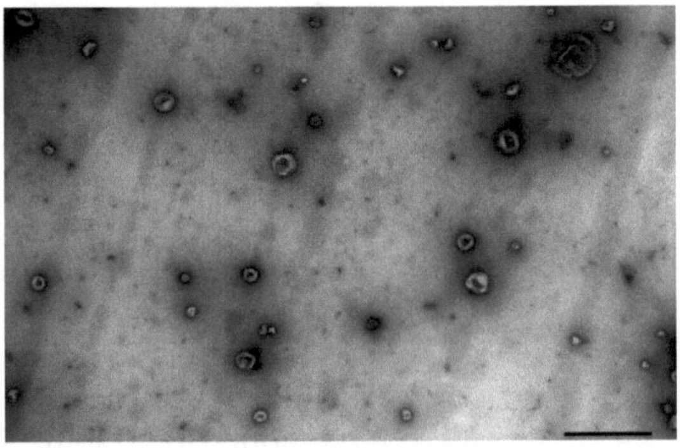

Fig. 1 Electron microscopy image of MSC-EVs. Briefly, EVs were harvested from confluent human Wharton's jelly derived-MSCs (WJ-MSCs) over 36 h in serum-free-media. Following differential centrifugation and tangential flow filtration (50-fold concentrate, 300 kDa membrane), MSC-EVs were purified from enriched culture medium via a gradient density cushion (iodixanol). Here, 6 ml of EV-enriched media was floated over 10% IDX (3 ml) and 55% IDX (3 ml). Twelve (1 ml) gradient fractions were taken. The representative image shows highly pure EVs from fraction 9 that correspond with a density of ~1.2 g/cm^3. Scale bar: 500 nm

intracellular) and predominant protein markers. However, with such a diverse and heterogeneous nature, it is increasingly difficult to uphold one terminology, thus as per ISEV suggestions [34], the term EV will be used herein to encompass both exosomes (vesicles <150 nm diameter) and MVs (>150 nm diameter).

Eukaryotic EV release is a process where the precise cellular and cytoskeletal events have been well studied, although certain crucial details remain unclear. Although cell membrane budding is acknowledged to contribute to EV production, typically, EV biogenesis, as relevant to the EV class considered to be the "true exosomes" is thought to involve the formation of intraluminal vesicles within multivesicular bodies (MVB). MVB are formed by the invagination (reverse budding) of the membrane. During the formation of the MVB, internalized membrane receptors/proteins are often silenced and degraded, a process which often requires the ubiquitination of the receptor/protein. This posttranslational modification leads to the subsequent interaction with the endosomal sorting complex required for transport (ESCRT) machinery. Mature MVBs fuse with the plasma membrane to release their content/secrete EVs to the extracellular environment. Intracellular trafficking and fusion of compartments classically involve small GTPases of the Rab family. The final step of EV release involves the fusion of MVBs with the plasma membrane and expulsion of EVs. This process is likely to involve a specific combination of soluble attachment protein receptor (SNARE) proteins (a membrane bound, multi-subunit protein complex), reviewed in Refs. [35–39].

The exact composition of a vesicle is variable and depends upon the cellular origin, cellular stimuli and biogenesis. During their biogenesis, EVs engulf an array of bioactive cargo from their parental cell. Although EV cargo has been reported to include genetic information in the form of DNA, ribosomal RNA, mRNA and noncoding RNAs such as microRNA, bioactive free fatty acids (FFAs) and protein, it is unclear if all such moieties are actively and specifically packaged into the EVs or if their detection merely indicates contamination of the EV preparations with diverse cellular detritus, in particular fragments of apoptotic bodies. Typical protein markers for EVs include flotillin1 (FLOT1), Alix (PDCD6IP), TSG101, and the tetraspanins CD9, CD63 and CD81, among others. The bioactive cargo can reflect the stimulus that triggered EV formation, suggesting specific packaging of "*message*" prior to export from the parent cell. In turn, EVs may represent a novel means of cell–cell communication [3, 36, 40–42].

The precise mechanism(s) by which EVs interact with cells, deliver their cargo and induce functional responses have not been well elucidated, but EV–target cell interaction is likely to be cell and EV specific. EV-target cell fusion and/or uptake may involve different endocytic pathways, including but not limited to direct membrane fusion but also clathrin-mediated and caveolin-mediated endocytosis, pinocytosis, and micropinocytosis, depending on the cell type [43]. Even after the uptake of EVs by a target cell, the fate of its bioactive cargo remains unclear and the clearance of vesicle components in target cells is also likely to depend on the route of uptake.

Emerging evidence has demonstrated integrin activation, phosphatidylserine (PS)-dependent internalization of EVs, IL-1 receptor activation, sonic hedgehog signaling, free fatty acid (FFA) adoption and microRNA transfer as possible mediators of the EV-induced biological effect [44–46]. Numerous studies have shown the transfer of EV-microRNA content elicits the beneficial EV mediated effects in targets cells [45, 47, 48]. Specifically, MSC-EV microRNA mediated intercellular communication was investigated using a cel-miR-67 luciferase reporter system in an in vitro MSC-astrocyte coculture model. Xin et al [49], found that microRNAs in MSCs were transferred to astrocytes via MSC-EVs (50 and 100 nm in diameter). They noted that cultured neurons treated with EVs significantly increased the neurite branch number and total neurite length. miR-133b levels were also increased in cultured neurons and astrocytes treated with MSC-EVs. Moreover, knockdown of miR-133b in MSCs confirmed that the increased miR-133b level in astrocytes is attributed to their transfer from MSCs, in turn providing evidence that MSCs communicate with brain parenchymal cells and may regulate neurons by EV-mediated transfer of microRNA. Although the mechanism of action needs to be addressed, with the potential broad application of MSC EV-based therapeutics, a single mechanism might not be definable and specific to all diseases. Clearly, a significant effort must be directed toward elucidating the detailed mechanisms of EV function and the active moieties responsible for their therapeutic action. This represents an urgent task, since EVs have been shown to be highly potent and to exert diverse effects in a spectrum of in vitro and in vivo models. On balance, the presence of heterogeneous EV populations, which is a consequence of significant diversity in EV biogenesis and the lack of standardized isolation protocols, coupled with potentially diverse mechanisms by which EVs interact with target cells, represents the most prominent challenges to overcome in the near term for the rational design and development of EV-based therapeutics.

5 Therapeutic Application of MSC-EVs

The interest in the therapeutic potential of EVs has exploded in recent years. Several lines of evidence suggest that MSCs act via a paracrine mechanism to protect the neonatal lung. In support, in vitro studies have shown that MSC-CM prevented alveolar epithelial cell (II) apoptosis, accelerated cell wound healing, and augmented endothelial cord formation following a hyperoxic insult [21]. Other in vitro studies have shown BMSC-CM protects cardiomyocytes via the mitochondria-mediated apoptotic pathway. Here, Xiang and coworkers found that the application of BMSC-CM to cardiomyocytes exposed to hypoxia–reoxygenation reduced apoptosis through inhibition of the release of cytochrome C from mitochondria and reduction of caspase-3 activation [50]. In vivo, our group, and others have inferred the therapeutic actions off MSCs are also paracrine. Previously, we reported that a bolus dose of concentrated mouse BMSC-CM (mBMSC-CM) was even more beneficial than live BMSC injections in a hyperoxia-induced neonatal mouse model of BPD, in improving alveolar simplification and reducing fibrosis [20]. In an extension of these studies, using neonatal mice exposed to 75% O_2 from post-natal day-1 (PND1) to PND14 to initiate lung injury, we demonstrated that administration of mBMSC-CM at a dosage of 10 μg per animal, reversed O_2-induced parenchymal fibrosis and peripheral devascularization (pruning), improved lung alveolar development and fully reversed pulmonary hypertension and right ventricular hypertrophy, compared to an equivalent dose mouse lung fibroblast CM [24]. Other groups have found similar effects using cord blood-derived MSC-CM, where the beneficial effects were still evident for >3 months. In a rat BPD model, other cellular sources of MSCs (including umbilical cord blood) have shown similar cytoprotective effects [25].

The clearance of such exogenous EVs in vivo is believed to involve mononuclear cells (such as macrophages). Interestingly, a spectrum of risk factors may act synergistically to induce an injurious inflammatory response in the airways and the pulmonary interstitium of preterm infants with BPD. This inflammatory response is characterized by an accumulation of macrophages as well as an arsenal of proinflammatory mediators that may be responsible for acute lung injury [51]. MSCs are capable of modulating the immune system and have been shown to modulate the macrophages response to injury, directing macrophage polarization states from M1 (classical, proinflammatory state) to the alternative M2 macrophage polarization which is associated with anti-inflammatory activity [52]. In support, our group has found that highly purified EVs produced by human Wharton's Jelly MSCs (WJ-MSCs) can readily modulate the polarization of primary mouse bone marrow-derived macrophages and an immortalized line of mouse alveolar macrophages (MHS cell line, Willis et al., manuscript in preparation). To supplement this line of thought, Tropea and coworkers have also demonstrated that mBMSC-CM increased the number of resident bronchoalveolar stem cells in neonatal mice exposed to hyperoxia [53]. This was interpreted to indicate that mBMSC-CM appears to possess potent proliferating and/or differentiating factors enhancing resident stem cell survival, a process which undeniably would support the development of the neonatal lung and restore homeostasis in a state of injury.

In addition to EVs, mBMSC-CM contains a plethora of soluble growth factors that have shown known to be cytoprotective in the lung, including, vascular endothelial growth factor (VEGF), adiponectin, staniocalcin-1 and TNF-6. Using mass spectroscopy analysis of mBMSC-CM, our group previously identified a high abundance of extracellular space proteins, including matrix components, with fibronectin (Fn1), heparan sulfate proteoglycan (Hspg2; perlecan), and collagen (Col6a3) that accounted for more than 20% of the peptides detected [20]. The remaining CM proteome is complex. The majority of the protein species detected could be classified into four overlapping functional groups, namely molecules associated with (1) proliferation and apoptosis, (2) cell–cell interactions and cell motility, (3) immune responses and inflammation and (4) respiratory disease. Other abundant proteins of interest identified in the BMSC-CM included osteopontin and macrophage colony stimulating factor 1 (M-csf1). However, although the therapeutic capacity of all MSC-CM constituents remains to be investigated, newly adopted biologic drugs in isolation have not afforded significant protection in BPD to date.

In other diseases where ischemia prevails, MSC-CM has also shown promise. For example, in a porcine model of myocardial infarction, CM treatment (IV dose: 1 ml/2.0 mg followed by a intracoronary dose: 4 ml/8 mg) resulted in a 60% reduction in infarct size coupled with an improved systolic and diastolic cardiac performance [54]. Later work from the same research group demonstrated that EVs (<200 nm vesicles) were the bioactive constituents of CM. EVs purified by sucrose density gradient coupled with HPLC were confirmed to contain CD9, CD81 and Alix. These purified EV fractions were cytoprotective in a murine model of myocardial ischemia–reperfusion [55]. Taken together this supports the notion that EVs harness the therapeutic capacity witnessed in the CM.

We have demonstrated that EVs mediate the cytoprotective effect of BMSCs in hypoxia-induced pulmonary hypertension [27]. Here, administration of MSC-EVs, identified through widely accepted EV markers and visualized by electron microscopy, protected against the elevation of right ventricular systolic pressure and the development of right ventricular hypertrophy after 3 weeks of hypoxic exposure. Importantly, EV-depleted CM had no effect. EV treatment was also able to abrogate early hypoxic macrophage influx and downregulate hypoxia-activated inflammatory pathways, thus mediating the anti-inflammatory properties of MSCs. Similarly, Cruz and colleagues found that the administration of either MSC-CM or isolated MSC-EVs derived from bone marrow and other sources are as effective as the MSCs themselves in mitigating inflammation and injury in an immunocompetent mouse model of allergic airway inflammation [31]. Compared to fibroblast derived-EVs, they found MSC-EVs (3×10^6 MSC equivalents, isolated by ultracentrifugation (UC), $100{,}000 \times g \times 1$ h) that were administered at the onset of antigen challenge, significantly ameliorated the inflammatory-provoked increases in airway hyper-responsiveness, lung inflammation, and the antigen-specific CD4 T-cell Th2 and Th17 phenotype. Interestingly, EVs from human MSCs were generally more potent than those from mouse MSCs for the majority of the outcome measures [31].

In vitro studies have noted MSC-EVs are beneficial in models of immune disease via immunomodulatory actions, regulating M2-like phenotypes in monocytes [56]. MSC-EVs have been shown to improve acute kidney injury in several preclinical models [57] and the transfer of MSC-EV microRNAs have shown promise in improving neurological disorders in rats subjected to middle cerebral artery occlusion [58]. Collectively, the above studies clearly identify that MSC-CM, and specifically isolated EV populations are highly effective in treating a diverse array of aliments. The pleiotropic effects of MSC-EVs represent an exciting opportunity to develop an innovative therapy for complex multifactorial diseases such as BPD. Strikingly, in an isolated clinical treatment attempt for a steroid-resistant grade IV GvHD patient, MSC-EVs suppressed symptoms over time without revealing any side effects [59]. Considering the array of diseases that MSCs and subsequent MSC-EVs may benefit, it is reasonable to suggest that rather than a singular mechanism of action, the rich collection of bioactive ingredients may act upon several cellular pathways, with the precise therapeutic action being target cell and disease specific. Indeed, it has been demonstrated that MSC-EVs favor migration to damaged tissue in comparison to healthy tissue [21, 60] and that in experimental control groups, MSCs have little effect. Thus, MSC-EV based-therapies may act as specific signaling vectors that aim to rescue damaged tissues and restore homeostasis.

6 Application and Regulation of EV Therapies

EV-based therapeutics may represent the next breakthrough for diseases of the newborn, however significant technological, analytical, safety and functional features are still to be addressed. Additionally, to bring EV-based therapies to clinic, their development and application is subject to compliance with regulatory frameworks concerning biological medicinal products.

7 Stem Cell Origin of EVs

Although primary MSC cells are commercially available, researchers often isolate MSCs from numerous species and sources. MSCs can be obtained from several tissues, including bone marrow, Wharton's jelly, umbilical cord blood, and adipose tissue, among others [61]. Historically bone marrow has been the default source for the derivation of MSCs for human stem cell engineering and EV harvesting. Human BMSCs show notable changes with donor age (these include a reduction in cell number, longer population doubling times, and lower differentiation potential in vitro) [62–66]. It remains unclear as to what effect bone marrow donor demographics have on subsequent EV production and should be addressed. Bone marrow extraction also comes with ethical constraint and problems associated with painful harvesting and donor site morbidity. With no known specific MSC marker, cells must adhere to plastic, demonstrate a baseline differentiation potential to osteoblast, chondroblast, and adipocyte lineage in vitro, and express the presence of widely excepted surface markers to be verified as MSCs [3]. However, there remain discrepancies in differentiation capacity and gene transcription programs depending on the tissue origin and species of these cells [67–70]. To what extent do these inconsistencies affect the immunomodulatory and regenerative capacity of MSCs is unclear, as is how such variations effect EV production.

Several questions remain unanswered: is the production of therapeutic EVs the same for all MSCs, independent of their tissue of origin? This is important for future large-scale production and clinical translation, as other MSC sources such as Wharton's jelly may afford multiple benefits over bone marrow-derived MSCs as they are often considered medical waste and conceptually provide an endless reservoir of primary human MSCs. Donor and patient safety is covered by guidelines from the International Council for Harmonization (ICH) of Technical Requirements for Pharmaceuticals for Human Use [71]. They provide guidance for the development and generation of novel biological medicinal products, with regard to product safety, efficacy and quality. For clinical application, in addition to an exclusion criteria (e.g., HIV and hepatitis positive patients), relevant donor demographics (including age, sex) should be documented. Special considerations should be in place if samples are pooled from multiple donors to account for possible EV–EV interactions. Thus, the manufacturing of EV-based therapeutics requires the appropriate infrastructure, quality assurance and compliance with GxP standards that put both donor and recipient safety at the forefront of importance.

8 EV Isolation

EVs have been isolated from various biological fluids and cell culture medium. Preanalytical protocols have been extensively discussed in a review by Théry and colleagues [72] and more recently in a position paper by ISEV [73]. In addition to using CM, most studies to date have employed ultra centifugation (UC) based protocols to isolate EVs. This procedure is based upon the ability of the centrifugation force to sediment matter based on size and density, where larger and denser components migrate away from the axis of the centrifuge and vice versa for smaller, less dense components. Differential ultra centrifugation involves sequential centrifugation steps with the aim of separating smaller and less dense constituents at each step. However, it has been proposed that the use of forces in excess of 100,000 × *g*, encourages the formation of EV aggregates [74]. Furthermore it is widely accepted that UC methods often co-isolate soluble factors and protein, contaminating EV preparations [75, 76]. For regulatory approval of biologic medicines (EV-therapies), nonactive components needed in the final formulation of a drug have to be characterized. Thus, recently, a general consensus in the EV community has shifted to more "*gentle*" isolation techniques, with the goal of further purifying EVs, reducing contaminants, maintaining bioactive integrity and separating bioactive vesicle populations. To date, parallel avenues of investigation include gradient density isolation [77] and size exclusion chromatography [78, 79]. These approaches have allowed EVs to be separated from proteins and soluble factors. With multiple EV isolation methods reported, how these different techniques affect the bioactive integrity of EVs remains unknown; however this question is vital when considering the efficacy of EVs.

9 Dose Evaluation

The administered EV dose represents a huge challenge in the research community. Current practices are difficult to follow with investigators using several different parameters to indicate EV dose [80]. With subtle changes in tissue culture conditions being shown to not only impact both EV amount but also composition, the need to standardize EV dosing is imperative [81, 82]. Current quantitative practices include reporting cell equivalents, protein concentration and/or specialized quantitative analytical measurements of EVs by instruments such as flow cytometry, tunable resistive pulse sensing (TRPS) and nanoparticle tracking analysis (NTA), with each technique harboring its own advantages and limitations [72, 83, 84]. Vesicle concentration maybe confounded by fragments, aggregates and contaminants, hence the importance of a rigorous and validated preanalytical isolation protocol. In vitro studies have found that EV-mediated signaling is dose dependent [80, 85], so optimizing EV dose may support the balancing of deleterious and therapeutic effects of EV administration. Thus, more rigorous characterization of EV-based formulation coupled with mathematical modeling and a combination of advanced analytical techniques is required. Moreover, vesicle integrity and stability is infrequently reported in current preclinical practice and should be addressed.

To ensure the integrity and purity of EV preparations, investigators may assess the protein–particle ratio as an indication of protein contamination [86]. Cell fragments can also skew vesicle quantification, thus membrane protein–total protein ratios coupled with staining for phalloidin (binds to actin filaments and used to detect fragmented cell debris) may indicate the degree of debris/fragments. With different quantitative parameters for dosing, inter--study comparisons are difficult. Preclinical studies to date have seldom undertaken several doses; thus, dose-dependent studies are lacking. Future studies may consider normalizing EV dose to body weight. Identification of the active substance in EV-based therapeutics is challenging. However, the regulatory classification of a biological medicine depends on a pharmaceutically active substance. Albeit this does not necessarily have to be a defined molecule, in terms of cellular therapeutics, this can be the cells themselves [87].

For future EV-based therapeutic applications understanding the in vivo fate of EVs is critical. Recently, several studies have investigated the biodistribution of EVs in mice. Wiklander et al., [88] and others [89, 90] have recently labeled EVs and tracked there in vivo fate after IV administration. Notably, Wiklander and coworkers found that the biodistribution pattern was dependent on route of administration. In contrast to IV injections, intraperitoneal (IP) and subcutaneous (SC) administration reduced EV accumulation in liver and spleen but increased EV accumulation in pancreas and gastrointestinal tract. However, in addition to dye dilution via target cell fusion, a potential limitation of using lipophilic dyes to track EVs is the disparity in half-life between EVs and the dye. Accurate trafficking of EVs in vivo would help in understanding the half-life of EVs and may allude to mechanisms of action.

10 Developing an EV Potency Assay

The definition of the bioactive substance(s) will remain a crucial question in the preclinical development of EV-based therapeutics. Whether the mechanism of action of MSC-EVs depends on the vesicles' cargo, the surface membrane or a combination of both is currently unclear. By FDA standards, potency is defined as the product(s) specific ability or capacity to effect a given result [87, 91, 92]. The capacity to evaluate specific loading could aid batch–batch variation. For example, the dose of EV preparations from two separate donors may be normalized via a given quantitative technique however the bioactive load may differ, subsequently the potency and degree of efficacy will be not be the same. Thus, investigators should consider employing a unique EV potency unit (EPU) to standardize practices and minimize variation between different samples. If established correlations are found between efficacy and EV markers (for example exosomal protein marker TSG101), surrogate markers of potency maybe considered. Jiao and colleague described an in vitro potency unit to for MSC-EVs [93]. The method describes the release of IL-10 (detected by ELISA) in mononuclear cells following the addition of MSC-EVs . Such tools would provide a powerful tool for analyzing the potency, however, the potency assessment must be intrinsically related to the disease model for which it is intended and be tuned to the mechanism of action. We propose a novel approach to standardize EV-therapeutics for inflammatory disorders, such as BPD and in preliminary studies, we use an in vitro macrophage polarization assay as a catalyst to assess the MSC-EV potency and subsequently the required dose for in vivo studies (Willis et al., manuscript in preparation).

11 Manufacturing, Scale-Up, Storage, and Reproducibility

Transitioning from preclinical to clinical studies introduces numerous challenges. EV preparations should be assessed for impurities and consistency with all materials used in the EV harvesting adhering to the required regulatory compliance. The manufacturing of EVs should adhere to current Good Manufacturing Practice (cGMP). Navabi and coworkers detailed the development of a method for the preparation of cGMP-grade EVs from the ascites fluid of ovarian cancer patients, and the methods used for the molecular and immunological characterization of these exosomes preceding their use in a clinical trial [94]. Moreover, in 2002 Lamarski and coworkers described a method for the production and characterization of clinical grade EVs derived from dendritic cells for their application in cancer vaccine clinical trials. They isolated particles (50–90 nm in diameter) containing antigen presenting molecules (MHC class I, MHC class II, and CD1), tetraspanin molecules (CD9, CD63 and CD81) and adhesion molecules (CD11b and CD54), via ultrafiltration of the clarified supernatant (500-kDa membrane) coupled with ultracentrifugation into a 30% sucrose/deuterium oxide (98%) cushion (density 1.210 g/cm^3) and used MHC class II concentration as a surrogate marker for quality control [95]. Since 2002, our analytical understanding of EV biology has widened and the development of specialized isolation and characterization methods has allowed investigators to more accurately prepare and comprehensively characterize pure EV populations. However, although the current consensus in the field is moving towards "gentle" EV isolation methods (such as size exclusion chromatography), an in-depth comparison of clinical grade EV isolation techniques is required.

Cells should be processed and cultured under strict laboratory conditions to minimize contamination and aid consistency. Previous in vitro studies have demonstrated that the protein and RNA levels in EVs secreted by endothelial cells reflects the culture and medium conditions and physiological/pharmacological/pathological stimulus that triggers EV release [96]. Moreover, under specific stimuli, EVs become enriched with an array of surface antigens and may harbor an altered composition. This questions if certain cell stimuli could be used to target the production of '*more*' therapeutic EVs? Indeed, Waszak preconditioned rat bone marrow MSCs ex vivo in hyperoxic conditions for 24 h [97]. The CM harvested from preconditioned cells exerted a more potent therapeutic effect in vivo on lung architecture compared to CM collected from cells grown under control conditions. Of course, in the absence of rigorous characterization in that system, one can only speculate where the observed augmentation of activity resides (EVs or soluble fraction of secretome).

Scaling up EV production to meet the needs of preclinical and clinical studies is of upmost importance. Although MSCs secrete significant numbers of EVs, demand for preclinical studies often requires massive cell cultures and consequently large CM volumes. Following differential centrifugation to remove cells/debris, current preanalytical protocols often reduce CM volume via tangential flow filtration (~25- to 50-fold), before further isolation of bioactive vesicles. More recently, the inevitable shift to using tissue culture bioreactors has been used to generate large-scale EV preparations, yet research into how such culture conditions effects EV production is required (Fig. 2).

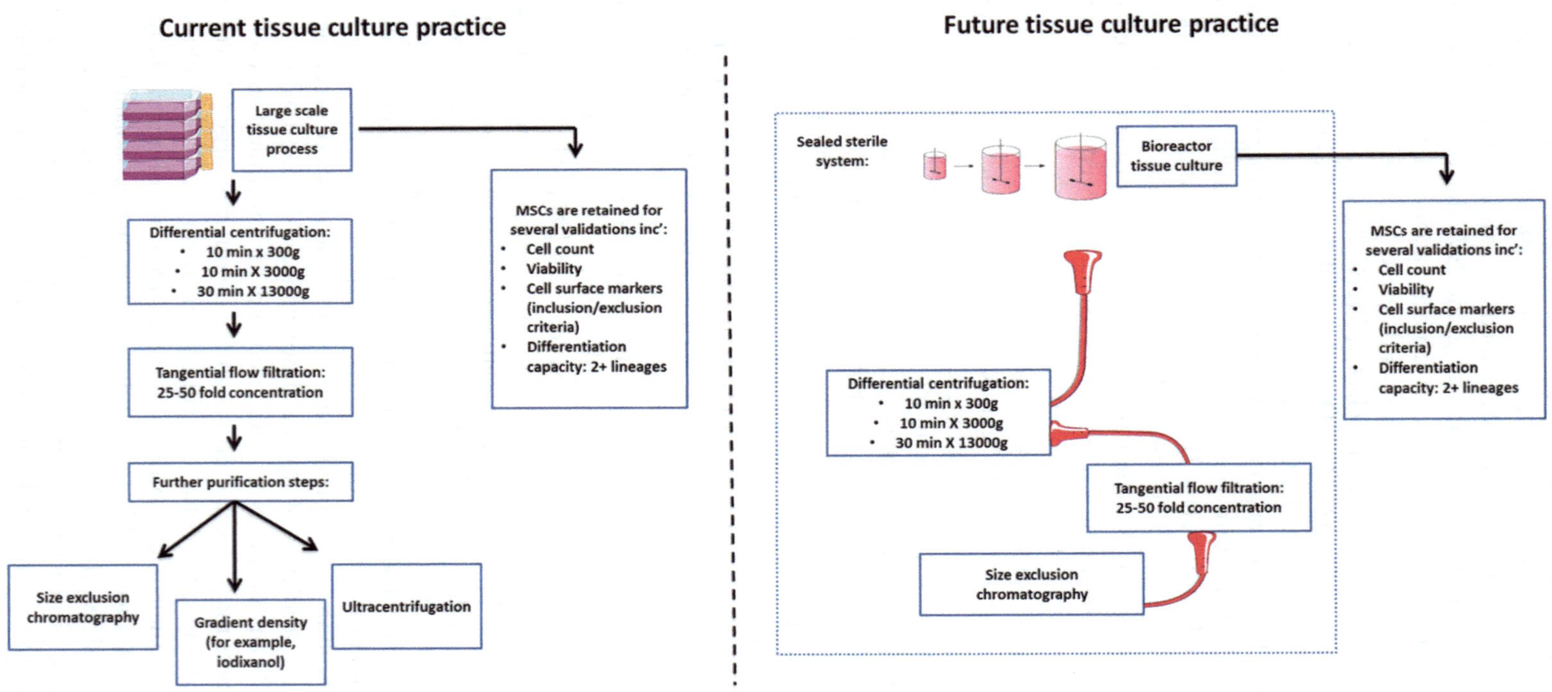

Fig. 2 Work flow for MSC-EV isolation. Preclinical studies require a substantial amount of *pure* EVs. Thus current practices often involve a large number of tissue culture flasks which is often labor intensive. Following conditioned media (CM) harvesting, several centrifugation steps and tangential flow filtration is readily used to remove cell debris, apoptotic bodies and concentrate volume, often in nonsterile conditions. Currently, with no "gold standard" method, preanalytical steps differ between studies, but ultracentrifugation (UC) has long been the most popular method for isolating EVs. Furthermore the purification of the active EV population from contaminants is currently lacking. Future tissue culture practices should aim to minimize the amount of purification steps. Firstly, bioreactors are becoming readily employed for large cell culture preparations. Unlike current practices which use calf serum, xeno-free media should be trialled at the preclinical development stage. Cell culture practice and purification steps should be conducted in an enclosed sterile environment. Attempts should be made to isolate the active EV population from contaminates (for example, size exclusion chromatography)

Lai et al., found that MSCs secrete many distinct sub types of vesicles, which differ in RNA and protein composition [98]. Although, this study utilized an immortal MSC cell line which is likely to secrete different EV subtypes to primary cells, the principle should still apply to other cell sources. Often, isolation techniques struggle to distinguish and separate such vesicle subtypes. Thus, it would be advantageous to separate and test the efficacy of different EV subtypes with the aim of identifying the bioactive/therapeutic vesicle population. Our laboratory has employed size exclusion chromatography to separate heterogeneous vesicle populations based on their size. Using CL-2B matrix, we can effectively distinguish 100 nm from 50 nm polystyrene beads. With biological samples, specifically MSC-EV populations, we are able to separate vesicles that are <130 nm, with smaller vesicles (40–100 nm) with minimal overlap (Willis et al., manuscript in preparation). Ongoing investigations aim to test the bioactivity between larger and smaller vesicles.

Efforts for standardizing EV storage protocols are being developed for biobanking, in which EVs are stored in their natural matrix (for example, body fluid or cell culture supernatant), following steps to remove contents that may influence the EV-pool (such as cells) [73, 99]. Unless delivering autologous compounds that require no in vitro culture/manipulation, such as EVs derived from the amniotic fluid, it remains impractical to always use '*fresh*' EV preparations. Thus, there is a need to isolate and store EVs of high purity. For EVs of such purity, no standardized protocol is currently available. In current practice, EV products are stored in isotonic buffers to prevent pH shifts during storage as well as during freeze–thaw cycles. Typically, dimethyl sulfoxide (DMSO) and glycerol are not used in the storage process as previous reports have shown they may impact EV integrity [100]. Storage vials can also affect the quality of EVs obtained, as EVs might unexpectedly and irreversibly bind to certain materials [99]. Thus, non-polypropylene vials are not advised as they may affect the concentration or integrity of stored EVs. The storage temperature of EVs is important. Reports have shown that EV integrity is greater at −80 °C in comparison to −190 °C or at 4 °C [100–102]. With a lack of studies addressing EV structural stability and functional efficacy after increased storing times, more studies which incorporate functional analyses in potency assays are needed. Such analyses will define a provisional shelf-life for EV-based products and help facilitate the manufacturing and distribution process.

12 Toxicology and Safety

Collectively, the above sections are all directed at improving the efficacy of EV-therapeutics, providing pitfalls and suggestions in the preclinical development of EV-based therapeutics with the over-arching goal of improving patient wellbeing. Central to this is the challenge of demonstrating the safety and efficacy of novel medicinal products. The ICH noted typical toxicology testing approaches are usually not considered appropriate for biologic medicines due to complex biological properties that may include species specificity, immunogenicity and unknown pleiotropic activities [87]. In the development of EV-based therapeutics, lessons can be drawn from the development and application of stem cell based-therapies. However, in contrast to the adult patient, additional precautions and long-term adverse consequences of EV-based therapeutics need to be considered when working with immunocompromised developing infant [3]. Thus, we propose that infants who receive MSC-EVs (or treatments akin to cell-based therapies) are closely monitored for several years for the theoretical risk of tumor formation. In a preclinical BPD study looking at the long term-outcomes of rodents administered with MSC-CM, no tumors were found at 6 months (2/3 year life expectancy); however, it would be of benefit to undertake well-designed preclinical studies in large animals for further supporting evidence [25]. Encouragingly, within excess of 500 MSC-based clinical trials that are currently registered, a recent meta-analysis looking at 1000 subjects has shown that there is no incidence of increased tumor risk up to 60 months post treatment [103].

13 Summary

As the incidence of chronic neonatal lung disease continues to rise, it is clear that EV therapeutics may represent the next major breakthrough for diseases of the newborn infant. Although EVs have previously been administered to humans for the treatment of cancer and, more recently, for a GvHD patient [59], no recommended standard techniques have been established for the clinical grade production and quality control of EV-based therapeutics . As current technology does not enable a comprehensive assessment at a single vesicle level, researchers should undertake a rigorous characterization of EV populations to reduce contaminants, and identify the bioactive moiety and mechanism of action. Determining the optimal dose response and number of doses to achieve efficacy without side effects is an important issue to resolve. The appropriate time window for EV administration, the best route of administration and the favorable cell source remain uncertainties and are yet to be investigated in a direct comparison (summarized in Fig. 3). Although each of the above issues require continued investigation, at the present time there is an urgency to develop novel therapeutic modalities for these as yet incurable diseases that pose a huge burden to the patient, family, and society.

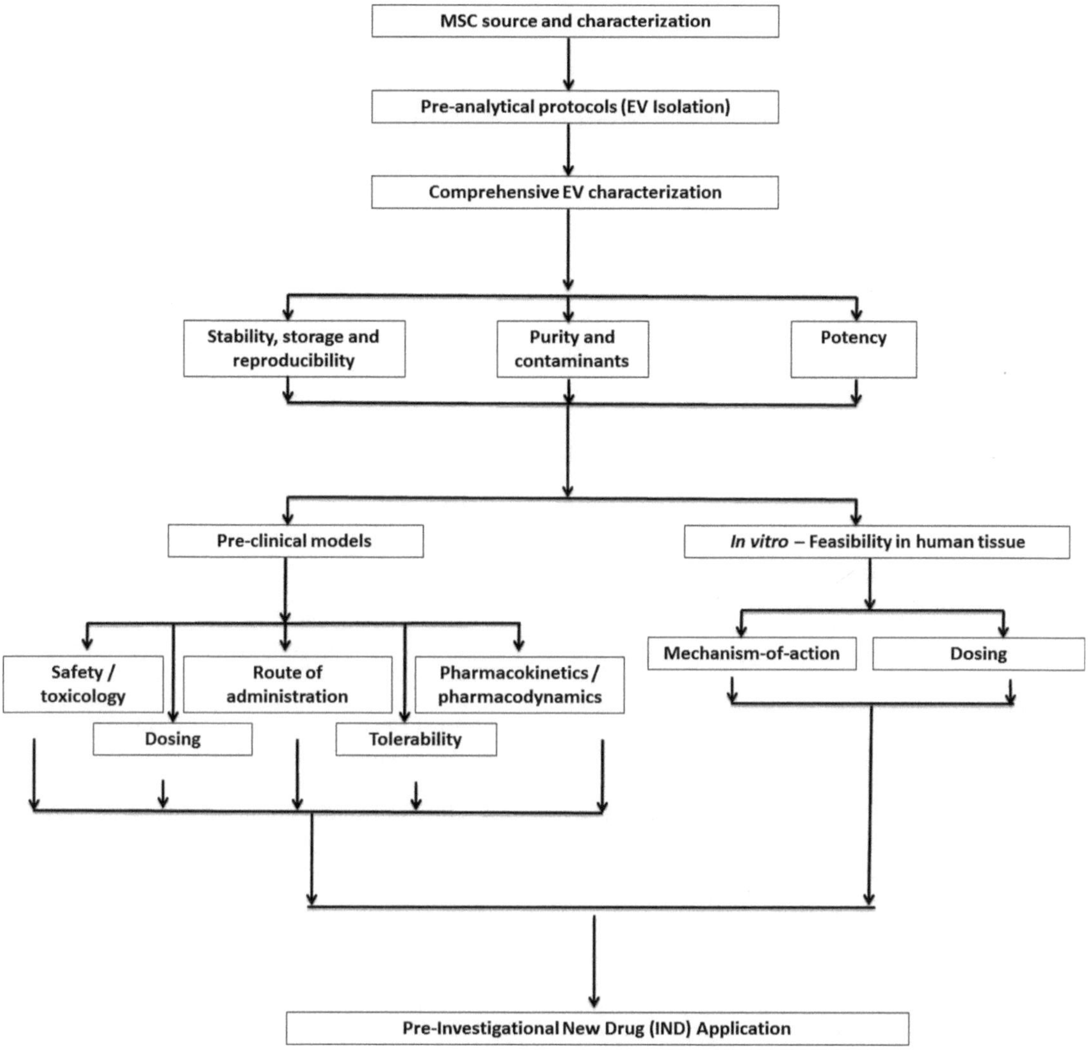

Fig. 3 Guidance on the overall scheme of testing of EV-based therapeutics in preclinical investigations. Adapted from [87]

EVs from MSCs, as well as other cell sources, hold significant potential as therapeutic vehicles of epigenetic immunomodulatory signals to treat a diverse array of diseases. To date, preclinical studies are very promising, but numerous uncertainties remain at large. Clearly, more work is required to better characterize the biology of MSC-EVs, and the molecular and epigenetic mechanisms of their action on inflammatory pathways of injury and repair that underlie lung diseases such as BPD. EV-based therapeutics offers new innovative possibilities to prevent and cure invalidating diseases such as BPD.

References

1. Borghesi A, Cova C, Gazzolo D, Stronati M (2013) Stem cell therapy for neonatal diseases associated with preterm birth. J Clin Neonatol 2(1):1–7
2. Stoll BJ, Hansen NI, Bell EF, Walsh MC, Carlo WA, Shankaran S et al (2015) Trends in care practices, morbidity, and mortality of extremely preterm neonates, 1993–2012. JAMA 314(10):1039–1051
3. Mitsialis SA, Kourembanas S (2016) Stem cell–based therapies for the newborn lung and brain: possibilities and challenges. Semin Perinatol 40(3):138–151
4. Baraldi E, Filippone M (2007) Chronic lung disease after premature birth. N Engl J Med 357(19):1946–1955
5. Hilgendorff A, O'Reilly MA (2015) Bronchopulmonary dysplasia early changes leading to long-term consequences. Front Med 2:2
6. Khemani E, McElhinney DB, Rhein L, Andrade O, Lacro RV, Thomas KC et al (2007) Pulmonary artery hypertension in formerly premature infants with bronchopulmonary dysplasia: clinical features and outcomes in the surfactant era. Pediatrics 120(6):1260–1269
7. del Cerro MJ, Sabaté Rotés A, Cartón A, Deiros L, Bret M, Cordeiro M et al (2014) Pulmonary hypertension in bronchopulmonary dysplasia: clinical findings, cardiovascular anomalies and outcomes. Pediatr Pulmonol 49(1):49–59
8. Borghesi A, Massa M, Campanelli R, Bollani L, Tzialla C, Figar TA et al (2009) Circulating endothelial progenitor cells in preterm infants with bronchopulmonary dysplasia. Am J Respir Crit Care Med 180(6):540–546
9. Baker CD, Balasubramaniam V, Mourani PM, Sontag MK, Black CP, Ryan SL et al (2012) Cord blood angiogenic progenitor cells are decreased in bronchopulmonary dysplasia. Eur Respir J 40(6):1516–1522
10. Titomanlio L, Kavelaars A, Dalous J, Mani S, El Ghouzzi V, Heijnen C et al (2011) Stem cell therapy for neonatal brain injury: perspectives and challenges. Ann Neurol 70 (5):698–712
11. Verina T, Fatemi A, Johnston MV, Comi AM (2013) Pluripotent possibilities: human umbilical cord blood cell treatment after neonatal brain injury. Pediatr Neurol 48 (5):346–354
12. Chicha L, Smith T, Guzman R (2014) Stem cells for brain repair in neonatal hypoxia–ischemia. Childs Nerv Syst 30(1):37–46
13. Gonzales-Portillo GS, Reyes S, Aguirre D, Pabon MM, Borlongan CV (2014) Stem cell therapy for neonatal hypoxic-ischemic encephalopathy. Front Neurol 5:147
14. Bennet L, Tan S, Van den Heuij L, Derrick M, Groenendaal F, van Bel F et al (2012) Cell therapy for neonatal hypoxia–ischemia and cerebral palsy. Ann Neurol 71(5):589–600
15. Baker CD, Seedorf GJ, Wisniewski BL, Black CP, Ryan SL, Balasubramaniam V et al (2013) Endothelial colony-forming cell conditioned media promote angiogenesis in vitro and prevent pulmonary hypertension in experimental bronchopulmonary dysplasia. Am J Physiol Lung Cell Mol Physiol 305(1):L73–L81
16. Ilancheran S, Michalska A, Peh G, Wallace EM, Pera M, Manuelpillai U (2007) Stem cells derived from human fetal membranes display multilineage differentiation potential. Biol Reprod 77(3):577–588
17. Hodges RJ, Jenkin G, Hooper SB, Allison B, Lim R, Dickinson H et al (2012) Human amnion epithelial cells reduce ventilation-induced preterm lung injury in fetal sheep. Am J Obstet Gynecol 206(5):448.e8–448.15
18. Vosdoganes P, Hodges RJ, Lim R, Westover AJ, Acharya RY, Wallace EM et al (2011) Human amnion epithelial cells as a treatment for inflammation-induced fetal lung injury in sheep. Am J Obstet Gynecol 205(2):156.e26–156.e33
19. Curley GF, Hayes M, Ansari B, Shaw G, Ryan A, Barry F et al (2012) Mesenchymal stem cells enhance recovery and repair following ventilator-induced lung injury in the rat. Thorax 67(6):496–501
20. Aslam M, Baveja R, Liang OD, Fernandez-Gonzalez A, Lee C, Mitsialis SA et al (2009) Bone marrow stromal cells attenuate lung injury in a murine model of neonatal chronic lung disease. Am J Respir Crit Care Med 180 (11):1122–1130
21. van Haaften T, Byrne R, Bonnet S, Rochefort GY, Akabutu J, Bouchentouf M et al (2009) Airway delivery of mesenchymal stem cells prevents arrested alveolar growth in neonatal lung injury in rats. Am J Respir Crit Care Med 180(11):1131–1142
22. Zhang X, Wang H, Shi Y, Peng W, Zhang S, Zhang W et al (2012) Role of bone marrow-derived mesenchymal stem cells in the prevention of hyperoxia-induced lung injury in newborn mice. Cell Biol Int 36(6):589–594
23. Chimenti L, Luque T, Bonsignore MR, Ramírez J, Navajas D, Farré R (2012) Pre-

treatment with mesenchymal stem cells reduces ventilator-induced lung injury. Eur Respir J 40(4):939–948

24. Hansmann G, Fernandez-Gonzalez A, Aslam M, Vitali SH, Martin T, Mitsialis SA et al (2012) Mesenchymal stem cell-mediated reversal of bronchopulmonary dysplasia and associated pulmonary hypertension. Pulm Circ 2(2):170–181
25. Pierro M, Ionescu L, Montemurro T, Vadivel A, Weissmann G, Oudit G et al (2013) Short-term, long-term and paracrine effect of human umbilical cord-derived stem cells in lung injury prevention and repair in experimental bronchopulmonary dysplasia. Thorax 68(5):475–484
26. Islam MN, Das SR, Emin MT, Wei M, Sun L, Westphalen K et al (2012) Mitochondrial transfer from bone marrow-derived stromal cells to pulmonary alveoli protects against acute lung injury. Nat Med 18 (5):759–765
27. Lee C, Mitsialis SA, Aslam M, Vitali SH, Vergadi E, Konstantinou G et al (2012) Exosomes mediate the cytoprotective action of mesenchymal stromal cells on hypoxia-induced pulmonary hypertension. Circulation 126(22):2601–2611
28. Lee JW, Fang X, Krasnodembskaya A, Howard JP, Matthay MA (2011) Concise review: Mesenchymal stem cells for acute lung injury: role of paracrine soluble factors. Stem Cells 29(6):913–919
29. Gu W, Song L, Li X-M, Wang D, Guo X-J, Xu W-G (2015) Mesenchymal stem cells alleviate airway inflammation and emphysema in COPD through down-regulation of cyclooxygenase-2 via p38 and ERK MAPK pathways. Sci Rep 5:8733
30. Bonfield T, Sutton M, Lennon D, Caplan A (2015) Mesenchymal stem cells: new directions in treating asthma (THER2P.955). J Immunol 194(1):67.6
31. Cruz FF, Borg ZD, Goodwin M, Sokocevic D, Wagner DE, Coffey A et al (2015) Systemic administration of human bone marrow-derived mesenchymal stromal cell extracellular vesicles ameliorates Aspergillus hyphal extract-induced allergic airway inflammation in immunocompetent mice. Stem Cells Transl Med 4(11):1302–1316
32. Ringden O, Keating A (2011) Mesenchymal stromal cells as treatment for chronic GVHD. Bone Marrow Transplant 46(2):163–164
33. Chang YS, Ahn SY, Yoo HS, Sung SI, Choi SJ, Oh WI et al (2014) Mesenchymal stem cells for bronchopulmonary dysplasia: phase 1 dose-escalation clinical trial. J Pediatr 164 (5):966–972.e6
34. Gould SJ, Raposo G (2013) As we wait: coping with an imperfect nomenclature for extracellular vesicles. J Extracell Vesicles. 15 Feb 2013. doi:10.3402/jev.v2i0.20389
35. van der Pol E, Böing AN, Harrison P, Sturk A, Nieuwland R (2012) Classification, functions, and clinical relevance of extracellular vesicles. Pharmacol Rev 64(3):676–705
36. Kourembanas S (2015) Exosomes: vehicles of intercellular signaling, biomarkers, and vectors of cell therapy. Annu Rev Physiol 77 (1):13–27
37. Colombo M, Raposo G, Théry C (2014) Biogenesis, secretion, and intercellular interactions of exosomes and other extracellular vesicles. Annu Rev Cell Dev Biol 30 (1):255–289
38. Abels ER, Breakefield XO (2016) Introduction to extracellular vesicles: biogenesis, RNA cargo selection, content, release, and uptake. Cell Mol Neurobiol 36(3):301–312
39. El Andaloussi S, Mager I, Breakefield XO, Wood MJA (2013) Extracellular vesicles: biology and emerging therapeutic opportunities. Nat Rev Drug Discov 12(5):347–357
40. Zaborowski MP, Balaj L, Breakefield XO, Lai CP (2015) Extracellular vesicles: composition, biological relevance, and methods of study. Bioscience 65(8):783–797
41. Sdrimas K, Kourembanas S (2014) MSC microvesicles for the treatment of lung disease: a new paradigm for cell-free therapy. Antioxid Redox Signal 21(13):1905–1915
42. Yáñez-Mó M, Siljander PR-M, Andreu Z, Zavec AB, Borràs FE, Buzas EI et al (2015) Biological properties of extracellular vesicles and their physiological functions. J Extracell Vesicles 4:27066
43. Mulcahy LA, Pink RC, Carter DRF (2014) Routes and mechanisms of extracellular vesicle uptake. J Extracell Vesicles 3. doi:10.3402/jev.v3.24641
44. Benameur T, Soleti R, Porro C, Andriantsitohaina R, Martínez MC (2010) Microparticles carrying sonic hedgehog favor neovascularization through the activation of nitric oxide pathway in mice. PLoS One 5(9):e12688
45. Valadi H, Ekstrom K, Bossios A, Sjostrand M, Lee JJ, Lotvall JO (2007) Exosome-mediated transfer of mRNAs and microRNAs is a novel mechanism of genetic exchange between cells. Nat Cell Biol 9(6):654–659
46. Subra C, Grand D, Laulagnier K, Stella A, Lambeau G, Paillasse M et al (2010) Exosomes account for vesicle-mediated

transcellular transport of activatable phospholipases and prostaglandins. J Lipid Res 51 (8):2105–2120

47. Zhang J, Li S, Li L, Li M, Guo C, Yao J et al (2015) Exosome and exosomal microRNA: trafficking, sorting, and function. Genomics Proteomics Bioinformatics 13(1):17–24
48. Stoorvogel W (2012) Functional transfer of microRNA by exosomes. Blood 119 (3):646–648
49. Xin H, Li Y, Buller B, Katakowski M, Zhang Y, Wang X et al (2012) Exosome-mediated transfer of miR-133b from multipotent mesenchymal stromal cells to neural cells contributes to neurite outgrowth. Stem Cells 30 (7):1556–1564
50. Xiang M-X, He A-N, Wang J-A, Gui C (2009) Protective paracrine effect of mesenchymal stem cells on cardiomyocytes. J Zhejiang Univ Sci B 10(8):619–624
51. D'Angio CT, Maniscalco WM (2006) Bronchopulmonary dysplasia in preterm infants: pathophysiology and management strategies. Paediatr Drugs 6(5):303–330
52. Cho D-I, Kim MR, Jeong H-Y, Jeong HC, Jeong MH, Yoon SH et al (2014) Mesenchymal stem cells reciprocally regulate the M1/M2 balance in mouse bone marrow-derived macrophages. Exp Mol Med 46:e70
53. Tropea KA, Leder E, Aslam M, Lau AN, Raiser DM, Lee J-H et al (2012) Bronchioalveolar stem cells increase after mesenchymal stromal cell treatment in a mouse model of bronchopulmonary dysplasia. Am J Physiol Lung Cell Mol Physiol 302(9):L829–L837
54. Timmers L, Lim SK, Arslan F, Armstrong JS, Hoefer IE, Doevendans PA et al (2008) Reduction of myocardial infarct size by human mesenchymal stem cell conditioned medium. Stem Cell Res 1(2):129–137
55. Lai RC, Arslan F, Lee MM, Sze NSK, Choo A, Chen TS et al (2010) Exosome secreted by MSC reduces myocardial ischemia/reperfusion injury. Stem Cell Res 4(3):214–222
56. Zhang B, Yin Y, Lai RC, Tan SS, Choo ABH, Lim SK (2013) Mesenchymal stem cells secrete immunologically active exosomes. Stem Cells Dev 23(11):1233–1244
57. Zhang G, Wang D, Miao S, Zou X, Liu G, Zhu Y (2016) Extracellular vesicles derived from mesenchymal stromal cells may possess increased therapeutic potential for acute kidney injury compared with conditioned medium in rodent models: a meta-analysis. Exp Ther Med 11(4):1519–1525
58. Xin H, Li Y, Chopp M (2014) Exosomes/miRNAs as mediating cell-based therapy of stroke. Front Cell Neurosci 8:377
59. Kordelas L, Rebmann V, Ludwig AK, Radtke S, Ruesing J, Doeppner TR et al (2014) MSC-derived exosomes: a novel tool to treat therapy-refractory graft-versus-host disease. Leukemia 8(4):970–973
60. Griffiths MJD, Bonnet D, Janes SM (2005) Stem cells of the alveolar epithelium. Lancet 366(9481):249–260
61. Romanov YA, Darevskaya AN, Merzlikina NV, Buravkova LB (2005) Mesenchymal stem cells from human bone marrow and adipose tissue: isolation, characterization, and differentiation potentialities. Bull Exp Biol Med 140(1):138–143
62. Choudhery MS, Badowski M, Muise A, Pierce J, Harris DT (2014) Donor age negatively impacts adipose tissue-derived mesenchymal stem cell expansion and differentiation. J Transl Med 12:8
63. Romanov YA, Svintsitskaya VA, Smirnov VN (2003) Searching for alternative sources of postnatal human mesenchymal stem cells: candidate MSC-like cells from umbilical cord. Stem Cells 21(1):105–110
64. Secco M, Zucconi E, Vieira NM, Fogaça LLQ, Cerqueira A, Carvalho MDF et al (2008) Multipotent stem cells from umbilical cord: cord is richer than blood! Stem Cells 26 (1):146–150
65. Musina RA, Bekchanova ES, Sukhikh GT (2005) Comparison of mesenchymal stem cells obtained from different human tissues. Bull Exp Biol Med 139(4):504–509
66. Chen J-Y, Mou X-Z, Du X-C, Xiang C (2015) Comparative analysis of biological characteristics of adult mesenchymal stem cells with different tissue origins. Asian Pac J Trop Med 8(9):739–746
67. Jansen BJH, Gilissen C, Roelofs H, Schaap-Oziemlak A, Veltman JA, Raymakers RAP et al (2009) Functional differences between mesenchymal stem cell populations are reflected by their transcriptome. Stem Cells Dev 19(4):481–490
68. Tsai M-S, Hwang S-M, Chen K-D, Lee Y-S, Hsu L-W, Chang Y-J et al (2007) Functional network analysis of the transcriptomes of mesenchymal stem cells derived from amniotic fluid, amniotic membrane, cord blood, and bone marrow. Stem Cells 25(10):2511–2523
69. Foudah D, Redaelli S, Donzelli E, Bentivegna A, Miloso M, Dalprà L et al (2009) Monitoring the genomic stability of in vitro cultured rat bone-marrow-derived mesenchymal stem cells. Chromosom Res 17(8):1025–1039
70. Ren Z, Wang J, Zhu W, Guan Y, Zou C, Chen Z et al (2011) Spontaneous transformation of adult mesenchymal stem cells from

cynomolgus macaques in vitro. Exp Cell Res 317(20):2950–2957
71. Harron DWG (2007) Technical requirements for registration of pharmaceuticals for human use: the ICH process. In: The textbook of pharmaceutical medicine. Blackwell Publishing Ltd, Oxford, pp 552–564
72. Théry C, Amigorena S, Raposo G, Clayton A (2006) Isolation and characterization of exosomes from cell culture supernatants and biological fluids. Curr Protoc Cell Biol. Chapter 3:Unit 3.22
73. Witwer KW, Buzás EI, Bemis LT, Bora A, Lässer C, Lötvall J et al (2013) Standardization of sample collection, isolation and analysis methods in extracellular vesicle research. J Extracell Vesicles 2. doi:10.3402/jev.v2i0.20360
74. Linares R, Tan S, Gounou C, Arraud N, Brisson AR (2015) High-speed centrifugation induces aggregation of extracellular vesicles. J Extracell Vesicles 4:29509
75. Franquesa M, Hoogduijn MJ, Ripoll E, Luk F, Salih M, Betjes MGH et al (2014) Update on controls for isolation and quantification methodology of extracellular vesicles derived from adipose tissue mesenchymal stem cells. Front Immunol 5:525
76. Jeppesen DK, Hvam ML, Primdahl-Bengtson B, Boysen AT, Whitehead B, Dyrskjøt L et al (2014) Comparative analysis of discrete exosome fractions obtained by differential centrifugation. J Extracell Vesicles 3:25011
77. Iwai K, Minamisawa T, Suga K, Yajima Y, Shiba K (2016) Isolation of human salivary extracellular vesicles by iodixanol density gradient ultracentrifugation and their characterizations. J Extracell Vesicles 5. doi:10.3402/jev.v5.30829
78. Lozano-Ramos I, Bancu I, Oliveira-Tercero A, Armengol MP, Menezes-Neto A, Del Portillo HA et al (2015) Size-exclusion chromatography-based enrichment of extracellular vesicles from urine samples. J Extracell Vesicles 4:27369
79. Baranyai T, Herczeg K, Onódi Z, Voszka I, Módos K, Marton N et al (2015) Isolation of exosomes from blood plasma: qualitative and quantitative comparison of ultracentrifugation and size exclusion chromatography methods. PLoS One 10(12):e0145686
80. György B, Hung ME, Breakefield XO, Leonard JN (2015) Therapeutic applications of extracellular vesicles: clinical promise and open questions. Annu Rev Pharmacol Toxicol 55:439–464
81. Beach A, Zhang H-G, Ratajczak MZ, Kakar SS (2014) Exosomes: an overview of biogenesis, composition and role in ovarian cancer. J Ovarian Res 7:14
82. Parolini I, Federici C, Raggi C, Lugini L, Palleschi S, De Milito A et al (2009) Microenvironmental pH is a key factor for exosome traffic in tumor cells. J Biol Chem 284 (49):34211–34222
83. Akers JC, Ramakrishnan V, Nolan JP, Duggan E, Fu C-C, Hochberg FH et al (2016) Comparative analysis of technologies for quantifying extracellular vesicles (EVs) in clinical cerebrospinal fluids (CSF). PLoS One 11(2): e0149866
84. Maas SLN, de Vrij J, van der Vlist EJ, Geragousian B, van Bloois L, Mastrobattista E et al (2015) Possibilities and limitations of current technologies for quantification of biological extracellular vesicles and synthetic mimics. J Control Release 200:87–96
85. Yu S, Liu C, Su K, Wang J, Liu Y, Zhang L et al (2007) Tumor exosomes inhibit differentiation of bone marrow dendritic cells. J Immunol 178(11):6867–6875
86. Webber J, Clayton A (2013) How pure are your vesicles? J Extracell Vesicles 2. doi:10.3402/jev.v2i0.19861
87. Batrakova EV, Kim MS (2016) Development and regulation of exosome-based therapy products. Wiley Interdiscip Rev Nanomed Nanobiotechnol 8(5):744–757
88. Wiklander OPB, Nordin JZ, O'Loughlin A, Gustafsson Y, Corso G et al (2015) Extracellular vesicle in vivo biodistribution is determined by cell source, route of administration and targeting. J Extracell Vesicles 4:26316
89. Takahashi Y, Nishikawa M, Shinotsuka H, Matsui Y, Ohara S, Imai T et al (2013) Visualization and in vivo tracking of the exosomes of murine melanoma B16-BL6 cells in mice after intravenous injection. J Biotechnol 165 (2):77–84
90. Ohno S-i, Takanashi M, Sudo K, Ueda S, Ishikawa A, Matsuyama N et al (2013) Systemically injected exosomes targeted to EGFR deliver antitumor microRNA to breast cancer cells. Mol Ther 21(1):185–191
91. Downing NS, Aminawung JA, Shah ND, Krumholz HM, Ross JS (2014) CLinical trial evidence supporting FDA approval of novel therapeutic agents, 2005–2012. JAMA 311 (4):368–377
92. U.S. Department of Health and Human Services Food and Drug Administration (1998)

Guidance for industry: providing clinical evidence of effectiveness for human drug and biological products

93. Jiao J, Milwid JM, Yarmush ML, Parekkadan B (2011) A mesenchymal stem cell potency assay. Methods Mol Biol 677:221–231
94. Navabi H, Croston D, Hobot J, Clayton A, Zitvogel L, Jasani B et al (2005) Preparation of human ovarian cancer ascites-derived exosomes for a clinical trial. Blood Cell Mol Dis 35(2):149–152
95. Lamparski HG, Metha-Damani A, Yao J-Y, Patel S, Hsu D-H, Ruegg C et al (2002) Production and characterization of clinical grade exosomes derived from dendritic cells. J Immunol Methods 270(2):211–226
96. de Jong OG, Verhaar MC, Chen Y, Vader P, Gremmels H, Posthuma G et al (2012) Cellular stress conditions are reflected in the protein and RNA content of endothelial cell-derived exosomes. J Extracell Vesicles 1. doi:10.3402/jev.v1i0.18396
97. Waszak P, Alphonse R, Vadivel A, Ionescu L, Eaton F, Thébaud B (2012) Preconditioning enhances the paracrine effect of mesenchymal stem cells in preventing oxygen-induced neonatal lung injury in rats. Stem Cells Dev 21 (15):2789–2797
98. Lai RC, Tan SS, Yeo RWY, Choo ABH, Reiner AT, Su Y et al (2016) MSC secretes at least 3 EV types each with a unique permutation of membrane lipid, protein and RNA. J Extracell Vesicles 5:29828
99. Lener T, Gimona M, Aigner L, Börger V, Buzas E, Camussi G et al (2015) Applying extracellular vesicles based therapeutics in clinical trials – an ISEV position paper. J Extracell Vesicles 4:30087
100. Lőrincz ÁM, Timár CI, Marosvári KA, Veres DS, Otrokocsi L, Kittel Á et al (2014) Effect of storage on physical and functional properties of extracellular vesicles derived from neutrophilic granulocytes. J Extracell Vesicles 3. doi:10.3402/jev.v3.25465
101. Sokolova V, Ludwig A-K, Hornung S, Rotan O, Horn PA, Epple M et al (2011) Characterisation of exosomes derived from human cells by nanoparticle tracking analysis and scanning electron microscopy. Colloids Surf B Biointerfaces 87(1):146–150
102. Kalra H, Adda CG, Liem M, Ang C-S, Mechler A, Simpson RJ et al (2013) Comparative proteomics evaluation of plasma exosome isolation techniques and assessment of the stability of exosomes in normal human blood plasma. Proteomics 13(22):3354–3364
103. Lalu MM, McIntyre L, Pugliese C, Fergusson D, Winston BW, Marshall JC et al (2012) Safety of cell therapy with mesenchymal stromal cells (SafeCell): a systematic review and meta-analysis of clinical trials. PLoS One 7 (10):e47559

Chapter 35

Therapeutic Use of Tumor Cell-Derived Extracellular Vesicles

Jing Liu, Jingwei Ma, Ke Tang, and Bo Huang

Abstract

Extracellular vesicles (EVs) are nanosized membrane vesicles, released by many cell types, which are an ideal drug-delivery system for antitumor therapy. Exosomes, microparticles (MP), and apoptotic bodies are the three main types of EVs. MPs are EVs with size ranges between 100 and 1000 nm. Tumor cell-derived EVs (T-EVs) can serve as a safe and efficient vehicle to specific deliver chemotherapeutic drugs or oncolytic adenoviruses for tumor treatment. In this section, we describe how to prepare EVs for therapeutic use.

Key words Extracellular vesicles, Tumor microparticles, Drug delivery, Chemotherapeutic drugs, Oncolytic adenoviruses, Tumor vaccines, Tumor antigens, Dendritic cells

1 Introduction

James E. Rothman, Randy W. Schekman, and Thomas C. Südhof won the 2013 Nobel Prize for Physiology or Medicine. Their work advanced our understanding of the vesicle transport system in eukaryotic cells; they focused on intracellular vesicles released by organelles such as endoplasmic reticulum and Golgi apparatus. Extracellular vesicles (EVs), and formerly named microparticles, are nanosized membrane vesicles, released by many cell types. Eukaryotic cells can release different types of EVs, including exosomes, microvesicles, and apoptotic bodies [1]. EVs are vesicles ranging from 40 to 100 nm. They originate from intraluminal budding of multivesicular endosomes (MVE) and are released upon fusion of MVEs with the plasma membrane. EVs are larger (diameter ranged from 100 to 1000 nm), and are formed through direct budding from the plasma membrane. Apoptotic bodies are the largest vesicles (50 nm–5 μm) which are released when apoptosis occurs.

Since the time microparticles were first identified by Peter Wolf in 1967, which were originally termed "platelet dust" by him, researchers have carried out a large number of studies on it which

Winston Patrick Kuo and Shidong Jia (eds.), *Extracellular Vesicles: Methods and Protocols*, Methods in Molecular Biology, vol. 1660, DOI 10.1007/978-1-4939-7253-1_35, © Springer Science+Business Media LLC 2017

have shown that microparticles play pivotal roles under physical or pathological conditions. We refer to microparticles as extracellular vesicles (EVs) going forward. Releasing EVs are primary ways for cells to discharge cellular waste. The cargo of EVs comprises small and long, coding and noncoding RNAs (e.g., mRNA, microRNA, and lncRNA), lipids and proteins [2, 3]. EVs may subsequently be internalized by target recipient cells, resulting in transfer of mRNAs and miRNAs. Such events can result in production or silencing of target proteins, respectively, and of proteins, including membrane proteins [4]. Generation and shedding of EVs occurs during biological processes of considerable diversity, including exogenous stimulation or endogenous reaction when cells were exposed to danger signal or other pathogen invasion including physical stress, UV radiation, bacterial or virus infection. Whereas EVs were originally considered as inert cellular debris, numerous studies provided now a rationale for the concept of EVs as veritable vehicles for the intercellular exchange of biological signals and information.

Pioneering work was carried out in a mouse model showing rejection of established tumors by EVs from autologous dendritic cells (DCs) loaded with tumor peptides [5].The use of EVs instead of DCs as an antigen-presenting unit may address current limitations of DC-based therapies, such as controlling the percentage of live DCs after freeze–thaw protocols and the unknown behavior or fate of DCs after in vivo injection. In a reported study, tumor derived EVs had been used for antitumor immunotherapy as a source of tumor antigen. Several clinical trials using EVs or EV plus established therapies have been shown as an effective vaccine for activation of specific cytotoxic T cells [6]. Nevertheless, very little study focused on the therapeutic use of EVs. Recently, our group has made great effects to study the therapeutic use of EVs.

The aim of this chapter is to highlight the concept of EVs as a novel vector for drugs or biomaterials delivery in tumor biotherapy and immunotherapy. Our studies indicate that tumor cell-derived EVs can be used as vectors to deliver chemotherapeutic drugs. In addition, it highlights a novel drug delivery strategy with potential clinical applications. EVs are a natural packaging system for drug delivery. When tumor cells were treated with chemotherapeutic drugs upon ultraviolet stimulation, tumor cell derived-EVs were found to have therapeutic effect for many tumor models. For example, MTX-packaging EVs inhibited growth of peritoneal H22 cells without typical side effects in one study. In another, Cisplatin-packaging EVs inhibited ovarian cancer growth in SCID mice [7]. It was also demonstrated, drug-packaging EVs could reverse drug resistance of tumor-repopulating cells (TRCs) [8]. TRCs, by virtue of being more deformable than differentiated cancer cells, preferentially take up T-EVs that release antitumor drugs, which in turn lead to death of TRCs. The underlying mechanisms include interfering with drug efflux and promoting

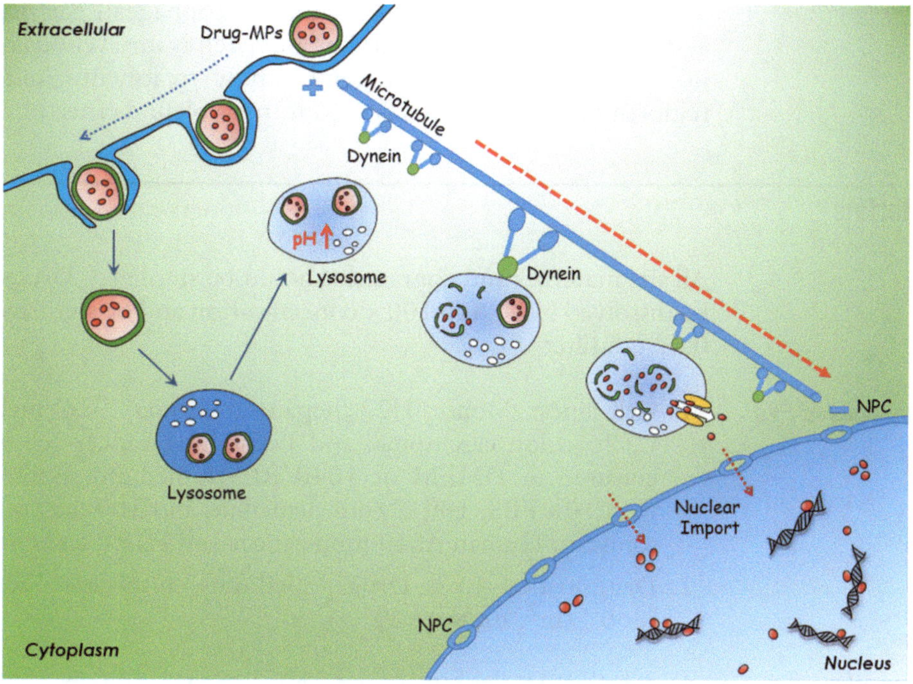

Fig. 1 The underlying mechanisms of reversal of drug resistance of TRCs by drug-packaging MPs

nuclear entry of the drugs. Interestingly, the results showed T-EVs enter and increase the pH value of lysosomes, leading to the migration of drug-loaded lysosomes along microtubule tracks toward the nucleus and discharging the drugs locally (Fig. 1). The unloaded drugs then readily enter the nucleus and exert its cytotoxicity more efficiently, leading to killing of the drug-resistant stem-like tumor cells.

During chemotherapeutic drug delivery, we also found that A549 lung cancer cell derived EVs could be used for loading oncolytic adenoviruses. Compared to oncolytic adenovirus alone, these EVs, which are loaded with oncolytic adenoviruses, showed numerous advantages on antitumor effect, including avoiding the diminishing of anti-adenovirus antibodies and a more efficient approach to enter tumor cell nucleus for a more effective oncolytic effect [9].

In addition to delivery, EVs can also act as a cell-free vaccine which has shown to be more effective in antitumor immune responses. EVs from allogeneic tumor cells (T-EV) containing both tumor antigens and innate DNA signals appear to have shared antigens that can be cross-presented by the host DCs, offering a possibility to simplify clinical application of T-EV-based vaccination therapies [10].

At present, there are three main EVs administration pathways. One is oral administration, which is ideal for patients who have digestive tract obstructions including esophageal, gastric, colon, and rectum carcinomas. Another approach is intravenous injection of EVs when applied for solid tumor treatment. And finally, regional perfusion of EVs is used to treat pleural effusion.

2 Materials

All the materials and containers should be sterilized. Drugs should be dissolved by PBS to 100× concentration and be sterilized with a 0.22 μm filter.

2.1 Drug Loaded EVs

1. Cell lines: Mouse H22 hepatocarcinoma, B16 melanoma, CT26 colon carcinoma, and Lewis lung cancer cell lines are cultured in DMEM or 1640 RPMI medium supplemented with 10% FBS, 100 U/mL penicillin, and 100 mg/mL streptomycin; Human A549 lung cancer and HL60 leukemia.
2. Drug-loaded EVs: Drugs, mediums, ultraviolet irradiation (300 J/m^2, *see* **Note 1**).

2.2 Oncolytic Adenovirus Loaded EVs

1. AdMaxTM recombinant adenovirus system (Microbix Canada): pDC515 (a shuttle plasma), pBGHfrtDE1, 3 FLP (adenoviral backbone).
2. pEGFP-C1 (ClonTech): a plasmid template for PCR amplification of EGFP gene.
3. pDC515 EGFP: a replication-defective adenovirus expressing enhanced green fluorescent protein (EGFP).
4. hTERTp-E1A: a recombinant DNA fragment that contains human telomerase promoter and adenoviral E1A cDNA fragment.
5. pBGHfrtDE1, 3 FLP.
6. Adenovirus purification and quantitation kits (Cell BioLabs Inc.).
7. OA-EV isolation: A549 cells, VP of OA.
8. Control EVs: uninfected tumor cells, ultraviolet irradiation (UBV, 300 Jm^2, *see* **Note 1**).

2.3 Mice Model

1. Mice: 6- to 8-week-old female BALB/c, BALB/c-nude mice, and C57BL/6 mice.
2. Tumor cells: H22 hepatocellular carcinoma cells, CT26 colon carcinoma cells, and B16 melanoma cells.

3 Methods

3.1 Preparation of Drug Loaded EVs

1. Drugs are added to mediums 1 h before ultraviolet irradiation.
2. Tumor cells are exposed to ultraviolet irradiation (300 J/m^2) for 1 h, and 16 h later, supernatants are centrifuged at 1000 × *g* for 10 min to remove whole cells.
3. The supernatants are centrifuged for 2 min at 14,000 × *g* to remove debris.
4. The supernatants are further centrifuged for 60 min at 14,000 × *g* to pellet EVs.
5. The pellets are washed three times and resuspended in culture medium for the subsequent experiments (*see* **Notes 2** and **3**).
6. EV counting: *see* **Note 4**.

3.2 Preparation of EVs Loaded Oncolytic Adenovirus

1. 1 × 10^7 A549 cells are infected with 1 × 10^8 VP of OA.
2. Forty-eight hours later, when the most cells detached from the culture dishes, the culture media are collected for OA-EV isolation.
3. The supernatants are first centrifuged for 10 min at 600 × *g* to get rid of cells and then centrifuged for 2 min at 14,000 × *g* to remove debris.
4. The supernatants are further centrifuged for 60 min at 14,000 × *g* to pellet OA-EVs.
5. To generate control EVs, uninfected tumor cells are exposed to ultraviolet irradiation (UBV, 300 J/m^2) for 1 h and continued 16 h culture.
6. The supernatants are used for control EV isolation.
7. EV counting: *see* **Note 4**.

3.3 Preparation of Oncolytic Adenovirus-Infected Cells EVs

1. The EGFP fragment is amplified from pEGFP-C1 by PCR and inserted into pDC515, to construct a replication-defective adenovirus expressing enhanced green fluorescent protein (EGFP). The construct is named as pDC515 EGFP.
2. To construct oncolytic adenovirus type 5, a recombinant DNA fragment that contains human telomerase promoter and adenoviral E1A cDNA fragment (hTERTp-E1A) is amplified from pDChTERTp-E1A-IRES-TK plasmid.
3. hTERTp-E1A is used to replace the CMV promoter of pDC515 to yield pDC-hTERTp-E1A.
4. For packaging Ad5 EGFP and Ad5 hTERTp-E1A, the shuttle plasmids pDC515 EGFP and pDC-hTERTp-E1A are cotransfected with backbone pBGHfrtDE1,3FLP into HEK 293 cells, respectively, according to the instruction of AdMaxTM recombinant adenovirus system.

5. Crude viruses from the HEK293 cells are purified and quantified by adenovirus purification and quantitation kits, respectively (Cell BioLabs Inc.).

3.4 Preparation of Mouse Tumor Model

1. For the human ovarian cancer murine peritoneal model, BALB/c-SCID mice are IP injected with 5×10^6 A2780 human ovarian cancer tumor cells.
2. For the subcutaneous lung cancer model, 6-week-old female BALB/c nude mice are subcutaneously (SC) injected with 1×10^6 A549 cells in 50 μl PBS.

3.5 Delivery of Drug Loaded EV to Mice

1. For the subcutaneous tumor model, 5 days after the tumor cells are injected, EVs (1×10^6 per mouse) or OA-EVs (1×10^6 per mouse) are intraperitoneally (IP) injected into mice once every 2 days.
2. For the intraperitoneal tumor model, 3 days after the tumor cells are injected, EVs (1×10^6 per mouse) or OA-EVs (1×10^6 per mouse) are IP injected into mice once every 2 days.
3. On day 16, all the mice are sacrificed to measure the tumor weight.

3.6 Delivery of Drug Loaded EV to Tumor Patients

EVs are administered orally, by intravenous injection, or by regional perfusion to tumor patients.

3.7 T-EV-Loaded DCs act as Therapeutic Vaccines

3.7.1 Isolation of Tumor Cell Derived EVs

1. Plate out tumor cells (1×10^7) in 100 mm culture dish in serum-depleted media.
2. The tumor cells are exposed to ultraviolet irradiation (300 J/m^2) for 1 h, and 16 h later, supernatants are centrifuged at $1000 \times g$ for 10 min to remove whole cells.
3. The supernatants are centrifuged for 2 min at $14{,}000 \times g$ to remove debris.
4. The supernatants are further centrifuged for 60 min at $14{,}000 \times g$ to pellet EVs.
5. The pellets are washed three times and resuspended in culture medium for the subsequent experiments (*see* **Note 2**).
6. EV counting: *see* **Note 4**.

3.7.2 Preparation of T-EVs-Loaded DCs

1. Bond marrow-derived dendritic cells are cultured as before.
2. The immature DCs are incubated with T-EVs overnight (12–16 h).
3. Harvest DCs and centrifuge at $400 \times g$ for 5 min at 4 °C.
4. The pellets are washed with PBS twice (remove the redundant T-EVs) for the subsequent experiments.

3.7.3 T-EV-Loaded DCs Exhibit Therapeutic Effects in Various Tumor Models

1. For H22 or CT26 tumor model: BALB/c mice are prechallenged with H22 or CT26 tumor cells (3×10^5 cells/mouse) in the lateral right flank. Five days after tumor inoculation, mice are treated SC with T-EVs or T-EV-loaded DCs three times at 5-day intervals.
2. For the B16 melanoma tumor model: C57BL/6 mice are pre-inoculated SC with 3×10^5 B16 melanoma cells receive therapy of T-EV-loaded DCs as above.
3. For the metastasis model: C57BL/6 mice are i.v. injected with 1×10^5 B16 melanoma cells. Five days later, mice are treated i.v. with T-EVs or T-EV-loaded DCs three times at 5-day intervals. 21 days after tumor incubation, mice are sacrificed and lung metastasis nodules are measured and calculated.

4 Notes

1. Always check the strength of ultraviolet irradiation because ultraviolet light weakens after long-time use.
2. EVs are stable when stored at 4 °C for 7 days.
3. The dose of chemotherapeutic drugs can be reduced to 1/10 of standard dosage when preparing for drug loaded EV.
4. EV counting: A flow cytometry-based method is used to count the number of EVs. After centrifugation, the EVs are suspended with PBS that is prefiltered through 0.1 mm filter and passed through 1 mm filter to further exclude background noise or nonspecific events. The EVs mixed evenly with 3 mm latex beads (LB-30, Sigma, St Louis, MO) with a denoted number. For flow cytometric analysis, 0.8 μm deep-blue dyed-latex beads (L1398, Sigma) are preused for gating and voltage adjustment, as such beads are fluorescent and can be detected on FL4 channel. When the mixture of LB30 and EVs were analyzed by flow cytometry, each LB30 bead formed a dot in the gate of the large-size population. If 10,000 counts of LB30 are collected, the number of EV could be calculated with the formula: $N = 10{,}000 \times$ (EV percentage/LB30 percentage).

Acknowledgments

This work was supported by National Basic Research Program of China (2015CB932600), National Science Fund for Distinguished Young Scholars of China (81225021), and National Natural Science Foundation of China (81472735, 81472653, and 81530080). J. Liu and J. Ma contributed equally to this work.

References

1. Lemoinne S, Thabut D, Housset C, Moreau R, Valla D, Boulanger CM et al (2014) The emerging roles of microvesicles in liver diseases. Nat Rev Gastroenterol Hepatol 11:350–361
2. Raposo G, Stoorvogel W (2013) Extracellular vesicles: exosomes, microvesicles, and friends. J Cell Biol 200:373–383
3. van Dommelen SM, Vader P, Lakhal S, Kooijmans SA, van Solinge WW, Wood MJ, Schiffelers RM (2012) Microvesicles and exosomes: opportunities for cell derived membrane vesicles in drug delivery. J Control Release 161:635–644
4. Skog J, Wurdinger T, van Rijn S, Meijer DH, Gainche L, Sena-Esteves M et al (2008) Glioblastoma microvesicles transport RNA and proteins that promote tumour growth and provide diagnostic biomarkers. Nat Cell Biol 10:1470–1476
5. Zitvogel L, Regnault A, Lozier A, Wolfers J, Flament C, Tenza D et al (1998) Eradication of established murine tumors using a novel cell-free vaccine: dendritic cell derived exosomes. Nat Med 4:594–600
6. Dai S, Wei D, Wu Z, Zhou X, Wei X, Huang H et al (2008) Phase I clinical trial of autologous ascites-derived exosomes combined with GM-CSF for colorectal cancer. Mol Ther 16:782–790
7. Tang K, Zhang Y, Zhang H, Xu P, Liu J, Ma J et al (2012) Delivery of chemotherapeutic drugs in tumour cell-derived microparticles. Nat Commun 3:1–11
8. Ma J, Zhang Y, Tang K, Zhang H, Yin X, Li Y et al (2016) Reversing drug resistance of soft tumor-repopulating cells by tumor cell-derived chemotherapeutic microparticles. Cell Res 26:713–727
9. Ran L, Tan X, Li Y, Zhang H, Ma R, Ji T et al (2016) Delivery of oncolytic adenovirus into the nucleus of tumorigenic cells by tumor microparticles for virotherapy. Biomaterials 89:56–66
10. Zhang H, Tang K, Zhang Y, Ma R, Ma J, Li Y et al (2015) Cell-free tumor microparticle vaccines stimulate dendritic cells via cGAS/STING signaling. Cancer Immunol Res 3:196–205

Erratum to: An Adaptable Polyethylene Glycol-Based Workflow for Proteomic Analysis of Extracellular Vesicles

Stephanie N. Hurwitz and David G. Meckes Jr.

Erratum to:
Chapter 25 in: Winston Patrick Kuo and Shidong Jia (eds.), *Extracellular Vesicles: Methods and Protocols*, Methods in Molecular Biology, vol. 1660, DOI 10.1007/978-1-4939-7253-1_25

The book was published with incorrect chapter title as "An Adaptable Polyethylene Glycol-Based Workflow for Proteomic Analysis of Extracellar Vesicles" and now the title has been corrected to "An Adaptable Polyethylene Glycol-Based Workflow for Proteomic Analysis of Extracellular Vesicles". The table of contents has been updated as well.

The updated original online version for this chapter can be found at http://dx.doi.org/10.1007/978-1-4939-7253-1_25

Winston Patrick Kuo and Shidong Jia (eds.), *Extracellular Vesicles: Methods and Protocols*, Methods in Molecular Biology, vol. 1660, DOI 10.1007/978-1-4939-7253-1_36, © Springer Science+Business Media LLC 2017

INDEX

Winston Patrick Kuo and Shidong Jia (eds.), *Extracellular Vesicles: Methods and Protocols*, Methods in Molecular Biology, vol. 1660, DOI 10.1007/978-1-4939-7253-1, © Springer Science+Business Media LLC 2017

D

E

F

G

H

I

MIX
Papier aus verantwortungsvollen Quellen
Paper from responsible sources
FSC® C105338

If you have any concerns about our products,
you can contact us on
ProductSafety@springernature.com

In case Publisher is established outside the EU,
the EU authorized representative is:
Springer Nature Customer Service Center GmbH
Europaplatz 3, 69115 Heidelberg, Germany

Printed by Libri Plureos GmbH
in Hamburg, Germany